ELECTRIC POWER SYSTEM COMPONENTS

TRANSFORMERS AND ROTATING MACHINES

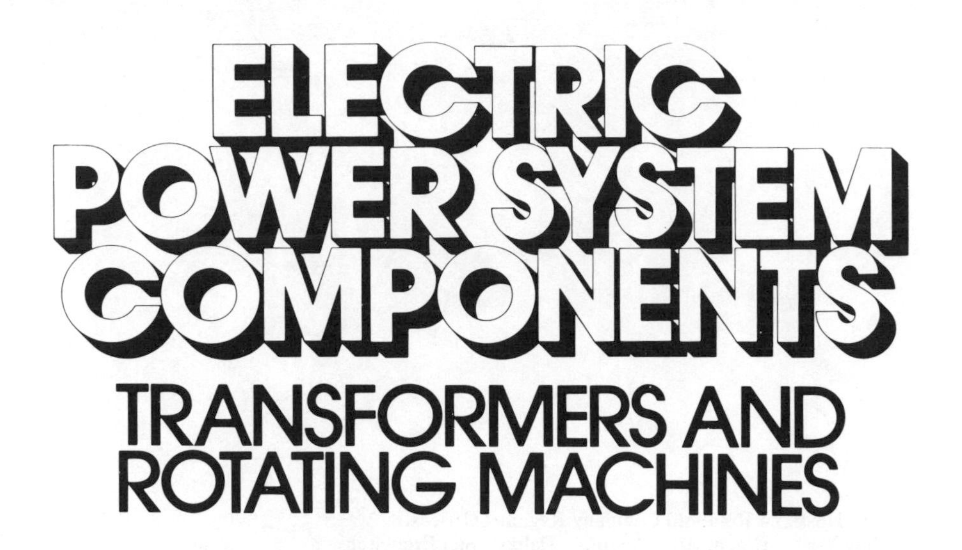

Robert Stein / William T. Hunt, Jr.

The City College of the City University of New York

VNR VAN NOSTRAND REINHOLD COMPANY
NEW YORK CINCINNATI ATLANTA DALLAS SAN FRANCISCO
LONDON TORONTO MELBOURNE

Van Nostrand Reinhold Company Regional Offices:
New York Cincinnati Atlanta Dallas San Francisco

Van Nostrand Reinhold Company International Offices:
London Toronto Melbourne

Library of Congress Catalog Card Number: 78-18398
ISBN: 0-442-17611-2

Manufactured in the United States of America

Published by Van Nostrand Reinhold Company
135 West 50th Street, New York, N.Y. 10020

Published simultaneously in Canada by Van Nostrand Reinhold Ltd.

15 14 13 12 11 10 9 8 7 6 5 4 3 2 1

Library of Congress Cataloging in Publication Data

Stein, Robert.
Electric power system components.

Includes index.
1. Electric machinery. 2. Electric transformers.
I. Hunt, William Thomas. II. Title.
TK2000.S73 621.31'4 78-18398
ISBN 0-442-17611-2

Preface

There are good reasons why the subject of electric power engineering, after many years of neglect, is making a comeback in the undergraduate curriculum of many electrical engineering departments. The most obvious is the current public awareness of the "energy crisis." More fundamental is the concern with social responsibility among college students in general and engineering students in particular. After all, electric power remains one of the cornerstones of our civilization, and the well-publicized problems of ecology, economy, safety, dependability and natural resources management pose ever-growing challenges to the best minds in the engineering community.

Before an engineer can successfully involve himself in such problems, he must first be familiar with the main components of electric power systems. This textbook will assist him in acquiring the necessary familiarity. The course for which this book is mainly intended can be taken by any student who has had some circuit analysis (using discrete elements, and including sinusoidal steady state) and elementary electromagnetic field theory. Most students taking the course will be in their junior or senior years. Once the course is completed, students may decide to go more deeply into the design and operation of these components and study them on a more advanced level, or they may direct their attention to the problems of the system itself, problems which are only hinted at briefly at various points herein.

Quite a few of the subjects covered here have been known and described by earlier writers. In fact, the older textbook literature in this field constitutes a valuable heritage. Nevertheless, and contrary to widespread belief, there also have been great changes in the practice of electric power engineering in recent years. Moreover, the electrical engineering students for whom this text is largely written bring a changed background of prerequisite training and knowledge to this study. All these changes call for a completely fresh start, not just a patchwork of adjustments and insertions to "update" obsolete treatments.

The authors, who have devoted half a lifetime to the teaching of this subject, have given much thought to the changing scene in electric power engineering and to the changing needs of their students. Some of the principles used in writing this text are given below.

a. To be at the right level for its readers, a textbook must be understandable to less-gifted students and yet be challenging enough to hold the interest of top-rated students.

b. An intelligent reader wants more than mere facts. He asks for explanations of such facts, for the logic behind them and the connections between them. A textbook should satisfy and encourage this attitude as much as possible.

c. Beware of time-honored traditions! It is true that many such traditions have earned respect because of their excellence, but many faulty explanations and illogical sequences have been perpetuated through generations of textbook authors. On the whole, it may be said that the more widely accepted the treatment of a subject, the more it should be scrutinized with suspicion before it is adopted.

d. In the past 10 to 20 years, the most significant change affecting electric power engineering has been the ascendency of electronic controls. This topic must be integrated into the text, appearing wherever it is pertinent, not just in an added chapter or appendix.

e. On the other hand, the digital computer, while having revolutionized the actual operation of power systems, as well as the design of their components, has only a minor influence on the basic theory and the understanding of these components.

f. The basic aim of this book is to impart to the reader an understanding of practical devices, not to give him additional practice in the use of the mathematical tools of linear systems theory or to preempt the teaching of related subjects such as control systems engineering.

The first two chapters are intended as an introduction, to clarify the purpose of the devices to be studied and the laws of nature on which their operation is based. As far as the further choice and sequence of subjects is concerned, magnetic circuits are a necessary prerequisite to all the devices to be studied, and among those, it is logical and helpful to analyze transformers before rotating machines. Two chapters (7 and 8) serve as a transition from transformers to motional devices. All the basic types of motors and generators are covered, with two or more chapters devoted to the more important ones. At the end of most chapters, there are sections containing illustrative examples and practice problems. Answers to problems are given at the end of the book.

Our thanks are due to the Department of Electrical and Computer Engineering of the University of Massachusetts in Amherst, for making the facilities of the University available.

ROBERT STEIN
WILLIAM T. HUNT, JR.

Contents

1
The Electric Power System

1-1 ELECTRIC POWER AND ITS COMPETITORS

For better or for worse, civilization is a goddess to whom we all pay tribute, and foremost among our payments is the use of electric power.

In engineering, power is a well-defined physical quantity (energy per unit of time) whose use is by no means restricted to its electric form. In fact, the power required for industrial processes is mostly mechanical or chemical; domestic power needs are largely for light and mechanical power; many means of transportation use some form of heat power. What distinguishes the electric form of power from all others is the fact that any form of power can be (and usually is) obtained by conversion from the electric form. (The only major exception is the familiar motor vehicle powered by an internal combustion engine where electric power is used in an auxiliary capacity only.)

So electric power has a predominant position among the various forms of power. The main reason is that all the enormous quantities of power needed for industry, homes, transportation, etc. can be centrally generated, transmitted over almost unlimited distances, and distributed over any desired area, all in electric form. At the location where it is to be used, power is converted into the desired form.

All the statements made above concerning power could equally well have been made in terms of *energy*. This makes it all the more imperative to observe the distinction between these two terms in the many cases when they are not interchangeable. For instance, a power plant could have a practically unlimited supply of energy and yet be unable to satisfy its consumers' demand for power, possibly due to one or more of its generators being under repair. Then again, a power plant fully capable of meeting all the power demands of its consumers could be short of energy if its fuel stocks were running too low.

1-2 POWER SYSTEM COMPONENTS

An electric power system consists basically of devices for generating, transmitting, distributing, and consuming electric power. The term *generating* means obtaining electric power by conversion from some other form of power. Similarly, *consuming* actually means converting electric power into some other form.

For the purpose of generating electric power, there is one almost unchallenged device: the *synchronous generator*, also called *alternator* (because its output is a-c power). It is a rotating machine, and a typical example of the kind of electromagnetic devices to whose study this book is devoted. There are other kinds of electromagnetic generators, which are only occasionally used, and there are other devices whose output is electric power, e.g., primary batteries, fuel cells, and so-called "direct" energy converters, but they have not (or at least not yet) been found suitable for major power systems.

The input power of an electromagnetic generator (synchronous or otherwise) is mechanical. The generator is driven by its *prime mover*. Any machine that produces mechanical power can be used as a prime mover. The most important one (in terms of power generated) is the *steam turbine*, which converts the heat power of steam into mechanical power. The steam in turn may be obtained by heating water in a boiler, a process that requires burning a "fossil" fuel like oil, gas, or coal. As an alternative, steam may be obtained by nuclear fission (or perhaps, at some time in the future, by nuclear fusion) in an atomic reactor and heat exchanger. In either case, it takes several steps to convert the chemical or nuclear energy stored in the fuel into electric energy.

Other important examples of prime movers are gas turbines and internal combustion engines (gasoline or Diesel), which convert the chemical power of their fuel directly into mechanical power, but which have been found more suitable for comparatively small, often auxiliary or portable, power systems. Finally, the power of moving water is used in hydroelectric power plants in which the prime movers are *waterwheels*. All these major ways of obtaining electric power are summarized schematically in Fig. 1-1.

It is worth noting that of all the prime sources of power discussed above, water power is the only one whose use does not constitute depletion of a natural resource. To be efficient, however, hydroelectric power plants usually require building a dam to store water, and this often encounters the rigorous opposition of conservationists. Even so, most available sites for the use of water power have already been utilized, at least on the North American continent. After all, hydroelectric power plants work only where there is enough water and enough difference of altitude, which means, broadly speaking, where there are mountains.

Thermal power plants, however, can be built anywhere, and the actual choice of a site is determined by the conflicting pressures of economy (where is the fuel least expensive?) and ecology (where are air and water pollution least objectionable?). For the case of nuclear power plants, there are additional problems of radiation leakage, radioactive waste disposal, and safety.

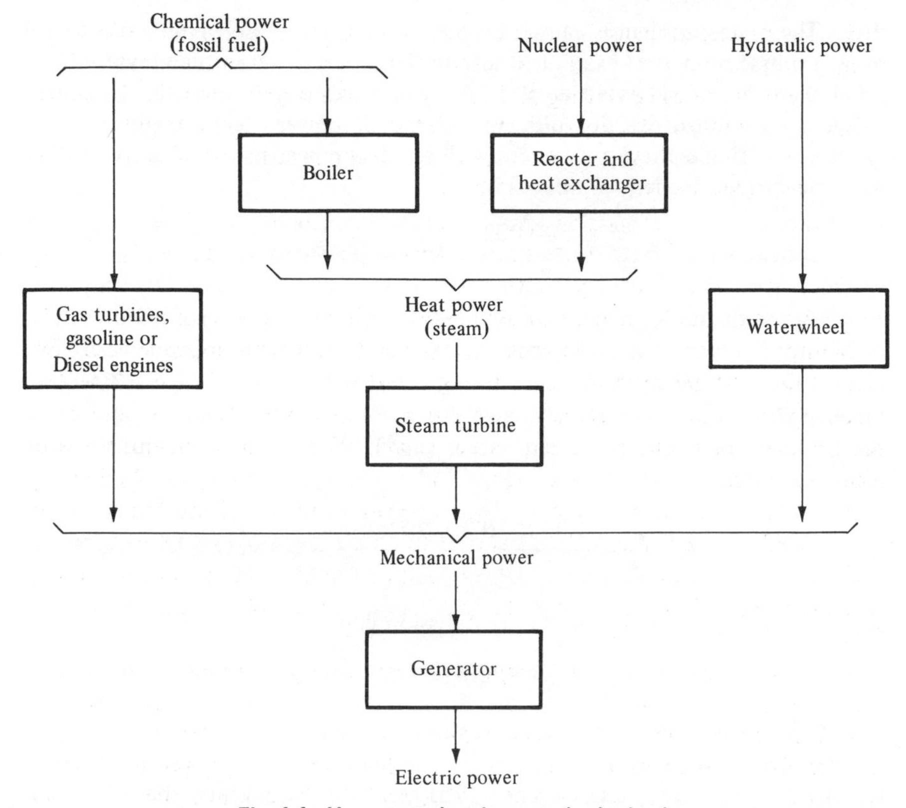

Fig. 1-1. How most electric power is obtained.

So much about the generation of power. The many devices *consuming* electric power are referred to as the *load* of the power system. For the purposes of this book, the only such devices of interest are those that convert electric power into mechanical power, the electric *motors*. Their operation constitutes a reversal of that of the generators; as a matter of fact, all electric generators can also operate as motors and vice versa, at least in principle. But even though motors are the same devices as generators, their different mode of operation raises different problems that require much additional study.

1-3 THE NEED FOR TRANSFORMERS

The transmission and distribution of electric power might be thought to require nothing but a suitable number of conducting wires, but there is more to it than that. The main problem is caused by the resistance of these wires.* It is useful to let a simple numerical example illustrate the orders of magnitude involved.

Let there be a load rated at 500 kilowatts (kw), barely enough for a small village, at a voltage of 250 volts (v). This little power system might be a d-c system, or a single-phase a-c system with an ideal power factor of unity. Either way, the current needed for this load is

$$I = \frac{P}{V} = \frac{500{,}000}{250} = 2000 \text{ amperes (amp)}$$

Now assume the power plant is located only 10 kilometers (km) or about 6 miles (mi) away from the load. To keep the resistance of the transmission wires low, choose the best material–copper, having a resistivity of $\rho = 1.73 \times 10^{-8}$ ohm-meters (Ωm)–and an extremely large cross-sectional area $A = 10$ square centimeters (cm^2) or about 1.6 square inches (in.^{2}). The resistance of this two-wire transmission line is

$$R = \frac{\rho l}{A} = \frac{1.73 \times 10^{-8} \times 20 \times 10^3}{10 \times 10^{-4}} = 0.346\ \Omega$$

and the full-load power loss in the transmission line is

$$P = RI^2 = 0.346\,(2000)^2 = 1.38 \times 10^6 \text{ watts (w)}$$

That is 2.76 times the power to be transmitted to the load! The generator must produce 1880 kw, of which 1380 are lost on their way to the load. Rather than sending power to the load efficiently, this generator would serve mainly to heat the countryside. An attempt to reduce this atrocious waste of power by further increasing the wire size would use enormous amounts of copper, making the entire project economically hopeless. And that for only a small load at a fairly short distance!

The only solution lies, as many readers may already know, in the use of higher transmission voltages. For instance, raising the voltage by a factor of 10 would reduce the current by the same factor, and thereby reduce the power loss by a factor of 100. Actually, transmission voltages for major power systems have climbed through the years up to as much as 750,000 v, and they will surely go even higher as progress is made in tackling such problems as insulation, corona, radio interference, etc. In this way, loads of over 10^9 w; that is 1000 megawatts (Mw), or 1 million kw can be transmitted over distances of many hundreds of miles without excessive power losses.

*The idea of using "superconducting" wires, i.e., wires cooled down to extremely low temperatures, should not be dismissed offhand as impractical or utopian. Serious feasibility studies have been made, although actual application of the idea still seems far away.

To operate *loads* at such high voltages is out of the question. In fact, considerations of insulation, spacing, and safety limit voltages for homes, offices, etc. to only a few hundred volts. Here again, there is only one way out: electric power must be transmitted at "high" voltages and consumed at "low" voltages. *Transformers*, which convert electric power between transmission and consumption with high efficiency, are indispensable to feasible power systems as we know them. It so happens that transformers are also electromagnetic devices. Their operation is simpler than that of generators and motors, and its study is a useful and even necessary prerequisite to the study of those machines.

The role of transformers in power systems is not limited to that of an intermediary between transmission lines and loads. For one thing, generators cannot be built (at least not economically) at voltages as high as those of the transmission lines. Transformers are used to "step up" the voltage between generator and transmission line, just as they are used to "step down" the voltage between transmission line and load. Furthermore, modern power systems consist of many generators, transmission lines, distribution networks, and loads, each of which may operate at its own most suitable voltage. Transformers are used to interconnect all parts of such a power system.

1-4 A WORD ABOUT RELIABILITY

Nothing man-made is perfect, but electric power systems have to come very close to perfection to do their job. Such is our dependence on an everready power supply that a power failure of more than the shortest duration has a catastrophic impact on our lives. (Readers living in the north-eastern United States and old enough to remember a certain night in November 1965 can testify to that observation.) Considering how many components built and operated by human beings are exposed to hazards such as extreme weather conditions, accidents, aging, and human error, the overall performance of our major power systems through the years must be called miraculous.

The major weapon used by power companies to maintain the reliability of their service is a certain amount of redundancy. Several generators always share their load and, thereby, are able to maintain service if one of them should break down. The same principle applies to transformers, transmission lines, and even distribution leads. Whenever any component suffers a failure, there should always be another way by which the power can be routed to every load.

As an extension of that idea, major power systems are not only integrated but also interconnected. Not only can faults be bypassed within a system; the various systems have also built connections so that power can be transmitted from one system to another. One system whose generators are temporarily unable to supply all its loads (e.g., due to a breakdown of one generator) can borrow or buy energy from another system.

It is true that such interconnections, intended to avoid blackouts, could actually have the opposite effect and spread them over wider areas, as in the above-mentioned 1965 event, but there is general agreement that, on the whole, interconnections do more good than harm. The situation is comparable to that of a group of mountain climbers who are tied to each other by a rope. The purpose of the rope is to save a member of the group who might slip or fall. Even though it has happened occasionally that one member of such a group pulled the others down to disaster, mountaineers continue to rely on ropes for their safety.

It should not be assumed that redundancy and interconnection by themselves are sufficient to maintain the entire power system in operation. There must also be elaborate sets of protective devices like lightning-arresters, as well as constant monitoring, to detect signs of component breakdowns, and circuit-breakers to open and close connections without shutting down the rest of the system.

1-5 THE QUALITY OF SERVICE

The term *maintaining service*, as used in the previous section, must include maintaining the waveshape of the voltage. Since voltages in power systems are practically all sinusoidal, what must be maintained are the magnitude and the frequency of the voltage.

It should be kept well in mind that basically all loads in a power system are connected in parallel to each other. The familiar power outlets in our rooms are pairs of "hot" terminals to which power-consuming devices can be connected. Therefore, the term *load* may mean, in a quantitative sense, the power or the current drawn from the terminals, but never the voltage. For instance, a generator is said to operate at *no-load* when the terminals are open and the current is zero.*

The fact that voltages have to be adjusted in order to be kept constant is probably well known to many readers. The resistance (and even more the inductive reactance) of a transmission line alone causes a voltage drop between generator and load; the magnitude of that drop depends on the current. Similarly, the output voltages of generators and transformers depend on the currents they carry, because of their own resistances and reactances. In our study of generators and transformers, we shall be very much concerned with the *regulation* of these devices, i.e., with the amounts by which their voltages change when their loads change.

*This *all-parallel* arrangement of power systems is the one universally chosen, but is is not the only possible one. In the early days of electric power engineering, there were experimental *all-series* systems in which all (d-c) generators and loads were connected in series. In such a system, *load* must mean voltage, and *no-load* means short-circuit.

To maintain the magnitude and frequency of voltages in the face of constantly changing loads is the main task in the operation of a power system. It involves the use of computers and automatic control systems, and it remains the subject of continuous study and search for improvement. A basic knowledge of the properties of synchronous generators and transformers is a necessary first step toward an understanding of such problems.

1-6 POWER LOSSES

Since the devices we shall study are all power converters, it makes sense to refer to their inputs and outputs in terms of power. It would be unrealistic to expect the *output power* of any such device to equal its input power. Inevitably, and for specific reasons we shall study later, the operation of all electromagnetic devices is accompanied by the loss of some power

$$P_{\text{loss}} = P_{\text{in}} - P_{\text{out}} \tag{1-1}$$

From the economic point of view, what matters is not so much the amount of power lost (in watts or kilowatts) as its relation to the power converted. This leads to the definition of *efficiency* as a dimensionless quantity

$$\eta = \frac{P_{\text{out}}}{P_{\text{in}}} \tag{1-2}$$

which can also be expressed in terms of input and losses

$$\eta = \frac{P_{\text{in}} - P_{\text{loss}}}{P_{\text{in}}} = 1 - \frac{P_{\text{loss}}}{P_{\text{in}}} \tag{1-3}$$

or, generally more usefully, in terms of output and losses

$$\eta = \frac{P_{\text{out}}}{P_{\text{out}} + P_{\text{loss}}} = 1 - \frac{P_{\text{loss}}}{P_{\text{out}} + P_{\text{loss}}} \tag{1-4}$$

In both Eqs. (1-3) and (1-4), the last expression is the most significant because it shows how the relative losses are subtracted from the ideal efficiency of unity.

The significance of power losses goes far beyond their economic aspect. All power "lost" is actually converted into *heat*,* and is thus lost for the purpose of the device. Now heat must inevitably raise the *temperature* of the object in which it is generated unless it can be dissipated, i.e., removed from that object. In the case of electromagnetic devices, there is a limit beyond which the temperature

*There is also a power loss by electromagnetic radiation, due to changing electric and magnetic fields, but this is negligibly small at the comparatively low frequencies of power systems.

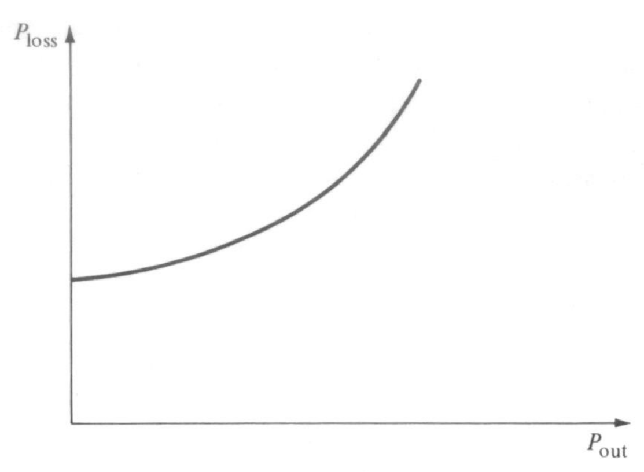

Fig. 1-2. Power losses.

must never rise. That is so because at high temperatures, all insulating materials deteriorate (first, their insulation resistance decreases and eventually they burn).

The dissipation of heat occurs naturally by three independent mechanisms: *heat radiation*, *heat conduction*, and *heat convection*; the last term means heat removal by the motion of the surrounding medium, e.g., air. The rate of heat removal by all three mechanisms depends on the surface area of the heated object and on the temperature difference between that object and the surrounding space (more strictly speaking, on the temperature gradient at the surface). The rate of heat convection in particular can be artificially increased by aiding the motion of the surrounding medium (e.g., by means of a fan), or by using a cooling medium, like water or oil, in place of air.

What happens then is that, when the device is turned on (made to convert power), the power losses cause its temperature to go up until it is hot enough to dissipate exactly as much heat as it generates. Thus a *steady-state* temperature (for a constant load) is reached asymptotically, similar to the steady-state current in an RL or RC circuit under d-c excitation, except that thermal time constants are generally reckoned in minutes, in contrast to the fractions of a second usually encountered as electric time constants.

As might be expected (and to be confirmed in later chapters), the magnitude of the power loss in an electromagnetic device depends on its output. Figure 1-2 shows a typical example of this relationship. As the load increases, more heat is generated, and the temperature rises. It follows that there is a certain amount of output power that leads to a steady-state temperature equal to the permissible limit. That is the *rated power* of the device, the power given on its nameplate.

The rated power of an electromagnetic device is thus seen to be determined by the power losses. It is not the maximum power that the device can deliver.

(Other energy-converting devices like automobile engines are often rated on such a basis.) For instance, an electric motor can be *overloaded*, i.e., it can be forced to deliver more than its rated output. If this is done (by an amount exceeding the small safety margin), the motor will eventually *overheat*, i.e., its insulation will deteriorate or even break down. It is, however, quite all right to overload a motor for a comparatively short time and then to operate it at reduced or zero output to let it cool off again. Such *load cycles* can be repeated indefinitely, and some motors intended for such use are actually given intermittent-duty ratings, which are higher than their steady-state ratings.

2

Faraday's Induction Law

2-1 A LITTLE HISTORY

Since all engineering devices discussed in this book (as well as many other devices of electrical engineering) are essentially based on the law of *electromagnetic induction*, it is appropriate to begin their study with a brief glance at the discovery of this fundamental law of nature.

Our story begins in 1819 when the Danish physicist *Oerstedt* found that the hitherto independently known phenomena of electricity and magnetism were related to each other. He was the first to show that an electric charge could produce magnetic effects, provided only that charge was in motion; in other words, that an electric current produces a magnetic field. Soon afterward, *Ampere* found a general mathematical formulation of this observation.

Their contemporary, *Faraday*, across the English Channel, thought it strange that this relationship between electricity and magnetism should be so one-sided. His intuition told him that it should also be possible to "produce electricity out of magnetism." After years of trying, he discovered, in 1831, the way to do it: the magnetic field must undergo a *change* in order to produce an electric effect.

The laws named after Ampere and Faraday have remained the two cornerstones of the science of electromagnetism. They are expressed in the full power of their basic simplicity and the beauty of their symmetry among *Maxwell's equations*.

2-2 FLUX AND FLUX LINKAGES

The reader should not expect the ideas discussed in this section (or even in the whole chapter) to be entirely new to him or her. It is always good to start with some familiar (though possibly not-too-well remembered) concepts and to use them as the first steps leading gradually into the realm of the unknown.

The magnetic quantity whose rate of change is responsible for an induced voltage (Faraday's electric effect) is called the magnetic *flux*. It may be defined mathematically in terms of the magnetic vector field quantity called the magnetic induction or (better) the magnetic *flux density* $\mathcal{B}$ by the equation

$$\phi = \int_A \mathfrak{B} \cdot d\mathfrak{A} \tag{2-1}$$

where A is a bounded surface consisting of infinitesimal area elements $d\mathfrak{A}$.

The whole concept is much more easily visualized if we can choose a surface A_p consisting of elements perpendicular to the flux density vector. In this case, we don't need vector notation and the dot product, and the equation becomes

$$\phi = \int_{A_p} B\, dA \tag{2-2}$$

A still simpler case we shall encounter frequently occurs when the flux distribution is uniform, i.e., when there is a region in space where the flux density has the same direction and the same magnitude at every point. Then we can choose a *plane* surface A_p, and the equation reads

$$\phi = B A_p \tag{2-3}$$

In any case, any mention of a flux implies the choice of surface A (or A_p). The main importance of a flux lies in the fact, found by Faraday, that a change of this flux results in an induced voltage around the boundary of its surface. It follows that, in order to define a certain flux, one needs to choose not the surface but only its boundary line. In other words, the choice of any surface bounded by the same closed line yields the same flux.

Another concept we owe to the genius of Faraday is the description of a vector field by its "lines of force." In addition to indicating the direction of the field vector at any point of space, these lines also indicate its magnitude by their density. A flux is proportional to the number of lines crossing its surface or linking its boundary line. The term "linking" aptly describes the relationship between the closed boundary line and the line of flux, comparing them to two links of a chain. (The reader knows that the lines of flux density must be closed, i.e., without beginning or end.)

As to that induced voltage, its magnitude is

$$e = \frac{d\phi}{dt} \tag{2-4}$$

while its sign (+ or −) depends on what may have been chosen as the positive direction of e and as the positive value of ϕ.

In place of our abstract boundary line, we can visualize a wire of negligibly small cross-sectional area, made of electrically conducting material. If this conducting loop links a magnetic flux, and if this flux is changing, then there will be an induced voltage and, therefore, a current in the conductor.

We shall study engineering devices that make use of this phenomenon. It can be expected that such devices will be better if larger voltages can be induced. At any rate, it has been found that voltages induced in a single loop are rarely large enough for their purposes. But, if we can wind several *turns* of our conducting wire next to each other to form a *coil*, then there is voltage according to Eq. (2-4) induced in each turn, and, since the coil consists of its turns connected in series, the voltage induced in a coil of N turns is

$$e = N \frac{d\phi}{dt} \tag{2-5}$$

It was tacitly assumed that all N turns are located so close to each other that they all link the same flux ϕ. Practical devices to be studied in this book mostly have coils consisting of many hundreds or even thousands of turns, and in many cases the turns link practically the same flux. Figure 2-1 shows a sketch of a coil of three turns linking a flux (shown as a bundle of lines). Note that the highest and the lowest of the lines drawn do not belong to the flux linked by the coil.

A slight simplification arises from the definition of *flux linkages*

$$\lambda = N\phi \tag{2-6}$$

Since N is a constant, a substitution of Eq. (2-6) into Eq. (2.5) results in

$$e = \frac{d\lambda}{dt} \tag{2-7}$$

which is useful in circuit analysis as the dual or counterpart of the familiar equation defining the electric current i as the derivative of the charge q.

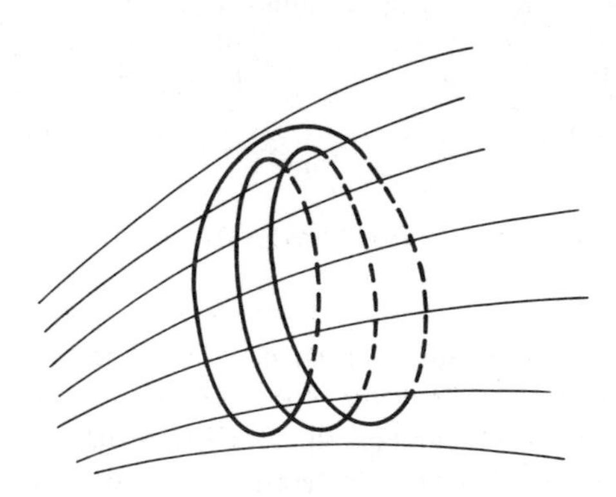

Fig. 2-1. A coil in a magnetic field.

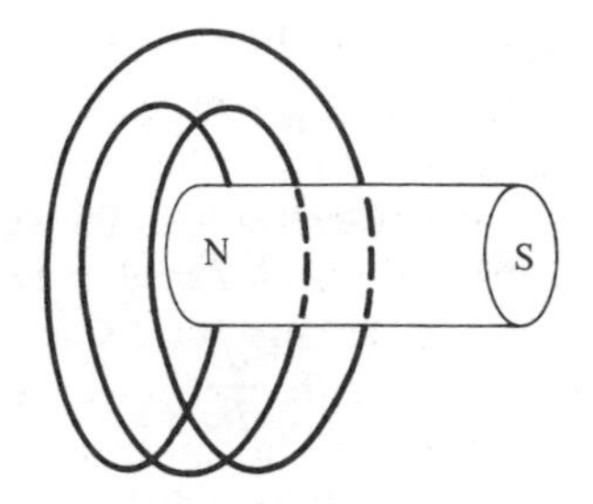

Fig. 2-2. Illustrating motional voltages.

2-3 TRANSFORMER VOLTAGES AND MOTIONAL VOLTAGES

The question arises: what could cause a flux to change? There are two distinct possibilities to be considered: a changing electric current and mechanical motion.

To begin with, magnetic fields in engineering devices are mostly produced by electric currents, the rarer alternative being the use of permanent magnets. Now, if an electric current changes, the change of the magnetic flux produced by this current must result in voltages being induced in any circuit that links any part of this flux. Such voltages account for the operation of transformers (see Chap. 5) and are known, therefore, as *transformer voltages*.

On the other hand, mechanical motion can change flux linkages in a variety of ways. Figure 2-2 illustrates one of the simplest cases. It shows a coil similar to that of Fig. 2-1, and a bar magnet with poles marked N and S (for north and south) at its left and right ends. (This could be either a permanent magnet, or an electromagnet whose "exciting" coil is not shown in the figure.) In the position shown, the coil links most if not all of the flux of the magnet. But if the magnet is moved far enough to the right, some of its lines of flux will bypass the coil. So the flux linked by the coil will change, and there must be a voltage induced in the coil during the motion. Such a voltage depends on the speed of the motion and is, therefore, called a *speed voltage*, or *motional voltage*. We shall see that such voltages form the basis of all electric motors and generators.

2-4 SELF- AND MUTUAL INDUCTANCE REVISITED

If we exclude permanent magnets from our consideration, every magnetic field is produced by electric currents, and its strength depends on the values of the currents that produced it. In linear circuit analysis, a science with which the reader must have some familiarity, the relation between current and flux is assumed to be one of strict proportionality.

Thus, if there is only one current path carrying $i(t)$ amperes at the time t, the flux linkages of this path are assumed to be

$$\lambda = Li \tag{2-8}$$

If the current changes, there is a transformer voltage induced, and its value is found by substituting Eq. (2-8) into Eq. (2-7):

$$e = L \frac{di}{dt} \tag{2-9}$$

so that the constant of proportionality introduced by Eq. (2-8) turns out to be the familiar *(self)-inductance*

$$L = \frac{\lambda}{i} = N \frac{\phi}{i} \tag{2-10}$$

In the more general case of several current paths (branches), linear circuit analysis assumes that the current in branch k contributes to the flux linkages of branch l the amount

$$\lambda_{kl} = M_{kl} \, i_k \tag{2-11}$$

Now, if the current i_k changes, the transformer voltage induced in branch l is

$$e_{kl} = \frac{d\lambda_{kl}}{dt} = M_{kl} \frac{di_k}{dt} \tag{2-12}$$

and we recognize the *mutual inductance* between branches k and l

$$M_{kl} = \frac{\lambda_{kl}}{i_k} \tag{2-13}$$

So we see that the reader has long been quite familiar with transformer voltages.

The trouble is that, with the kind of electromagnetic devices to be studied in this book, the assumption of linearity does not hold, as we shall see in the next chapter. We shall be constantly faced with the need to choose either *linear approximations* or other approaches that, although far more accurate and also more physically meaningful, deny us the use of the many powerful and often convenient tools of linear algebra.

If the idea of circuit parameters like L and M is to be extended to nonlinear circuits, these parameters must not be constants but functions of all currents. Furthermore, if motional voltages are to be introduced into circuit analysis, *time-varying* parameters L and M must be used.

3

Magnetic Circuits

3-1 AMPERE'S LAW

There are several ways to express Ampère's law, the relation between a magnetic field and the electric current or currents producing it. The following equation will turn out to be the most significant one:

$$\oint \mathcal{H} \cdot dl = \sum i \tag{3-1}$$

$\mathcal{H}$ is the field intensity vector. The left side of the equation suggests that we choose an arbitrary closed line consisting of an infinite number of infinitesimal line elements dl and form the line integral of $\mathcal{H}$ along this line. The right side of the equation is the algebraic sum of all electric currents whose paths link the chosen line.

This equation can be written in a much simpler and more easily visualized form (without vector notation and dot product) if the integration path is chosen to have the direction of the field intensity at every point of its course; in other words, if the integration path coincides with a line of force. In that case,

$$\oint Hdl = \sum i \tag{3-2}$$

In addition, if the field is uniform, the equation becomes further simplified:

$$Hl = \sum i \tag{3-3}$$

The simplest example is a single current in a straight linear path and a circular path of integration surrounding the current path at the distance (radius) r, as sketched in Fig. 3-1. As the reader presumably knows, this chosen integration path coincides with a line of force, and the field intensity has the same magnitude at every point of this path. Thus, Eq. 3-3 is applicable:

$$H(2\pi r) = i, \quad \text{or} \quad H = \frac{i}{2\pi r} \tag{3-4}$$

Much stronger fields are obtained by forming *coils* consisting of many turns of conducting wire wound side by side. In that case the lines of force link all turns of the coil, and the right side of Eq. 3-1 or 3-2 or 3-3 becomes the product

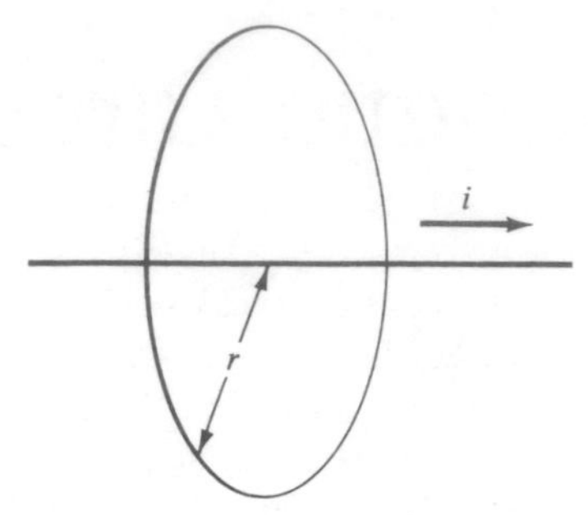

Fig. 3-1. Illustrating Ampere's law.

of i, the current, and N, the number of turns. This product is a significant quantity in all electromagnetic devices, and it is given the symbol $\mathcal{F}$ and the name *magnetomotive force* (mmf for short).*

$$\mathcal{F} = Ni \tag{3-5}$$

The unit of $\mathcal{F}$, the ampere-turn (At), is dimensionally the same as the ampere.

3-2 IRON CORES

The description of a magnetic field requires the use of two vector quantities, the field intensity $\mathcal{H}$ and the flux density $\mathcal{B}$. As we have seen in the previous section, the field intensity is directly related to the current or currents producing the field; on the other hand, we know that the effects of a magnetic field (mechanical forces or induced voltages, as described in the previous chapter) are directly related to the flux density. The "constituent" relation between these two vectors is

$$\mathcal{B} = \mu\mathcal{H} \tag{3-6}$$

where the *permeability* μ (normally a scalar quantity) depends on the material. If magnetic effects are to be increased, materials with larger values of μ should be used.

Most materials have practically the same permeability as free space:

$$\mu_0 = 4\pi \times 10^{-7} \text{ rationalized mks units}^\dagger$$

But there is a distinct group of materials whose permeability exceeds that of free space by several orders of magnitude. Iron and steel, the materials whose magnetic behavior is so obviously "special," are the principal members of this group,

*We use the symbol $\mathcal{F}$ although the mmf is not a vector quantity because the letters f and F are already overworked (frequency, force, etc.)

†The units of $\mathcal{B}$ and $\mathcal{H}$ are determined by their relations to electric quantities. By Eq. 2-4, the unit of flux is a volt-second, and it is called a weber. Thus, the unit of flux density is a weber per square meter. By Eqs. 3-3 and 3-5, the unit of field intensity is an ampere-turn per meter. These units are called rationalized, in contrast to other systems in which the factor 4π appears in Eqs. 3-1 through 3-5, and not in the value of μ_0.

(others are nickel and many oxides and alloys of iron), and they account for the name *ferromagnetic* materials (*ferrum* is iron in Latin).

To use this property of ferromagnetic materials–namely, their high permeability–for electromagnetic devices, the basic idea consists of winding coils around iron *cores*, as shown schematically in Fig. 3-2, for instance, except that instead of four turns, most devices have many hundreds or thousands of turns.

Let such a coil carry an electric current, and visualize the magnetic field in terms of lines of force. All lines must link the coil, and, if we consider the permeability of the surrounding space as negligible compared to that of the core, all lines of flux density must be closed through the core. So the use of an iron core results not only in a considerably larger flux, but it also forces the lines of flux density to follow a prescribed path, comparable to the path a closed electric conductor provides for an electric current. This *analogy* introduces the concept of *magnetic circuits* in which a magnetic flux is seen as the counterpart of an electric current.

Figure 3-3 is a simplified redrawing of the core and coil of Fig. 3-2, which also indicates the approximate path of the *mean line* of flux density. We visualize the flux linking the coil as a bundle of such lines, and we use their average density as the approximate value of the flux density throughout the core. This permits us to use Eq. 2-3. So we write

$$B = \frac{\phi}{A} \tag{3-7}$$

where A is the cross-sectional area of the core. (Since this area is perpendicular to the flux, the subscript p used in the original equation is omitted as unnecessary.) Next, we obtain the field intensity from Eq. 3-6

$$H = \frac{B}{\mu} \tag{3-8}$$

where script letters have once again been replaced by ordinary italic capitals to indicate a relation of magnitudes. This value of H is valid for every point

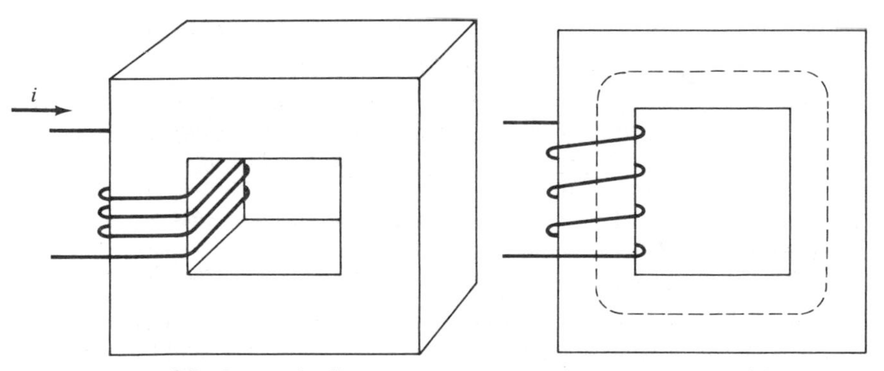

Fig. 3-2. Core and coil.

Fig. 3-3. Mean line of flux.

along the mean line of flux so that we can use Eq. 3-3 to obtain

$$\mathcal{F} = Hl \tag{3-9}$$

What we have accomplished by these three steps (the last three equations) is to find the mmf required to produce a desired flux. We can also combine the three steps and obtain

$$\mathcal{F} = Hl = \frac{B}{\mu} l = \phi \frac{l}{\mu A} \tag{3-10}$$

In the analogy between magnetic and electric circuits, the mmf is the counterpart of the voltage (for which the old term *electromotive force*, or emf for short, is still occasionally used). In that sense, Eq. 3-10 is the magnetic counterpart of *Ohm's law*, and the fraction on the right side of the equation is the counterpart of resistance. It is given the symbol $\mathcal{R}$ and the name *reluctance*.

$$\mathcal{R} = \frac{\mathcal{F}}{\phi} \tag{3-11}$$

For the case of a core with a uniform cross-section, it is

$$\mathcal{R} = \frac{l}{\mu A} \tag{3-12}$$

The reader can see the perfect analogy to the expression for the resistance of a cylindrical conductor (length l, cross-sectional area A) where the conductivity appears in place of the permeability. This should come as no surprise, since it is the higher permeability of the core (compared to its surrounding space) that forces the flux to follow a prescribed path, just as the high conductivity of a metallic wire provides a prescribed path for the electric current.

Just as, in electric circuit analysis, it is often more convenient to use the reciprocal of resistance (or of impedance), so the reciprocal of reluctance is a useful quantity in the study of magnetic circuits. Its name is *permeance* and its symbol is $\mathcal{P}$

$$\mathcal{P} = \frac{\phi}{\mathcal{F}} \tag{3-13}$$

There is nothing inaccurate in Eqs. 3-11 and 3-13, which are actually nothing but definitions. But any numerical calculation based on Eqs. 3-7 and 3-9 can only be an approximation, for several reasons. To begin with, we assumed that the flux is strictly confined to the core as it would be if the permeability of the core were an infinitely high multiple of that of the surrounding space. Actually, the ratio of permeabilities is only of the order of 10^3, and we shall sometimes be concerned with the existence of *leakage fluxes*, which are simply fluxes outside the intended ferromagnetic path. The same consideration is valid for an electric current in a conductor, but the ratio of conductivities (metallic conductor to

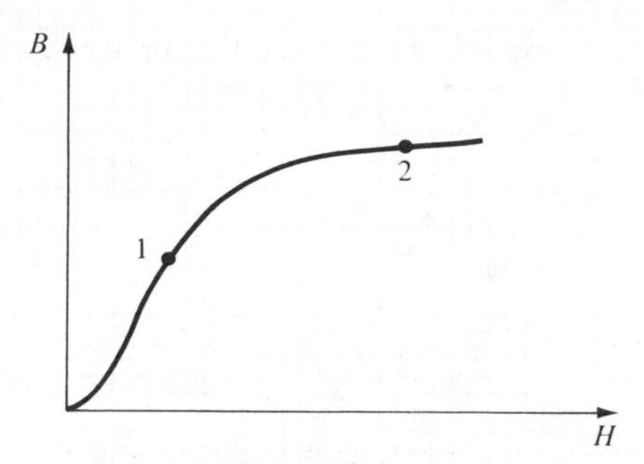

Fig. 3-4. Magnetization curve.

surrounding insulating material) is usually of the order of 10^{15} or even more, which makes the concept of electric circuits a much better model of true conditions than that of magnetic circuits. Furthermore, the use of a mean line and the assumption of a uniform field within the core are clearly nothing but approximations, especially with respect to the corners of the core. Again, the same method is much more nearly correct for electric conductors, because their ratio of length to cross-sectional area is normally much higher than the corresponding ratio for magnetic cores.

Altogether, the whole concept of a circuit is never anything more than an approximation, needed to circumvent the complexities of a rigorous application of Maxwell's equations to an engineering problem. Fortunately, magnetic circuit calculations based on our simplifying assumptions have results with satisfactory accuracy, and have proved to be useful in the study of electromagnetic devices.

3-3 SATURATION

Ferromagnetic materials are practically indispensable for the cores of most electromagnetic devices, because of their high values of permeability. They have other properties, however, that cause great complications in the study, design, and operation of such devices.

The relation between field intensity and flux density can be obtained empirically, and, for all ferromagnetic materials, the nature of this relation is shown in Fig. 3-4. One significant aspect of this curve is the shape of its upper part where it becomes nearly horizontal. Once the flux density has been raised to a certain value, it becomes insensitive to further changes of the field intensity; the material is then said to be *saturated*. The curve is called the *saturation curve* or the *magnetization curve* of the material.

We see that no single value of permeability can be assigned to such a material. For instance, the permeability at point 2 in Fig. 3-4 is lower than at point 1. The more the material is saturated beyond the "knee" of the curve, the lower the permeability becomes, and the more the advantage of using the material

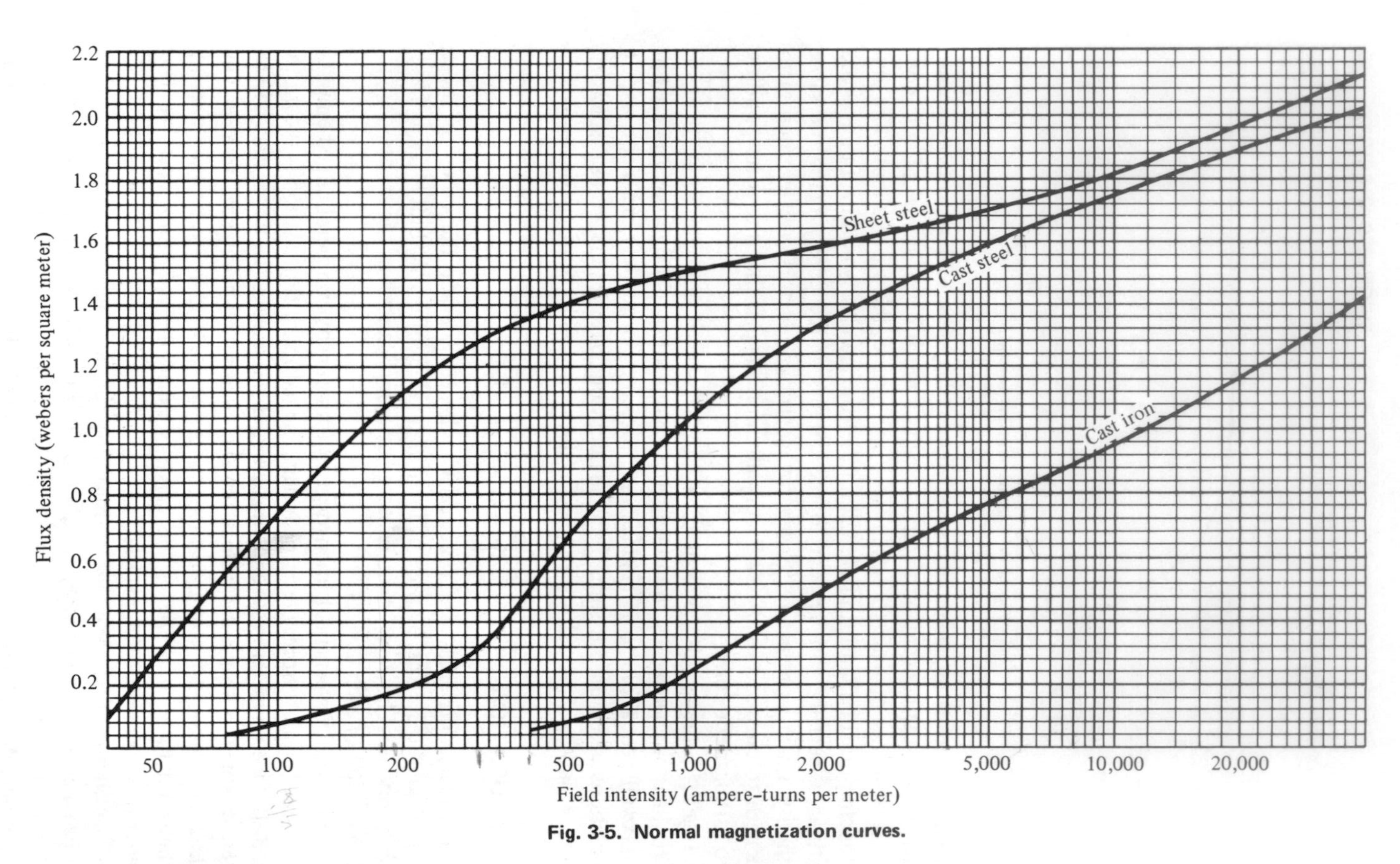

Fig. 3-5. Normal magnetization curves.

diminishes. The fact that the relation between B and H is nonlinear lies at the root of many complications we shall encounter in our further studies.

The calculating procedures shown in the previous section remain correct. Still, the use of Eq. 3-8 (or 3-6) would require that the appropriate value of μ be found first, from the magnetization curve of the material. Instead, it is simpler to look up the values of H directly from that curve. Similarly, we can find the flux produced by a given mmf in three steps: $H = \mathfrak{F}/l$, look up the corresponding value of B, and $\phi = BA$.

Figure 3-5 is a set of magnetization curves for commercially used materials, for use in illustrative examples and practice problems. The abscissa scale is logarithmic to cover a wider range of values.

3-4 SERIES MAGNETIC CIRCUITS

When a core does not have a uniform cross-section, it must be considered as consisting of several *parts*, each with a different value of cross-sectional area A_1, A_2, etc. Application of Eq. 3-7 then yields several values of B, one for each part. To each of these values, we can look up the corresponding value of H on the magnetization curve. But then, instead of using Eq. 3-9, we have to go back to Ampère's law (Eq. 3-2) and approximate the integral by the sum:

$$\mathfrak{F} = \sum_k H_k l_k \tag{3-14}$$

This equation is the magnetic counterpart of *Kirchhoff's voltage law*, as can be shown by the following series of substitutions:

$$\mathfrak{F} = \sum_k H_k l_k = \sum_k \frac{B_k}{\mu_k} l_k = \sum_k \frac{\phi_k}{\mu_k A_k} l_k = \sum_k \mathfrak{R}_k \phi_k \tag{3-15}$$

Figure 3-6 shows a typical magnetic circuit consisting of four parts. Since the flux is the same in each part, this core is considered a *magnetic series circuit*, and the last equation can be written as

$$\mathfrak{F} = \phi \sum_k \mathfrak{R}_k = \mathfrak{R}\phi \tag{3-16}$$

where the total reluctance appears logically as the sum of the reluctances of the parts. Figure 3-7 shows the electric analog to the magnetic circuit of Fig. 3-6. Notice, however, that the voltage source V has to be inserted into the electric circuit by cutting into that circuit. In the magnetic circuit, the corresponding quantity $\mathfrak{F}$ requires only the presence of a current in the coil surrounding the magnetic core.

How to solve a typical problem concerning such a core, is shown in Example 3-1.

While it is a straightforward procedure (accepting the limited accuracy inherent

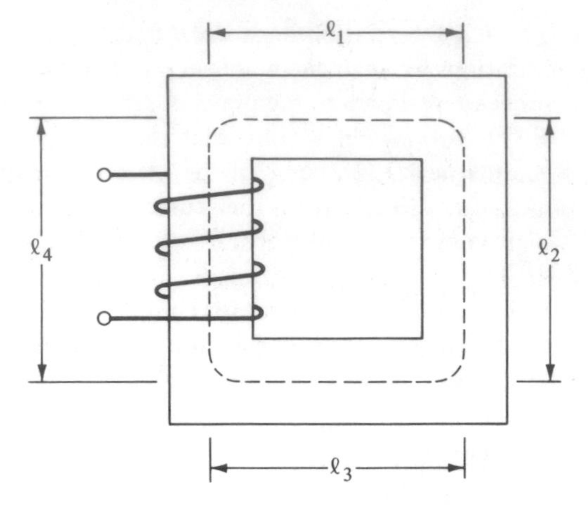

Fig. 3-6. Magnetic series circuit.

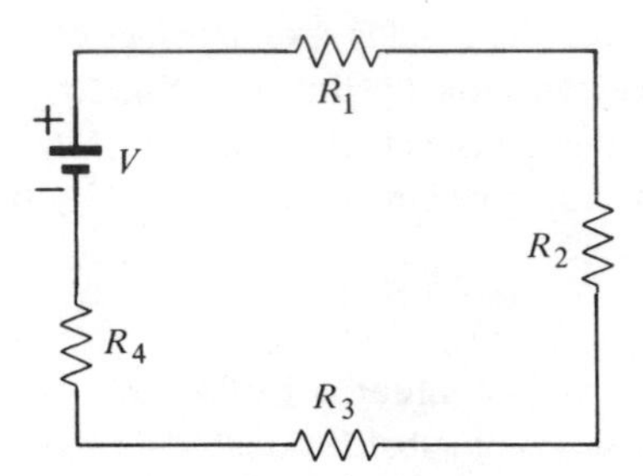

Fig. 3-7. Electric series circuit.

in the circuit concept) to find the mmf needed to produce a given flux in a series magnetic circuit, the opposite problem, that of finding the flux produced by a given mmf in such a core, leads to a peculiar difficulty that is a direct consequence of the nonlinearity of the relation between B and H. There is simply no straightforward way to break up the total mmf into its parts according to Eq. 3-14, since the values of H and B (and, therefore, of μ) are different for each part of the magnetic circuit. The problem can be solved only by an iterative process of *trial and error*, beginning with an arbitrary guess, as illustrated by Example 3-2.

3-5 PARALLEL BRANCHES

Figure 3-8 shows a typical magnetic circuit of three parts where parts 2 and 3 are in parallel. Since lines of flux must be closed lines, and since all lines of flux

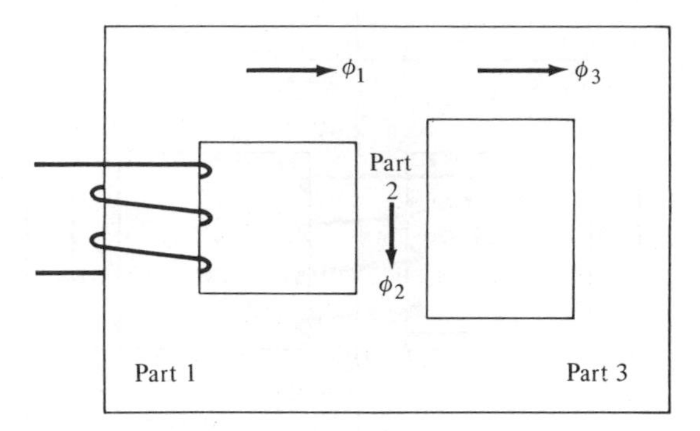

Fig. 3-8. Magnetic circuit with parallel branches.

are inside the magnetic core, we can write

$$\phi_1 = \phi_2 + \phi_3 \tag{3-17}$$

Figure 3-9 shows the electric circuit analog to the core of Fig. 3-8. Equation 3-17 is clearly the magnetic counterpart to *Kirchhoff's current law*. Assumed positive directions for fluxes are assigned arbitrarily, similar to branch currents in an electric network.

In the three-legged core of Fig. 3-8, there are three different ways to apply Ampère's law (Eq. 3-1, approximated by Eq. 3-14), just as there are three loops in the electric circuit of Fig. 3-9 to which Kirchhoff's voltage law can be applied. For instance, for the loop formed by legs 1 and 2,

$$H_1 l_1 + H_2 l_2 = Ni, \text{etc.}$$

A typical sample of a numerical problem for such a core is given in Example 3-3.

In later chapters dealing with rotating machines, we shall encounter symmetrical magnetic circuits. Figure 3-10 illustrates the principle. We can see that, in that core,

$$\phi_1 = \phi_3 = \phi_2/2, \quad \text{and} \quad Ni = \phi_1 \mathcal{R}_1 + \phi_2 \mathcal{R}_2 = \phi_2(\mathcal{R}_2 + \mathcal{R}_1/2)$$

where $\mathcal{R}_1/2$ is the combined reluctance of the two equal parallel legs.

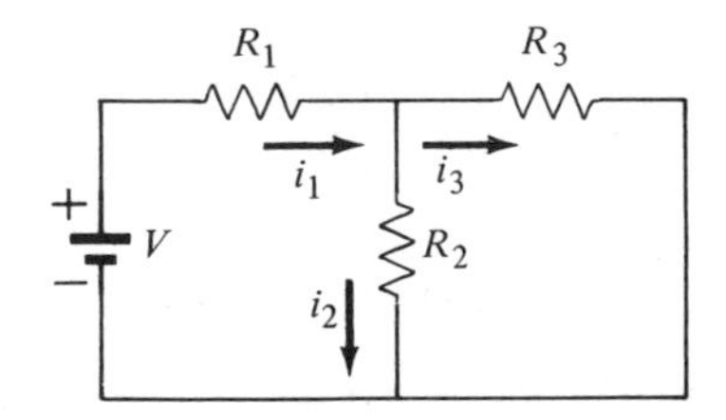

Fig. 3-9. Electric circuit with parallel branches.

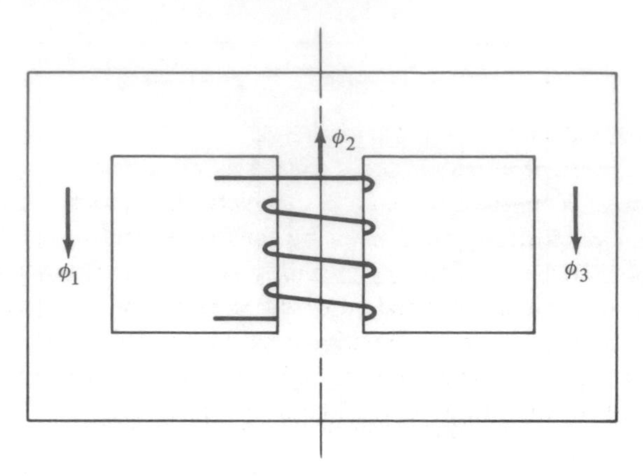

Fig. 3-10. Symmetrical core.

3-6 AIR GAPS

Many electromagnetic devices have air gaps in the core to allow access to the magnetic field or to permit motion of some parts of the magnetic circuit. Figure 3-11 shows such a core. If the length l_g of the air gap (i.e., the distance across the gap) is short compared to the dimensions associated with the cross-sectional area, then the lines of flux will be confined within a region that allows us to apply our magnetic circuit assumptions. The fact that the lines of flux tend to bulge in the air gap (a fact usually referred to as "fringing") can be taken into consideration in an empirical way, if necessary, by using an equivalent cross-sectional area A_g, which slightly exceeds the area of the adjacent iron part, so that the flux density in the air gap is assumed to be a little less than in these iron parts. The reluctance of the air gap thereby becomes

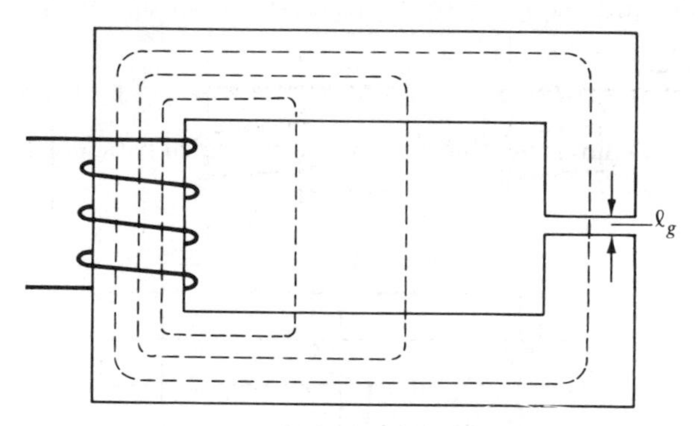

Fig. 3-11. Magnetic circuit with air gap.

$$\mathcal{R}_g = \frac{l_g}{\mu_0 A_g} \tag{3-18}$$

For a sample calculation, see Example 3-4.

Although the length of an air gap is usually no more than a tiny fraction of the entire length of a magnetic circuit, the reluctance of such a gap is almost invariably a major portion, often even by far the largest one, of the total reluctance. This is so because the permeability of air is so much smaller than the permeability of the core material. For the same reason, the presence of an air gap also tends to increase the relative importance of the leakage flux. This is illustrated in Fig. 3-11 where the dotted lines are suggestive of the flux paths, but should not be interpreted as the literal flux distribution.

3-7 SATURATION CURVES

Magnetic circuit calculations are concerned with relations between mmfs and fluxes, much as electric circuit calculations are all about voltages and currents. If it were not for the curvature of the B versus H curve, fluxes would be proportional to mmfs, and vice versa. As it is, this relation is characterized by *saturation*.

In a uniform magnetic core (i.e., a magnetic circuit consisting of only one part), we shall see that the ϕ versus $\mathcal{F}$ curve is identical with the B versus H curve, with appropriate scale changes. If the mmf is that of a single coil, the curve can also be changed by further scale changes into one describing the relation between the flux linkages and the current in the coil.

Figure 3-12 demonstrates all these scale changes. The original curve (B versus

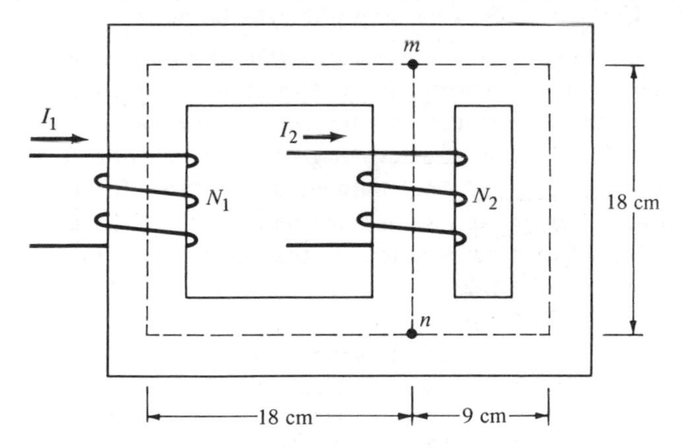

Fig. 3-12. Scale changes.

H) describes the property of the core *material*. For any point of the curve, the ratio of ordinate over abscissa is the *permeability* $\mu = B/H$.

By multiplying the abscissa scale by the length of the mean line of flux, we obtain the mmf $\mathcal{F} = Hl$. By multiplying the ordinate scale by the cross-sectional area, we obtain the flux $\phi = BA$. Now the curve describes the property of the *magnetic circuit*. For any of its points, the ratio of ordinate over abscissa is the permeance $\mathcal{P}$.

Furthermore, by dividing our new abscissa scale by the number of turns of the coil wound around the core, we obtain the current $i = \mathcal{F}/N$, and by multiplying our new ordinate scale by the same number of turns, we obtain the flux linkages $\lambda = \phi N$. This time, the curve describes the property of the *electric circuit* (of the coil). For any of its points, the ratio of ordinate over abscissa is the inductance $L = \lambda/i$ (see Eq. 2-10).*

We note that all these three ratios, μ, $\mathcal{P}$, and L, have values that depend on the chosen point on the curve. In other words, since B is not proportional to H, ϕ is not proportional to $\mathcal{F}$, and λ is not proportional to i. Or, due to the saturation of the ferromagnetic material, both the magnetic circuit of the core and the electric circuit of the coil are *nonlinear*. As was mentioned earlier, this fact of nonlinearity will confront us throughout our studies of electromagnetic devices.

The conclusions just reached about the nonlinearity of magnetic and electric circuits are not limited to the case of uniform cores. For a nonuniform core, the ϕ versus $\mathcal{F}$ curve cannot be obtained by simple scale changes. What one has to do, instead, for all parts in series, is to add the abscissas of their scale-changed saturation curves (BA versus Hl). Similarly, for a core with parallel branches, the ordinates must be added. In each case, the result is a composite ϕ versus $\mathcal{F}$ curve, which still looks very much like its component curves, nonlinear with saturation.

It must be mentioned that the terms *saturation curve* or *magnetization curve* can be applied to all such curves. In most cases, when such a term is used, it is necessary to specify (unless it is clear from the context) what is meant by it: a description of the behavior of a ferromagnetic material, or of a core made of such a material, or of a coil wound around such a core. It should be pointed out, furthermore, that the terms *linear* magnetic circuit, *linear* inductance, etc., as used in this book, are meant to refer to the case where the saturation curve is a straight line through the origin.

*Strictly speaking, $L = \lambda/i$ is valid only for the case where the graph of λ versus i is a straight line through the origin. The general expression for L is found from $e = d\lambda/dt = (d\lambda/di)\,(di/dt) = L\,di/dt$; thus $L = d\lambda/di$, which is the *slope* of the λ versus i curve while the ratio λ/i can be used as an equivalent or average value of L for a current varying between zero and i. The point remains: the inductance, whether defined as λ/i or as $d\lambda/di$, takes on different values for different values of the current i.

3-8 EXAMPLES

Example 3-1 (Section 3-4)

The core of Fig. E-3-1 is made of cast steel. The dimensions are given in centimeters. The depth into the page is 8 cm. The coil has 300 turns. Find the current that produces a flux of 0.0064 webers.

The core is seen to be uniform except for the right leg, which is thinner. Thus, the core is considered as a magnetic series circuit consisting of two parts whose mean lengths l_1 and l_2 are indicated in the figure.

The pertinent dimensions are $l_1 = (0.05 + 0.22 + 0.05) + 2(0.05 + 0.3 + 0.04) = 0.32 + 0.78 = 1.1$ m, $l_2 = (0.05 + 0.22 + 0.05) = 0.32$ m, $A_1 = 0.1 \times 0.08 = 0.008$ m^2, and $A_2 = 0.08 \times 0.08 = 0.0064$ m^2.

The following tabulation is useful. Fill in all the given data (flux, areas, and lengths), and proceed from left to right until all quantities are known.

Part	ϕ(webers)	A(m^2)	B(webers/m^2)	H(At/m)	l(m)	Hl(At)
1	0.0064	0.008	0.8	620	1.1	682
2	0.0064	0.0064	1.0	900	0.32	288
						970

By way of explanation, here are the computations used for line 1 of the tabulation:

$B_1 = \phi_1/A_1 = 0.0064/0.008 = 0.8$ webers/m^2,
H_1 is then found from the graph (Fig. 3-5) for cast steel,
$H_1 l_1 = 620 \times 1.1 = 682$ At.

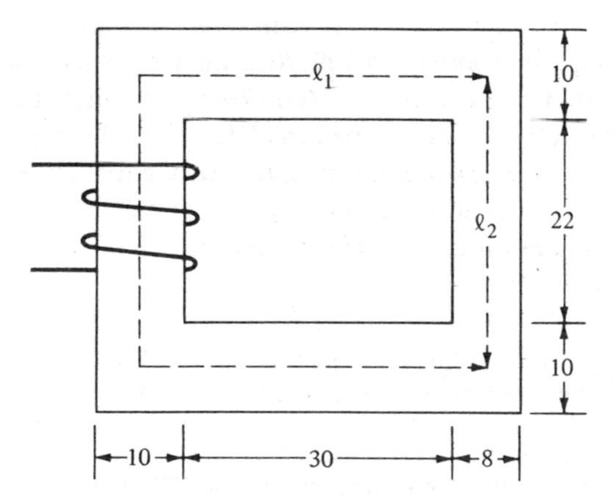

Fig. E-3-1. Core of Example 3-1.

The same procedure is used in line 2. The result is $I = NI/N = \Sigma(Hl)/N = 970/300 = \mathit{3.233\ amp}$.

Familiarity with this procedure will permit computing and entering results in the tabulation without writing out all details. This will result in some saving of writing, as well as conveniently summarizing all quantities that pertain to a magnetic circuit problem.

Example 3-2 (Section 3-4)

For the core of the previous example, find the flux produced by an mmf of 2000 At.

This is a typical problem to be solved by trial and error. The question is how to make the first guess. It should be assumed that the previous problem has not been solved before; otherwise, it would already be known that the result must be substantially higher than 0.0064 webers.

Ignoring the solution of Example 3-1, however, leaves even the order of magnitude of the result completely unknown. Instead of making a "wild" first guess of ϕ, it is preferable to guess at a value of $(Hl)_1$ or $(Hl)_2$, since it is known that each of them must be less than 2000 At.

The following procedure is based on a first guess of $(Hl)_1 = 1000$ At. Use the tabulation introduced for the previous example, and enter the known dimensions and the first guess in their proper places. Then follow line 1 from right to left.

Part	ϕ	A	B	H	l	Hl
1	0.008	0.008	1.0	909	1.1	1000
2	0.008	0.0064	1.25	1600	0.32	512
						1512

Thus, $H_1 = (Hl)_1/l_1 = 1000/1.1 = 909$ At. Find B_1 from the curve for cast steel, and get $\phi = B_1 A_1 = 1.0 \times 0.008 = 0.008$ webers. Enter this result in line 2, and follow it from left to right, leading to $(Hl)_2 = 512$ At. Since the total mmf came out lower than the given value of 2000 At, the first guess must have been too low.

Repeat the procedure with, say, $(Hl)_1 = 1500$ At.

Part	ϕ	A	B	H	l	Hl
1	0.00944	0.008	1.18	1364	1.1	1500
2	0.00944	0.0064	1.475	3200	0.32	1024
						2524

This time, the guess was too high. Try one in between, perhaps $(Hl)_1 = 1250$ At.

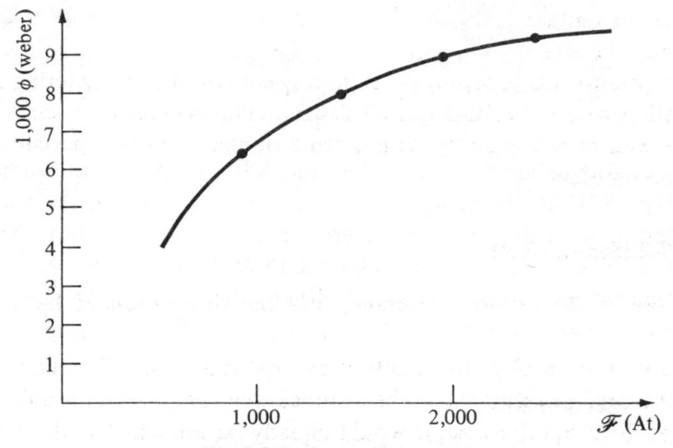

Fig. E-3-2. Saturation curve.

Part	ϕ	A	B	H	l	Hl
1	0.00888	0.008	1.11	1136	1.1	1250
2	0.00888	0.0064	1.39	2400	0.32	768
						2018

Clearly, this can be repeated to any extent of accuracy desired. In most cases, a result like that obtained from the third guess in this example would be considered close enough.

It is of some interest to plot the results of all guesses (plus that of Example 3-1) to obtain the composite curve of ϕ versus $\mathcal{F}$. (See Section 3-7.) This is done in Fig. E-3-2. Note the pronounced saturation for high values of $\mathcal{F}$.

Example 3-3 (Section 3-5)

The three-leg magnetic circuit shown in Fig. 3-8 is made of cast steel and has effective dimensions as shown in the following table:

Part	$A(\text{m}^2)$	l(m)
1 left	0.0090	0.56
2 center	0.0032	0.26
3 right	0.0045	0.51

The coil has 300 turns. Find the value of current that will make the flux in leg 3 be 0.005 weber.

Part	ϕ	A	B	H	l	Hl
1	0.00935	0.009	1.04	980	0.56	548
2	0.00435	0.0032	1.36	2200	0.26	571
3	0.005	0.0045	1.11	1120	0.51	571

Since ϕ_3 is given, start there; complete line 3 from left to right, leading to the result $(Hl)_3 = 571$ At. Since legs 2 and 3 are in parallel, this is also the value of $(Hl)_2$. Starting with this value, complete line 2 from right to left. Next, compute $\phi_1 = \phi_2 + \phi_3 = 0.00435 + 0.005 = 0.00935$ weber. Beginning with this value, complete line 1 from left to right. Then, obtain i from Ampère's law $Ni = (Hl)_1 + (Hl)_2 = 548 + 571 = 1119$ At and $i = 1119/300 = 3.73$ amp.

Example 3-4 (Section 3-6)

Return to the core of Fig. E-3-1, but alter it by cutting an air gap 0.1 cm long into the right-side leg. Find the current needed to produce a flux of 0.0064 weber. Neglect leakage and fringing.

Part	ϕ	A	B	H	l	Hl
1	0.0064	0.008	0.8	620	1.1	682
2	0.0064	0.0064	1	900	0.32	288
Gap	0.0064	0.0064	1	796,000	0.001	796
						1766

Since fringing is to be neglected, the cross-sectional area for the air gap is the same as for Part 2 of the magnetic circuit. The length of Part 2 could be corrected to 0.319 m without making any appreciable difference. The whole procedure is the same as in Example 3-1, except only that the value of H for the air gap is not found from a curve, but rather by dividing the value of B by $\mu_0 = 4\pi \times 10^7$. The result of the problem is 1766 At.

3-9 PROBLEMS

3-1. Find the value of permeability of sheet steel for flux densities of $B_1 = 0.4$, $B_2 = 0.8, B_3 = 1.2, B_4 = 1.6$ webers/m^2.

3-2. Find the value of relative permeability ($\mu_r = \mu/\mu_0$) for sheet steel, cast steel, and cast iron for a flux density of 0.8 webers/m^2.

3-3. A ring of ferromagnetic material has a rectangular cross-section. The inner diameter is 19 cm, the outer diameter is 23 cm, and the thickness is 2 cm. There is a coil of 500 turns wound on the ring. When the coil has a current of 3 amp, the flux in the core is 0.0006 weber. On the basis of the magnetic circuit concept, find the following quantities, all in rationalized mks units: (a) the mmf $\mathfrak{F}$; (b) the field intensity H; (c) the flux density B; (d) the reluctance $\mathfrak{R}$; (e) the permeability μ; (f) the relative permeability $\mu_r = \mu/\mu_0$.

3-4. The core shown in Fig. P-3-4 is made of cast iron. The depth into the page is 7.5 cm. Coil N_1 has 537 turns. Find the current I_1 that will produce a flux of 0.003 weber.

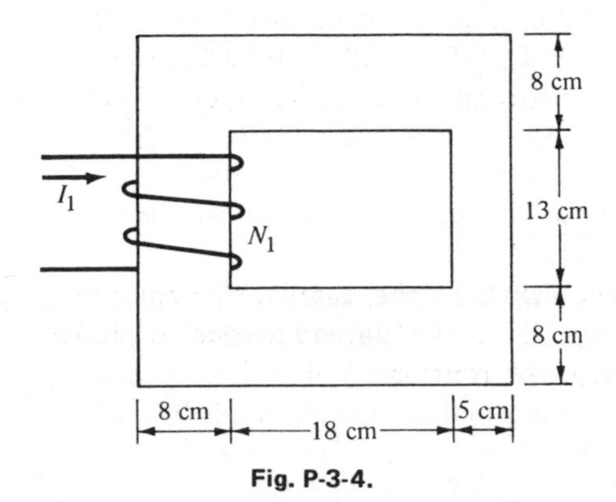

Fig. P-3-4.

3-5. The three-leg magnetic circuit shown in Fig. P-3-5 is made of sheet steel. Effective dimensions are shown in the following table.

Part	$A(\text{m}^2)$	$l(\text{m})$
1 left	0.0014	0.46
2 right	0.0012	0.40
3 center	0.0024	0.14

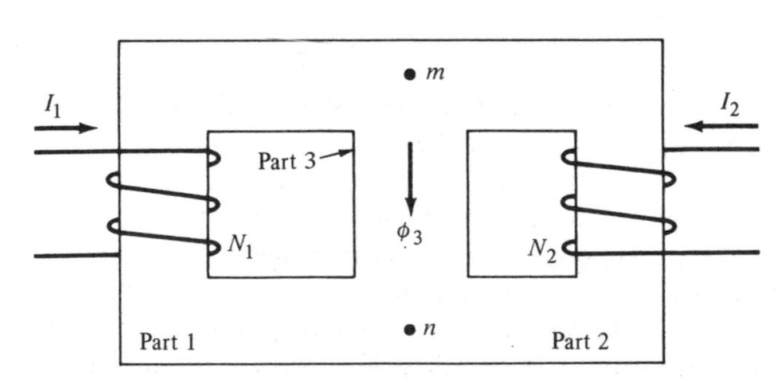

Fig. P-3-5.

Coil N_1 has 160 turns. Current I_1 is fixed at 1.1 amp. Coil N_2 has 180 turns. Find the current I_2 that will make the flux in the center leg, ϕ_3, be 0.0034 weber.

3-6. The three-leg magnetic circuit shown in Fig. P-3-5 is made of cast steel. Effective dimensions are given in the following table.

Part	A(m^2)	l(m)
1 left	0.0018	0.52
2 right	0.0016	0.50
3 center	0.0026	0.16

Coil N_1 has 230 turns. Coil N_2 has 300 turns. Current I_2 is fixed at 2 amp. Find the current I_1 that will make the flux in the center leg, ϕ_3, be 0.0032 weber.

3-7. The magnetic circuit shown in Fig. P-3-7 is made of laminated sheet steel with a stacking factor of 0.94. (The stacking factor is the ratio of the effective area to the gross area.) The gross dimension for depth into the page is 8 cm. The length of the air gap is 0.4 cm. The other dimensions are shown in the figure. The coil N_1 has 800 turns. Find the current in the coil that will make the flux be 0.0086 weber. Neglect fringing.

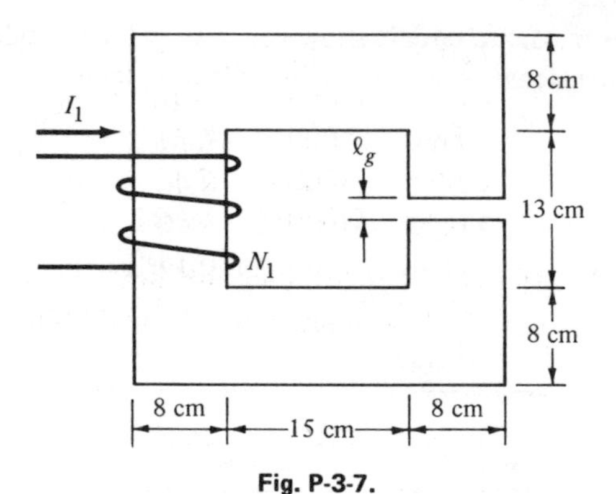

Fig. P-3-7.

3-8. For the magnetic circuit in Problem 3-7 and Fig. P-3-7, find the flux produced by a current of 4.5 amp. Neglect fringing.

3-9. The magnetic circuit of Problem 3-7 is to be energized with 10 amp in the 800-turn coil. The flux is to be 0.0086 weber. Find the length of the air gap that is required for these conditions. Neglect fringing.

3-10. The cast iron core shown in Fig. P-3-10 has effective dimensions as follows:

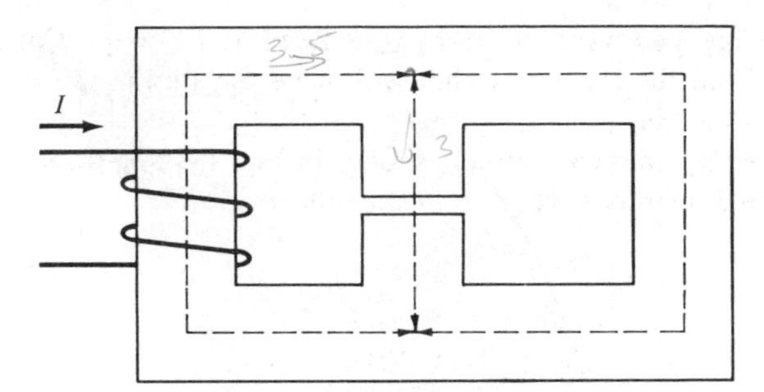

Fig. P-3-10.

Part	$A(\text{m}^2)$	$l(\text{m})$
1 left	0.0038	0.36
2 right	?	0.52
3 center	0.0032	0.14
4 air gap	0.0035	0.0025

The coil has 8600 At. The flux through the air gap is to be 0.003 weber. Find the cross-sectional area of the right side of the core.

3-11. The cast steel core shown in Fig. P-3-11 has effective dimensions as shown in the following table.

Part	$A(\text{m}^2)$	$l(\text{m})$
1 left	0.0006	0.40
2 right	0.0007	0.50
3 center	0.0008	0.12
4 air gap	0.00086	0.003

The coil on the right has 2950 At. Find the mmf of the other coil for the following three cases: (a) The air gap flux is 0.0009 weber. (b) The air gap flux is zero. (c) The left-side flux is zero.

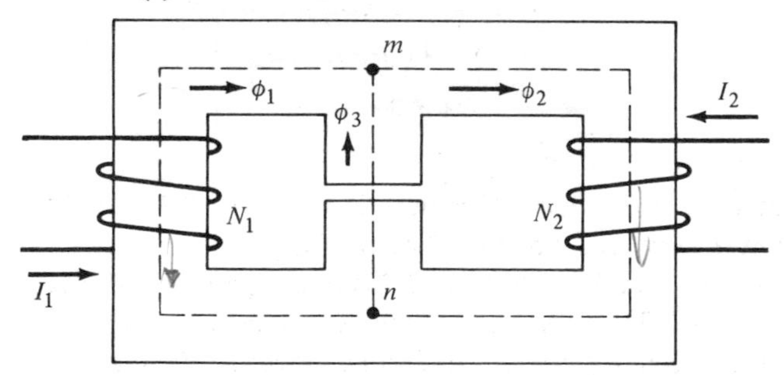

Fig. P-3-11.

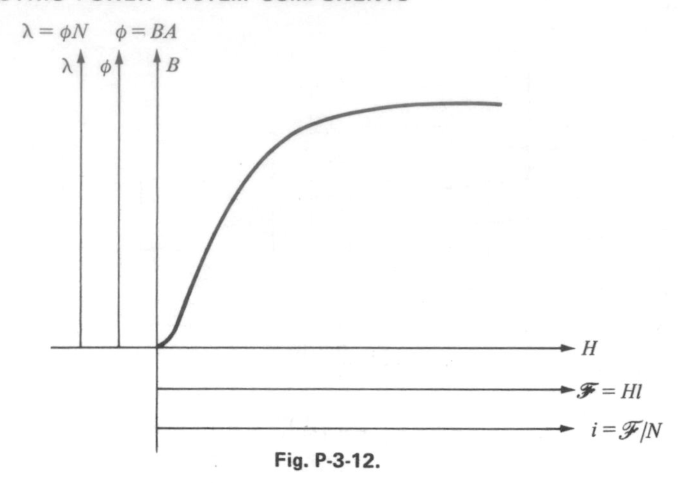

Fig. P-3-12.

3-12. For the core shown in Fig. P-3-12, all legs have the same cross-sectional area. Distances are given in the figure. Find the mmf ratio N_1I_1/N_2I_2 for the following three cases: (a) The flux in the left leg is zero. (b) The flux in the center leg is zero. (c) The flux in the right leg is zero.

3-13. A magnetic core is made with two parts as shown in Fig. P-3-13(a). Effective dimensions are as shown.

Part	$A(\mathrm{m}^2)$	$l(\mathrm{m})$
1 left	0.0005	0.23
2 right	0.0003	0.08

The core material has a magnetization curve that is approximated in idealized form as shown in Fig. P-3-13(b). The coil supplies 100 ampere-turns (At). Find the flux, the flux density, and the field intensity in each part of the core.

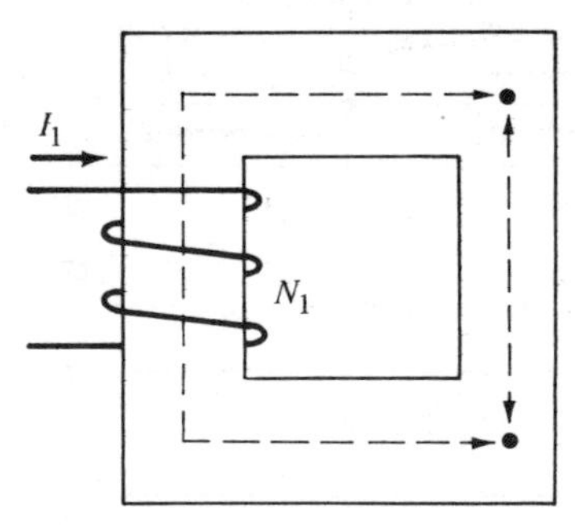

Fig. P-3-13(a).

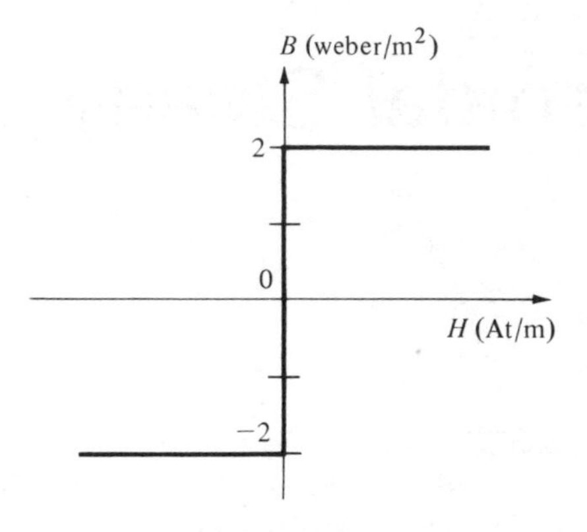

Fig. P-3-13(b).

3-14. The core material in Fig. P-3-14 has a magnetization curve with idealized form as given in Fig. P-3-13(b). Effective dimensions are shown.

Part	$A(\text{m}^2)$	$l(\text{m})$
1 core	0.0038	0.62
2 air gap	0.0042	0.0032

The coil N_1 has 850 turns. Find the flux across the air gap and the field intensity in the core for the following values of current in the coil: (a) 2 amp, (b) 4 amp, (c) 6 amp, and (d) 8 amp.

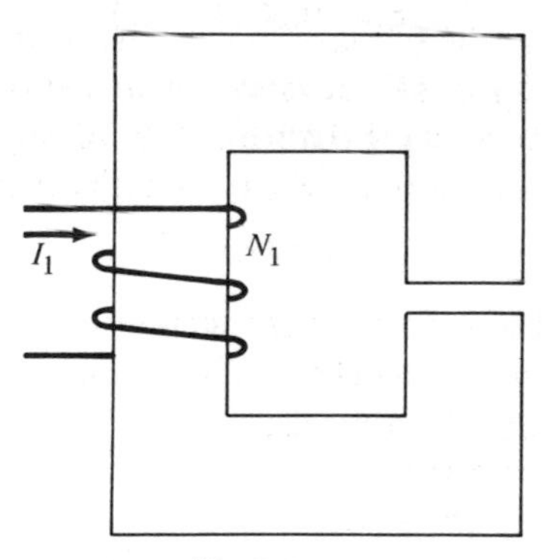

Fig. P-3-14.

4

Sinusoidal Steady State

4-1 REVIEW

In many of the engineering devices to be studied in this book, there are time-varying or moving magnetic fluxes. For continuous operation, a time variation must be *periodic*, and this means usually *sinusoidal*. Any departure from this ideal waveshape can be, if necessary, taken into account by the principle of the Fourier series. That is, periodic waveshapes must be composed of sinusoids, called fundamentals and harmonics, and the harmonics can be disregarded if they are small enough, or else they can be considered separately.

For most of the devices to be considered in this book, sinusoidal steady state is the most important mode of operation.* Although the reader is expected to be reasonably familiar with the applicable methods of analysis, a brief review might be welcome to many, at least to establish the notation and terminology to be used.

For instance, let a current be a sinusoidal function of time

$$i = I_{\max} \cos(\omega t + \alpha) \tag{4-1}$$

where the *amplitude* $I_{\max}$ (the subscript stands for *maximum value*) is $\sqrt{2}$ times the *rms* (root-mean-square) or *effective value* I, and the *radian frequency* ω (in radians per second) in 2π times the *frequency* f (in Hertz). This current is represented by the *phasor*

$$\mathbf{I} = I \underline{/\alpha} \tag{4-2}$$

where the boldface symbol is used to distinguish the phasor, which is a complex number, from its magnitude. The phasor thus introduced is an rms phasor. It is also possible to use amplitude phasors like

$$\mathbf{I}_{\max} = I_{\max} \underline{/\alpha} \tag{4-3}$$

The value of phasor representation is largely based on the fact that the phasor representing a *sum* of sinusoids (of the same frequency) is the complex sum of the phasors representing the component sinusoids to be added. Therefore, Kirchhoff's laws can be written as phasor equations for sinusoidal steady state.

*The popular term *a-c* (alternating current) is less specific but acceptable.

Furthermore, the differential equations by which voltages across passive circuits are related to currents through the circuits, can be replaced by complex equations in which voltage and current phasors are proportional to each other, the factor of proportionality being the complex *impedance* or *admittance* of the circuit.

A sinusoidal time function whose phase angle is zero is said to be in the *axis of reference*. Its phasor appears in a *phasor diagram* as a horizontal line with an arrow pointing in the positive (right-hand) direction. The choice of the voltage or current to be the axis of reference is arbitrary since it amounts to the same thing as choosing the origin of the time scale, i.e., the instant when $t = 0$.

When current and voltage are time-varying, their product, the power, must also be a function of time. The *average value* of power (over one period or any whole multiple thereof) for sinusoidal waveshapes is

$$P = VI \cos \theta \tag{4-4}$$

where V and I are the rms values of voltage and current, θ is the phase difference between them, and its cosine is called the *power factor*. For pure energy-consuming elements, $\theta = 0$ and $\cos \theta = 1$, whereas for pure energy-storing elements (inductance and capacitance), $\theta = \pm 90^\circ$ and $\cos \theta = 0$. The *average power* P is also often called the *real power*, *active power*, or simply *power* (in the context of sinusoidal steady state).

The *reactive power*

$$Q = -VI \sin \theta \tag{4-5}$$

accounts for the presence of energy-storing elements in the same way in which P accounts for the presence of energy-consuming elements. The minus sign is an arbitrary choice adopted by many (though not all) authors. To make it meaningful, a more specific definition of the angle θ is required. In this book (as in many others), θ is always to be understood as the angle by which the voltage leads the current. Thereby, inductive (*lagging*) reactive power is negative, and capacitive (*leading*) reactive power is positive. Reactive power has the same dimension as active power, but its established unit is the *var* (volt-ampere reactive), in contrast to the *watt* by which active power is measured.

The *apparent power* is the product of voltage and current rms values

$$P_a = VI \tag{4-6}$$

This quantity is used to express the *rating* (see the last two paragraphs of Section 1-6) of a power device intended to operate in the sinusoidal steady state, because the power *losses* in such a device depend on voltage and current and not on their phase relation. The unit of apparent power is the *volt-ampere*.

All these quantities can be combined into a single complex number, the *complex power*. In Cartesian (rectangular) form it is

$$\mathbf{P}_a = P + jQ \tag{4-7}$$

By substituting Eqs. 4-4 and 4-5 into this definition, we obtain the complex power in polar form

$$\mathbf{P}_a = VI(\cos\theta - j\sin\theta) = VI\,e^{-j\theta} = VI\,\underline{/-\theta} = P_a\,\underline{/-\theta} \tag{4-8}$$

So complex power contains them all: average power is its real part, reactive power is its imaginary part, apparent power is its magnitude, and the negative of the phase angle is its angle.

4-2 HYSTERESIS

When a coil wound around a ferromagnetic core carries an alternating current i, then the mmf $\mathfrak{F} = Ni$ also alternates, and so does the flux ϕ produced by it. Nevertheless, the instantaneous values of ϕ corresponding to the instantaneous values of $\mathfrak{F}$ cannot be taken from the saturation curve ϕ versus $\mathfrak{F}$. That is so because, for an alternating field, the relation between the instantaneous values of B and H does not follow a curve like Fig. 3-4, but rather a loop like Fig. 4-1.

What this diagram indicates is that the values of B depend not only on the

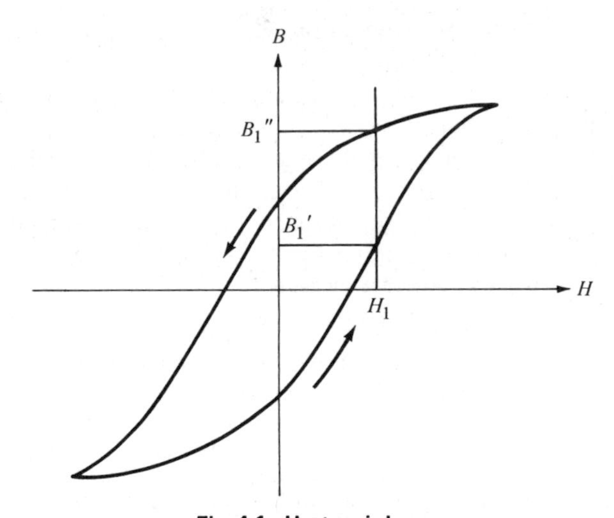

Fig. 4-1. Hysteresis loop.

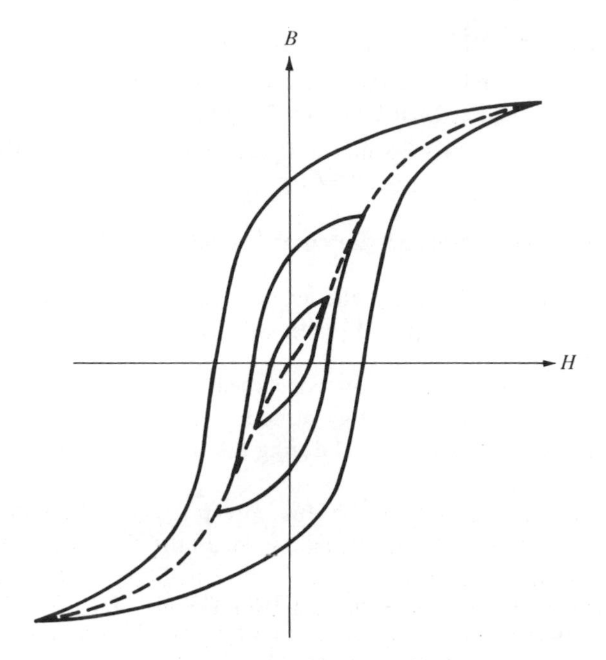

Fig. 4-2. Family of hysteresis loops.

values of H, but also on the previous history of the material. For instance, on the *ascending* (lower) branch of the loop, the value corresponding to H_1 is B_1' whereas, on the *descending* (upper) branch, the value corresponding to H_1 is B_1''. The whole loop is called a *hysteresis loop*.*

The size and shape of such a loop depends (for a given material) on the maximum value of H. Figure 4-2 shows a family of hysteresis loops for several values of H_{max}. The dotted line drawn through the end points of the loops is called the *normal* saturation (or magnetization) curve. This is the curve that is used for magnetic circuit calculations like those in the previous chapter; in fact, the curves of Fig. 3-5 are such normal magnetization curves; they are also approximately valid for d-c magnetization.

Thus, the phenomenon of hysteresis may be bypassed in magnetic circuit calculations, but it is very important nonetheless. To understand its significance, we must first study the energy stored in a magnetic circuit.

*Hysteresis is Greek for *lagging*. What is meant is that B lags H (except at the tip of the loop). For instance, when H is positive and reaches its zero value, B is still positive (at its "residual" value B_r; or, when B reaches zero, then H is already negative, etc.

4-3 ENERGY STORAGE

Let a coil wound around a ferromagnetic core be connected to a voltage source (of arbitrary wave shape). Then a current flows through the coil, and

$$v = Ri + \frac{d\lambda}{dt} \tag{4-9}$$

where R is the resistance of the coil circuit. The source delivers energy to the circuit at the rate

$$p = vi = Ri^2 + i\frac{d\lambda}{dt} \tag{4-10}$$

The first term represents the power consumed by the resistance. The second term is the rate of change of energy stored in the magnetic field. Thus, the energy delivered to the magnetic field during the time interval from t_1 to t_2 is

$$\Delta W_m = \int_{t_1}^{t_2} i\frac{d\lambda}{dt}\,dt = \int_{\lambda_1}^{\lambda_2} i\,d\lambda \tag{4-11}$$

where λ_1 and λ_2 are the flux linkages at the instants t_1 and t_2, respectively. In the case of a linear circuit (actual or approximated), this expression may be resolved into the form familiar from electric circuit analysis, by substituting $\lambda = Li$:

$$\Delta W_m = \int_{i_1}^{i_2} i\,d(Li) = \frac{L}{2}(i_2^2 - i_1^2) \tag{4-12}$$

In general (i.e., without assuming linearity), the energy may also be expressed in magnetic circuit terms since the magnetic field is confined to the core. Thus

$$\Delta W_m = \int_{\phi_1}^{\phi_2} \frac{\mathfrak{F}}{N}\,d(N\phi) = \int_{\phi_1}^{\phi_2} \mathfrak{F}\,d\phi \tag{4-13}$$

where ϕ_1 and ϕ_2 are the values of the flux at the instants t_1 and t_2.

Now the core can always be subdivided into parts within which the field is uniform (even if these parts have to be infinitesimal). For each such part, the substitutions $\mathfrak{F} = Hl$ and $\phi = BA$ can be made. Then the energy supplied to the entire core between instants t_1 and t_2 is the sum

$$\Delta W_m = \sum_k \int_{B_{k_1}}^{B_{k_2}} H_k l_k A_k\,dB_k \tag{4-14}$$

where B_{k_1} and B_{k_2} are the flux densities in part k at t_1 and t_2. The product $l_k A_k$ is the volume of part k of the core. Thus, we arrive at an expression for the increase of *energy density* (energy per volume) during the time interval from t_1 to t_2 for part k. Using the lower case symbol w for energy *density*, and omitting the subscript k,

$$\Delta w_m = \int_{B_1}^{B_2} H\,dB \tag{4-15}$$

This integral appears in Fig. 4-3 as the shaded area to the left of the saturation curve. Incidentally, if the magnetic circuit is *linear*, the substitution of $B = \mu H$ or $H = B/\mu$ makes it possible to evaluate the integral analytically:

$$\Delta w_m = \int_{H_1}^{H_2} H d(\mu H) = \frac{\mu}{2}(H_2^2 - H_1^2) = \frac{1}{2\mu}(B_2^2 - B_1^2) = \frac{1}{2}(B_2 H_2 - B_1 H_1) \tag{4-16}$$

the last of which can be recognized as the area of the trapezium into which the shaded area of Fig. 4-3 changes for the linear case.

Now return to the hysteresis loop, which is redrawn in Fig. 4-4. Let the state of the core material at a certain part of the core be given by the coordinates of point *a* of the diagram for the instant t_1 and by those of point *b* for the instant t_2. Thus, the energy density increases between t_1 and t_2 by an amount represented by the area *abca*. Continue along the descending branch until point *d* is reached at the instant t_3. During the time interval from t_2 to t_3 the energy density decreases by an amount represented by the area *bcdb*. Note that more energy has been supplied to the core than was returned by it even though H has the same value (zero) at t_1 and t_3. The excess energy per unit volume for one-half cycle, represented by the area *abda*, is converted into *heat* inside the ferro-

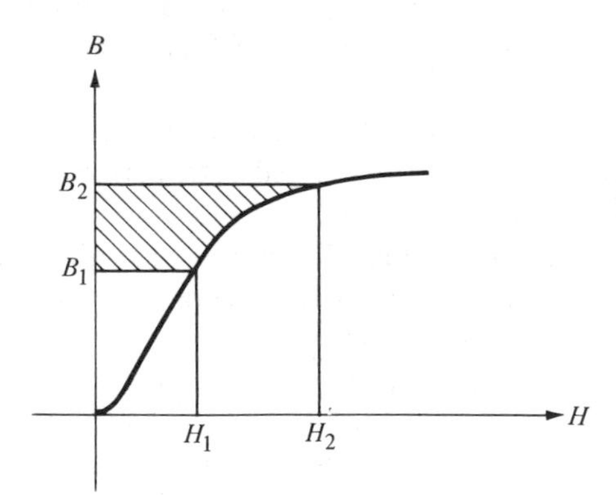

Fig. 4-3. Energy density.

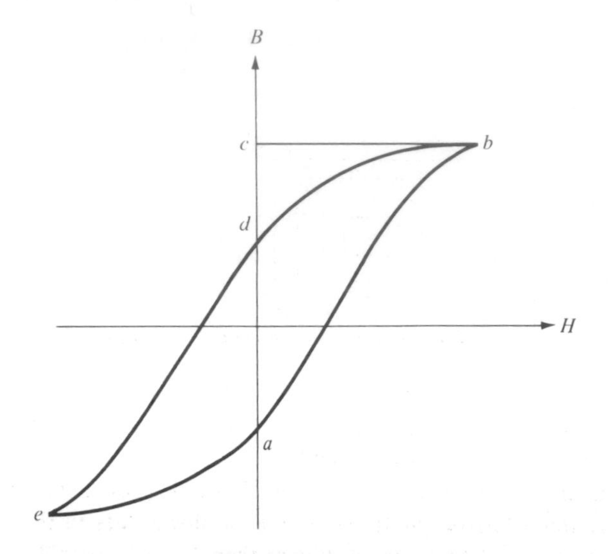

Fig. 4-4. Hysteresis loss.

magnetic material. From the symmetry of the hysteresis loop it may be concluded that the energy per unit volume per cycle converted into heat due to hysteresis is represented by the whole *loop area abdea*.

The size of this area depends on the material and on the extent to which it is magnetized. For a given material, it may be expressed as a function of the maximum flux density

$$\oint H\,dB = k_h(B_{\max})^n \tag{4-17}$$

where the factor k_h varies quite widely but the exponent n ranges only between 1.5 and 2.5, depending on the material. Both constants, k_h and n, are found empirically.

4-4 CORE LOSSES

The previous section has shown that, in a power-converting device using an iron core, a certain amount of power must be lost due to hysteresis if the flux in the core alternates. The term *hysteresis loss* must be understood to mean the average *power* loss. It is obtained by multiplying the energy loss per cycle by the number of cycles per second (i.e., by the frequency f expressed in Hertz). For a uniform core (or part of a core), the hysteresis loss in watts is

$$P_h = k_h \mathcal{V} f(B_{\max})^n \tag{4-18}$$

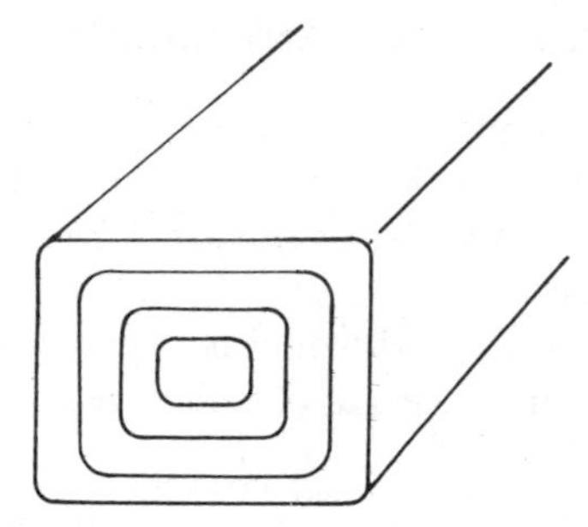

Fig. 4-5. Core cross-section and eddy-current paths.

where $\mathcal{V}$ is the *volume* of the core (or of the part). For a nonuniform core, the losses are computed for each part and added.

If an experiment is conducted in which an alternating source is connected to a coil wound around a ferromagnetic core and the power loss in the core is measured, this loss turns out to be much higher than Eq. 4-18 would indicate. The explanation is that there are additional power losses attributed to currents in the core. These currents are called *eddy currents*, and they are caused by electromagnetic induction since their paths, as sketched in Fig. 4-5, link alternating fluxes. In fact, power losses due to eddy currents would be altogether too high for efficient power-converting devices if they were not sharply reduced by the standard remedy of breaking down the eddy-current paths. This is done by using not solid but *laminated* cores for alternating fluxes. Such cores consist of sheets separated from each other by thin insulating layers.* The *eddy current loss* in a laminated core (or, again, part of a core) can be approximated by

$$P_e = k_e \mathcal{V} \tau^2 f^2 (B_{\max})^2 \tag{4-19}$$

where τ is the thickness of the laminations.

Since both hysteresis loss and eddy-current loss produce heat in the core, they may be lumped together, under the name *core losses* (or, less accurately, iron losses)

$$P_c = P_h + P_e = k_h' f (B_{\max})^n + k_e' f^2 (B_{\max})^2 \tag{4-20}$$

In the last form, the core losses are expressed in terms of variable quantities only (volume and thickness of laminations being constants for a given core).

*As an inevitable consequence, the effective cross-sectional area of the core is slightly smaller than the gross area. The ratio of effective area to gross area is called the *stacking factor*.

4-5 WAVE SHAPES AND EQUIVALENT CIRCUITS

Once again, we visualize connecting a voltage source to a coil wound around a ferromagnetic core. This time, let the source be ideal and of sinusoidal waveshape, and let the effect of the circuit resistance be negligible. Our purpose is to find the steady-state current.

If the circuit were linear, the coil would be a pure linear inductance L. Thus, if the source voltage is taken as the axis of reference,

$$v = V_{max} \cos \omega t \tag{4-21}$$

then the current could be written as

$$i = \frac{V_{max}}{\omega L} \sin \omega t \tag{4-22}$$

which indicates, among other things, that it lags the voltage by 90°.

Actually, the circuit is nonlinear. This has the consequence that, before the current can be found, it is necessary to determine the steady-state flux by means of Faraday's induction law. (Without circuit resistance, the induced voltage equals the source voltage.)

$$\phi = \frac{1}{N} \int v\, dt = \frac{V_{max}}{N\omega} \sin \omega t \tag{4-23}$$

In this last equation, we can recognize the *maximum flux*

$$\phi_{max} = \frac{V_{max}}{N\omega} \tag{4-24}$$

Using the substitutions $V_{max} = \sqrt{2}\, V$ and $\omega = 2\pi f$, the relation between voltage and flux in the sinusoidal steady state can be written

$$V = \frac{2\pi}{\sqrt{2}} N f \phi_{max} = 4.44\, N f \phi_{max} \tag{4-25}$$

with an easily remembered numerical coefficient. Note that the voltage magnitude is expressed by the rms value whereas, for the flux, the maximum value is more significant.

More importantly, the last equation indicates that a sinusoidal source of given magnitude and frequency dictates the steady-state flux magnitude, regardless of any property of the core, or even (as long as the assumption of a negligible resistance holds) in the absence of a core! What the core determines is the size and waveshape of the *current*.

As a first approach, let core losses be neglected. The wave shape of the current may be obtained, point by point, from a *normal saturation curve* ($N\phi$ versus i).

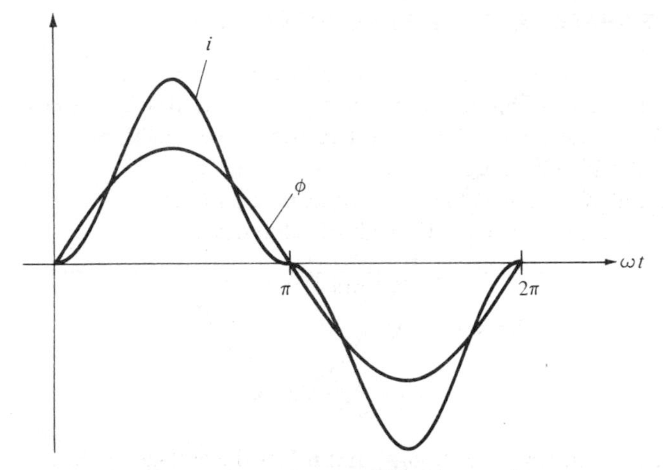

Fig. 4-6. Current wave without hysteresis.

The more saturated the maximum flux, the more sharply peaked must the current wave be, and the less can it resemble a sinusoid. Figure 4-6 shows a sinusoidal flux (according to Eq. 4-23) and the corresponding current wave. Distorted though it is, it can still be considered to be *in phase* with the flux (strictly speaking, the current *fundamental* is in phase with the flux) and thereby lagging the voltage by 90°, as it must in any pure inductance, whether linear or not. Thus, the power factor and the power consumed by the circuit are zero (no wonder, when resistance and core losses have been neglected).

Figure 4-7 is a phasor diagram for this lossless circuit. Since phasors can only represent sinusoids and the current is not a sinusoid, we should understand the current phasor as representing the fundamental of the current wave. The result $P = 0$ remains valid since the current harmonics do not produce any average power in the absence of voltage harmonics.

Now replace the normal saturation curve by a *hysteresis loop*, and use *that* to

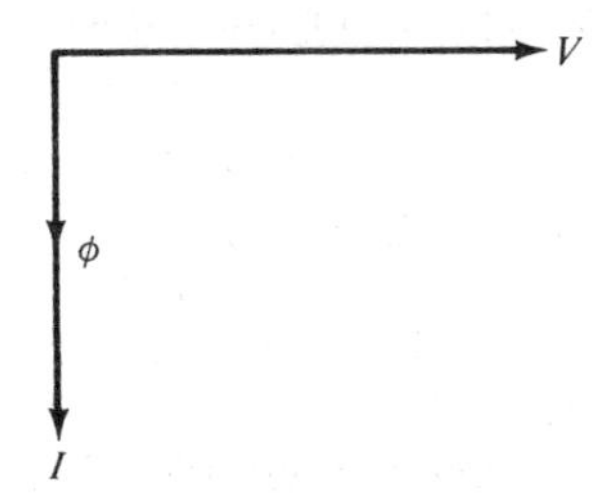

Fig. 4-7. Phasor diagram of lossless coil.

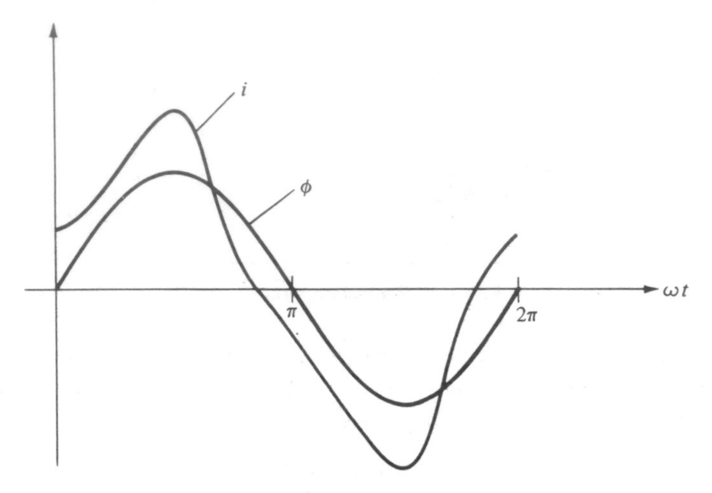

Fig. 4-8. Current wave leading flux.

get the waveshape of the current. The result is shown in Fig. 4-8. This time, the current wave (strictly speaking, again it is fundamental) is seen to lead the flux. More insight can be gained by splitting the current into two fictitious components. The one in phase with the flux is called the *magnetizing current* i_m. The other, i_h, leads the flux by 90° and thereby is in phase with the voltage, as if it were flowing through a pure energy-consuming element. Note that this circuit does consume power, namely the hysteresis loss

$$P = VI \cos \theta = VI_h \tag{4-26}$$

To consider eddy currents as part of this concept, all that is needed is to add another current component i_e in phase with the voltage. Representation of each of these currents (again, omitting their harmonics) by its rms phasor leads to the equation

$$\mathbf{I} = \mathbf{I}_m + \mathbf{I}_h + \mathbf{I}_e \tag{4-27}$$

which can be interpreted as Kirchhoff's current law applied to the *equivalent circuit* of Fig. 4-9. It is also possible to combine the last two terms into one, the *core loss current*

$$\mathbf{I}_c = \mathbf{I}_h + \mathbf{I}_e \tag{4-28}$$

Note that the circuit of a coil wound around a core cannot, if core losses are to taken into consideration, be represented by an inductance alone. Energy-consuming elements must be added to obtain a circuit that is valid for sinusoidal steady state analysis. Since all elements of the equivalent circuit appear in

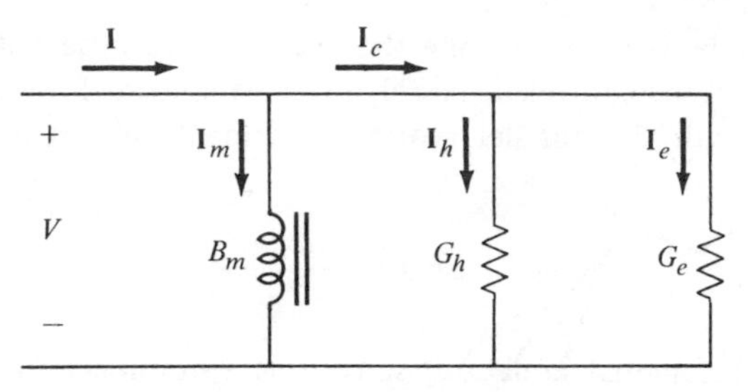

Fig. 4-9. Equivalent circuit.

parallel, they are more conveniently expressed in mhos than in ohms. B_m is called the *magnetizing susceptance*, and G_h and G_e are conductances that can be combined to a single *core loss conductance*

$$G_c = G_h + G_e \tag{4-29}$$

which carries the core loss current $\mathbf{I}_c$. All the elements can be combined to form a complex admittance

$$\mathbf{Y} = G_c + j B_m \tag{4-30}$$

Note, incidentally, that B_m is negative because the current $\mathbf{I}_m$ lags the voltage by 90°. Also, B_m is not a linear circuit element because I_m is not proportional to V, due to saturation. The parallel lines drawn in Fig. 4-9 next to the inductance symbol (making it an "iron-cored" inductance) are suggestive of this fact. Nevertheless, the current $\mathbf{I}$ can always be expressed as

$$\mathbf{I} = \mathbf{YV} \tag{4-31}$$

and the power consumed by the circuit is the core loss

$$P_c = VI \cos \theta = VI_c \tag{4-32}$$

A phasor diagram corresponding to this equivalent circuit is shown in Fig. 4-10.

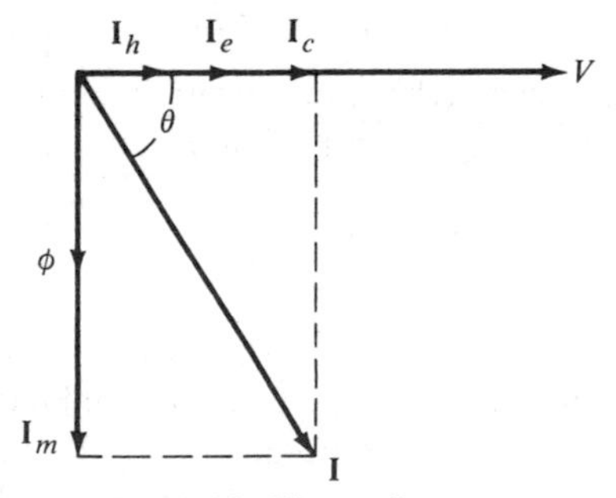

Fig. 4-10. Phasor diagram.

Finally, it may be required to take the resistance R of the coil into consideration. In this case, the equivalent circuit must be amended to contain this resistance in series with the parallel group of elements shown in Fig. 4-9. The equation

$$v = Ri + \frac{d\lambda}{dt} \tag{4-33}$$

is a nonlinear differential equation, since λ is a nonlinear function of i (the saturation curve). A solution (for i) requires an iterative procedure of digital computation.

4-6 EXAMPLES

Example 4-1 (Section 4-3)

The cast steel core in Fig. E-4-1 is assumed to have constant permeability of 1.1×10^{-3} henrys per meter. The coil has 1200 turns. Effective dimensions are: $A_s = 0.003\ \text{m}^2$, $l_s = 0.5$ m, $A_g = 0.0034\ \text{m}^2$, $l_g = 0.0004$ m. The flux in the air gap is 0.003 weber. (a) Find the current in the coil. (b) Find the energy stored in the air gap. (c) Find the energy stored in the steel. (d) Find the self-inductance.

Solution

(a) Solve the magnetic circuit problem using a table.

Part	ϕ(weber)	A(m²)	B(weber/m²)	H(At/m)	l(m)	Hl(At)
steel core	0.003	0.003	1.0	910	0.5	455
air gap	0.003	0.0034	0.884	704,000	0.0004	281

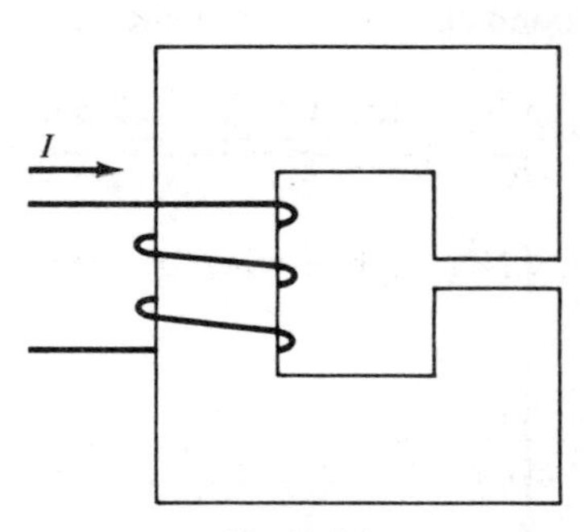

Fig. E-4-1.

The mmf required in the coil is $NI = \Sigma Hl = 736$ At. The current in the coil is $I = 736/1200 = 0.613$ amp.

(b) Use Eq. 4-14 with $B_{k_1} = 0$, and restrict it to just one portion of the magnetic circuit. For the air gap

$$\begin{aligned} W_g &= (l_g A_g)\ \tfrac{1}{2}\ (H_g B_g) \\ &= 0.0004 \times 0.0034 \times \tfrac{1}{2} \times 704{,}000 \times 0.884 \\ &= 0.422 \text{ joules} \end{aligned}$$

(c) For the steel portion, the stored energy is

$$\begin{aligned} W_s &= (l_s A_s)\ \tfrac{1}{2}\ (H_s B_s) \\ &= 0.5 \times 0.003 \times \tfrac{1}{2} \times 910 \times 1 \\ &= 0.683 \text{ joules} \end{aligned}$$

(d) The self-inductance is given by

$$L = \lambda/i = N\phi/i = (1200 \times 0.003)/0.613 = 5.87 \text{ henrys}$$

Example 4-2 (Section 4-4)

A sample of iron having a volume of 33 cm^3 is subjected to a magnetizing force varying sinusoidally at a frequency of 400 Hz. The hysteresis loop is plotted using the following scales: 1 cm represents 300 At/m, and 1 cm represents 0.2 weber/m^2. The area of the hysteresis loop is 57.5 cm^2. Find the hysteresis loss in watts.

Solution

Let P_h' denote the energy loss represented by Eq. 4-17. This is the energy loss per unit volume for one cycle.

$$P_h' = \oint H\,dB = \left(57.5\ \frac{\text{cm}^2}{\text{cycle}}\right)\left(\frac{300 \text{ At/m}}{1 \text{ cm}}\right)\left(\frac{0.2 \text{ weber/m}^2}{1 \text{ cm}}\right) = 3450\ \frac{\text{joules}}{\text{m}^3 \text{ cycle}}$$

The hysteresis loss of Eq. 4-18 is P_h' multiplied by the volume and the frequency.

$$\begin{aligned} P_h &= P_h' \mathcal{V} f \\ &= \left(3450\ \frac{\text{joules}}{\text{m}^3 \text{ cycle}}\right)\left(400\ \frac{\text{cycles}}{\text{sec}}\right)(33 \text{ cm}^3)\left(\frac{1 \text{ m}}{100 \text{ cm}}\right)^3 \\ &= 45.5 \text{ watts} \end{aligned}$$

Example 4-3 (Section 4-4)

The flux in a magnetic core is alternating sinusoidally with a frequency of 400 Hz. The maximum flux density is 0.6 weber/m^2. The eddy-current loss is 28 w. Find the eddy-current loss in this core when the frequency is 300 Hz and the maximum flux density is 0.7 weber/m^2.

Solution

Let $k_e' = k_e \mho \tau^2$

Use Eq. 4-19 to find k_e'.

$$k_e' = \frac{P_{e1}}{f_1^2 B_1^2} = \frac{28}{(400 \times 0.6)^2} = 4.86 \times 10^{-4} \ \frac{\text{watts m}^4}{\text{Hz}^2 \text{ weber}^2}$$

For the new frequency and new maximum flux density, we can find

$$P_{e2} = k_e' f_2^2 B_2^2 = (4.86 \times 10^{-4})(300)^2 (0.7)^2 = 21.4 \text{ w}$$

Example 4-4 (Section 4-4)

The total core losses (hysteresis plus eddy current) for a sheet steel core are found to be 500 w at 25 Hz. When the frequency is increased to 50 Hz and the maximum flux density is kept constant, the total core loss becomes 1400 w. Find the hysteresis and eddy-current losses for both frequencies.

Solution

Since $B_{\max}$ is constant, Eq. 4-20 can have the following form:

$$P_c = Af + Bf^2 \text{ where } A = k_h'(B_{\max})^n \text{ and } B = k_e'(B_{\max})^2$$

For a frequency of 25 Hz, the core loss is

$$P_{c1} = 500 = A(25) + B(25)^2$$

For a frequency of 50 Hz, the core loss is

$$P_{c2} = 1400 = A(50) + B(50)^2$$

Solve the two equations to find $A = 12$ and $B = 0.32$. Now, we can find the individual losses.

$$P_{h1} = Af_1 = 300 \text{ w} \qquad P_{e1} = Bf_1^2 = 200 \text{ w}$$

$$P_{h2} = Af_2 = 600 \text{ w} \qquad P_{e2} = Bf_2^2 = 800 \text{ w}$$

Example 4-5 (Section 4-5)

A magnetic core is made of sheet steel laminations. Effective dimensions are: length of 0.6 m and cross-section area of 0.0022 m^2. The density of steel is 7700 kg/m^3. The coil of 125 turns is energized with a 60-Hz voltage that makes the flux be $\phi(t) = 0.003 \sin 377\, t$ weber. Find (a) the applied voltage, (b) the peak current, (c) the rms current, (d) the core loss, (e) the core loss current, and (f) the magnetizing current.

Solution

(a) Use Eq. 4-25 to find the rms voltage.

$$V = 4.44\, Nf\phi_{max} = 4.44 \times 125 \times 60 \times 0.003 = 100 \text{ v}$$

(b) The maximum flux density is found from

$$B_{max} = \phi_{max}/A = 0.003/0.0022 = 1.36 \text{ weber/m}^2$$

From Fig. 3-5 find $H_{peak} = 400$ At/m. (The subscript *max* is reserved to sinusoidal functions.) The peak current then is

$$I_{peak} = H_{peak}\, l/N = (400 \times 0.6)/125 = 1.92 \text{ amp}$$

(c) The apparent power, P_a, is obtained from empirical data in Fig. E-4-5. For $B_{max} = 1.36$ weber/m^2, we find $P_a^* = 11.7$ volt-amps/kg. The weight of the core is

$$\text{Weight} = (0.0022 \text{ m}^2)(0.6 \text{ m})(7700 \text{ kg/m}^3) = 10.1 \text{ kg}$$

The exciting apparent power is

$$P_a = (11.7 \text{ volt-amps/kg})(10.1 \text{ kg}) = 118 \text{ volt-amps} = VI$$

The rms current is $I = P_a/V = 118/100 = 1.18$ amps. Notice that $I_{peak} > \sqrt{2}\, I$ because of the nonlinear relation between B and H.

(d) The coré losses are obtained from empirical data in Fig. E-4-5. For B_{max} = 1.36 weber/m^2, we find $P^* = 2.2$ w/kg. The core loss power is

$$P_c = (2.2 \text{ w/kg})(10.1 \text{ kg}) = 22 \text{ w}$$

(e) We can represent an iron cored reactor with an equivalent circuit as shown in Fig. 4-9. This neglects the resistance of the wire in the coil. The core losses are accounted for by a current I_c that is in phase with the voltage function. See Fig. 4-10.

$$I_c = P_c/V = 22/100 = 0.22 \text{ amp}$$

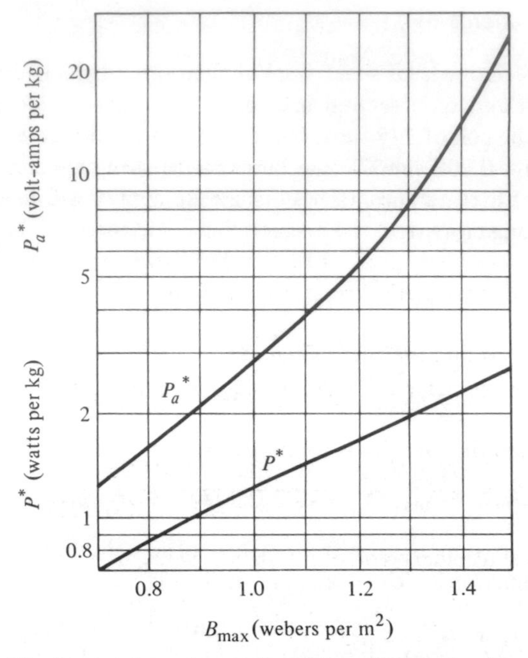

Fig. E-4-5. Curves of core loss and exciting volt-amps for sheet steel.

(f) The exciting current I has been found from the apparent power. Use the voltage function as the reference.

$$I_c = I \cos \theta$$

$$\theta = \cos^{-1} (I_c/I) = \cos^{-1} (0.22/1.18) = 79.3°$$

$$\mathbf{I} = 1.18 \underline{/-79.3°} = 0.22 - j\,1.16$$

The magnetizing current, I_m, can be found from

$$\mathbf{I} = \mathbf{I}_c + \mathbf{I}_m$$

$$\mathbf{I}_m = (0.22 - j\,1.16) - (0.22 + j\,0) = 0 - j\,1.16 = 1.16 \underline{/-90°}$$

The iron cored reactor differs from an ideal linear inductance in that core losses are present and the currents I_c, I_m, and I are nonsinusoidal when the flux function is a sinusoid.

Example 4-6 (Section 4-5)

The core of Example 4-5 has an air gap cut in it. The effective dimensions of the air gap are an area of 0.0025 m^2 and a length of 0.002 m. The effective dimensions of the core are the same as given in Example 4-5. For the same flux, $\phi(t) = 0.003 \sin 377\, t$ weber, find the rms current.

Solution

The voltage in the coil and all conditions in the steel core are the same as in Example 4-5. The reluctance of the magnetic circuit will be increased by the presence of the air gap, and the magnetizing component of the current will have to supply more ampere-turns. The core-loss component of the current will be the same. For this magnetic circuit, we can write

$$Ni_m = N(i_{mc} + i_{mg}) = \phi(\mathcal{R}_c + \mathcal{R}_g)$$

where $\mathcal{R}_c$ is the reluctance of the steel core and $\mathcal{R}_g$ is the reluctance of the air gap.

$$I_{mg\,\max} = \frac{\phi_{\max}\,\mathcal{R}_g}{N} = \frac{0.003 \times 0.002}{125 \times 4\pi \times 10^{-7} \times 0.0025} = 15.3 \text{ amp}$$

This current is sinusoidal. Therefore, its rms value is

$$I_{mg} = 15.3/\sqrt{2} = 10.8 \text{ amp}$$

Now we can find I_m. From Example 4-5, $I_{mc} = 1.16$ amp.

$$I_m = I_{mc} + I_{mg} = 1.16 + 10.8 = 12 \text{ amp}$$

The exciting current I is

$$\mathbf{I} = \mathbf{I}_c + \mathbf{I}_m = (0.22 + j\,0) + (0 - j\,12) = 0.22 - j\,12 = 12\underline{/-89^\circ} \text{ amp}$$

Comparing the answers from Examples 4-5 and 4-6, the presence of the air gap results in a closer approximation to an ideal linear inductance, but at the price of reduced value of inductance (and reactance).

4-7 PROBLEMS

4-1. Refer to the magnetization curve for cast steel in Fig. 3-5. Neglect hysteresis in this problem. (a) If the field intensity is the time function $H(t) = 2500 \sin 3.14\, t$ At/m, draw a graph of the time function $B(t)$. (b) If the flux density is the time function $B(t) = 1.4 \sin 3.14\, t$ webers/m^2, draw a graph of the time function $H(t)$.

4-2. A ring of ferromagnetic material has a toroidal winding with 700 turns. This magnetic circuit has cross-sectional area of 0.0004 m^2 and mean length of 0.7 m. Assume the permeability is constant at 1.1×10^{-3} henry per meter. (a) Find the self-inductance and the energy stored in the magnetic field for a current of 0.12 amp. (b) For the same toroidal winding, but with an air core, find the self-inductance and the energy stored in the magnetic field for a current of 0.12 amp.

4-3. A sheet steel core similar to Fig. E-4-1 is assumed to have constant permeability of 6.7×10^{-3} henry per meter. The coil has 1400 turns. Effective dimensions are $A_s = 0.0052$ m^2, $l_s = 0.7$ m, $A_g = 0.0064$ m^2, $l_g = 0.0005$ m. The flux in the air gap is 0.006 weber. (a) Find the current in the coil. (b) Find the energy stored in the air gap. (c) Find the energy stored in the steel. (d) Find the self-inductance.

4-4. In plotting a hysteresis loop, the following scales are used: 1 cm represents 400 At/m, and 1 cm represents 0.1 weber/m^2. For a certain material, the area of the loop is 28 cm^2. For a volume of 450 cm^3, calculate the hysteresis loss in joules per cycle for the specimen tested.

4-5. The flux in a magnetic core is alternating sinusoidally with a frequency of 60 Hz. The maximum flux density is 0.6 weber/m^2. The eddy-current loss is 16 w. Find the eddy-current loss in this core when the frequency is 90 Hz and the maximum flux density is 0.5 weber/m^2.

4-6. The total core loss (hysteresis plus eddy current) for a sheet steel core is found to be 1200 w at 100 Hz. If the maximum flux density is kept constant and the frequency is reduced to 60 Hz, the total core loss is found to be 528 w. Find the separate hysteresis and eddy-current losses for both frequencies.

4-7. A sheet steel core has a coil with 1200 turns. The resistance of the coil may be neglected. The frequency is held constant at 60 Hz. The core losses are 1250 w when energized from a sinusoidal voltage source with rms voltage of 150 v. The core losses are 603 w when the voltage is changed to 100 v. The core losses are 175 w when the voltage is 50 v. Find the exponent n in Eq. 4-20.

4-8. A sheet steel core has a coil with 70 turns. The resistance of the coil may be neglected. The sinusoidal voltage is held constant at rms value of 100 v. With a frequency of 30 Hz, the hysteresis loss is 90.5 w and the eddy-current loss is 64.5 w. With a frequency of 60 Hz, the total core loss is 116.5 w. Find the exponent n in Eq. 4-20.

4-9. Data for one-half of the symmetrical hysteresis loop for the steel in a core are given.

B weber/m^2	0	0.45	0.80	1.00	0.90	0.77	0.60	0.30	0
H At/m	190	200	245	300	100	0	−80	−155	−190

Effective dimensions of the core are area of 0.04 m^2 and length of 1.7 m. The flux density function is $B(t) = 1.00 \sin 314\, t$ webers/m^2. Find the hysteresis loss in watts.

4-10. A coil with 62 turns is wound around the core of Problem 4-9. A sinusoidal voltage of 50-Hz frequency is impressed on this coil. (a) Find the rms magnitude of the voltage to make the given hysteresis loop applicable. (b) Find the peak value of the current. (c) Draw neat curves of the voltage and the current for one cycle.

4-11. A sinusoidal voltage is impressed on a reactor coil. Assume the exponent n in the hysteresis loss to be 2. Resistance of the coil is to be neglected. Determine the percent change in flux, hysteresis loss, eddy-current loss, and coil current for the following conditions: (a) magnitude of voltage is increased 10 percent with frequency unchanged; (b) frequency is increased 10 percent with the magnitude of voltage unchanged; (c) both frequency and magnitude of voltage are increased 10 percent.

4-12. Two reactor cores are made of the same material with laminations of the same thickness. When Core 1 is operated with 60 Hz and B_{max} at 1.3 webers/m^2, the voltage is 570 v, the current is 1.35 amp, the core loss power is 184 w, and the power factor is 0.24 lag. Core 2 has all linear dimensions 30 percent larger than Core 1. When Core 2 is operated with 60 Hz and B_{max} at 1.3 webers/m^2, find the voltage, current, core loss power, and power factor.

4-13. A magnetic core is made of sheet steel laminations. Effective dimensions are mean length of 0.8 m, and cross-section area of 0.0037 m^2. The density of steel is 7700 kg/m^3. The coil of 86 turns is energized from a 60-Hz voltage source that makes the flux be $\phi(t) = 0.0048 \sin 377\, t$ weber. Find (a) the applied voltage, (b) the peak current, (c) the rms current, (d) the core loss, and (e) the magnetizing current.

4-14. An air gap is added to the core of Problem 4-13. Effective dimensions of the air gap are length of 0.002 m and cross-section area of 0.0042 m^2. All given conditions in Problem 4-13 still apply. Find (a) the rms current, and (b) the power factor.

4-15. A magnetic core is made of sheet steel laminations. Effective dimensions are cross-section area of 0.0041 m^2 and mean length of 0.82 m. The coil of 140 turns is energized from a 60-Hz voltage source. The input power is 80 w. Find the voltage and current. Neglect coil resistance in this problem.

4-16. A magnetic core is made of sheet steel laminations. There are no air gaps. The core has two parts with effective dimensions given.

Part	A(m^2)	l(m)
1	0.0044	0.2
2	0.0036	0.6

A coil with 220 turns is energized from a 60-Hz voltage source. The flux is $\phi(t) = 0.00546 \sin 377\, t$ weber. Find (a) the core loss in each part of the core, (b) the applied voltage, and (c) the rms current.

5
Transformers

5-1 THE IDEAL TRANSFORMER

The purpose of a power transformer, as explained in Section 1-3, is to enable different parts of a power system to operate at different voltages. Such a transformer may be interposed between a generator and a transmission line, between a distribution network and a load, etc. It should neither consume nor store any energy. Thus, its input power should equal its output power at any instant. Referring to Fig. 5-1, this requirement can be written as

$$v_1 i_1 = v_2 i_2 \tag{5-1}$$

Since the purpose of a transformer calls for a constant voltage ratio, this means

$$\frac{v_1}{v_2} = a \tag{5-2}$$

and

$$\frac{i_1}{i_2} = \frac{1}{a} \tag{5-3}$$

where the constant a is called the ratio of transformation.*

A device that satisfies Eqs. (5-2) and (5-3) is called an *ideal transformer*. Of course, nothing man-made can ever be ideal, but it is possible, in the case of the transformer, to come close to the ideal by remarkably simple means.

Consider once again a ferromagnetic core, this time one with *two* coils wound around it, as shown schematically in Fig. 5-2.† Call their turns numbers N_1 and N_2, respectively. Let there be a flux ϕ in the core. Then the voltages induced in the coils are

$$v_1 = N_1 \frac{d\phi}{dt} \tag{5-4}$$

*This is a traditional but arbitrary definition. It would be equally valid to give a name and a symbol to the ratio $1/a = v_2/v_1 = i_1/i_2$.

†The figure is drawn only to illustrate the principle of a transformer, not its physical appearance. On an actual transformer, the two coils are arranged to make full use of the core and to provide close magnetic coupling at the same time.

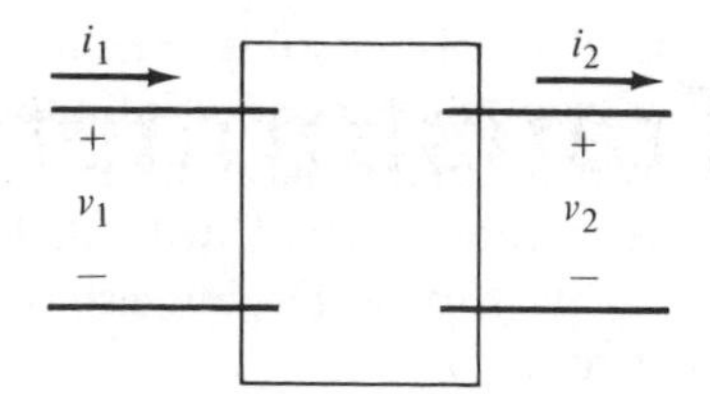

Fig. 5-1. Assumed positive directions and polarities.

$$v_2 = N_2 \frac{d\phi}{dt} \tag{5-5}$$

Dividing Eq. 5-4 by Eq. 5-5 leads to the desired constant *voltage ratio*

$$\frac{v_1}{v_2} = a = \frac{N_1}{N_2} \tag{5-6}$$

But how about the *current ratio*? The key to the answer is that the flux must be produced by the combined mmf of the two currents

$$\mathfrak{F} = N_1 i_1 - N_2 i_2 \tag{5-7}$$

where the minus sign is the result of having used the same current arrows in Fig. 5-2 as in Fig. 5-1. Note that Eq. 5-3 can be satisfied only if $\mathfrak{F} = 0$ at all times. On the other hand, surely the flux cannot be zero at all times if nonzero

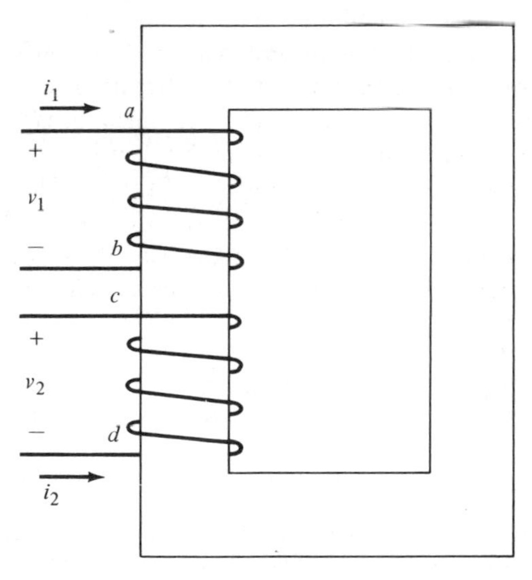

Fig. 5-2. Core with two coils.

voltages are to be induced. The only possible conclusion is that an ideal transformer must have a core of zero reluctance (or infinite permeance). Or, to use electric circuit terms, the two coils must have infinite inductances. This shows that the use of ferromagnetic core material, with its high permeability, serves not only to link two coils to one flux better, but also to come close to satisfying the condition of zero reluctance.

Now the principle of operation of a transformer can be explained. Connecting coil 1 (called the *primary*) to a voltage source dictates the flux in the core, in accordance with Eq. 5-4, and thereby the voltage induced in coil 2 (the *secondary*), in accordance with Eq. 5-5. As long as nothing is connected to the secondary terminals (*open-circuit*), the primary coil draws zero current from the source (ideally) because it is an infinite inductance connected to a finite source voltage. But when a *load* is connected to the secondary terminals, the current in the secondary circuit would produce an infinite flux in the core if this were not counteracted by a primary current satisfying the ideal current ratio. As a result, power can be transferred from the source to the load even though the two circuits are not connected to each other.

There remains a question of polarities to be resolved. Interchanging one pair of terminals in Fig. 5-2 would introduce a minus sign into Eq. 5-2, and reversing one current arrow would do the same thing to Eq. 5-3. The best way to handle that is to use the same *dot notation* that most readers should have encountered in circuit analysis in the study of mutual inductance. One dot is placed at one terminal of each coil, indicating that

(a) voltages from dotted to undotted terminals have the same sign, and
(b) currents entering at the dotted terminals produce mmfs in the same direction (i.e., aiding each other in producing a flux).

This means that, in Fig. 5-2, dots could be placed at terminals *a* and *c* (or *b* and *d*). With the aid of dots, schematic diagrams can be drawn in a much simpler way. For instance, Fig. 5-3 contains exactly the same information as Fig. 5-2.

5-2 IMPERFECTIONS

Since a core cannot really have zero reluctance, the total mmf in an actual transformer is not really zero. Its value can be determined for every instantaneous value of flux by the methods studied for magnetic circuits in Chapter 3. After that, Eq. 5-7 may be solved for the current i_1

$$i_1 = \frac{\mathcal{F}}{N_1} + \frac{N_2}{N_1} i_2 \qquad (5\text{-}8)$$

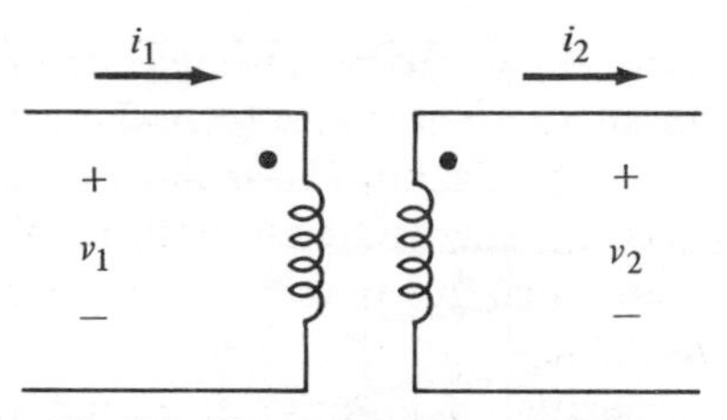

Fig. 5-3. Illustrating dot notation.

splitting this current into two fictitious but meaningful components. The first term is what the current i_1 would be if the secondary circuit were opened (or even removed). It is the magnetizing current

$$i_m = \frac{\mathcal{F}}{N_1} \tag{5-9}$$

The second term is what i_1 would be if the transformer were ideal ($\mathcal{F} = 0$).

$$i_2' = \frac{N_2}{N_1} i_2 \tag{5-10}$$

It is called the *secondary current referred to the primary*. With this terminology, Eq. 5-8 can be rewritten

$$i_1 = i_m + i_2' \tag{5-11}$$

It was shown in the previous chapter that the magnetizing current, in order to produce a time-varying flux, must be accompanied by a core loss component. So there is a second imperfection to be considered, changing the current equation to

$$i_1 = i_m + i_c + i_2' = i_\phi + i_2' \tag{5-12}$$

Thus, two imperfections, the finite core reluctance and the core losses, act to spoil what would otherwise be the ideal *current ratio*. For an actual transformer, Eq. 5-3 must be replaced by Eq. 5-10 in the form

$$\frac{i_2'}{i_2} = \frac{1}{a} \tag{5-13}$$

Additional imperfections are equally effective in spoiling the *voltage ratio*. They are the coil *resistances* and the *leakage fluxes*. To take them into consideration, refer to Fig. 5-4, which is actually Fig. 5-2 redrawn, but with resistances added and typical flux lines sketched. It shows that three distinctive

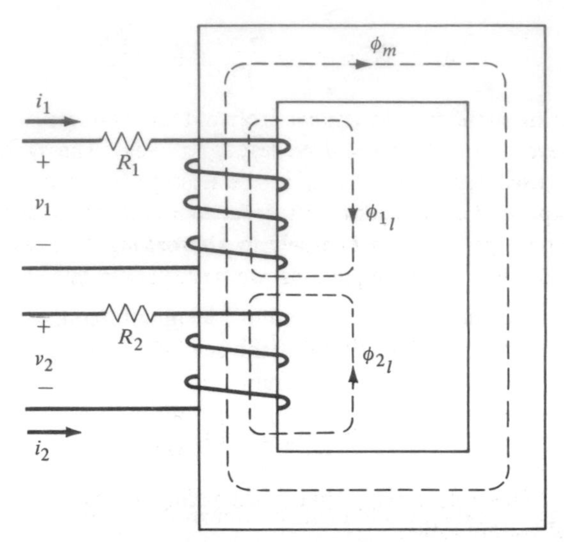

Fig. 5-4. Resistances and leakage fluxes.

fluxes can be identified. The *main flux* ϕ_m, which follows the magnetic circuit and thereby links both coils, is considered positive when the mmf $\mathfrak{F}$ is positive according to Eq. 5-7. Each of the two leakage fluxes ϕ_{1_l} and ϕ_{2_l} links only the one current that produces it, and each leakage flux is considered positive when its own current is positive. (Current directions were assigned earlier.)

The total fluxes linking the two coils are no longer the same. Call them

$$\phi_1 = \phi_m + \phi_{1_l} \tag{5-14}$$

and

$$\phi_2 = \phi_m - \phi_{2_l} \tag{5-15}$$

The voltage equations for each coil now are written as follows:

$$v_1 = R_1 i_1 + N_1 \frac{d\phi_1}{dt} \tag{5-16}$$

$$v_2 = -R_2 i_2 + N_2 \frac{d\phi_2}{dt} \tag{5-17}$$

Now substitute Eqs. 5-14 and 5-15 into Eqs. 5-16 and 5-17, and arrange them to read

$$v_1 = R_1 i_1 + N_1 \frac{d\phi_{2_l}}{dt} + N_1 \frac{d\phi_m}{dt} \tag{5-18}$$

$$v_2 = -R_2 i_2 - N_2 \frac{d\phi_{2l}}{dt} + N_2 \frac{d\phi_m}{dt} \tag{5-19}$$

The last terms in these two equations are what the voltages would be in each coil if there were no resistances and no leakage fluxes. The symbols e_1 and e_2 are commonly used for them. As far as the leakage flux terms are concerned, they can be expressed as voltages across linear inductances L_{1_l} and L_{2_l}, since the leakage fluxes have paths that are largely in air, making them less susceptible to saturation. So the two voltage equations can be rewritten once more to read

$$v_1 = e_1 + R_1 i_1 + L_{1_l} \frac{di_1}{dt} \tag{5-20}$$

$$v_2 = e_2 - R_2 i_2 - L_{2_l} \frac{di_2}{dt} \tag{5-21}$$

See that the voltages e_1 and e_2, that is, the *voltages induced by the main flux*, satisfy the ideal voltage ratio

$$\frac{e_1}{e_2} = a \tag{5-22}$$

whereas the *terminal voltages* v_1 and v_2 differ from them by the effects of imperfections.

5-3 SINUSOIDAL STEADY-STATE EQUATIONS

All equations relating time-varying voltages and currents to each other can be "translated" into sinusoidal steady-state "language," i.e., into *phasor equations*. In the case of equations stating the voltage or current ratio, all that is needed is to replace the time function symbols by the corresponding phasor symbols, since the ratio a is a constant so that there are no phase differences to be considered. Thus Eqs. 5-13 and 5-22 become

$$\frac{\mathbf{I}_2'}{\mathbf{I}_2} = \frac{1}{a} \tag{5-23}$$

$$\frac{\mathbf{E}_1}{\mathbf{E}_2} = a \tag{5-24}$$

Next, the voltage equations, Eqs. 5-20 and 5-21, are to be changed into phasor form. Remember that the derivative of a sinusoid always leads that sinusoid by 90°, and that the magnitude (amplitude or rms value) of the derivative is that of the original sinusoid multiplied by the radian frequency ω. In other words, the phasor representing the derivative is the phasor representing the original sinusoid multiplied by $j\omega$. Furthermore, the product of the radian frequency and an inductance is a reactance. So the voltage equations in phasor form contain the leakage reactances called X_1 and X_2 (The subscript l for leakage is omitted as unnecessary because there will be no other reactances considered in this context). These equations are

$$\mathbf{V}_1 = \mathbf{E}_1 + R_1\mathbf{I}_1 + jX_1\mathbf{I}_1 \tag{5-25}$$

$$\mathbf{V}_2 = \mathbf{E}_2 - R_2\mathbf{I}_2 - jX_2\mathbf{I}_2 \tag{5-26}$$

Finally, there is the current equation, Eq. 5-12, to be written in phasor form. The (fundamental of the) pure magnetizing current i_m is in phase with the flux it produces, which is the main flux ϕ_m, and the value of i_m depends on that of ϕ_m. Since i_m is in phase with ϕ_m, it lags the voltage e_1 by 90° since that voltage is induced by ϕ_m. Similarly, the core loss current i_c is in phase with e_1 and its value also depends on the flux ϕ_m.* As a result of all this, the current equation can be written

$$\mathbf{I}_1 = \mathbf{I}_c + \mathbf{I}_m + \mathbf{I}_2' = G_c\mathbf{E}_1 + jB_m\mathbf{E}_1 + \mathbf{I}_2' \tag{5-27}$$

where G_c and B_m are concepts familiar from Section 4-5. It should be remembered that B_m (and also G_c, to a lesser extent) is not a linear circuit element, since a flux in a ferromagnetic core is not proportional to the mmf that produces it.

5-4 THE BASIC EQUIVALENT CIRCUIT

The five equations of the previous section are statements of the voltage and current ratios of a fictitious *ideal transformer* plus Kirchhoff's voltage and current equations of a circuit that relates the voltages and currents of the *actual transformer* to those of the ideal one. The *equivalent circuit* drawn in Fig. 5-5 shows the actual voltages V_1 and V_2 and the actual currents I_1 and I_2 as its terminal quantities, and it satisfies the five equations. What is more, all passive elements of that circuit represent imperfections of the actual transformer. In other words, if these imperfections were zero, the entire circuit would be reduced to that of an ideal transformer. Note that the series elements (those expressed in ohms) are those that spoil the ideal voltage ratio, whereas the elements across the voltage E_1 (expressed in mhos) are those that spoil the ideal current ratio.

The equations, and also the diagram, may be somewhat simplified by the use of complex circuit elements. Thus, the resistances and leakage reactances may be combined to form complex impedances

*This neglects the small contribution the leakage fluxes make to the core losses.

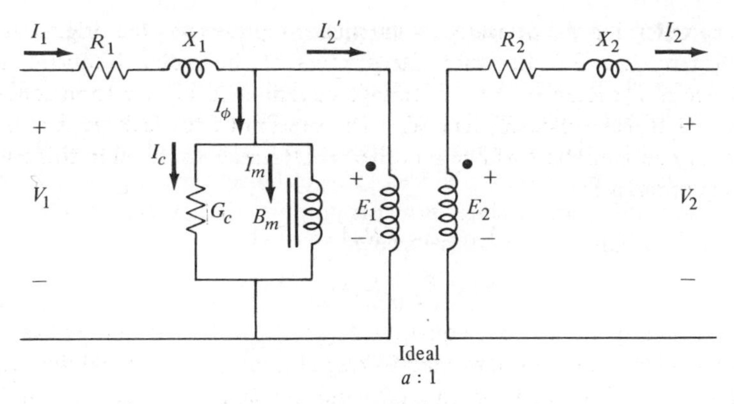

Fig. 5-5. Basic equivalent circuit.

$$R_1 + jX_1 = \mathbf{Z}_1 \tag{5-28}$$

$$R_2 + jX_2 = \mathbf{Z}_2 \tag{5-29}$$

and the magnetizing susceptance (which, by the way, is negative) and the core loss conductance form a (nonlinear) complex admittance already encountered in Section 4-5

$$G_c + jB_m = \mathbf{Y}_{\phi_1} \tag{5-30}$$

The voltage and current equations (the last three equations in the previous section) now read

$$\mathbf{V}_1 = \mathbf{E}_1 + \mathbf{Z}_1\mathbf{I}_1 \tag{5-31}$$

$$\mathbf{V}_2 = \mathbf{E}_2 - \mathbf{Z}_2\mathbf{I}_2 \tag{5-32}$$

$$\mathbf{I}_1 = \mathbf{I}_2' + \mathbf{Y}_{\phi_1}\mathbf{E}_1 \tag{5-33}$$

The equivalent circuit in its simplified form appears in Fig. 5-6. It still shows the actual transformer as consisting of an ideal transformer plus elements representing its imperfections.

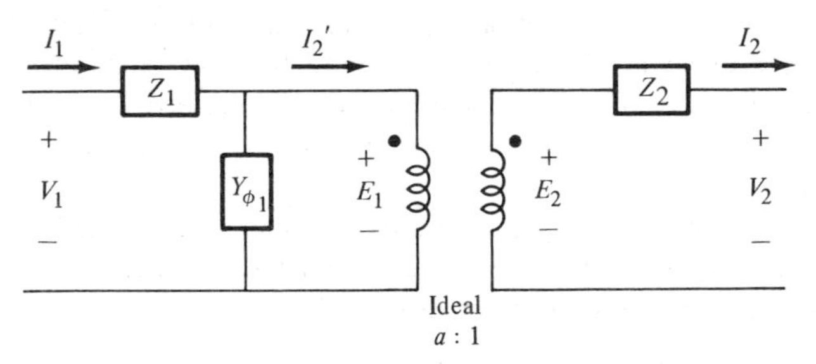

Fig. 5-6. Complex circuit elements.

5-5 A CHOICE OF EQUIVALENT CIRCUITS

Before further studies of a transformer, or of the power system of which it is a part, are undertaken, the equivalent circuit will be simplified some more. As a first step, all elements can be made to appear on the same side of the ideal transformer, instead of being on both sides. For instance, the impedance $\mathbf{Z}_2$ can be transferred to the primary side. For this purpose, the secondary voltage equation will be multiplied by the turns ratio a

$$a\mathbf{V}_2 = a\mathbf{E}_2 - a\mathbf{Z}_2\mathbf{I}_2 \tag{5-34}$$

which, with the aid of the ratio equations, Eqs. 5-23 and 5-24, can be written as

$$a\mathbf{V}_2 = \mathbf{E}_1 - a^2\mathbf{Z}_2\mathbf{I}_2' \tag{5-35}$$

To satisfy this modified secondary voltage equation, as well as the other four equations of Section 5-3, the equivalent circuit must be redrawn as shown in Fig. 5-7.

Comparing this diagram to the previous one shows that the element $\mathbf{Z}_2$ was transferred from the right side to the left side of the ideal transformer. For that purpose, $\mathbf{Z}_2$ had to be multiplied by a^2, the square of the turns ratio. This is significant, and not too surprising. After all, the primary voltage of an ideal transformer is a times the secondary voltage, and its primary current is $1/a$ times the secondary current. This leads logically to the rule that an impedance can be transferred from the secondary to the primary side by being multiplied by a^2, and from the primary to the secondary side by being multiplied by $(1/a)^2$, while the opposite is true for admittances.

Figure 5-7 shows all elements on the primary side of the ideal transformer. They can just as well all appear on the secondary side, as shown in Fig. 5-8. This can be obtained simply by following the rule formulated in the last paragraph, or more rigorously as follows: multiply the primary voltage equation

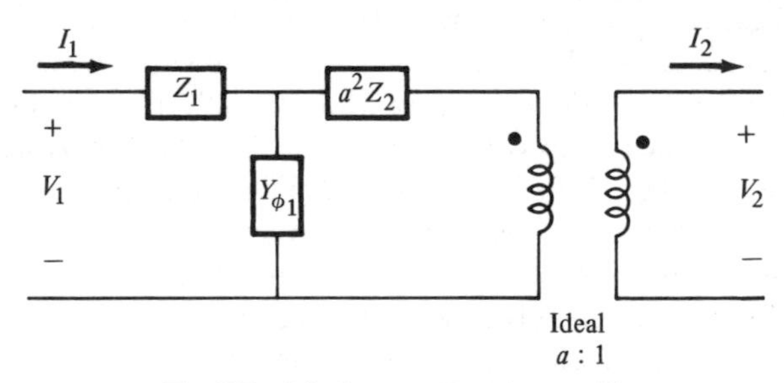

Fig. 5-7. All elements on primary side.

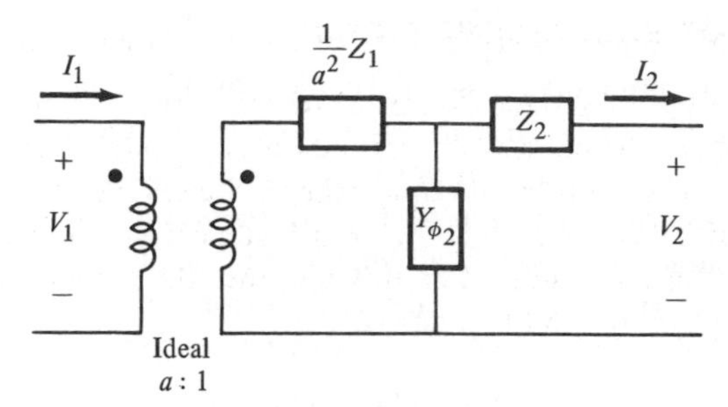

Fig. 5-8. All elements on secondary side.

(Eq. 5-31) by $1/a$ and write it in the form

$$\frac{1}{a}\mathbf{V}_1 = \mathbf{E}_2 + \left(\frac{1}{a^2}\mathbf{Z}_1\right)(a\mathbf{I}_1) \tag{5-36}$$

and multiply the current equation (Eq. 5-33) by a in the form

$$a\mathbf{I}_1 = \mathbf{I}_2 + a^2\mathbf{Y}_{\phi_1}\mathbf{E}_2 = \mathbf{I}_2 + \mathbf{Y}_{\phi_2}\mathbf{E}_2 \tag{5-37}$$

which will confirm that the rule works correctly.

In Eq. 5-37, a new symbol was introduced

$$\mathbf{Y}_{\phi_2} = a^2\mathbf{Y}_{\phi_1} \tag{5-38}$$

This was done to stress the point that there is no preference between the last two equivalent circuits. Which of the two coils of a transformer is to be called the primary and which the secondary depends entirely on which one is connected to the source and which one to the load; in other words, on the direction of power flow. The same transformer can equally well be used to step up a voltage or to step it down. Nor is it necessary ever to memorize any rule in terms of multiplying or dividing by a^2. All that is needed is to realize that, when an impedance is to be transferred from the low voltage side to the high voltage side, it must be multiplied by the factor that will make it larger, etc.

Having now collected all elements on one side of the ideal transformer leads to a T-shaped connection of three complex elements, all of which are referred to the same side. The next step is to change the T into an L by transferring the admittance branch either to the left or to the right, at some expense of accuracy. The error committed by this operation is small enough to be permissible, provided the imperfections represented by the complex elements are as small as they must be if the transformer is to serve its purpose.

For instance, let the circuit of Fig. 5-7 be altered by moving the admittance $\mathbf{Y}_{\phi_1}$ to the left terminals. The voltage across $\mathbf{Z}_1$ is now changed because the current through it is now $\mathbf{I}_1 - \mathbf{I}_\phi$ instead of $\mathbf{I}_1$. Nevertheless, the voltage across $\mathbf{Z}_1$ could never be more than a small fraction of $\mathbf{V}_1$, even in the worst case, namely when the current I_1 has its rated magnitude compared to which I_ϕ is only a small fraction. So the error in the voltage across Z_1 cannot be more than a small fraction of a small fraction of V_1 and thereby it remains within reasonable limits of accuracy inherent in all engineering calculations. The same thing is true about the current error, or about the errors resulting from shifting the admittance to the right side of the T, in either one of the two equivalent circuits with Ts.

But what was accomplished by sacrificing accuracy to change Ts into Ls? First of all, the impedance elements are now connected in series and can, therefore, be combined into one:

$$\mathbf{Z}_1 + a^2 \mathbf{Z}_2 = \mathbf{Z}_{e_1} \tag{5-39}$$

$$\frac{1}{a^2} \mathbf{Z}_1 + \mathbf{Z}_2 = \mathbf{Z}_{e_2} \tag{5-40}$$

These are called the *equivalent impedances* of the transformer, referred to the primary and secondary side, respectively. They combine the resistances and leakage reactances of the two coils. Their real and imaginary parts are

$$\Re e\ \mathbf{Z}_{e_1} = R_{e_1} = R_1 + a^2 R_2 \tag{5-41}$$

$$\Im m\ \mathbf{Z}_{e_1} = X_{e_1} = X_1 + a^2 X_2 \tag{5-42}$$

$$\Re e\ \mathbf{Z}_{e_2} = R_{e_2} = \frac{1}{a^2} R_1 + R_2 \tag{5-43}$$

$$\Im m\ \mathbf{Z}_{e_2} = X_{e_2} = \frac{1}{a^2} X_1 + X_2 \tag{5-44}$$

Note that each of these resistances, reactances, and impedances can be transferred from the secondary to the primary side by means of the factor a^2, and vice versa. Figure 5-9 shows the four *approximate equivalent circuits* obtained by the method discussed. The idea is that results obtained from any of these circuits are practically identical, so that, for any problem to be solved, the most convenient circuit can be chosen.

5-6 VOLTAGE REGULATION

The study of a power transformer is largely the study of the effects of its imperfections, and the approximate equivalent circuits are the best tools for that purpose.

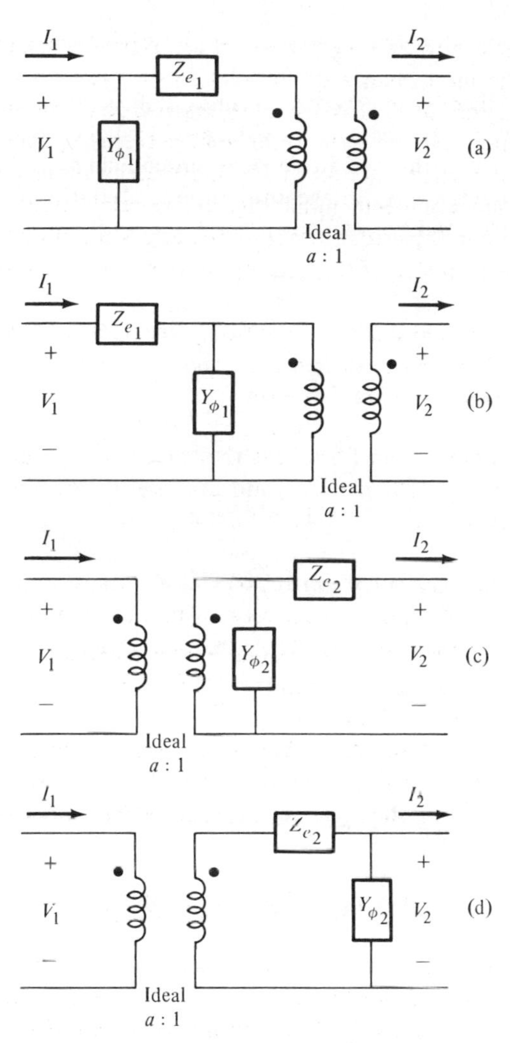

Fig. 5-9. Four approximate equivalent circuits.

If the imperfections were negligibly small, the rms value of the primary voltage V_1 would be a times the rms value of the secondary voltage V_2. Due to the imperfections represented by the impedance Z_{e_1}, there is a difference between the magnitudes of V_1 and aV_2. That difference depends on what is connected to the two pairs of terminals, or, as it is called, on the *operating condition* of the transformer. Since there is an infinite multitude of possible operating conditions, certain definitions have been chosen to make it possible to describe or choose a transformer on the basis of how close it comes to the ideal voltage ratio.

To begin with, there is an accepted definition of the term *rated conditions* for a power transformer. It means rated voltage and rated current at the *secondary* terminals. Notice that this definition does not contain any statement about the phase relation between this voltage and current. A transformer is thus said to operate under rated conditions when

$$V_2 = V_{2\ \text{rated}} \tag{5-45}$$

and

$$I_2 = I_{2\ \text{rated}} = \frac{P_{a\ \text{rated}}}{V_{2\ \text{rated}}} \tag{5-46}$$

The ratings of a power transformer, as they appear on its name plate, also contain a primary voltage. The meaning and purpose of this rating are to establish the exact turns ratio

$$a = \frac{V_{1\ \text{rated}}}{V_{2\ \text{rated}}} \tag{5-47}$$

Nevertheless, the primary terminal voltage under rated conditions is, in general, *not* $V_{1\ \text{rated}}$. Its actual value depends on the operating condition, and can be calculated from any equivalent circuit.

The sensible choice of a circuit for a voltage calculation is one that does not require any current calculations. Thus, to find V_1 from given values of V_2 and I_2, we may either choose the diagram of Fig. 5-9a and write the voltage phasor equation

$$\mathbf{V}_1 = a\mathbf{V}_2 + \mathbf{Z}_{e_1} \frac{\mathbf{I}_2}{a} \tag{5-48}$$

or we may choose Fig. 5-9c and write

$$\mathbf{V}_1 = a(\mathbf{V}_2 + \mathbf{Z}_{e_2} \mathbf{I}_2) \tag{5-49}$$

which comes to exactly the same thing. In both cases, the rated values are to be substituted for V_2 and for I_2, and the phase difference between these two quantities must be known. Note, incidentally, that the use of Fig. 5-9b or Fig. 5-9d would have required solving a current equation first, and yet the result would have been no more accurate.

The established standard for the departure from the ideal voltage ratio of a power transformer is the *difference between the magnitudes* of V_1 and aV_2, normalized by being divided by the rated value, and called the *regulation* (short for voltage regulation).

$$\epsilon = \frac{V_1 - aV_2}{aV_2} = \frac{V_1/a - V_2}{V_2} \tag{5-50}$$

where V_2 is the rated value, and V_1 (or V_1/a) is calculated from Eqs. 5-48 or 5-49. Note that the regulation depends on the phase difference between V_2 and I_2, but that the voltages in Eq. 5-50 are expressed as magnitudes, not phasors. For an ideal transformer, $\epsilon = 0$, regardless of the load power factor.

Regulation can also be defined in terms of *no-load* and *full-load* voltages. Let the transformer operate under full-load (i.e., rated) conditions, as stated by Eqs. 5-45 and 5-46. This requires that the input voltage be adjusted to the value obtained from Eq. 5-48 or Eq. 5-49. Now disconnect the load, thereby make $I_2 = 0$ and observe the change in output voltage V_2 to its new value V_1/a. This leads to the expression

$$\epsilon = \frac{V_{2\,\mathrm{NL}} - V_{2\,\mathrm{FL}}}{V_{2\,\mathrm{FL}}} \tag{5-51}$$

whereby the regulation is seen as the normalized difference between the output voltage at no-load and at full-load, when the input voltage is held at the value that makes $V_{2\,\mathrm{FL}} = V_{2\,\mathrm{rated}}$.

The three phasor diagrams of Fig. 5-10 illustrate how the regulation of a given transformer depends on the power factor of the load. In diagram (a), the current is *in phase* with the voltage (the load is resistive, its power factor is unity), whereas in diagram (b), the current *lags* the voltage (the load is inductive, the power factor is called *lagging*), and in diagram (c), the current *leads* the voltage (the load is capacitive, its power factor is said to be *leading*). In each case, the voltage $Z_{e_2} I_2$ must be small compared to V_2 (in fact, much smaller than in the diagrams where its size is much exaggerated for clarity's sake). The regulation in case (b) is much worse (larger) than in case (a), whereas in case (c) it is even negative.

A sample calculation, using realistic numerical values, is given in Example 5-1. The concept of regulation will reappear when generators are discussed in later chapters.

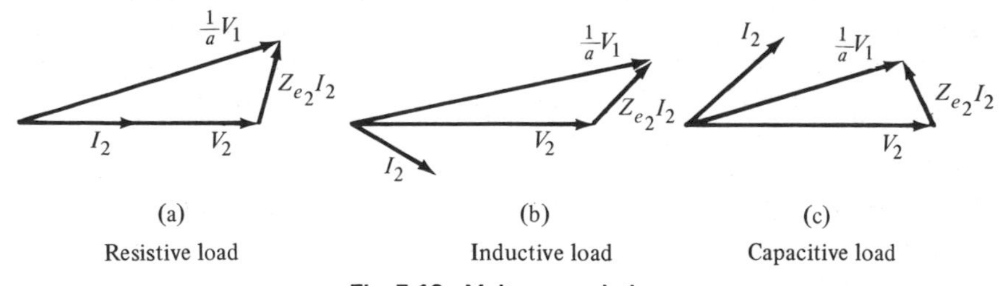

Fig. 5-10. Voltage regulation.

5-7 EFFICIENCY

A glance at any equivalent transformer circuit shows that it contains both energy-consuming and energy-storing elements. This means that the transformer, due to its imperfections, "consumes" both real power and reactive power. Of all the effects of its imperfections, the consumption of real power is the most serious one because it means that there is continuously some energy to be supplied (and paid for) that gets lost instead of reaching the load for which it is intended.

The concepts of power losses and efficiency were introduced in Section 1-6, and their relations are given by Eqs. 1-1 through 1-4. In the case of the transformer, the power losses are clearly identified as the *core losses* (hysteresis and eddy currents) in the ferromagnetic core and the so-called *copper losses* caused by the energy-consuming property (i.e., the resistance) of the conducting material (usually copper) of the coils. In terms of elements of the approximate equivalent circuits, the core losses are represented by the real part of the admittance $\mathbf{Y}_{\phi_1}$ or $\mathbf{Y}_{\phi_2}$, and the copper losses by the real part of the impedance $\mathbf{Z}_{e_1}$ or $\mathbf{Z}_{e_2}$.

To obtain mathematical expressions of the power losses, the most convenient of the various equivalent circuits should be chosen in each case. Thus, the *copper losses* may be written from Fig. 5-9 (b) or (c):

$$P_R \approx R_{e_1} I_1^2 \approx R_{e_2} I_2^2 \tag{5-52}$$

It is true that these two expressions are not identical, since I_2 does not equal aI_1 whereas $R_{e_1} = a^2 R_{e_2}$. In fact, both expressions are only approximations while the truth lies somewhere in the middle (see the "exact" equivalent circuit of Fig. 5-6). The point is that, such inaccuracies are permissible because the power losses are small in comparison to rated output power.

The *core losses* are best obtained from Fig. 5-9 (a) or (d):

$$P_c \approx G_{c_1} V_1^2 \approx G_{c_2} V_2^2 \tag{5-53}$$

where the symbols G_{c_1} and G_{c_2} are introduced for the real parts of $\mathbf{Y}_{\phi_1}$ and $\mathbf{Y}_{\phi_2}$, respectively. Appropriate comments on the accuracy of these expressions run similar to those about copper losses.

The efficiency of a given transformer depends on its operating condition. In most cases of interest, values of efficiency are based on *rated output voltage*, but they cannot be limited to rated output current because efficiency varies with the magnitude and the phase angle of the load current.

Using Eq. 1-4, the efficiency of a transformer can be written:

$$\eta = \frac{V_2 I_2 \cos\theta_2}{V_2 I_2 \cos\theta_2 + G_{c_2} V_2^2 + R_{e_2} I_2^2} \tag{5-54}$$

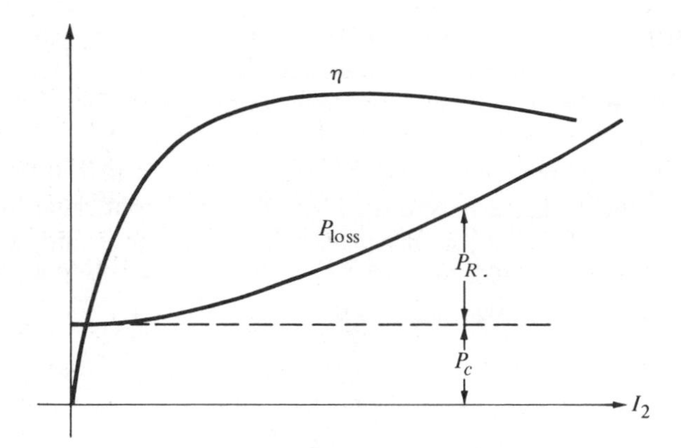

Fig. 5-11. Losses and efficiency.

where $V_2 = V_{2\ \text{rated}}$, and θ_2 is the phase difference between V_2 and I_2. Note that only output quantities appear in this equation, and only I_2 and θ_2 (both determined by the load) are variable parameters. Note also that the core losses are constant (i.e., independent of the load) while the copper losses are proportional to the square of the load current. Sample calculations are given in Example 5-2.

Figure 5-11 shows how the power losses and the efficiency of a transformer vary with its load, for a constant value of the load power factor. The abscissa can be the load current I_2 or, by means of scale changes, the apparent power $V_2 I_2$ or the power $V_2 I_2 \cos\theta_2$ delivered to the load. At no-load, the efficiency must be zero because the output is zero while the input is not (it equals the core losses). With increasing load, the efficiency curve rises for a while but ultimately, it must go down and approach zero asymptotically because of the rapidly increasing copper losses. (Equation 5-54 shows that the limit of η for $I_2 \to \infty$ is zero.)

So the efficiency curve must have a maximum that can be calculated by differentiating Eq. 5-54 with respect to I_2 (remembering that V_2 and $\cos\theta_2$ are constants). Thus, for maximum efficiency,

$$\frac{d\eta}{dI_2} = 0 \tag{5-55}$$

which results in

$$G_{c_2} V_2^2 = R_{e_2} I_2^2 \tag{5-56}$$

indicating that the efficiency is highest at that load at which the constant losses (the core losses) equal the losses that vary with the load (the copper losses). The same statement is also applicable, in principle, to motors and generators.

If the curves of Fig. 5-11 are drawn for a different power factor, the losses curve remains unchanged, but the efficiency at any value of load current is highest at unity power factor and goes down as the power factor decreases, as can be seen from Eq. 5-54.

If a transformer is intended to operate continuously at its full rated load, it can and should be designed to have its highest efficiency at rated load. Many power transformers, however, supply loads that change from time to time, but follow the same pattern day after day. For such a transformer, the efficiency at one specific load is not as significant as the *energy efficiency* for the whole *load cycle* (usually, a day).

$$\eta_W = \frac{\int_{t_1}^{t_2} P_{\text{out}}\, dt}{\int_{t_1}^{t_2} P_{\text{out}}\, dt + \int_{t_1}^{t_2} P_c\, dt + \int_{t_1}^{t_2} P_R\, dt} \tag{5-57}$$

where the integration limits t_1 and t_2 are the beginning and end of a load cycle. See the sample calculations of Example 5-3.

5-8 A LOOK AT DESIGN PRINCIPLES

At this point, it is instructive to look at the *physical* dimensions of a transformer and their relation to power losses. Suppose a given transformer design for given ratings were changed by multiplying every linear dimension by a factor $k > 1$. The cross-section of the core would then be multiplied by k^2, and the flux density by $1/k^2$. The volume of the core would be increased by the factor k^3, and the *core losses* (see Eqs. 4-18 and 4-19) would thereby be approximately (by giving the hysteresis loss exponent n the reasonable value of 2) reduced by the factor k. The same is true for the *copper losses*, since the cross-section of all conductors would be multiplied by k^2, their lengths by k, and, thus, their resistances by $1/k$. So each of the power losses can be reduced as much as the designer wants, at the expense of increased size, weight, and cost of the transformer. The actual design is the result of a trade-off between operating expenses (i.e., the cost of lost energy) and purchasing price. The same kind of considerations apply to the design of electric motors and generators.

Another pertinent question in this context is how a given transformer design ought to be changed for a different *rating*. Some insight can be gained by making the arbitrary but plausible assumption that the flux density B in the core and the current density J in the conductors be maintained at the same values. Now if all linear dimensions are multiplied by a factor K, both the flux and the current are multiplied by K^2. Since flux is proportional to voltage (for constant frequency, see Eq. 4-25), it follows that the volt-ampere rating of the transformer, is multiplied by K^4. On the other hand, the core losses are multiplied by K^3 (see again Eqs. 4-18 and 4-19), and so are the copper losses, because $RI^2 = (\rho l/A)(JA)^2 = \rho lAJ^2$ where A is the cross-sectional area of a conductor and l is its length. To summarize: the rating increases with the fourth power of a factor $K > 1$, and the losses only with the third power. The larger the transformer, the better is its efficiency.

This kind of reasoning explains what is behind the trend toward ever increasing ratings of individual transformers. (Essentially the same can be said about generators.) One larger unit is more efficient than several smaller ones for the same total output. But there is a penalty to be paid for the advantage of the larger unit size. While the power losses grow with the third power of the scale factor K, the *surface* on whose magnitude the dissipation of these losses (in the form of heat) depends, increases only with the second power of K. The larger unit with the more favorable efficiency presents a more difficult *cooling* problem.

There are many techniques for its solution. The surface can be increased by cooling ribs. When this becomes insufficient, heat convection can be improved by various artificial means. Fans can be used to increase the motion of the ambient air, and more effective cooling media, like water, can be used. Large transformers are frequently placed in tanks filled with cooling *oil*; many large generators use *hydrogen* as a cooling medium. Sizes of transformers and generators have kept growing through the years just as fast as improved cooling techniques became available and economically worthwhile.

5-9 TESTING

The behavior of a transformer under any set of operating conditions can be predicted from its equivalent circuit. Both the manufacturer and the user of a power transformer have a need to know the elements of its equivalent circuit. There is no simple and reliable way, however, to obtain all these elements from the dimensions or design drawings. Fortunately, the elements of the approximate equivalent circuit can be determined from fairly simple tests that do not require the transformer to be loaded.

In an *open-circuit test* a voltage source is connected to either one of the terminal pairs, and the other terminal pair is kept open, i.e., no current is drawn from it. Suppose the source is connected to the primary side and the secondary side is opened. The most convenient diagram for this case is that of Fig. 5-9(a) with $I_2 = 0$ and, therefore, $I_2' = 0$, because of the current ratio of the ideal transformer. Thus the impedance $\mathbf{Z}_{e_1}$ carries no current, and the entire circuit is reduced to the admittance

$$\mathbf{Y}_{\phi_1} = \frac{\mathbf{I}_{1\text{ oc}}}{\mathbf{V}_{1\text{ oc}}} \tag{5-58}$$

Three measurements are needed. *Ammeter* and *voltmeter* readings are the rms values of current and voltage, and a *wattmeter* reading (which equals the core losses) makes it possible to calculate the phase angle of the admittance. The procedure is straightforward:

$$Y_{\phi_1} = \frac{I_{1\text{ oc}}}{V_{1\text{ oc}}} \tag{5-59}$$

$$\cos\theta_\phi = \frac{P_{\text{oc}}}{V_{1\text{ oc}} I_{1\text{ oc}}} \tag{5-60}$$

$$\mathbf{Y}_{\phi_1} = Y_{\phi_1} \underline{/\theta_\phi} \tag{5-61}$$

where the negative angle (in the fourth quadrant) must be chosen. If desired, the real and imaginary parts of $\mathbf{Y}_{\phi_1}$ can be obtained by

$$G_{c_1} = Y_{\phi_1} \cos\theta_\phi \tag{5-62}$$

$$B_{m_1} = Y_{\phi_1} \sin\theta_\phi \tag{5-63}$$

The test may equally well be taken from the secondary side (i.e., with a source connected to the secondary terminals and the primary terminals left open). In that case, the best diagram to use is that of Fig. 5-9(d), and the circuit is reduced to the admittance $\mathbf{Y}_{\phi_2}$. Since $\mathbf{Y}_{\phi_2} = a^2\mathbf{Y}_{\phi_1}$, either one of the two procedures may be used. It is important in either case, however, to use a voltage source of rated magnitude because the admittances are *nonlinear* circuit elements that would have different values at different voltages.

There remains the impedance $\mathbf{Z}_{e_1}$ (or $\mathbf{Z}_{e_2}$) to be found. This can be done by means of a *short-circuit* test. Again, a source is connected to one pair of terminals but, this time, the other pair is short circuited. For instance, let the source be connected to the secondary terminals and make $V_1 = 0$ by means of a short-circuit connection. The diagram of Fig. 5-9(c) shows that, in this case, the voltage across $\mathbf{Y}_{\phi_2}$ is also zero, and the circuit is reduced to the impedance

$$\mathbf{Z}_{e_2} = \frac{\mathbf{V}_{2\text{ sc}}}{\mathbf{I}_{2\text{ sc}}} \tag{5-64}$$

Again, voltage, current, and power are measured and evaluated:

$$Z_{e_2} = \frac{V_{2\text{ sc}}}{I_{2\text{ sc}}} \tag{5-65}$$

$$\cos \theta_e = \frac{P_{sc}}{V_{2\,sc} I_{2\,sc}} \tag{5-66}$$

$$\mathbf{Z}_{e_2} = Z_{e_2} \underline{/\theta_e} \tag{5-67}$$

where the angle θ_e is in the first quadrant. Then, if desired:

$$R_{e_2} = Z_{e_2} \cos \theta_e \tag{5-68}$$

$$X_{e_2} = Z_{e_2} \sin \theta_e \tag{5-69}$$

As before, the test may just as well be taken from the primary side, whereby the impedance $\mathbf{Z}_{e_1}$ is obtained. For sample calculations, see Example 5-4.

The impedances $\mathbf{Z}_{e_1}$ and $\mathbf{Z}_{e_2}$ are considered linear circuit elements, which means they do not depend on the value of the voltage used for the short-circuit test. The coils must not be overloaded too much, however, and, therefore, the source voltage for the short-circuit test is limited to a small fraction of rated voltage. This is so because the voltage drop across the impedance Z_{e_1}, even with currents in excess of rated values, is only a small fraction of the rated value of V_1, and the voltage drop across Z_{e_2} is only a small fraction of the rated value of V_2.

It is worth noting that an important advantage of using *approximate* equivalent circuits for power transformers is, in addition to simplifying the analysis of operating conditions, that two simple standard tests (open- and short-circuit tests) are sufficient to obtain all the values of the elements of such circuits.

5-10 NORMALIZED QUANTITIES

Physical quantities can be expressed in terms of their units, or in comparison to a reference or *base quantity*. For instance, people's height might be stated as 1.7 meters (m), 1.6 m, etc. Or, a height of, say, 2 m might be chosen as the base quantity, and these people's height would then be expressed as 0.85 per unit, 0.8 per unit, etc. (or as 85 percent, 80 percent, etc.). A quantity is thus "expressed in *per unit*" when it is divided by the base quantity, which must have the same dimension. By this operation, known as *normalization*, quantities are made dimensionless, and their per-unit values are often more significant than their absolute values. Many readers have undoubtedly encountered this idea before.

For power transformers, normalization is particularly useful. Rated apparent power and rated voltages are chosen as base quantities. Since power, reactive power, and apparent power are all of the same dimension, the volt-ampere rating is used as the base for all these quantities.

$$P_{\text{base}} = Q_{\text{base}} = P_{a\ \text{base}} = P_{a\ \text{rated}} \tag{5-70}$$

$$V_{1\ \text{base}} = V_{1\ \text{rated}} \tag{5-71}$$

$$V_{2\ \text{base}} = V_{2\ \text{rated}} \tag{5-72}$$

Consistently, the base currents must be

$$I_{1\ \text{base}} = \frac{P_{a\ \text{rated}}}{V_{1\ \text{rated}}} \tag{5-73}$$

$$I_{2\ \text{base}} = \frac{P_{a\ \text{rated}}}{V_{2\ \text{rated}}} \tag{5-74}$$

Finally, base impedances (and base admittances) are chosen to be consistent with the previous choices:

$$Z_{1\ \text{base}} = \frac{V_{1\ \text{base}}}{I_{1\ \text{base}}} \tag{5-75}$$

$$Z_{2\ \text{base}} = \frac{V_{2\ \text{base}}}{I_{2\ \text{base}}} \tag{5-76}$$

These bases are also valid for resistances and reactances, just as admittance bases are valid for conductances and susceptances.

When all quantities are divided by their proper bases and thereby normalized, all equations can be written in normalized form. As an example, take Eq. 5-25. With all quantities appearing in that equation properly normalized, each term of the equation is actually divided by V_1, which leaves the equation intact. Example 5-5 illustrates the procedure, and shows its special advantages for transformer calculations:

(a) The elements of an equivalent circuit represent imperfections, and, therefore should be small, but that is a relative term. Just how small is "small"? Actually, an impedance of, say, $Z_{e_1} = 1\ \Omega$ of a transformer may quite well be more "imperfect" than one of $2\ \Omega$ of another transformer with different ratings. The answer becomes entirely clear when normalized values are used. All imperfections must be much less than *unity*, and the lesser the better, regardless of ratings.

(b) Due to the different bases on the two sides, the ideal transformer that is part of every equivalent circuit becomes a 1:1 transformer when normalized values are used. So it may be simply omitted (except when the physical separation of the two sides is relevant). No quantity needs to be referred to one or the other side any more; their normalized values are the same no matter which side they are on. For instance, in per-unit values, $\mathbf{Z}_{e_1} = \mathbf{Z}_{e_2}$, etc.

(c) Some quantities take on added significance when they are normalized. From Fig. 5-9(a) or (c) (the two are identical for normalized quantities), we see that Z_e (notice there is no longer a subscript 1 or 2) is also the voltage drop across that impedance at full load (i.e., when $I_2 = 1.0$ per unit), and its real part R_e also equals the full-load copper losses. Similarly, using the diagram of Fig. 5-9(b) or (d), the admittance Y_ϕ is also the no-load current, and its real part G_c also equals the core losses (both for $V_2 = 1.0$ per unit).

(d) Regulation and efficiency (themselves two dimensionless quantities) are more easily calculated when the circuit elements and the load current are given in normalized form. For the calculation of both, the output voltage is assumed to be at rated value (thus, $V_2 = 1.0$ per unit), and for the calculation of regulation (and of full-load efficiency), the output current has rated value ($I_2 = 1.0$ per unit).

The use of normalization is not limited to the study of one single transformer. In fact, entire power systems, consisting of many generators, transformers, transmission lines, and loads, can be analyzed in normalized form, with one base power chosen and consistently used.

5-11 AN ALTERNATIVE: MUTUAL INDUCTANCE

In the study of electromagnetic devices, in this book and elsewhere, extensive use is made of equivalent circuits that are derived from *physical descriptions* of these devices. Such descriptions consider such phenomena as energy storage and electromagnetic induction, and they involve not only electric quantities (voltages and currents) but also magnetic ones (mmfs and fluxes).

It is also possible to describe electric circuits coupled by magnetic fields by the abstract tool of *circuit analysis*, in terms of self- and mutual inductances. It is as if there were two entirely different languages in which the same facts could be stated. Surely an intelligent reader is entitled to know how to translate from one such language into the other, and what should determine a preference for one or the other.

In the approach of abstract circuit analysis, a transformer is described by a diagram like that of Fig. 5-12, with arbitrarily chosen current arrows, polarity marks, and dots. The equations that correspond to this diagram are

$$v_1 = R_1 i_1 + L_1 \frac{di_1}{dt} - M \frac{di_2}{dt} \tag{5-77}$$

$$v_2 = M \frac{di_1}{dt} - R_2 i_2 - L_2 \frac{di_2}{dt} \tag{5-78}$$

On the other hand, the physical approach leads to an equivalent circuit like that of Fig. 5-5. Since the notation in that diagram is for sinusoidal steady state

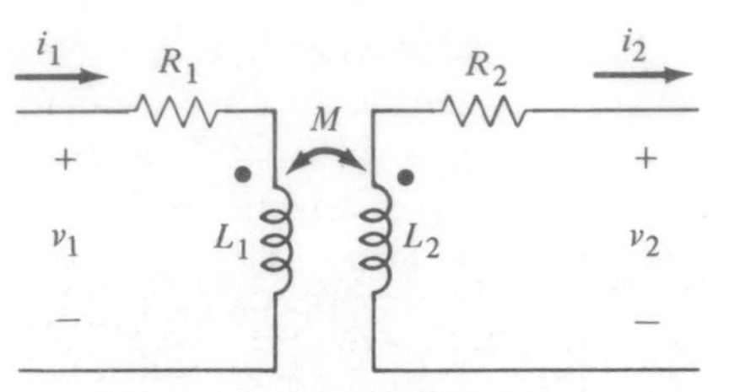

Fig. 5-12. Mutual inductance.

only, the diagram is redrawn in Fig. 5-13 with appropriate notation, and with one change: the core loss conductance is omitted because the circuits approach does not consider core losses. The inductive element related to the main flux is given a new symbol L_ϕ. This inductance is related to the susceptance B_m by the equation

$$B_m = -\frac{1}{\omega L_\phi} \tag{5-79}$$

Now the induced voltage e_1 can be expressed in terms of currents:

$$e_1 = L_\phi \frac{d}{dt}\left(i_1 - \frac{i_2}{a}\right) \tag{5-80}$$

and the voltage equations can be read from the diagram

$$v_1 = R_1 i_1 + L_{1_l}\frac{di_1}{dt} + e_1 = R_1 i_1 + (L_{1_l} + L_\phi)\frac{di_1}{dt} - \frac{L_\phi}{a}\frac{di_2}{dt} \tag{5-81}$$

$$v_2 = \frac{e_1}{a} - R_2 i_2 - L_{2_l}\frac{di_2}{dt} = \frac{L_\phi}{a}\frac{di_1}{dt} - R_2 i_2 - \left(L_{2_l} + \frac{L_\phi}{a^2}\right)\frac{di_2}{dt} \tag{5-82}$$

These two equations are identical with Eq. 5-77 and 5-78 if the following substitutions are made:

$$M = \frac{L_\phi}{a} \tag{5-83}$$

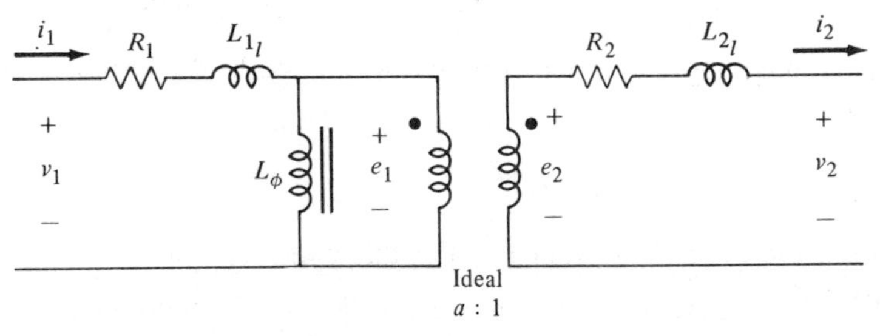

Fig. 5-13. Back to the basic equivalent circuit.

$$L_1 = L_{1l} + L_\phi \tag{5-84}$$

$$L_2 = L_{2l} + \frac{L_\phi}{a^2} \tag{5-85}$$

This constitutes the promised "translation" between the two approaches, which are thereby shown to be identical, with the exception of the core losses for which the circuit approach is not well suited.

And yet, there is a big difference. It lies in the *nonlinearity* of the magnetic circuit. The three inductances in Eqs. 5-77 and 5-78 are nonlinear circuit elements, so much so that their values may change by several orders of magnitude for changing currents. In other words, for coupled circuits using ferromagnetic cores, these two equations are not suitable for numerical evaluations.

It is indeed a big advantage of the descriptive approach that it leads to an equivalent circuit like that of Fig. 5-13, in which the only nonlinear element, the fictitious inductance L_ϕ, is shunted into an auxiliary branch, away from the main current paths. Even better, for a power transformer, the current through this element has an amplitude practically independent of the load, so that a constant value of L_ϕ (or B_m) can be assigned to that element. So this method is by far superior to that of abstract circuit analysis for the study of power transformers or other devices based on similar principles. Nevertheless, for the study of devices *without* ferromagnetic cores, sets of linear equations resulting from the circuits method are more systematic and often easier to handle.

5-12 EXAMPLES

Example 5-1 (Section 5-6)

A transformer is rated 10 kva, 2400/240 v, 60 Hz. The parameters for the approximate equivalent circuit of Fig. 5-9(a) are:

$$\mathbf{Y}_{\phi_1} = G_{c_1} + j\,B_{m_1} = 12.5 - j\,28.6 = 31.2\underline{/-66.4^\circ}\ \mu\text{mhos}$$

$$\mathbf{Z}_{e_1} = R_{e_1} + j\,X_{e_1} = 8.4 + j\,13.7 = 16.1\underline{/58.5^\circ}\ \text{ohms}$$

Find the voltage regulation for operation with rated load and power factor of 0.8 lagging.

Solution

We plan to use Eq. 5-48 to find V_1. For this purpose, we first have to find the rated current.

$$I'_{2\,\text{rated}} = I_{1\,\text{rated}} = 10{,}000/2400 = 4.17\ \text{amp}$$

Choose I_2' as the reference. Either this or V_2 may be chosen. It is a matter of personal preference.

$$\mathbf{I}_2/a = \mathbf{I}_2' = 4.17\underline{/0^\circ} = 4.17 + j\,0 \text{ amp}$$

The power factor angle is

$$\theta = \cos^{-1}(0.8) = 36.9^\circ$$

The voltage V_2 leads the current I_2 by this angle.

$$a\mathbf{V}_2 = 2400\underline{/36.9^\circ} = 1920 + j\,1440 \text{ volts}$$

The voltage across the equivalent impedance is

$$\mathbf{Z}_{e_1}\mathbf{I}_2' = (8.4 + j\,13.7)(4.17 + j\,0) = 35 + j\,57 \text{ volts}$$

The terminal voltage on side 1 is found from Eq. 5-48.

$$\mathbf{V}_1 = a\mathbf{V}_2 + \mathbf{Z}_{e_1}\mathbf{I}_2' = 1955 + j\,1497 = 2462\underline{/37.4^\circ} \text{ volts}$$

Use only the magnitudes in Eq. 5-50 to find the voltage regulation.

$$\epsilon = \frac{V_1 - aV_2}{aV_2} = \frac{2462 - 2400}{2400} = 0.0258$$

Example 5-2 (Section 5-7)

The transformer of Example 5-1 is operated with rated load and power factor of 0.8 leading. Find the efficiency.

Solution

We plan to use Eq. 1-4 to find the efficiency. We do not solve for all of the values in Fig. 5-9(a). We work with rated quantities as being close enough for this computation of efficiency. The rated load current is

$$aI'_{2\,\text{rated}} = I_{2\text{rated}} = 10{,}000/240 = 41.7 \text{ amp}$$

The output power is

$$P_{\text{out}} = V_2 I_2 \cos\theta = (240)(41.7)(0.8) = 8000 \text{ w}$$

Use Eq. 5-52 to find the copper losses. Since R_{e_1} is given, it is convenient to use

$$I_1 \approx I_2' = I_2/a = 4.17 \text{ amp}$$

$$P_R = I_1^2 R_{e_1} = (4.17)^2\,(8.4) = 146 \text{ w}$$

A measured value of core losses is often available and can be used. In this problem, G_{c_1} is given so we use Eq. 5-53 to find the core losses. We use $V_1 \approx V_{1\,\text{rated}}$.

$$P_c = G_{c_1} V_1^2 = (12.5 \times 10^{-6})(2400)^2 = 72 \text{ w}$$

The summation of losses is

$$P_{\text{loss}} = P_R + P_c = 146 + 72 = 218 \text{ w}$$

Use Eq. 1-4 to find the efficiency.

$$\eta = 1 - \frac{P_{\text{loss}}}{P_{\text{out}} + P_{\text{loss}}} = 1 - \frac{218}{8000 + 218} = 0.973$$

Example 5-3 (Section 5-7)

All-day efficiency is defined to be the ratio of energy output to energy input for a 24-hour period. The 10-kva transformer of Example 5-1 is operated for 24 hours a day. Loads during the day are: 10 kva at 1.0 P.F. for 3 hours; 6 kva at 0.8 P.F. lag for 5 hours; no load for 16 hours. Find the all-day efficiency.

Solution

The given conditions of this problem make it simple to evaluate the integrals in Eq. 5-57. The output energy is

$$W_{\text{out}} = (10 \times 1.0) \times 3 + (6 \times 0.8) \times 5 + 0 \times 16 = 54 \text{ kwh}$$

The core losses are constant for the full 24 hours. The energy into core losses is

$$W_c = (72/1000) \times 24 = 1.728 \text{ kwh}$$

During 3 hours, the copper loss is

$$P_R = \left(\frac{10{,}000}{2400}\right)^2 (8.4)/1000 = 0.146 \text{ kw}$$

During 5 hours, the copper loss is

$$P_R = \left(\frac{6000}{2400}\right)^2 (8.4)/1000 = 0.053 \text{ kw}$$

During 16 hours, the copper loss is

$$P_R = 0$$

The energy into copper losses is

$$W_R = (0.146) \times 3 + (0.053) \times 5 + (0) \times 16 = 0.703 \text{ kwh}$$

The denominator of Eq. 5-57 is the energy input during the period.

$$W_{\text{in}} = W_{\text{out}} + W_R + W_c = 54 + 1.728 + 0.703 = 56.431 \text{ kwh}$$

The energy efficiency for the full 24 hours is

$$\eta_W = W_{\text{out}}/W_{\text{in}} = 54/56.431 = 0.957$$

Example 5-4 (Section 5-9)

Test data for a 10-kva, 2400/240-v, 60-Hz transformer are as follows: open-circuit test, with input to the low side, 240 v, 0.75 amp, 72 w; short-circuit test, with input to the high side, 80.5 v, 5.0 amp, 210 w. Find the parameters for the approximate equivalent circuit in Fig. 5-9(a).

Solution

Turns ratio = a = 2400/240 = 10

$$I_{1\text{rated}} = 10{,}000/2400 = 4.17 \text{ amp}$$

$$I_{2\text{rated}} = 10{,}000/240 = 41.7 \text{ amp}$$

Since the input for the open-circuit test is measured into the low side, it is convenient to work with Fig. 5-9(d). The admittance is

$$Y_{\phi_2} = I_{2\text{oc}}/V_{2\text{oc}} = 0.75/240 = 3.12 \times 10^{-3} \text{ mho}$$

The angle of the admittance is found by

$$\cos\theta_\phi = P_{\text{oc}}/(V_{2\text{oc}} I_{2\text{oc}}) = 72/(240 \times 0.75) = 0.4$$

$$\theta_\phi = \cos^{-1}(0.4) = -66.4°$$

The conductance is

$$G_{c_2} = Y_{\phi_2} \cos\theta_\phi = 1.25 \times 10^{-3} \text{ mho}$$

The susceptance is

$$B_{m_2} = Y_{\phi_2} \sin\theta_\phi = -2.86 \times 10^{-3} \text{ mho}$$

In Fig. 5-9(a) these quantities are referred to the high side.

$$\mathbf{Y}_{\phi_1} = \mathbf{Y}_{\phi_2}/a^2 = 31.2\underline{/-66.4°} = 12.5 - j\,28.6\ \mu\text{mho}$$

Since the measurements for the short-circuit test are made into the high side winding, Fig. 5-9(b) is applicable. The impedance is

$$Z_{e_1} = V_{1\text{sc}}/I_{1\text{sc}} = 80.5/5 = 16.1\ \Omega$$

The angle of the equivalent impedance is found by

$$\cos\theta_e = P_{sc}/V_{1sc}I_{1sc} = 210/(80.5)(5) = 0.522$$

$$\theta_e = \cos^{-1}(0.522) = 58.5°$$

The equivalent resistance is

$$R_{e_1} = Z_{e_1}\cos\theta_e = 8.4\ \Omega$$

The equivalent reactance is

$$X_{e_1} = Z_{e_1}\sin\theta_e = 13.7\ \Omega$$

Example 5-5 (Section 5-10)

(a) Repeat Example 5-4, using per-unit values. (b) Find the regulation for a power factor of 0.8, this time *leading*. (c) Find the efficiency at *half* load and a power factor of 0.8.

Solution

First establish the bases: $P_{base} = P_{a\,base} = 10{,}000$ va.

$$V_{1\,base} = 2400\text{ v},\ V_{2\,base} = 240\text{ v}$$

$$I_{1\,base} = 10{,}000/2400 = 4.17\text{ amp}$$

$$I_{2\,base} = 10{,}000/240 = 41.7\text{ amp}$$

Convert all test data into per-unit values

$$V_{oc} = 240/240 = 1\text{ pu}$$

$$I_{oc} = 0.75/41.7 = 0.018\text{ pu}$$

$$P_{oc} = 72/10{,}000 = 0.0072\text{ pu}$$

$$V_{sc} = 80.5/2400 = 0.0335\text{ pu}$$

$$I_{sc} = 5.0/4.17 = 1.2\text{ pu}$$

$$P_{sc} = 210/10{,}000 = 0.021\text{ pu}$$

Note that there are no subscripts indicating "primary" or "secondary" for per-unit values.

(a) The procedures of Example 5-4 are repeated with pu values.

$$Y_\phi = 0.018/1 = 0.018\text{ pu}$$

$$\cos\theta_\phi = 0.0072/(1 \times 0.018) = 0.4$$

Being a dimensionless number, it must be the same as that obtained in the previous example. So are

$$\theta_\phi = -66.4^\circ \text{ and } \sin\theta_\phi = -0.9165$$

$$G_c = 0.018 \times 0.4 = 0.0072 \text{ pu}$$

$$B_m = -0.018 \times 0.9165 = -0.0165 \text{ pu}$$

$$\mathbf{Y}_\phi = 0.018\underline{/-66.4^\circ} = 0.0072 - j\,0.0165 \text{ pu}$$

$$Z_e = 0.0335/1.2 = 0.0279 \text{ pu}$$

$$\cos\theta_e = 0.021/(0.0335 \times 1.2) = 0.522$$

$$\theta_e = 58.5^\circ, \sin\theta_e = 0.853$$

$$R_e = 0.0279 \times 0.522 = 0.01456 \text{ pu}$$

$$X_e = 0.0279 \times 0.853 = 0.0238 \text{ pu}$$

$$\mathbf{Z}_e = 0.0279\underline{/58.5^\circ} = 0.01456 + j\,0.0238 \text{ pu}$$

These values can be checked against those obtained in the previous example by using the admittance and/or impedance bases. For instance

$$Z_{1\,\text{base}} = 2400/4.17 = 576\ \Omega$$

and

$$Z_{e_1} = 16.1/576 = 0.0279, \text{etc.}$$

(b) Use the method of Example 5-1

$$\mathbf{I}_2 = 1\underline{/0^\circ} \text{ pu}$$

$$\mathbf{V}_2 = 1\underline{/-36.9^\circ} \text{ pu}$$

(For a "leading" power factor, the voltage lags the current), thus $\mathbf{V}_2 = 0.8 - j\,0.6$ pu.

$$\mathbf{Z}_e\mathbf{I}_2 = 0.0146 + j\,0.0238 \text{ pu}$$

Add these two phasors to get $\mathbf{V}_1 = 0.8146 - j\,0.576 = 0.9876\underline{/-35.3^\circ}$. So the regulation is $(0.9876 - 1)/1 = -0.0124$ (negative, due to the leading load).

(c) At half-load,

$$I_2 = 0.5 \text{ pu}$$

$$P_{\text{out}} = 1 \times 0.5 \times 0.8 = 0.4 \text{ pu}$$

$$P_c = 0.018 \times 1^2 = 0.018 \text{ pu (equals } G_c)$$

$$P_R = 0.01456 \times 0.5^2 = 0.00364 \text{ pu}$$

$$P_{\text{loss}} = 0.018 + 0.00364 = 0.02164 \text{ pu}$$

So the efficiency is

$$\eta = 0.4/(0.4 + 0.02164) = 0.949$$

5-13 PROBLEMS

5-1. The transformer in Fig. 5-2 has a sheet steel core. The mean length of the flux path is 0.7 m. The effective cross-sectional area is 0.005 m^2. Windings a-b and c-d are connected to electric circuits. $N_{1(a-b)} = 80$ turns; $N_{2(c-d)} = 160$ turns. At the instant when flux is 0.006 webers, the current flowing into terminal a is 63 amp. Neglect core losses. (a) Find the magnetizing mmf at this time. (b) Find the magnitude and direction of the current in coil c-d at this time. (c) Find the ratio of I_2/I_1. Compare this to the turns ratio N_1/N_2.

5-2. The parameters of a transformer are: $a = 2.5, R_1 = 0.07\ \Omega, X_1 = 0.20\ \Omega$, $R_2 = 0.01\ \Omega$, $X_2 = 0.025\ \Omega$, $G_{c_1} = 0.0005$ mho, $B_{m_1} = -0.0025$ mho. This transformer is operated with $V_2 = 220$ v, $I_2 = 91$ amp, and the power factor is 0.85 lagging. Find $\mathbf{I}_1$ and $\mathbf{V}_1$, using $\mathbf{I}_2$ as the reference. Use the basic equivalent circuit of Fig. 5-5.

5-3. Solve Problem 5-2 using the approximate equivalent circuit of Fig. 5-9(a).

5-4. Test data for a 10-kva, 2300/230-v, 60-Hz transformer are as follows: open-circuit test, with input to the low side, 230 v, 0.62 amp, 69 w; short-circuit test, with input to the high side, 127 v, 4.9 amp, 263 w. Find the equivalent circuit parameters $R_{e_1}, X_{e_1}, G_{c_1}, B_{m_1}$.

5-5. For the transformer in Problem 5-4, find the voltage regulation when operated with load on the low voltage side of 230 v, 43.5 amp, power factor of 0.8 lagging.

5-6. Find the efficiency for the operating conditions in Problem 5-5.

5-7. Test data for a 30-kva, 2400/240-v, 60-Hz transformer are as follows: open-circuit test, with input to the low side, 240 v, 3 amp, 230 w; short-circuit test, with input to the high side, 100 v, 18.8 amp, 1050 w. Find the equivalent circuit parameters $R_{e_1}, X_{e_1}, G_{c_1}, B_{m_1}$.

5-8. For the transformer in Problem 5-7, find the voltage regulation when operated with 240 v, 125 amp, power factor of 0.6 leading, load on the low voltage side.

5-9. Find the efficiency for the operating conditions in Problem 5-8.

5-10. A 30-kva transformer has core losses of 230 w and copper losses of 320 w when operated at rated voltage and rated kva. This transformer is connected for 24 hours a day. Loads during the day are: 30 kva at 0.8 P.F. lagging for 3 hours; 20 kva at 0.6 P.F. lagging for 6 hours; and no load for 15 hours. Find the all-day efficiency.

5-11. A transformer is rated 5 kva, 220/110 v, 50 Hz. For rated operating conditions it has hysteresis loss of 30 w, eddy-current loss of 30 w, and copper loss of 80 w. If this transformer is used in a 60-Hz installation, 220/110 v, find the available load kva such that total losses remain at 140 w.

5-12. The transformer in Problem 5-11 is to be used in a 50-Hz system with voltage of 110/55 v. Find the permissible load current such that total losses remain at 140 w.

5-13. The transformer in Problem 5-7 is operated at rated voltage and rated frequency. Find the maximum efficiency.

5-14. A 5-kva, 330/220-v, 60-Hz transformer has the 220-v winding shorted. With 6.3 v impressed on the 330-v winding, the input power is 50 w and the input current is 15 amp. If a short-circuit test is performed with the input to the 220-volt winding with the 330-volt winding shorted, find the impressed voltage and input power for a current of 24 amp.

5-15. A 5-kva, 330/220-v, 60-Hz transformer is tested with the 330-v winding open. With 220-v impressed, the input power is 40 w and the current is 0.4 amp. If this transformer is operated with the 220-v winding open and with 330 v impressed on the high-voltage winding, find the input power and the no-load current.

5-16. For the transformer in Problem 5-7, find $R_{e_1}, X_{e_1}, G_{c_1}, B_{m_1}$ in per-unit.

5-17. For the transformer in Problem 5-7, use the per-unit method to find the voltage regulation for a power factor of 0.9 lagging.

5-18. For the transformer in Problem 5-7, use the per-unit method to find the efficiency for rated load conditions and with power factor of 0.9 lagging.

5-19. A transformer, rated 50 kva, has copper losses of 400 w when operated at 80 percent of rated current. (a) Find the value of the equivalent resistance in per-unit. (b) If maximum efficiency occurs for operation at 70 percent of rated current, find the core loss in per-unit.

5-20. For the transformer in Problem 5-7, assume $X_1 = 2.20\ \Omega$ and $X_2 = 0.022\ \Omega$. Find the inductances L_1, L_2, and M. Find the coefficient of coupling, $k = M/\sqrt{L_1 L_2}$.

5-21. A transformer has the following parameters in per-unit: $\mathbf{Z}_1 = 0.01 + j\,0.02$ pu, $\mathbf{Z}_2 = 0.01 + j\,0.04$ pu, $\mathbf{Y}_{\phi_1} = 0.01 - j\,0.05$ pu. (a) Find the coefficient of coupling. (b) For a load current of 0.7 pu, with power factor of 0.8 lagging, and rated load voltage, find the input voltage in per-unit.

5-22. The purpose of this problem is to compare the transmission of electric energy at two different voltages. It is desired to deliver 24 kw at 240 v into a resistance load that is at a distance of 2 km from the available source. Transformers may be considered to be ideal. Two systems are

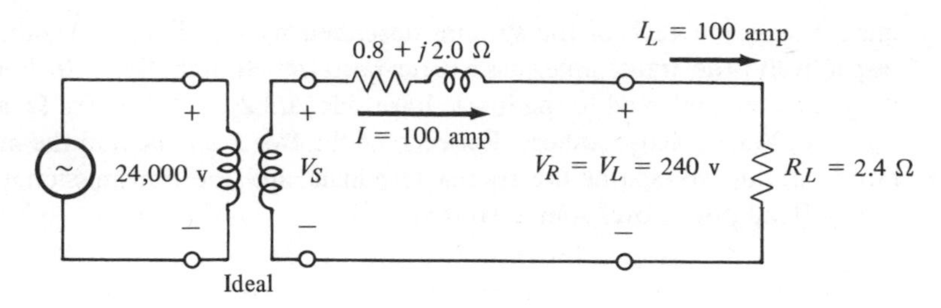

Fig. P-5-22(a).

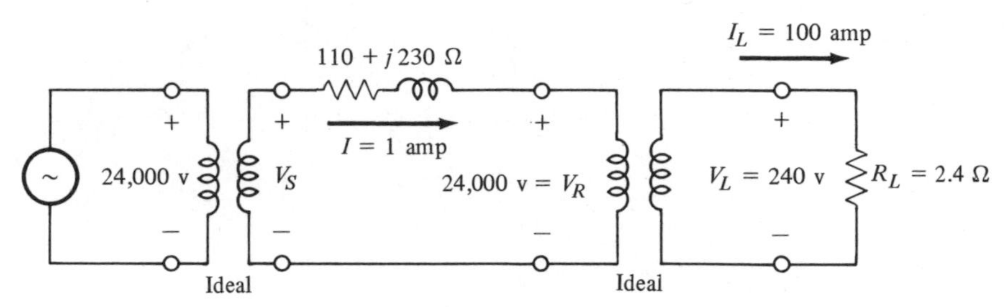

Fig. P-5-22(b).

shown in Fig. P-5-22(a) and (b). For each system, find the efficiency and the value of no-load voltage at the load terminals.

5-23. The load resistance of 2.4 Ω is supplied through a transformer and a trans-

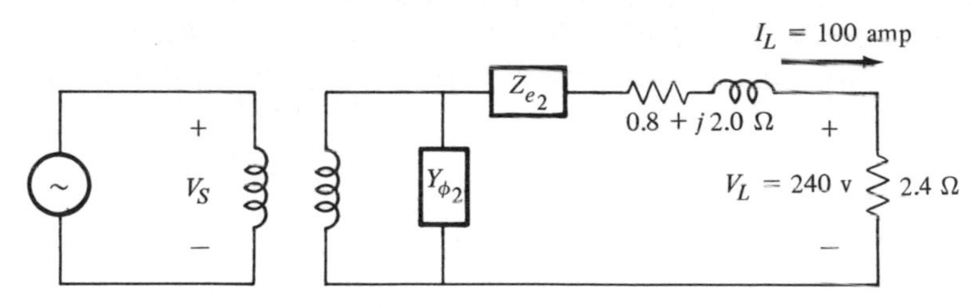

Fig. P-5-23(a).

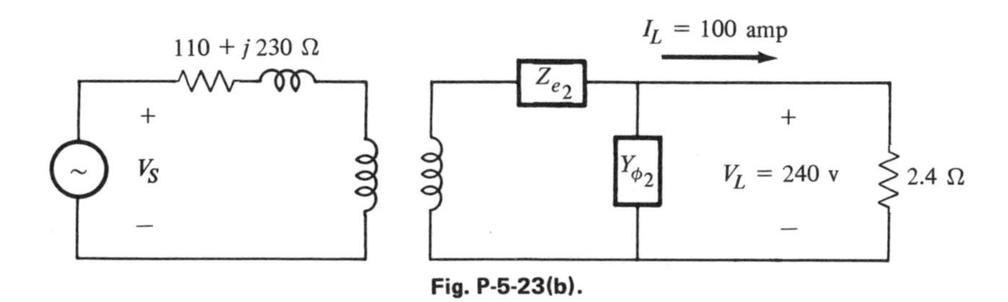

Fig. P-5-23(b).

mission line. In each of the systems described by Fig. P-5-23(a) and (b), respectively, the transformer has a turns ratio (left to right) of 100 : 1, and its parameters, referred to the low-voltage side, are $\mathbf{Z}_{e_2} = 0.1 + j\,0.2\ \Omega$, and $\mathbf{Y}_{\phi_2} = 0.005 - j\,0.015$ mhos. For each of the two systems, find the magnitude of the voltage at the source terminals, and the transmission efficiency (load power over source power).

6

Transformer Connections

6-1 AUTOTRANSFORMERS

The transformer described in Chapter 5 accomplishes not only the transfer of power between power system components operating at different voltages, but also the *electrical isolation* of these components from one another. There are many cases when this latter feature is quite important. For instance, considerations of safety in case of accidental failure demand that low-voltage load circuits be physically separated from high-voltage transmission lines. But in many other cases, particularly when the two different voltages are in the same order of magnitude, that separation is not needed, and in these cases, considerable savings in both cost and operating expenses can be accomplished by using transformers in which the primary and secondary circuits are connected to each other to form an *autotransformer* circuit.

Consider the circuit diagram of Fig. 6-1. It can be considered either as two coils connected to each other, or as a single coil with a tap (i.e., a connection from inside the coil to a third terminal). The latter view is the one that gives rise to the name autotransformer. *Auto* is Greek for *self* (as in automobile—that strange horseless carriage that seems to move by itself), and the single winding acts as a transformer all by itself. For purposes of analysis, however, it is preferable to view the autotransformer as consisting of two coils, called S (for *series*) and C (for *common*). Coil S is in *series* with the source (or with the load, in Fig. 6-2), and coil C is *common* to both the primary and the secondary circuits. Of course, the two coils are wound around the same ferromagnetic core.

To study the principles of the autotransformer, all of its imperfections (mag-

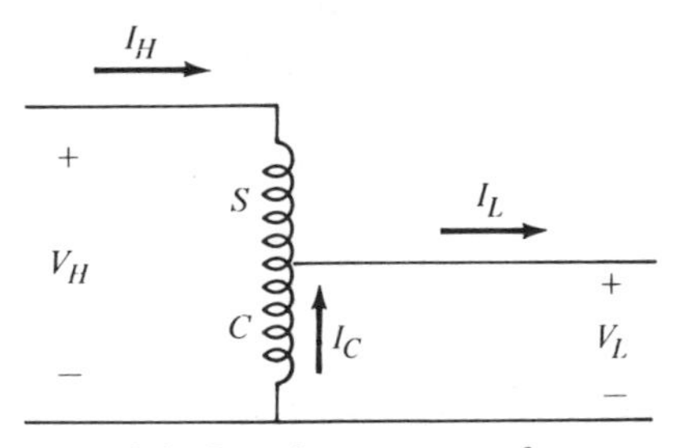

Fig. 6-1. Step-down autotransformer.

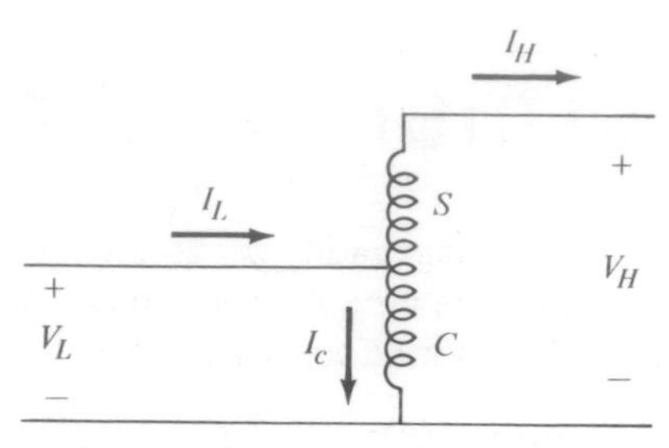

Fig. 6-2. Step-up autotransformer.

netizing current, core losses, leakage fluxes, and coil resistances) will at first be disregarded. Also, the discussion will be limited to sinusoidal steady state.

Connecting the two left-side (primary) terminals of either Fig. 6-1 or Fig. 6-2 to a voltage source determines the value of the flux in the core, in accordance with Eq. 4-25, with the number of turns between these two terminals substituted for N. This alternating flux causes the same voltage to be induced in every turn that is wound around the core. Therefore, calling the turns numbers of the two coils N_S and N_C, respectively, the ratio of voltage phasors is

$$\frac{\mathbf{V}_H}{\mathbf{V}_L} = \frac{N_S + N_C}{N_C} \tag{6-1}$$

where the subscripts H and L stand for *higher voltage* and *lower voltage*. In the case of Fig. 6-1, the primary (input) voltage is higher than the secondary (output) voltage, but the autotransformer can just as well be used to step the voltage up rather than down, as in Fig. 6-2. Equation 6-1 is valid in either case.

Now, to get the *current ratio*, remember that an ideal transformer must have zero ampere-turns for a finite flux. In the following equations, as well as in the two diagrams, the subscripts H and L are consistently used to indicate the higher voltage and lower voltage side. Thus, the current I_H is the current on the high voltage side (and is identical with the current through coil S), regardless of the fact that it is actually smaller than I_L. Use either Fig. 6-1 or Fig. 6-2.

$$\mathfrak{F} = \mathbf{I}_H N_S - \mathbf{I}_C N_C = 0 \tag{6-2}$$

and eliminate the current I_C by Kirchhoff's current law

$$\mathbf{I}_H + \mathbf{I}_C = \mathbf{I}_L \tag{6-3}$$

This leads to

$$\mathbf{I}_H N_S - (\mathbf{I}_L - \mathbf{I}_H) N_C = \mathbf{I}_H(N_S + N_C) - \mathbf{I}_L N_C = 0 \tag{6-4}$$

and, thus, to the ratio of current phasors

$$\frac{\mathbf{I}_H}{\mathbf{I}_L} = \frac{N_C}{N_S + N_C} \tag{6-5}$$

which is the reciprocal of the voltage ratio, just as in the regular transformer with two separate coils. Note that the ratio is a real number, which means $\mathbf{I}_H$ and $\mathbf{I}_L$ are in phase. Thus Eq. 6-3 is also valid for magnitudes

$$I_C = I_L - I_H \tag{6-6}$$

Example 6-1 illustrates all these relationships.

To demonstrate the advantages of the autotransformer, let Fig. 6-3 be understood to represent a two-coil transformer for the same voltages and currents as the autotransformer of Fig. 6-2. For the sake of easier comparison, assume both transformers have identical iron cores.

The *common* coil of the autotransformer requires as many turns as the low-side coil of the two-coil transformer, but it may be made of thinner wire because it has to carry I_C amperes, which, as Eq. 6-6 shows, is less than the current I_L carried by the low-side coil of the two-coil transformer. The *series* coil of the autotransformer carries as much current as the high-side coil of the two-coil transformer, but it consists of fewer turns since its voltage $V_H - V_L$ is less than V_H. So the two coils of an autotransformer are smaller, lighter, and thus, cheaper than those of an equivalent two-coil transformer. In addition, due to their smaller size, they require less space; so the length of the core may also be reduced for a further saving.

It is also possible to compare an autotransformer to a two-coil transformer made of the same core and coils, without the interconnection of the coils. Again, the autotransformer comes out ahead in value. The reader is referred to Example 6-2 for a numerical illustration.

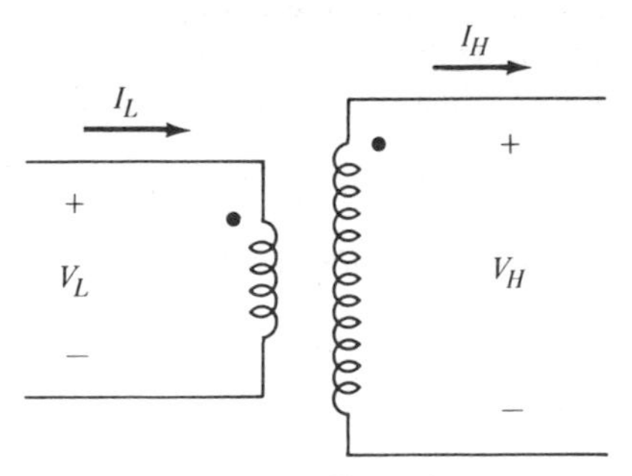

Fig. 6-3. Two-coil transformer.

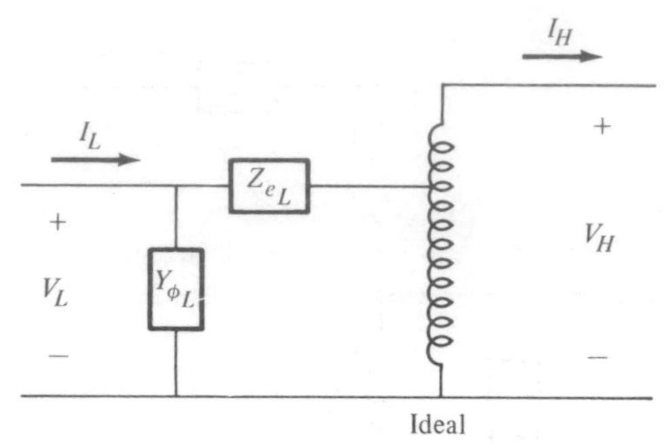

Fig. 6-4. Approximate equivalent circuit of autotransformer.

The differences are particularly pronounced when $N_C > N_S$, i.e., when the voltage and current ratios are not too far from unity. It is rather fortunate that the autotransformer is most advantageous in those cases in which the separation of the primary from the secondary side is usually not required. Typical examples are transformers that serve to adjust load voltages that would otherwise vary with load changes (the number of turns on one side is slightly altered by means of switches). By contrast, there are cases when the primary and secondary voltages are of different orders of magnitude. In these cases, the separation is often necessary, but they are just the cases when the differences in cost are much smaller anyway.

The imperfections of an actual (not ideal) autotransformer can be considered just like those of a two-coil transformer, in terms of an approximate equivalent circuit. For example, the circuit of Fig. 6-4 is similar to that of Fig. 5-9a, but with an ideal *auto*transformer. The parameters of such a circuit can be obtained from open- and short-circuit tests.

It is also of interest to compare these parameters to those that describe the two-coil transformer consisting of the same core and coils. For this purpose, Fig. 6-5 gives the equivalent circuit of this two-coil transformer, with the ad-

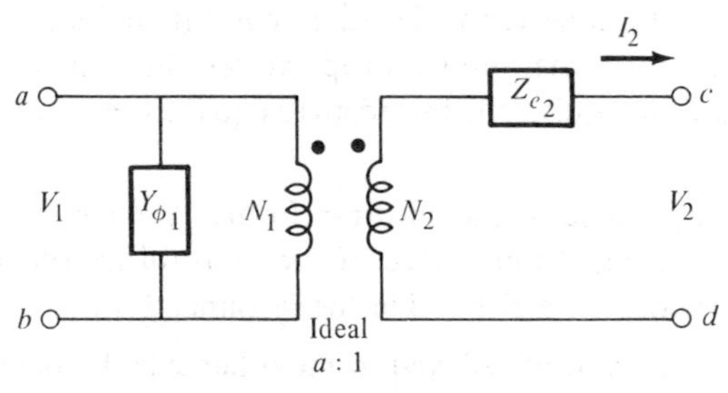

Fig. 6-5. Transformer with imperfections.

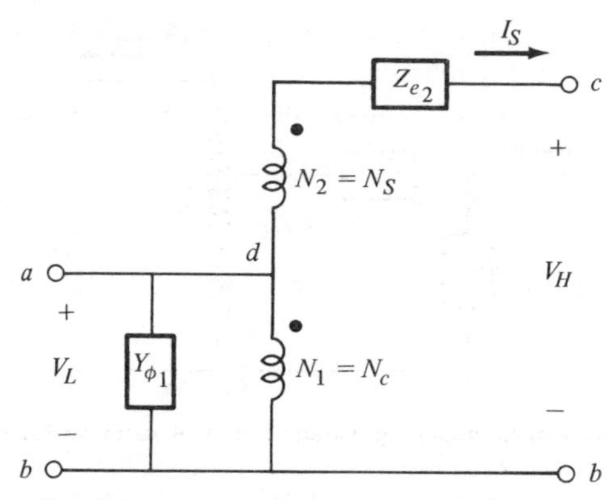

Fig. 6-6. Autotransformer with imperfections.

mittance Y_ϕ referred to the primary side and the impedance Z_e to the secondary side. In Fig. 6-6, the same coils are reconnected to form an autotransformer, the original primary coil becoming the common coil, and the original secondary coil becoming the series coil.

Let the low-side terminal voltage V_L of the autotransformer be the same as the primary terminal voltage V_1 of the two-coil transformer. This requires that the flux be the same, which in turn calls for the same magnetizing current and the same core losses. Therefore, the admittances Y_{ϕ_1} in the two diagrams must also be the same. To formulate this as a general rule,

> the magnetizing admittance of the autotransformer, referred to its *low-voltage* side, equals the magnetizing admittance of the two-coil transformer referred to that side that becomes the *common* coil in the reconnection.

Now consider the equivalent impedance in Fig. 6-5. It represents the combined resistances and leakage reactances of the two windings. The voltage across it is the phasor difference between the actual secondary terminal voltage $\mathbf{V}_2$ and its ideal value $\mathbf{V}_1/a$. When the load of the autotransformer of Fig. 6-6 is adjusted so that its current I_S is equal to the current I_2 drawn by the load of the two-coil transformer, then both coils carry the same current in Fig. 6-6 as they do in Fig. 6-5. Consequently, the voltage drops across their imperfections are the same, which makes the equivalent impedances equal to each other. As a general rule,

> the equivalent impedance of the autotransformer, referred to its *high-voltage* side, equals the equivalent impedance of the two-coil transformer, referred to that side that becomes the series coil in the reconnection.

These parameters can also be referred to the other side, by being multiplied by the square of the proper turns ratio. Observe that this is not the turns ratio a of the two-coil transformer, but the ratio appearing in Eqs. 6-1 or 6-5. For instance, to refer the equivalent impedance Z_{e_2} of Fig. 6-6 to the low-voltage side, it must be multiplied by the voltage ratio and divided by the current ratio. Thus

$$\mathbf{Z}_{e_1} = \left(\frac{N_C}{N_S + N_C}\right)^2 \mathbf{Z}_{e_2} \tag{6-7}$$

Similarly, to refer the admittance from the low- to the high-voltage side

$$\mathbf{Y}_{\phi_2} = \left(\frac{N_C}{N_S + N_C}\right)^2 \mathbf{Y}_{\phi_1} \tag{6-8}$$

The reader is advised never to memorize such equations, but rather to consider that impedances must be larger when referred to the high-voltage side, and admittances must be larger when referred to the low-voltage side.

The equivalent circuit of Fig. 6-6 (or that of Fig. 6-4) permits a comparison of the *regulation* and *efficiency* of the autotransformer to those of the two-coil transformer made of the same core and coils. The full-load voltage drop across Z_{e_2} is the same in Fig. 6-5 and 6-6, but in Fig. 6-6 it is a smaller fraction of the output voltage. Similarly, the power losses for a given fraction of rated load are the same in both circuits, but those for the autotransformer are a smaller fraction of the output power. On both counts, the autotransformer wins the competition. Its advantage is the more pronounced, the more the turns number N_C exceeds the turns number N_S.

6-2 SINGLE-PHASE POWER

This section and the following two are inserted here as a review of subjects with which the reader should be familiar before he reaches the discussion of three-phase power transformation.

Since power systems need transformers, (see Section 1-3), they must use *alternating* voltages and currents. This raises the question of how a uniform flow of power can be obtained. The *instantaneous value of power* is the product of the instantaneous values of voltage and current

$$p(t) = v(t)\, i(t) \tag{6-9}$$

Let voltage and current be sinusoids, and form the product

$$p(t) = V_{\max} \cos(\omega t + \alpha)\, I_{\max} \cos(\omega t + \beta) \tag{6-10}$$

The product of two cosines can be converted according to the formula

$$\cos x \cos y = \tfrac{1}{2}\left[\cos(x + y) + \cos(x - y)\right] \tag{6-11}$$

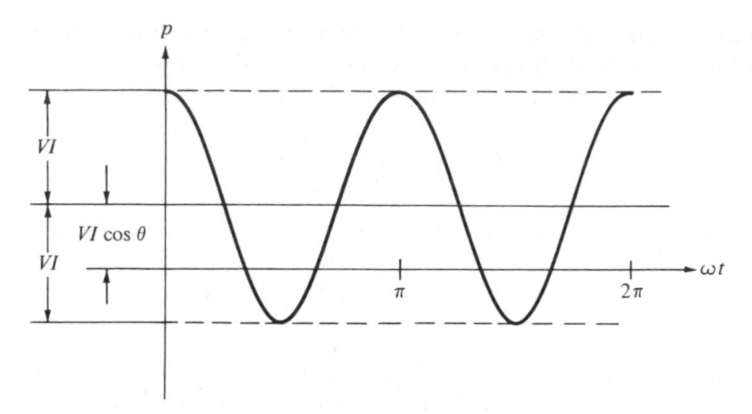

Fig. 6-7. Single-phase power.

resulting in the expression

$$p(t) = \frac{V_{\max} I_{\max}}{2} [\cos(2\omega t + \alpha + \beta) + \cos(\alpha - \beta)] \tag{6-12}$$

Finally, substitute the *rms values* $V = V_{\max}/\sqrt{2}$ and $I = I_{\max}/\sqrt{2}$, and the phase difference $\theta = \alpha - \beta$

$$p(t) = VI \cos(2\omega t + \alpha + \beta) + VI \cos\theta \tag{6-13}$$

Figure 6-7 shows the wave shape of the function $p(t)$ for arbitrary values of VI and θ, and with the axis of reference chosen for the sake of convenience to make $\alpha + \beta = 0$. The power is seen to be pulsating at twice the frequency of voltage and current, thus becoming negative twice during every period. This is particularly undesirable for electromechanical power conversion where the pulsating action causes vibration and noise. None of this happens when the electric power is that of a balanced three-phase power system. The idea is to add three curves like that of Fig. 6-7, displaced against each other by one-third of a period but otherwise identical. The three pulsating components add up to zero, and the total power is a constant. The rotating magnetic field that will be encountered in later chapters is a manifestation of this fact.

6-3 REVIEW OF THREE-PHASE CIRCUITS

The authors hope that the contents of this section are familiar to the readers, but they believe that some will find a brief review helpful.

A three-phase source is a set of three single-phase sources whose voltages are sinusoids of equal amplitude and frequency, displaced against each other by 120° which is one-third of a period. For instance:

$$\begin{aligned} v_1(t) &= V_{max} \cos \omega t \\ v_2(t) &= V_{max} \cos(\omega t - 120°) \\ v_3(t) &= V_{max} \cos(\omega t - 240°) \end{aligned} \tag{6-14}$$

where the phase sequence is 1-2-3 because v_2 follows v_1, etc. In terms of rms phasors

$$\begin{aligned} \mathbf{V}_1 &= V\underline{/0°} \\ \mathbf{V}_2 &= V\underline{/-120°} \\ \mathbf{V}_3 &= V\underline{/-240°} \end{aligned} \tag{6-15}$$

It should be mentioned right here that the sum of three such symmetrical quantities—voltages, currents, or whatever else—is always zero. This can be seen by adding the three Eqs. 6-14 or the three Eqs. 6-15, either analytically or graphically.

There is no difficulty in obtaining three such voltages in one generator, as later chapters will make clear. But it would be far too uneconomical to transmit three-phase power to the loads if each phase required its own two transmission wires. Instead, the three circuits must be *interconnected*. There are two basic connections, the *delta* and the *wye*. Figure 6-8 shows that, by interconnecting three-phase sources, in both cases the number of line terminals is reduced from six to three (a, b, c) although, in the case of the wye connection, a fourth terminal, the *neutral*, is also available, if desired.

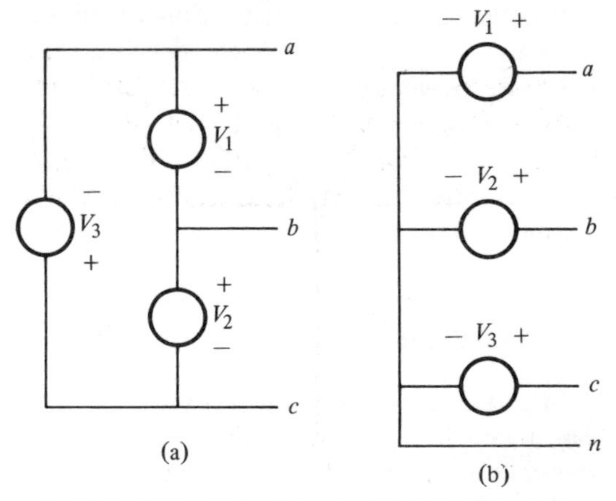

Fig. 6-8. Three-phase sources: (a) delta; (b) wye.

An important distinction is to be made between the *phase voltages* (for sources, these are the voltages across each source) and the *line voltages* (these are the voltages between each pair of line terminals). For a delta connection, this distinction is trivial because the two are the same (see Fig. 6-8a)

$$\mathbf{V}_1 = \mathbf{V}_{ab}$$
$$\mathbf{V}_2 = \mathbf{V}_{bc} \tag{6-16}$$
$$\mathbf{V}_3 = \mathbf{V}_{ca}$$

which may be written as a single equation with subscripts P for phase and L for line

$$\mathbf{V}_L = \mathbf{V}_P \tag{6-17}$$

For a wye connection, however, Fig. 6-8b shows that the phase voltages are the *voltages to neutral*

$$\mathbf{V}_1 = \mathbf{V}_{an}$$
$$\mathbf{V}_2 = \mathbf{V}_{bn} \tag{6-18}$$
$$\mathbf{V}_3 = \mathbf{V}_{cn}$$

and the relation between each line voltage and the nearest phase voltage, found by application of Kirchhoff's voltage law (Fig. 6-9) is

$$\mathbf{V}_L = \sqrt{3}\,\mathbf{V}_P \underline{/\pm 30^\circ} \tag{6-19}$$

where the choice of the plus or minus sign depends on the phase sequence.

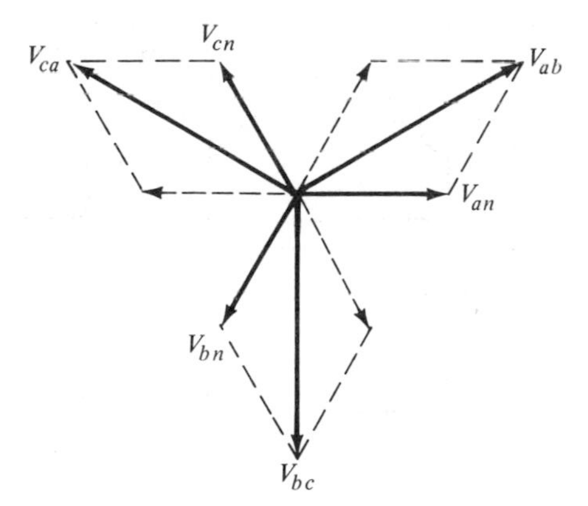

Fig. 6-9. Line and phase voltages for wye connection.

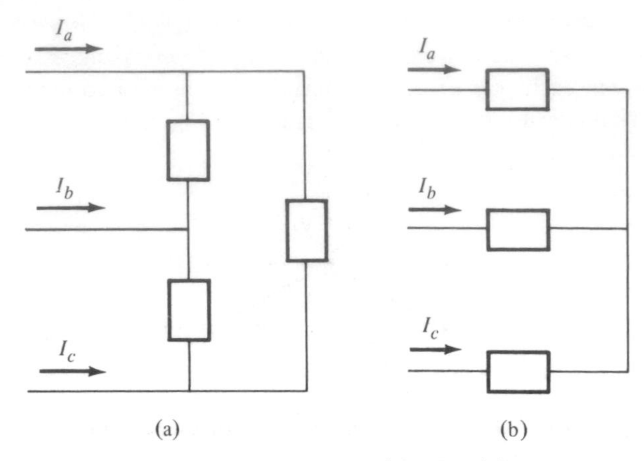

Fig. 6-10. Three-phase loads: (a) delta; (b) wye.

All of this has concerned voltages. *Currents* depend on loads. Figure 6-10 depicts three-phase loads in delta and wye connections, using steady state notation. Either one of the two configurations can be connected to the source terminals a, b, c regardless of whether the sources are delta- or wye-connected. If both the sources and the loads are wye-connected, their neutrals may be (and sometimes are) connected to each other, forming a four-wire system, as shown in Fig. 6-11.

Similar to the distinction between line voltages and phase voltages, there are *line currents* (in the three lines a, b, c connecting the source terminals to the load terminals) and *phase currents* (in each of the three loads) to be kept apart. In a

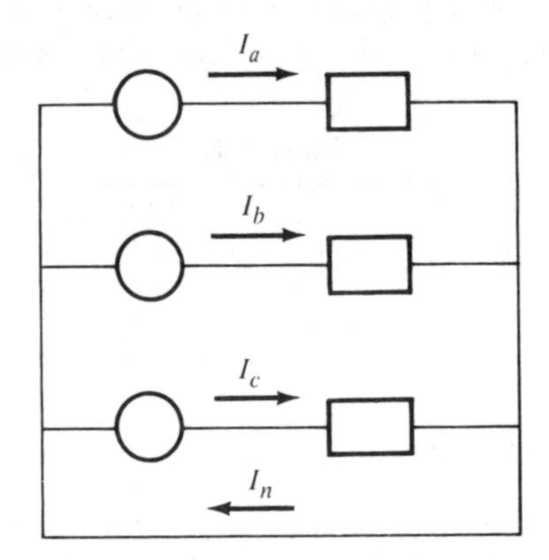

Fig. 6-11. Four-wire system.

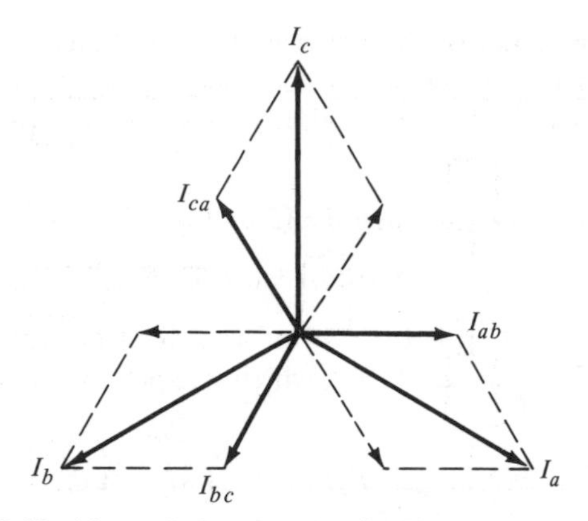

Fig. 6-12. Line and phase currents in delta-connected loads.

wye-connected load, line currents are the same as phase currents (see Figs. 6-10b or 6-11).

$$\mathbf{I}_L = \mathbf{I}_P \tag{6-20}$$

If there is a neutral conductor as in Fig. 6-11, it carries the "return current" I_N which, in the case of a *balanced load* (three identical leads in delta or wye) is zero.

In a delta-connected load, the three line currents are related to the three phase currents in accordance with Kirchhoff's current law. For a balanced load, the steady state relation between any line current and the nearest phase current (see Fig. 6-12) is

$$\mathbf{I}_L = \sqrt{3}\,\mathbf{I}_P \underline{/\mp 30^\circ} \tag{6-21}$$

with the plus or minus sign depending on the phase sequence.

Table 6-1 is a recapitulation of the magnitude relations in balanced three-phase systems.

Table 6-1.

	Δ	Y
V_L	V_P	$\sqrt{3}\,V_P$
I_L	$\sqrt{3}\,I_P$	I_P

6-4 THREE-PHASE POWER

Let the instantaneous power expressed by Eq. 6-13 and depicted by Fig. 6-7 represent the power in phase 1 of a balanced three-phase circuit, with V and I

representing the rms values of the phase voltages and phase currents. Then the power for the other two phases can be obtained by substituting $\omega t - 120°$ and $\omega t - 240°$, respectively, for ωt, and the total power for the three phases becomes

$$p(t)_{3\,\text{phase}} = V_P I_P [\cos(2\omega t + \alpha + \beta) + \cos(2\omega t + \alpha + \beta - 240°) + \cos(2\omega t + \alpha + \beta - 480°)] + 3V_P I_P \cos\theta \quad (6\text{-}22)$$

The bracket in this equation is the sum of a symmetrical set of three sinusoids (because $-480°$ is the same as $-120°$), which was previously recognized to equal zero. Thus

$$p(t)_{3\,\text{phase}} = P_{3\,\text{phase}} = 3V_P I_P \cos\theta \quad (6\text{-}23)$$

which means that this power is not pulsating. In a balanced three-phase system, energy is generated, transmitted, and consumed at a uniform rate.

Equation 6-23 expresses power in terms of *phase quantities* (V_P and I_P), which is not really useful when the connection (delta or wye) of sources or loads is not known. But power can equally well be expressed in terms of *line quantities*. The product $V_P I_P$ equals, for a delta connection, $V_L(I_L/\sqrt{3})$, and, for a wye connection, $(V_L/\sqrt{3})I_L$, which is exactly the same thing. Substitution into Eq. 6-23 yields

$$P_{3\,\text{phase}} = \sqrt{3}\, V_L I_L \cos\theta \quad (6\text{-}24)$$

which is usually the more convenient expression to use. It must be kept in mind, though, that θ remains the angle between phase quantities, never that between line quantities.

Similar reasoning leads to expressions for reactive power and apparent power in terms of either phase or line quantities for balanced three-phase circuits:

$$Q_{3\,\text{phase}} = -3V_P I_P \sin\theta = -\sqrt{3}\, V_L I_L \sin\theta \quad (6\text{-}25)$$

and

$$P_{a\,3\,\text{phase}} = 3V_P I_P = \sqrt{3}\, V_L I_L \quad (6\text{-}26)$$

Finally, complex power in a balanced three-phase circuit remains

$$\mathbf{P}_a = P + jQ = P_a \underline{/-\theta} \quad (6\text{-}27)$$

6-5 TRANSFORMATION OF THREE-PHASE POWER

The transmission of three-phase power from generators to loads requires the use of transformers, just as that of single-phase power. Transformation of three-phase power can be accomplished with three transformers (also called a *bank* of

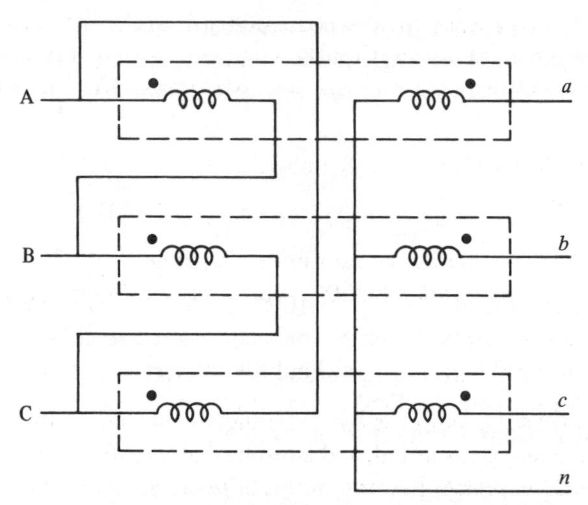

Fig. 6-13. Primaries in delta, secondaries in wye.

transformers) whose coils must be interconnected to permit their connection to the line terminals. Using the basic configurations, delta and wye, for the transformer primaries and secondaries, provides a choice among *four* possibilities.

There are several factors to consider in choosing. First, there may be the desire to use a *neutral conductor*. For instance, there may be single-phase loads calling for voltages of different magnitudes, like domestic lights and appliances for 120 v, and tools or large air conditioners for $\sqrt{3}\ 120 = 208$ v. This would necessitate the use of wye-connected transformer secondaries, as shown in Fig. 6-13, where the primaries are delta-connected. In this diagram, each of the three transformers is drawn inside a broken outline.

A neutral conductor may also be needed on the primary side, e.g., for protection of an overhead transmission line against lightning. In this case, the transformer primaries must be wye-connected. Further purposes of neutral conductors can be to provide a path for the "return current," in the case of unbalanced loads, or for certain harmonics, in the case of nonsinusoidal currents drawn by nonlinear load elements.

The choice of transformer connections also has a bearing on the ratio of line voltages. The turns ratio of each of the three transformers is the ratio of their *phase* voltages, and a wye connection raises the line voltages to $\sqrt{3}$ times as much. Therefore, a transformer bank with primaries in wye and secondaries in delta helps to step down the line voltages, etc.

A bank of three transformers can be replaced by a *three-phase transformer*, which consists of three primary and three secondary coils wound around different legs of a three-leg magnetic circuit. That permits a saving of core material and is thus, cheaper than a transformer bank; although it is cheaper, of course, to replace one out of three single-phase transformers than an entire three-phase unit. Also, for extremely large ratings, it becomes impractical to transport three-phase units.

6-6 EXAMPLES

Example 6-1 (Section 6-1)

Two coils wound around a core have 1000 and 3000 turns, respectively. Connect a 400-v source across the series connection of these coils, and a 5-Ω resistance across the 1000 turns coil. (a) Find the voltages across each coil. (b) Find the currents in each coil. (c) Find the input power and the output power of this autotransformer. Consider the core and coils ideal.

Solution

Refer to Fig. 6-1. The 1000 turns coil is the common coil, the other the series coil.

$$V_C = (1000/4000) \times 400 = 100 \text{ v} = V_L$$

$$V_S = (3000/4000) \times 400 = 300 \text{ v}$$

$$I_L = 100/5 = 20 \text{ amp}$$

$$I_H = I_S = (1000/4000) \times 20 = 5 \text{ amp (from Eq. 6-5)}$$

$$I_C = 20 - 5 = 15 \text{ amp (from Eq. 6-6)}$$

Because of the resistive load, all power factors are unity. Thus

$$P_{\text{in}} = V_H I_H = 400 \times 5 = 2000 \text{ w}$$

$$P_{\text{out}} = V_L I_L = 100 \times 20 = 2000 \text{ w}$$

In an ideal transformer, these two quantities must be equal.

Example 6-2 (Section 6-1)

A 20-kva load is to be supplied at 500 v. An ideal step-up autotransformer is used to connect this load to a 400-v source. Find (a) the voltage and current of the series winding, (b) the voltage and current of the common winding, (c) the kva rating of this transformer if it were used as a two-winding transformer.

Solution

(a) The circuit of Fig. 6-2 is used. The autotransformer must have high side voltage, V_H = 500 v, and low side voltage, V_L = 400 volts. The series winding voltage is

$$V_S = V_H - V_L = 500 - 400 = 100 \text{ v}$$

The series winding current is the load current

$$I_S = I_H = I_{\text{Load}} = 20{,}000/500 = 40 \text{ amp}$$

(b) The voltage of the common winding is the low side voltage

$$V_C = V_L = 400 \text{ v}$$

The current in the common winding is

$$I_C = (N_S/N_C)\,I_S = (V_S/V_C)\,I_S = (100/400)\,(40) = 10 \text{ amp}$$

An alternative method to find I_C is first to find I_L

$$I_L = 20{,}000/400 = 50 \text{ amp}$$

Then

$$I_C = I_L - I_H = 50 - 40 = 10 \text{ amp}$$

(c) The apparent power associated with the two-winding transformer is the product of the voltage and current of one winding

$$P_{aT} = V_C I_C = (400 \text{ v})\,(10 \text{ amp}) = 4000 \text{ volt-amp} = 4 \text{ kva}.$$

If it were used as a two-winding transformer, its rating would be 4 kva, 400/100 volts.

Example 6-3 (Section 6-1)

A two-winding transformer is rated 5 kva, 440/110 v, 60 Hz. Referred to the 440-v winding, the equivalent impedance is $\mathbf{Z}_e = 0.5 + j\,0.8$ ohms, and the shunt admittance is $\mathbf{Y}_\phi = 0.00026 - j\,0.001$ mho. This transformer is connected as a 440/550-v step-up autotransformer. (a) Find the kva rating. (b) Find the voltage regulation for operation with power factor of 0.8 lagging. (c) Find the efficiency for operation with rated load and power factor of 0.8 lagging.

Solution

(a) Refer to Fig. 6-4. The 110-v winding is the series winding. The output current can be the rated current of the series winding.

$$I_H = I_S = 5000/110 = 45.5 \text{ amp}$$

The rated high side voltage is $V_H = V_C + V_S = 440 + 110 = 550$ v. The autotransformer rating $= V_H I_H = 550 \times 45.5/1000 = 25$ kva.

(b) The given equivalent impedance is referred to the 440-v winding. This would be Z_{e_1} in Fig. 5-9(a). For the autotransformer in Fig. 6-4, the equivalent impedance is Z_{e_L}.

$$\mathbf{Z}_{e_L} = \left(\frac{N_S}{N_S + N_C}\right)^2 \mathbf{Z}_{e_1} = \left(\frac{110}{550}\right)^2 (0.5 + j\,0.8) = 0.02 + j\,0.032$$

Let V'_H and I'_H designate the input values to the ideal autotransformer.

$$\mathbf{V}'_H = \left(\frac{N_C}{N_C + N_S}\right) \mathbf{V}_H = \left(\frac{440}{550}\right) 550 \underline{/0^\circ} = 440 \underline{/0^\circ} \text{ v}$$

The power factor angle is $\theta = \cos^{-1}(0.8) = 36.9^\circ$. The high side current is $\mathbf{I}_H = 45.5 \underline{/-36.9^\circ}$ amp.

$$\mathbf{I}'_H = \left(\frac{N_C + N_S}{N_C}\right) \mathbf{I}_H = \left(\frac{550}{440}\right) 45.5 \underline{/-36.9^\circ} = 56.9 \underline{/-36.9^\circ} \text{ amp}$$

Write Kirchhoff's voltage equation to find the actual low side voltage.

$$\mathbf{V}_L = \mathbf{V}'_H + \mathbf{Z}_{e_L}\mathbf{I}'_H = (440 \underline{/0^\circ}) + (0.038 \underline{/58^\circ})(56.9 \underline{/-36.9^\circ}) = 442 \underline{/0.1^\circ} \text{ v}$$

The voltage regulation is

$$\epsilon = \frac{V'_{HNL} - V'_{HFL}}{V'_{HFL}} = \frac{442 - 440}{440} = 0.0045$$

(c) The given shunt admittance is referred to the 440-v winding. This would be Y_{ϕ_1} in Fig. 5-9(a). For the autotransformer in Fig. 6-4, this is also Y_{ϕ_L}

$$\mathbf{Y}_{\phi_L} = \mathbf{Y}_{\phi_1} = 0.00026 - j\,0.001 \text{ mho}$$

In finding the core loss, we use $V_{L\,\text{rated}} = 440$ v.

$$P_c = G_{cL} V_L^2 = (0.0026)(440)^2 = 50.3 \text{ w}$$

In finding the copper losses, we use $I_{L\,\text{rated}} = I'_{H\text{rated}} = 56.9$ amp

$$P_R = R_{e_L}(I'_H)^2 = (0.02)(56.9)^2 = 64.8 \text{ w}$$

The summation of losses is

$$P_{\text{Loss}} = P_R + P_c = 64.8 + 50.3 = 115.1 \text{ w}$$

The output power is

$$P_{\text{out}} = V_H I_H \cos\theta = (550)(45.5)(0.8) = 20{,}000 \text{ w}$$

Use Eq. 1-4 to find the efficiency

$$\eta = 1 - \frac{P_{\text{Loss}}}{P_{\text{out}} + P_{\text{Loss}}} = 1 - \frac{115.1}{20{,}000 + 115.1} = 0.9943$$

Example 6-4 (Section 6-5)

Three single-phase transformers are connected Δ - Y as shown in Fig. 6-13. Each transformer is rated 100 kva, 2300/13,800 v, 60 Hz. The total three-phase load is 289.8 kva with P.F. = cos 20° = 0.94 lagging. The input voltages are:

$$\mathbf{V}_{AB} = 2300 \underline{/0^\circ}, \mathbf{V}_{BC} = 2300 \underline{/-120^\circ}, \mathbf{V}_{CA} = 2300 \underline{/+120^\circ}$$

Find all phasor voltages and currents for this transformer bank. Consider the transformers to be ideal.

Solution

The turns ratio is a = 2300/13,800 = 1/6. The dot-marked terminals determine voltages that are in phase. The line-to-neutral voltages on the secondary are

$$\mathbf{V}_{an} = (1/a)\,\mathbf{V}_{AB} = 13{,}800 \underline{/0^\circ}$$

$$\mathbf{V}_{bn} = (1/a)\,\mathbf{V}_{BC} = 13{,}800 \underline{/-120^\circ}$$

$$\mathbf{V}_{cn} = (1/a)\,\mathbf{V}_{CA} = 13{,}800 \underline{/+120^\circ}$$

Use Kirchhoff's voltage law to find the line-to-line voltages on the secondary

$$\mathbf{V}_{ab} = \mathbf{V}_{an} - \mathbf{V}_{bn} = 23{,}900 \underline{/+30^\circ}$$

$$\mathbf{V}_{bc} = \mathbf{V}_{bn} - \mathbf{V}_{cn} = 23{,}900 \underline{/-90^\circ}$$

$$\mathbf{V}_{ca} = \mathbf{V}_{cn} - \mathbf{V}_{an} = 23{,}900 \underline{/+150^\circ}$$

Observe the 30° phase shift of the secondary line voltages with respect to the primary line voltages. The line current on the high side is found from the information about the load

$$I = (289{,}800/3)/13{,}800 = 7 \text{ amp}$$

This current lags behind the line-to-neutral voltage by 20°. The currents through the secondary windings are

$$\mathbf{I}_{na} = 7 \underline{/0^\circ - 20^\circ} = 7 \underline{/-20^\circ}$$

$$\mathbf{I}_{nb} = 7 \underline{/-120^\circ - 20^\circ} = 7 \underline{/-140^\circ}$$

$$\mathbf{I}_{nc} = 7 \underline{/+120^\circ - 20^\circ} = 7 \underline{/+100^\circ}$$

The primary current flowing into a dot-marked terminal must be in phase with the secondary current flowing out of a dot-marked terminal. The current in the primary windings are

$$\mathbf{I}_{AB} = (1/a)\,\mathbf{I}_{na} = 42 \underline{/-20^\circ}$$

$$\mathbf{I}_{BC} = (1/a)\,\mathbf{I}_{nb} = 42\,\underline{/-140^\circ}$$

$$\mathbf{I}_{CA} = (1/a)\,\mathbf{I}_{nc} = 42\,\underline{/+100^\circ}$$

Use Kirchhoff's current law to find the input line currents

$$\mathbf{I}_A = \mathbf{I}_{AB} - \mathbf{I}_{CA} = 72.7\,\underline{/-50^\circ}$$

$$\mathbf{I}_B = \mathbf{I}_{BC} - \mathbf{I}_{AB} = 72.7\,\underline{/-170^\circ}$$

$$\mathbf{I}_C = \mathbf{I}_{CA} - \mathbf{I}_{BC} = 72.7\,\underline{/+70^\circ}$$

If a neutral point were available on the primary side, the line-to-neutral voltage would be $\mathbf{V}_{AN} = 1328\,\underline{/-30^\circ}$. Observe that the line current $\mathbf{I}_A$ lags behind this line-to-neutral voltage by the power factor angle.

Example 6-5 (Section 6-5)

Two transformers, each rated 20 kva, 440/220 v, 60 Hz, are operated in open-delta as shown in Fig. E-6-5. The input voltages are $\mathbf{V}_{ab} = 440\,\underline{/0^\circ}$, $\mathbf{V}_{bc} = 440\,\underline{/-120^\circ}$, $\mathbf{V}_{ca} = 440\,\underline{/+120^\circ}$. The load on the secondary is a balanced, wye-connected resistance load of 1.41 Ω per phase. Consider the transformers to be ideal. (a) Find the voltages on the secondary. (b) Find all currents.

Solution

(a) The turns ratio $= a = 440/220 = 2$. The dot-marked terminals determine voltages that are in phase.

$$\mathbf{V}_{BC} = (1/a)\,\mathbf{V}_{bc} = (1/2)\,440\,\underline{/-120^\circ} = 220\,\underline{/-120^\circ}$$

$$\mathbf{V}_{CA} = (1/a)\,\mathbf{V}_{ca} = 220\,\underline{/+120^\circ}$$

Use Kirchhoff's voltage law to find $\mathbf{V}_{AB}$

$$\mathbf{V}_{AB} = \mathbf{V}_{AC} + \mathbf{V}_{CB} = -\mathbf{V}_{CA} - \mathbf{V}_{BC} = 220\,\underline{/0^\circ}$$

The secondary voltages form a symmetrical three-phase set.

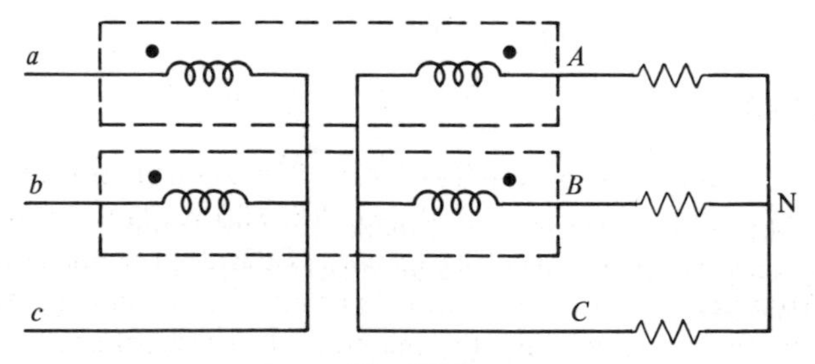

Fig. E-6-5.

(b) The load current will be in phase with line-to-neutral voltage. The magnitude of the load current is

$$I_{\text{Load}} = (220/\sqrt{3})/(1.41) = 90 \text{ amp}$$

$$\mathbf{I}_{AN} = \mathbf{V}_{AN}/\mathbf{Z}_L = 90 \underline{/-30^\circ} = \mathbf{I}_{CA}$$

$$\mathbf{I}_{BN} = \mathbf{V}_{BN}/\mathbf{Z}_L = 90 \underline{/-150^\circ} = -\mathbf{I}_{BC}$$

$$\mathbf{I}_{CN} = \mathbf{V}_{CN}/\mathbf{Z}_L = 90 \underline{/+90^\circ} = \mathbf{I}_{BC} - \mathbf{I}_{CA}$$

The primary current flowing into a dot-marked terminal must be in phase with the secondary current flowing out of a dot-marked terminal. The currents in the transformer primaries are

$$\mathbf{I}_{bc} = (1/a)\,\mathbf{I}_{CB} = 45 \underline{/-150^\circ} = \mathbf{I}_b$$

$$\mathbf{I}_{ac} = (1/a)\,\mathbf{I}_{CA} = 45 \underline{/-30^\circ} = \mathbf{I}_a$$

The remaining line current is found by using Kirchhoff's current law.

$$\mathbf{I}_c = \mathbf{I}_{cb} + \mathbf{I}_{ca} = 45 \underline{/+90^\circ}$$

Observe that the input line currents form a symmetrical three-phase set. These results show that two transformers can be used for symmetrical three-phase. We expect that including the equivalent impedances would result in some slight deviation from perfect symmetry.

It will be interesting to investigate the power and phase angle for each transformer. For transformer 2, we have $\mathbf{V}_{BC} = 220 \underline{/-120^\circ}$ and $\mathbf{I}_{BN} = 90 \underline{/-150^\circ}$. The complex power is $\mathbf{P}_{a_2} = P_2 + j\,Q_2 = 17.1 - j\,9.9$ kva. For transformer 3, we have $\mathbf{V}_{CA} = 220 \underline{/+120^\circ}$ and $\mathbf{I}_{AC} = -\mathbf{I}_{AN} = 90 \underline{/+150^\circ}$. The complex power is $\mathbf{P}_{a_3} = P_3 + j\,Q_3 = 17.1 + j\,9.9$ kva. The load power factor is unity, but each transformer is operating with a different power factor. The total three-phase load is 34.2 kva, while the load on each transformer is 19.8 kva. If three (instead of two) transformers had been used to supply the same load, each of them would have needed a rating of only $34.2/3 = 11.4$ kva.

6-7 PROBLEMS

6-1. A resistive load is to be operated at 100 v. The available source is 500 v. A two-winding transformer is rated 4 kva, 400/100 v. This transformer is to be used as a step-down autotransformer. Consider the transformer to be ideal. Without exceeding the current ratings of the windings, find (a) the current in the series winding, (b) the current in the common winding, (c) the maximum load kva. Compare this result with Example 6-2.

6-2. An autotransformer is used to supply a 400-kva load at 2640 v and power factor of 0.8 lagging. The available source voltage is 2400 v. Consider the autotransformer to be ideal. Find (a) the turns ratio between the series and the common windings, (b) the current in the series winding, (c) the current in the common winding, (d) the kva rating of this transformer if it were used as a two-winding tranformer.

6-3. An ideal autotransformer has 3 to 4 for the ratio of series turns to common turns. Loads are connected as shown in Fig. P-6-3. $\mathbf{Z}_1 = (0 - j\,9)\ \Omega$. $\mathbf{Z}_2 = (16 + j\,0)\ \Omega$. Find the driving point impedance at terminals *a-b* for two cases: (a) switch *k* is open, (b) switch *k* is closed.

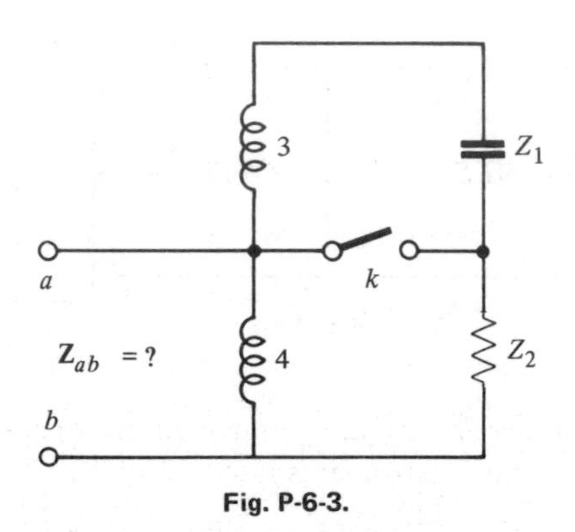

Fig. P-6-3.

6-4. A two-winding transformer is rated 10 kva, 220/110 v, 60 Hz. Referred to the 220-volt winding, $\mathbf{Z}_e = 0.06 + j\,0.16\ \Omega$ and $\mathbf{Y}_\phi = 0.002 - j\,0.01$ mho. This transformer is connected as a 330/110-v, step-down autotransformer. (a) What is the kva rating? (b) Find the voltage regulation for unity power factor load. (c) Find the efficiency for rated load and unity power factor.

6-5. The transformer in Problem 6-4 is connected as a 330/220 v, step-down autotransformer. (a) What is the kva rating? (b) Find the voltage regulation for unity power factor. (c) Find the efficiency for rated load and unity power factor.

6-6. Three identical single-phase transformers are to be used to supply a balanced three-phase load of 90 kva at a line-to-line voltage of 220 v. The power source has line-to-line voltage of 1320 v. No winding is to have more than 1000 v. No winding is to carry more than 150 amp. Find the correct connection for the transformer bank. Find the voltage and current ratings for the transformer windings.

6-7. Interchanging any two leads to the input to a three-phase device will impress voltages with the opposite phase sequence. Solve Example 6-4 with the lines to A and B interchanged. The input voltages will be $V_{AB} = 2300 \angle +180°$, $\mathbf{V}_{BC} = 2300 \angle -60°$, $\mathbf{V}_{CA} = 2300 \angle +60°$.

6-8. Two transformers are connected in open delta as shown in Fig. P-6-8. The balanced three-phase load requires 104 kva with power factor of 0.8 lagging. The transformer voltage ratings are 4400/440 v. Voltages on the load side are $\mathbf{V}_{AB} = 440 \angle -60°$, $\mathbf{V}_{BC} = 440 \angle -180°$, $\mathbf{V}_{CA} = 440 \angle +60°$. (a) Find the kva rating of each transformer. (b) Find the average power transferred through each transformer. (c) If a third transformer with the same rating were used with the other two to make a delta-delta connection, what would be the total kva capacity of this bank for balanced loading?

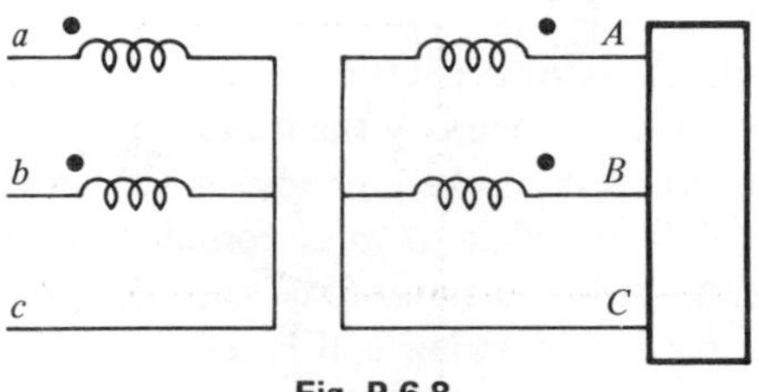

Fig. P-6-8.

6-9. Two ideal autotransformers are connected in open-delta as shown in Fig. P-6-9. Each autotransformer is rated 220/88 v. The input voltages are $\mathbf{V}_{ab} = 220 \angle 0°$, $\mathbf{V}_{bc} = 220 \angle +120°$, $\mathbf{V}_{ca} = 220 \angle -120°$. A balanced, Y-connected, three-phase load has impedance of $5.08 \angle 15°\ \Omega$ per phase. Find all phasor voltages and currents for this circuit.

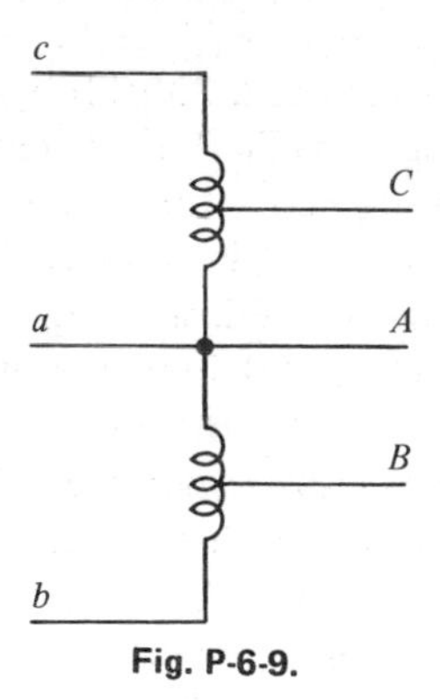

Fig. P-6-9.

7
Electromechanical Energy Conversion

7-1 MOTIONAL VOLTAGES

The great leap from the study of the static transformer to that of rotating generators and motors must be undertaken by the reader in several slow steps.

What would be the simplest model embodying the principles of a motional electromagnetic device? First of all, it must contain a magnetic circuit with an air gap. In Fig. 7-1a, only the portion of the core nearest the air gap is shown. The rest of the core with its exciting coil is left to the reader's imagination. Inside the air gap, an electric conductor must be located, and, since electric currents flow in closed circuits, the figure shows a closed rectangular loop. Only one of its four sides (the "active" conductor) is in a magnetic field.

In terms of the dimensions indicated in the figure, the loop, or one-turn coil, links the flux

$$\lambda = Blx \tag{7-1}$$

where B is the flux density in the air gap which depends on the mmf of the exciting coil and the dimensions and magnetic properties of the core. Now let the loop *move* horizontally, to the right or left (referring to the upper part of Fig. 7-1a). This changes the flux linkages at a rate

$$\frac{d\lambda}{dt} = \frac{d(Blx)}{dt} = Bl\,\frac{dx}{dt} \tag{7-2}$$

and by Faraday's induction law, this constitutes an induced voltage. Specifically, it is a motional voltage (see Section 2-3). Using the symbol u for the velocity dx/dt, the very basic result is

$$e = Blu \tag{7-3}$$

The reader familiar with electromagnetic field theory will recognize this equation as a special case of the vector equation for electric field intensity

$$\mathfrak{E} = \mathfrak{U} \times \mathfrak{B} \tag{7-4}$$

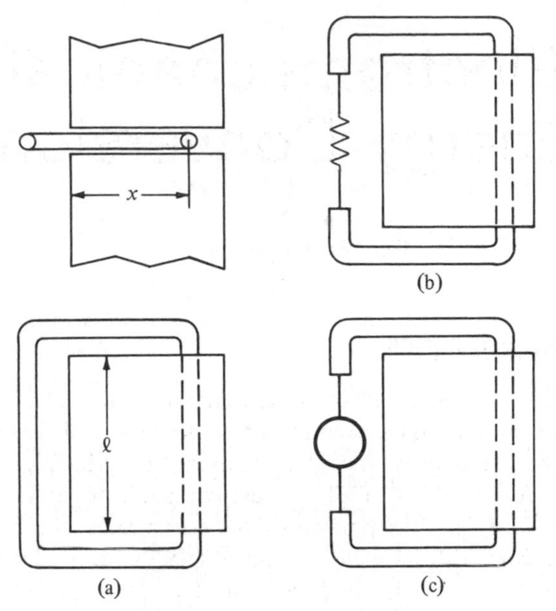

Fig. 7-1. Conductor in magnetic field: (a) plan and elevation; (b) resistance inserted; (c) source inserted.

The scalar equation, Eq. 7-3, is sufficient whenever, as in the arrangement of Fig. 7-1a (and of most engineering devices based on it), the three pertinent directions (those of the motion, the field, and the conductor) are all at right angles to each other.

The cross product of Eq. 7-4 also determines the direction of the electric field intensity. If Eq. 7-3 is to give the correct sign (plus or minus), the positive directions of motion, flux density, and voltage (in this sequence) must be chosen to form a right-hand system of rectangular coordinates.* The term *direction of voltage* is to be understood the same way as the commonly used arrow in circuit analysis. For instance, in Fig. 7-2, the voltage e is positive when terminal a has a higher potential then terminal b. In other words, $e = v_{ab}$.

Equation 7-3 also illustrates a different way to visualize the induction of motional voltages. Instead of being considered the result of a change of flux

*A right-hand system of rectangular (Cartesian) coordinates x, y, and z may be defined in terms of the first three fingers of the right hand, held at right angles to each other. The thumb points in the positive x-direction, the forefinger in the positive y-direction, and the middle finger in the positive z-direction. Applying this definition to Eq. 7-3 means that, when the thumb points in the direction of the motion, and the forefinger in the direction of the flux, the middle finger indicates the direction of the induced voltage.

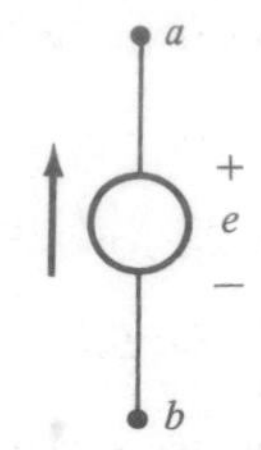

Fig. 7-2. Voltage polarity.

linkages, such voltages can also be attributed to the *relative motion* of an electric conductor and a magnetic field. They have been described as being due to the "cutting" of lines of flux, a graphic expression that some readers may find very helpful.

It must be kept in mind, however, that the expression Blu refers to the voltage induced in one conductor, whereas $d\lambda/dt$ is the voltage induced in an entire coil. For instance, if the whole loop (one-turn coil) of Fig. 7-1 were located inside the air gap, the voltage induced in it would be zero. This could be equally well explained by stating that the flux linkages of the loop would not change with motion, or that the two equal Blu voltages induced in the two active conductors would be opposed to each other. (No voltage would be induced in the other two conductors whose direction is the same as that of the motion.)

7-2 GENERATOR ACTION

The primitive device model of Fig. 7-1 can be used for much more than the given derivation of Eq. 7-3. The next step, easily visualized, is to cut one of the three conductors that are not in the magnetic field, so as to obtain two open terminals, and then to connect an electric circuit, for instance just a resistance, to these terminals (Fig. 7-1b).

Whenever the active conductor (the one in the air gap) moves in the direction indicated before, the induced voltage must cause a current to flow through this resistance which, consequently, consumes power "generated" by the motion of the active conductor in the magnetic field. A device based on the model sketched in Fig. 7-1 can operate as a *generator*. In the terminology to be used extensively in later chapters, the active conductor is called the *armature*, the unseen coil wound around the core the *field winding*, and the resistance connected to the armature terminals constitutes the *load* of the generator.

A generator cannot possibly deliver electric power to a load unless the same amount of power is made available to the generator in mechanical form. In the model of Fig. 7-1, that mechanical power is what is needed in order to move the armature against the *force* exerted on the current-carrying active conductor by the magnetic field.

Equating the mechanical input power to the electric output power,

$$Fu = ei \tag{7-5}$$

substituting Eq. 7-3, and solving for the force F results in

$$F = Bli \tag{7-6}$$

This equation of the force exerted on a current-carrying conductor in a magnetic field was derived on the basis of two premises: Faraday's law of electromagnetic induction, and the principle of conservation of energy. But many readers have undoubtedly recognized Eq. 7-6 as a special form of the magnetic component of the *Lorentz force*

$$\mathfrak{f} = i(\mathfrak{L} \times \mathfrak{B}) \tag{7-7}$$

where $\mathfrak{f}$ and $\mathfrak{L}$ are the vectors representing force and length. Again, the scalar equation, Eq. 7-6 (similar to the scalar equation, Eq. 7-3), is valid whenever the three pertinent directions are at right angles to each other, and the correct sign (plus or minus) is obtained from the equation if the positive directions of current, flux density, and force (in that sequence) are chosen to form a right-hand system of coordinates.*

The two directional rules (for the induced voltage and the magnetic force) should be taken together to reveal their full significance with respect to conservation of energy. Suppose, for instance, that the flux in the air gap (referring to the upper view of Fig. 7-1a) is directed upward and considered positive. Let the conductor move to the right, and call that direction positive. Then the induced voltage must be positive, which, in order to satisfy the rule that motion, flux density, and induced voltage form a right-hand system of coordinates, must mean "out of the paper," i.e., toward the viewer. If the circuit is closed through a load resistance, the current has the same direction as the voltage. Now to satisfy the rule that current, flux density, and force are to form a right-hand system of coordinates, the force must be directed to the left, *opposing the motion*. This is necessary in order to be in agreement with the principle of conservation of energy.

So a generator must be driven by its prime mover whose force has to overcome the magnetic force between field and armature. Note, however, that this force becomes zero when the armature circuit is open (at *no-load*). A generator can be driven essentially without expenditure of energy as long as it delivers no energy at its armature terminals.

*See the previous footnote. Accordingly, with the first three fingers of the right hand held at right angles to each other, let the thumb point in the direction of the current, and the forefinger in the direction of the flux. The middle finger then indicates the direction of the magnetic force exerted on the conductor.

7-3 MOTOR ACTION

Being aware of the force between a magnetic field and a current-carrying conductor leads to considering how to make use of that force. An electric current must be obtained in the armature by connecting its terminals to a source. Figure 7-1 can still be used as a model. The source now takes the place that was occupied by a load resistance in the previous section. (See Fig. 7-1c)

Let the current and the flux density have the same directions as before (i.e., current out of the paper, flux density upward, always referring to the upper view in Fig. 7-1a). Application of the rule for the direction of the force results in establishing that the force is directed to the left. Consequently, the conductor tends to move to the left. So the device modeled by Fig. 7-1 can be used to overcome a force that opposes the motion. In this case, the device operates as a *motor*, and the force to be overcome constitutes the mechanical *load* of the motor.

The principle of conservation of energy, expressed by Eq. 7-5, must again be satisfied but, this time, the electric power ei is the input and the mechanical power Fu the output. The fact that the device depicted by Fig. 7-1 can operate both as a generator and as a motor (in other words, that the process of energy conversion is *reversible*) is typical for most electromagnetic energy-converting devices.

Equations 7-3 and 7-6 remain valid, together with their directional rules. Application of these rules leads to the result (inevitable in order to agree with energy conservation) that the induced voltage *opposes* the voltage of the source connected to the terminals. That is why the voltage induced in a motor armature is sometimes referred to by the old name "counter-emf" or "back-emf."

Figure 7-3 illustrates the difference between generator and motor operation in terms of current and voltage directions. In the diagrams of Fig. 7-3a the direction of the current (and of the force) is the same. More significant from a practical viewpoint are the diagrams of Fig. 7-3b in which the direction of the voltage (and of the motion) is the same. Whichever pair of diagrams is used, it can be seen that,

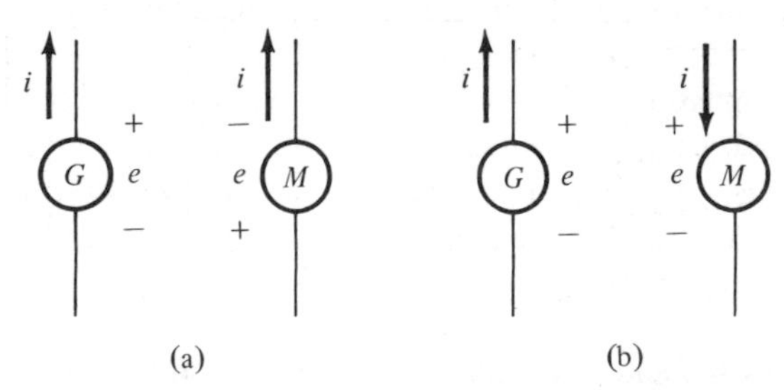

Fig. 7-3. Generator and motor operation: (a) same current direction; (b) same voltage direction.

when the device operates as a generator, the current flows through the armature from the terminal with the lower potential to the terminal with the higher potential. In a motor, by contrast, the current must flow from the higher to the lower potential (as it does in any circuit that receives power, e.g., in a resistance).

A motor might be blocked, i.e., forcibly kept at standstill. This happens when the opposing force is just too large to be overcome by the magnetic force. In this condition, the output of the motor is zero, and so is its input no matter how much current it may draw from the source since, without motion, there is no induced voltage.

7-4 SINGLY EXCITED DEVICES

The device modeled in Fig. 7-1 involves two electric circuits, one on its fixed part and one on its movable part. The same is true for most actual generators and motors. One of these circuits, that of the field winding, could be omitted by using a permanent magnet as the core, but that has not been found to be practical for most power-converting devices. It is possible, however, to build devices capable of motor and generator action and having only one electric circuit without taking recourse to permanent magnetism. Such devices are used for many other purposes (including instruments, microphones, phonograph pickups, and relays), but they will be discussed here because some further concepts useful in the study of generators and motors can be learned thereby.

Figure 7-4 shows a model of such a device. There is again a magnetic circuit with an air gap but, this time, part of the core itself is movable (the right-side top part). The movable part is meant to have one degree of freedom: it can move horizontally, to the left or right. There is a coil wound around the fixed part of the core. Positive directions are assigned in the diagram to the voltage and current

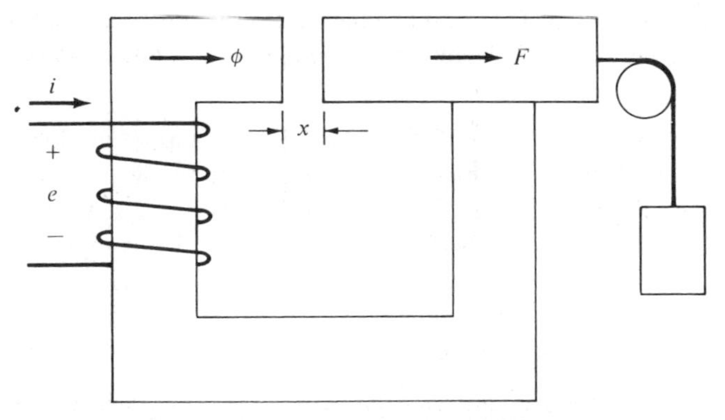

Fig. 7-4. A singly excited electromagnetic device.

of the coil and to the flux in the core. The choice of current and flux directions is consistent, i.e., such that a positive current produces a positive flux.

When the movable part actually moves, several changes take place. To begin with, the distance x changes, and since x is the length of the air gap, the reluctance of the magnetic circuit changes. This brings about a change of flux (unless the change of reluctance is counteracted by a simultaneous change of mmf), and a change of energy storage. Therefore, the simple *power balance* of Eq. 7-5 must be replaced by

$$ei = \frac{dW_m}{dt} + F\frac{dx}{dt} \tag{7-8}$$

where F is the electromagnetic force exerted on the movable part, and W_m is the energy stored in the magnetic field. Note that the positive direction assigned to the force F is the direction of a motion with a positive velocity dx/dt. (As a consequence of this choice, the reader may well anticipate that the actual values of F must be negative, since the electromagnetic force tends to pull the movable part to the left.) Also, the positive directions of the voltage e and the current i are chosen so that the product ei is positive when the device receives electric power.

The equation is written in such a way that the left side represents electric input power. Whenever both the product ei and the product $F(dx/dt)$ are positive, the device has an electric input and a mechanical output, which is the essence of *motor* action. If these two products are at any time both negative, the device operates as a *generator*. These statements are valid regardless of the sign of the other term, i.e., regardless of whether the magnetic field energy increases or decreases during the motion.

Magnetic field energy has been investigated in Section 4-3. The pertinent result, quoted from Eq. 4-13, is

$$\Delta W_m = \int_{\phi_1}^{\phi_2} \mathcal{F}\, d\phi \tag{7-9}$$

To evaluate this integral, the mmf $\mathcal{F}$ must be expressed as a function of the flux ϕ. The relation between these two quantities is the familiar saturation curve, which gives the flux as a function of the mmf (by tradition) instead of the other way around. Therefore, the energy *increment* of Eq. 7-9 is represented in the graph of Fig. 7-5 by the area *abcda* between the curve and the vertical axis. The total stored energy, obtained by setting the lower limit of the integral equal to zero, is represented by the area 0*bc*0 to the left of the saturation curve.

For a device like that of Fig. 7-4, however, the saturation curve itself depends on the position of the movable part, i.e., on the distance x. The larger x, the

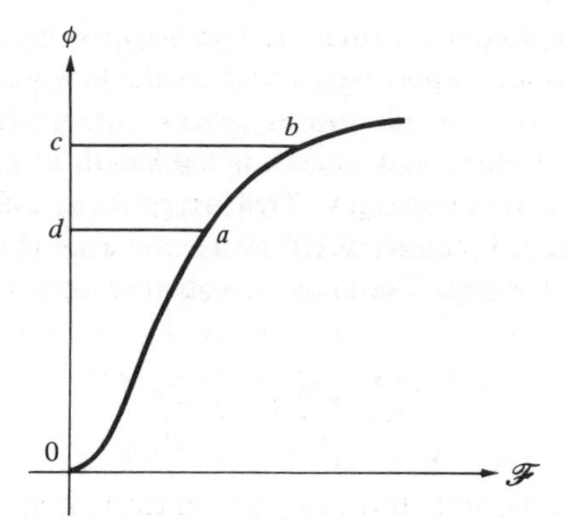

Fig. 7-5. Magnetic field energy.

larger the reluctance of the magnetic circuit, the larger the mmf required for a given flux. Instead of one curve, there is a family of saturation curves, one for every value of x. Figure 7-6 shows two curves of that family, belonging to the values x_1 and x_2. Note that x_1 must be larger than x_2.

Suppose that, at a certain instant, the movable part is located at $x = x_1$ and the flux is ϕ_1. Then the area $0ac0$ in Fig. 7-6 represents the energy stored at that instant, regardless of whether the movable part was always in the same position or not. In other words, the energy storage depends on the condition or *state* of the device, and not on its past history (This is true only because hysteresis is ignored in this discussion. From a practical viewpoint, the importance of hysteresis lies not in its effect on energy storage, but rather on the energy *loss* that it causes under alternating conditions.)

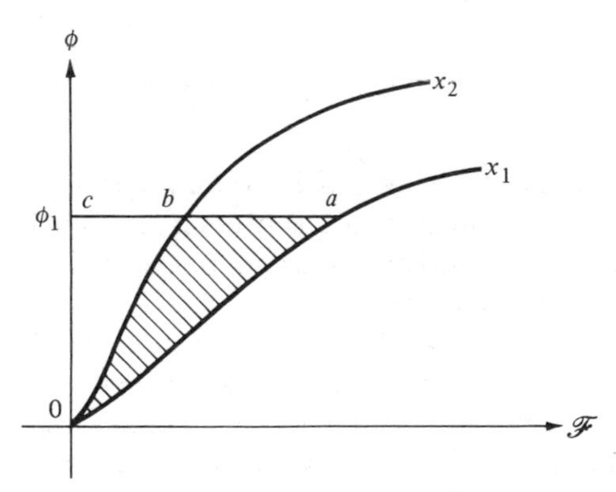

Fig. 7-6. Motion at constant flux.

Now consider what happens *when motion occurs*, changing the position x from x_1 to x_2. The state of the device is described by a point (in Fig. 7-6) that wanders from some place on the lower curve to some place on the upper one. The actual location of these two places and the path leading from one to the other depend on several external factors: the circuit connected to the coil (is it a resistive voltage source, or a pure current source, or what else?), and the assisting or resisting external force (for instance a weight attached to a rope around a pulley, as suggested in Fig. 7-4) plus the inevitable forces of inertia and friction. There is an infinite variety of possibilities out of which two cases are of special interest.

(a) Let the motion occur at *constant flux*. This could be done by adjusting the current during the motion, or it could be approximated by having the motion occur so fast that the flux cannot change appreciably until after the motion is completed. In this case, the electric power is zero as can be seen from

$$ei = \left(N \frac{d\phi}{dt}\right) i = \mathcal{F} \frac{d\phi}{dt} \tag{7-10}$$

Thus, the power balance of Eq. 7-8 becomes

$$0 = \frac{dW_m}{dt} + F \frac{dx}{dt} \tag{7-11}$$

which means that mechanical work is being done solely at the expense of stored magnetic energy. Referring to Fig. 7-6, keeping the flux constant at the value ϕ_1, the magnetic field energy is being reduced during the motion, from the area $0ac0$ to the smaller area $0bc0$. The difference, the area $0ab0$, represents the mechanical work done (for instance, the potential energy added to the weight

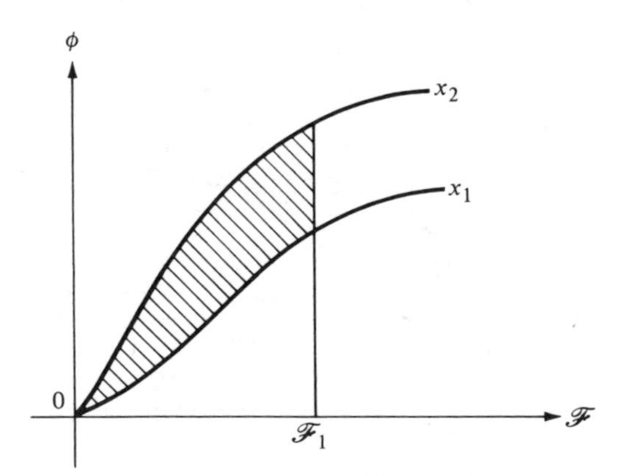

Fig. 7-7. Motion at constant mmf.

lifted). Equation 7-11 can also be solved for the force, which comes out as a partial derivative

$$F = -\left(\frac{\partial W_m}{\partial x}\right)_{\phi=\text{const}} \tag{7-12}$$

(b) The motion could be made to occur at *constant current* (constant mmf is the same thing), either by using a pure constant current source, or approximately by making the motion sufficiently slow. Figure 7-7 suggests that what is pertinent for such a process is not the area between a saturation curve and the vertical axis (the magnetic field energy), but rather the area between that curve and the horizontal axis. The quantity represented by this area has no physical meaning, but it is a useful function of the state of the device, and it is called the *coenergy* W'_m. Figure 7-8 shows the relation

$$W_m + W'_m = \mathcal{F}\phi \tag{7-13}$$

which can be differentiated:

$$\frac{dW_m}{dt} + \frac{dW'_m}{dt} = \mathcal{F}\frac{d\phi}{dt} + \phi\frac{d\mathcal{F}}{dt} \tag{7-14}$$

Solving this for dW_m/dt, and substituting it, as well as Eq. 7-10, into Eq. 7-8 (the power balance), leads to

$$\mathcal{F}\frac{d\phi}{dt} = -\frac{dW'_m}{dt} + \mathcal{F}\frac{d\phi}{dt} + \phi\frac{d\mathcal{F}}{dt} + F\frac{dx}{dt} \tag{7-15}$$

The purpose of all this was to be able to cancel the $\mathcal{F}(d\phi/dt)$ terms, and to introduce the term $\phi(d\mathcal{F}/dt)$, which is zero in the case under consideration (motion

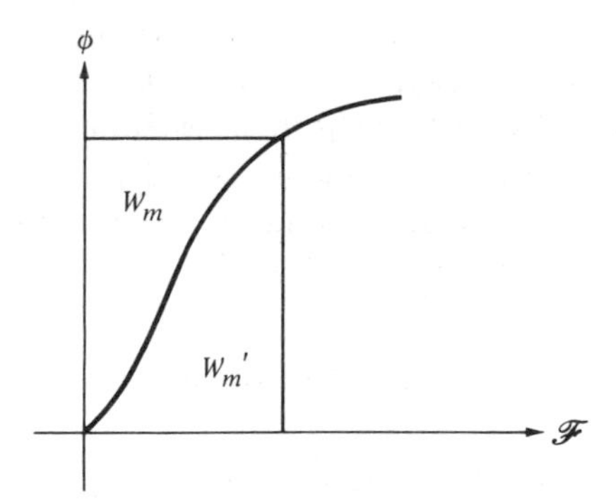

Fig. 7-8. Energy and coenergy.

with constant mmf). Thus, for this case

$$0 = -\frac{dW'_m}{dt} + F\frac{dx}{dt} \tag{7-16}$$

which says that the mechanical work done is equal to the increase in coenergy. It also leads to another expression for the force:

$$F = \left(\frac{\partial W'_m}{\partial x}\right)_{i=\text{const}} \tag{7-17}$$

Observe that in both cases (motion with constant flux and motion with constant current), the mechanical work done is represented by the area between two saturation curves. Since any arbitrary contour leading from one saturation curve to another can be broken up into (infinitesimal) horizontal and vertical steps, it follows that the mechanical work done by the motion of a device like that of Fig. 7-4 is always described by the area bounded by the initial and final saturation curves and the contour. From this, it may be further concluded that motion from right to left (referring to Fig. 7-4) must result in a mechanical output, or that motion from left to right requires a mechanical input.

7-5 LINEAR ANALYSIS

There is no difficulty in evaluating Eq. 7-12 or Eq. 7-17 for the force across an air gap of a magnetic circuit, provided that a family of saturation curves is given or has been calculated. There are algorithms for the use of Simpson's rule or of several other methods to evaluate definite integrals. Then, the slope of the function W_m or W'_m versus x can be found, either by graphical or by computational techniques. But for the purpose of obtaining general *qualitative* results rather than numerical answers to specific numerical problems, the use of *linear approximations* becomes preferable.

Figure 7-9 depicts a saturation curve that is a straight line through the origin. From this diagram, it is seen that, in the linear case, energy and coenergy are equal to each other, and that they can be found as the area of a triangle.

$$W_m = W'_m = \frac{\mathfrak{F}\phi}{2} \tag{7-18}$$

Other useful expressions are found by substitutions that introduce the reluctance $\mathfrak{R} = \mathfrak{F}/\phi$, the permeance $\mathfrak{P} = \phi/\mathfrak{F}$, or the inductance $L = \lambda/i = N^2\phi/\mathfrak{F}$. They are

$$W_m = \frac{(\mathfrak{R}\phi)\phi}{2} = \frac{\mathfrak{R}}{2}\phi^2 \tag{7-19}$$

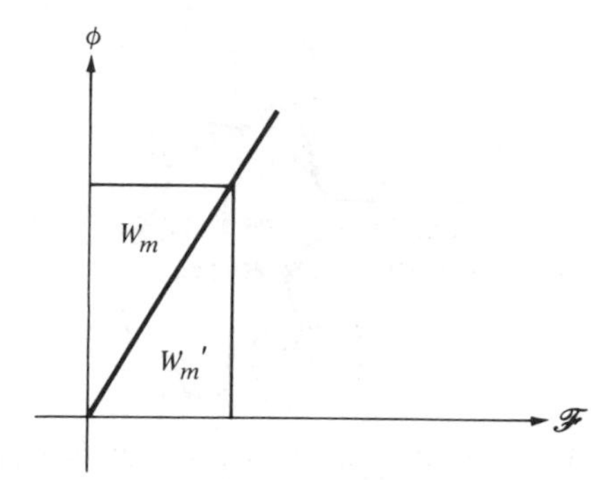

Fig. 7-9. Linear analysis.

$$W_m' = \frac{\mathcal{F}(\mathcal{P}\mathcal{F})}{2} = \frac{\mathcal{P}}{2}\mathcal{F}^2 \tag{7-20}$$

and the familiar

$$W_m' = \frac{\mathcal{F}(\mathcal{F}L/N^2)}{2} = \frac{L}{2}i^2 \tag{7-21}$$

These expressions can, in turn be substituted into the force equations, Eqs. 7-12 or 7-17, respectively, with the results

$$F = -\frac{\psi^2}{2}\frac{d\mathcal{R}}{dx} \tag{7-22}$$

$$F = \frac{\mathcal{F}^2}{2}\frac{d\mathcal{P}}{dx} \tag{7-23}$$

$$F = \frac{i^2}{2}\frac{dL}{dx} \tag{7-24}$$

Here are, then, three new equations for the magnetic force across an air gap, and they all tell the same story in three different ways: the magnetic force acts so that the motion it would tend to produce would decrease the reluctance or increase the permeance of the magnetic circuit, or increase the inductance of the electric circuit. The three statements are redundant: each of them could be derived from any of the two others. It also follows from them that there is no force in a direction in which a motion would not change the reluctance (permeance, inductance).

In the device model of Fig. 7-4, the force on the movable piece is directed to the left, according to any of these three equations, and also in agreement with

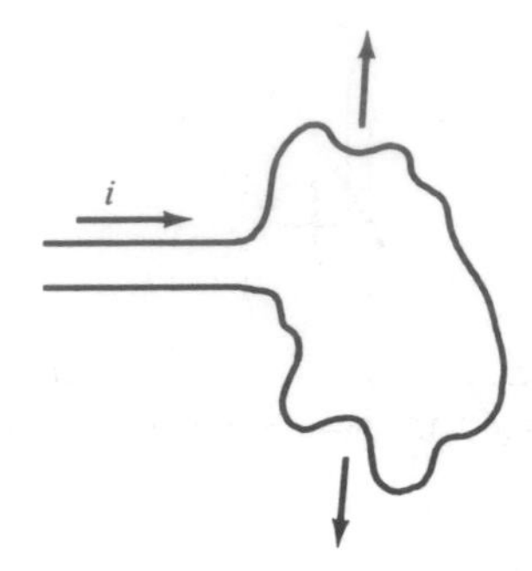

Fig. 7-10. Irregular loop.

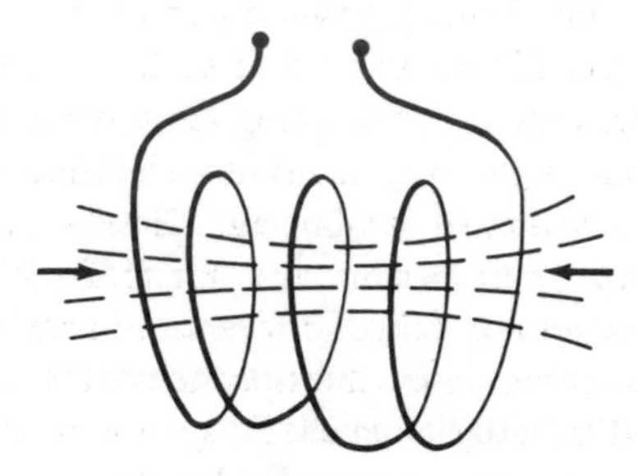
Fig. 7-11. Coil.

the results of the previous section. Many other phenomena, some of them well known, can also be explained in terms of reluctance, permeance or inductance.

(a) A current-carrying loop of wire (Fig. 7-10), loosely lying on the ground or on a table, will assume a circular shape to maximize the area it encloses.

(b) A coil (Fig. 7-11) tends to shorten itself to minimize the length for which the lines of flux are crowded into a limited area. These two examples can also be taken as illustrations of the magnetic force between two current-carrying conductors: they attract each other when their currents are in the same direction, etc.

(c) Consider a movable piece of iron in the air gap of a magnetic circuit (Fig. 7-12). Assume the effect of fringing to be negligible. Then, theoretically, there is no force in the horizontal direction because a horizontal motion would not change the reluctance of the magnetic circuit. But there is definitely a force pulling the piece downward toward the centered position. This can also be explained in terms of the attraction of induced magnetic poles.

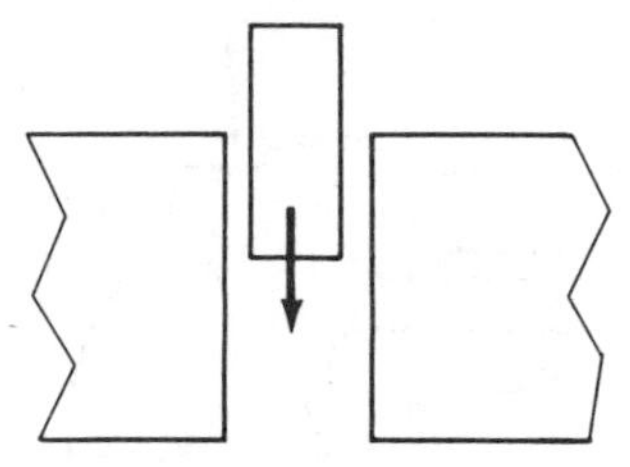

Fig. 7-12. Piece of iron in air gap.

7-6 ROTATING MOTION

Devices based on the models of Figs. 7-1 and 7-4 exhibit the main characteristics of motor or generator action during the time in which motion actually occurs. Practical power-converting devices, however, must be able to operate continuously. Therefore, devices like those studied so far in this chapter could be used as motors or generators only by performing *reciprocating motion*—moving back and forth. This is indeed the way many nonelectrical machines run: the old-fashioned steam engine, and most gasoline and Diesel engines.

Wherever *rotating motion* is feasible, it is far preferable to reciprocating motion. It avoids the constant rise, fall, and reversal of accelerating and decelerating forces and the stresses, strains, and vibrations they cause, with their further consequences of noise and material fatigue. This is the reason why the old reciprocating steam engines have been superseded by steam turbines, and it is one of the reasons why the invention of a rotating internal combustion engine (the Wankel motor) has caused such a stir. Electric motors and generators are well suited for rotating motion, and they are always built for it.

It is easy enough to change the model of Fig. 7-1 from a reciprocating to a rotating device. As Fig. 7-13 shows, it is only natural that, in the process, the magnetic core is provided with two air gaps instead of one. As a consequence, the one-turn armature coil of the figure now has two active conductors, a further advantage compared to the model of Fig. 7-1. The entire magnetic core is shown, including the exciting coil on the left leg. The circle drawn as a broken line indicates the path of the two active conductors which are themselves depicted as small circles representing their cross-sections.

This model can also be modified to become a rotating counterpart of the one of Fig. 7-4 (the singly-excited one). Just omit the armature coil and let the middle piece of the core rotate instead, around the same axis. This causes the reluctance of the magnetic circuit to change (it has its minimum value in the position of the figure), which makes electromechanical energy conversion possible.

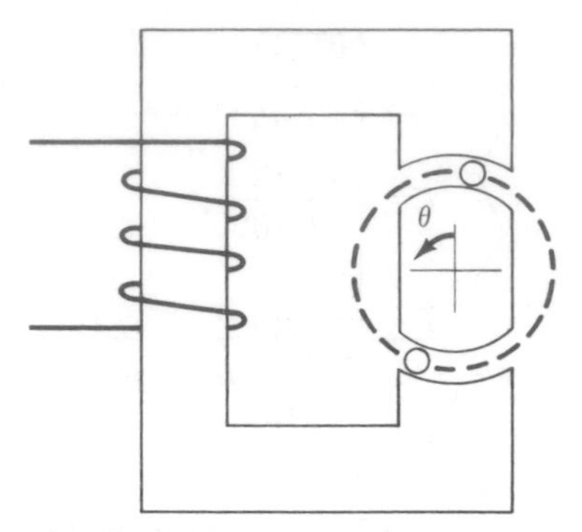

Fig. 7-13. Rotating device.

Many equations of this chapter have to be adapted to rotating motion. In place of the distance or position variable x, there is now an angular position θ, with an arbitrarily chosen reference position for which θ is zero. Consequently, the velocity $u = dx/dt$ is replaced by the ***angular velocity*** $\omega = d\theta/dt$. This is a more significant quantity, since all points of a rotating body have the same angular velocity whereas most of them do not have a velocity of the same magnitude and direction. When the angle θ is expressed in radians, the velocity of any point of a rotating body has a magnitude $u = dx/dt = r(d\theta/dt) = r\omega$ where r is the radius or distance from the axis of rotation.

Likewise, equations describing a rotating motion should not contain a force F but rather a *torque* $T = Fr$. This is so because, when forces are acting on several points of a rotating body, it is the torques whose sum causes the rotating motion, not the forces. For instance, in the device of Fig. 7-13, the sum of the forces exerted on the two active conductors is zero, but the total torque is twice that of a single conductor.

The "translation" of the pertinent equations into the "language" of rotating motion is done simply by replacing x by θ, u by ω, and F by T. That is so because the mechanical power for rotating motion is the product $T\omega$. Thus the straight electromechanical power balance of Eq. 7-5 becomes $ei = T\omega$, and the power balance involving a change of magnetic energy storage (Eq. 7-8) becomes

$$ei = \frac{dW_m}{dt} + T\omega \tag{7-25}$$

The various force equations for singly excited devices (Eqs. 7-12, 7-17, and 7-22 to 7-24) become torque equations:

$$T = -\left(\frac{\partial W_m}{\partial \theta}\right)_{\phi=\text{const}} \tag{7-26}$$

$$T = \left(\frac{\partial W'_m}{\partial \theta}\right)_{i=\text{const}} \tag{7-27}$$

$$T = -\frac{\phi^2}{2}\frac{d\mathfrak{R}}{d\theta} \tag{7-28}$$

$$T = \frac{\mathfrak{F}^2}{2}\frac{d\mathfrak{P}}{d\theta} \tag{7-29}$$

$$T = \frac{i^2}{2}\frac{dL}{d\theta} \tag{7-30}$$

all of which can be derived from Eq. 7-25 exactly the same way as the original equations were derived from Eq. 7-8.

Apart from reciprocating and rotating motion, there is yet another kind of motion that deserves to be mentioned. It is called *linear* motion (unfortunately, using a word that has different meanings), and it refers to a vehicle moving along its road or track. There exists the possibility of having a continuous system of electric conductors placed on the track, facing a similar set of conductors on the vehicle, just as the stationary and moving conductors of a conventional motor are facing each other. These conductors form a *linear* motor that drives the vehicle. The idea has been used for limited purposes (the launching of planes from an aircraft carrier) but at this writing, it is the subject of theoretical and experimental schemes for railroads.

7-7 TORQUES IN MULTIPLY EXCITED DEVICES

Electric motors and generators consist of a rotating part called the *rotor* and a stationary part called the *stator*. In most cases, rotor and stator are each equipped with one set of interconnected electric conductors called a *winding*. The two circuits indicated on the model of Fig. 7-13 may serve as an example. Some machines carry more than two windings.

To determine the electromagnetic torque exerted on the rotor will be one of the major subjects in the chapters dealing with the operation of each of the main types of motors and generators. At this point, however, a *linear approximation* is sufficient to provide insight into the general principles involved, and it will be limited to devices with double excitation, i.e., with two electric circuits. The derivation will be based on the same two basic laws of nature that were used to find the torque (or force) in singly excited devices; the principle of *conservation of energy*, and *Faraday's induction law*.

To begin with the conservation law: the power balance of Eq. 7-25 must now have two terms on its left side, one each for the stator and the rotor. With the induction law used for both induced voltages,

$$i_1 \frac{d\lambda_1}{dt} + i_2 \frac{d\lambda_2}{dt} = \frac{dW_m}{dt} + T\omega \tag{7-31}$$

where λ_1 and λ_2, the flux linkages of the two circuits, are produced by both currents. Thus, they can be written, for a linear device, in terms of *self- and mutual inductances* (see Section 2-4) as follows:

$$\lambda_1 = L_1 i_1 + M i_2 \tag{7-32}$$

$$\lambda_2 = M i_1 + L_2 i_2 \tag{7-33}$$

provided negative as well as positive values are allowed for M.

The inductances L_1, L_2, and M depend on the reluctance of the magnetic circuit. For a motional device, the reluctance, in turn, may depend on the angular position θ of the rotor. In order first to obtain an expression for the energy storage W_m, however, the rotor is held at an arbitrary *fixed position*, making the angular velocity zero and keeping all the inductances constant. So the power balance equation for a fixed rotor becomes

$$L_1 i_1 \frac{di_1}{dt} + M i_1 \frac{di_2}{dt} + M i_2 \frac{di_1}{dt} + L_2 i_2 \frac{di_2}{dt} = \frac{dW_m}{dt} \tag{7-34}$$

leading to the result

$$W_m = \frac{L_1}{2} i_1^2 + \frac{L_2}{2} i_2^2 + M i_1 i_2 \tag{7-35}$$

which is easily verified by differentiating it with respect to t.

On the other hand, when the rotor is free to move, the flux linkages can change for two different reasons (see Section 2-3): because of a change of current magnitudes, and because of the motion. Therefore, in differentiating flux linkages, the inductances can no longer be treated as constants. Thus, for instance,

$$e_1 = \frac{d\lambda_1}{dt} = L_1 \frac{di_1}{dt} + i_1 \frac{dL_1}{dt} + M \frac{di_2}{dt} + i_2 \frac{dM}{dt} \tag{7-36}$$

The reader can recognize that two of these four terms are transformer voltages, and the other two (characterized by the derivatives of inductances) *motional voltages*. Similarly, the other voltage e_2 is expressed as the sum of four terms, and then both voltages are substituted into the left side of Eq. 7-31. On the right side, there is the derivative of the magnetic energy. This means that Eq. 7-35 has to be differentiated but, this time, *not* for a fixed rotor, resulting in seven terms to which the mechanical power is added. So the power balance equation becomes rather lengthy:

$$i_1\left(L_1\frac{di_1}{dt}+i_1\frac{dL_1}{dt}+M\frac{di_2}{dt}+i_2\frac{dM}{dt}\right)$$

$$+i_2\left(M\frac{di_1}{dt}+i_1\frac{dM}{dt}+L_2\frac{di_2}{dt}+i_2\frac{dL_2}{dt}\right)$$

$$=L_1i_1\frac{di_1}{dt}+\frac{i_1^2}{2}\frac{dL_1}{dt}+L_2i_2\frac{di_2}{dt}+\frac{i_2^2}{2}\frac{dL_2}{dt}$$

$$+Mi_1\frac{di_2}{dt}+Mi_2\frac{di_1}{dt}+i_1i_2\frac{dM}{dt}+T\frac{d\theta}{dt} \qquad (7\text{-}37)$$

Fortunately, many terms can be cancelled, with the final result

$$T=\frac{i_1^2}{2}\frac{dL_1}{d\theta}+\frac{i_2^2}{2}\frac{dL_2}{d\theta}+i_1i_2\frac{dM}{d\theta} \qquad (7\text{-}38)^*$$

As was to be expected, this equation includes the case of the singly excited device. Just set one of the two currents equal to zero, and the expression reverts to that of Eq. 7-30. So the first two terms of the torque equation for a doubly excited device are *reluctance torques* as previously encountered. But the third term is new, and it is of particular interest because it is possible for the mutual inductance to be changed by motion even if the self-inductances are not affected by it. This happens in a device in which motion does not change the reluctance of the magnetic circuit but only the position of the two windings relative to each other, as the next section will show.

7-8 SALIENT AND NONSALIENT POLES

In this section, the model of Fig. 7-13 will be changed in various ways to come closer to the actual design of engineering devices. For this purpose, it is helpful first to introduce another pictorial representation of a winding. Figure 7-14 shows (a) the coil symbol used in previous chapters, and (b) the new one consisting of small circles that indicate the cross-section of conductors located perpendicular to the paper, and not showing the (horizontal) connecting wires in front and back of it. In place of arrows, this representation permits the indication of circuit directions by means of dots (the points of arrows facing the

*It might be thought that this result could have been obtained more quickly by substituting Eq. 7-35 into Eq. 7-27 (with $W_m' = W_m$). But it would have been necessary to prove that Eq. 7-27 is valid for devices with double excitation.

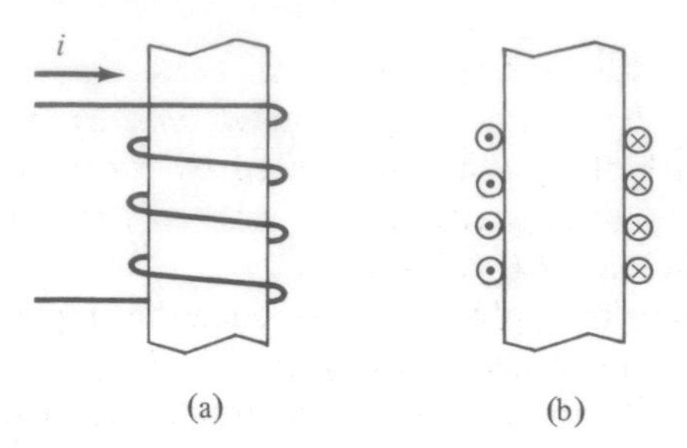

Fig. 7-14. Two equivalent representations of a winding.

viewer, meaning "out of the paper") and *crosses* (meaning "into the paper"). These symbols will be used consistently from here on.

The first change to be undergone by the model of Fig. 7-13 is to let the middle part of the right leg rotate with the coil. The reason is that the coil by itself could not possibly be made strong enough to withstand the magnetic and centrifugal forces acting on it. It must be firmly attached to, and supported by, that part of the core that thereby becomes the rotor.

This leads to a design whose basic features are depicted in Fig. 7-15, which shows two rotor positions. In these two sketches, the stator winding is placed near the air gaps, and the other legs of the core are not shown. The rotor position is described by the angle θ, which is meant to be the angle between the larger rotor axis of symmetry and the vertical direction. So, $\theta = 0$ in Fig. 7-15a, and 90° in Fig. 7-15b.

Attention must be drawn to the magnetic circuit and the flux configuration. When $\theta = 0$, the stator and rotor mmfs aid each other in producing a flux two

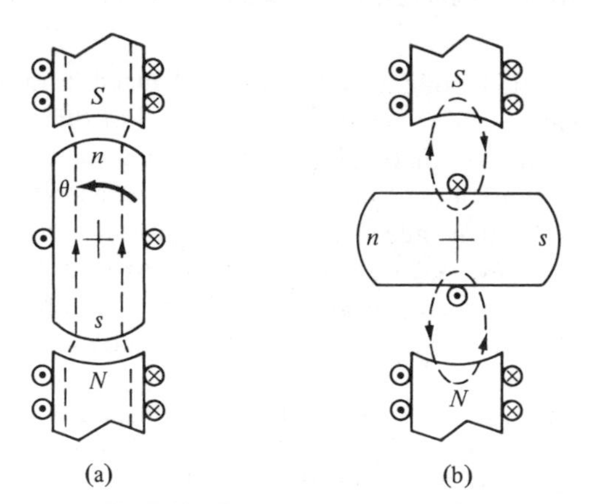

Fig. 7-15. Device with salient poles.

lines of which are sketched as broken lines (note that they cross the air gaps in *radial* direction). When $\theta = 90°$, the stator mmf by itself would still produce a vertical flux as before, although reduced by the reluctance of the larger air gaps. The rotor mmf by itself, however, can only produce some flux according to the two broken lines in Fig. 7-15b.

We are now ready to apply the torque equation, Eq. 7-38, with its derivatives of inductances, to the model device of Fig. 7-15. (This equation is based on a linear approximation, to be sure, but the results are at least qualitatively valid. They may even be close quantitatively to the true values for an actual machine due to the effect of the air gaps that tend to reduce the nonlinearity of the magnetic circuit.)

How do the inductances vary as functions of the angle θ? Self-inductances, or flux linkages per ampere, are proportional to the permeance (inversely proportional to the reluctance) of the magnetic circuit. Thus, both the stator inductance (say, L_1) and the rotor inductance (L_2) can be seen to have maximum values at $\theta = 0$ or $\theta = 180°$, and minimum values at $\theta = \pm 90°$. Mutual inductance, on the other hand, refers to the flux produced by one current and linking the other circuit. Figure 7-15b shows that, for $\theta = 90°$, the flux produced by the rotor current does not link the stator winding at all and vice versa; in other words, in this rotor position, $M = 0$. In the position of Fig. 7-15a, when $\theta = 0$, the mutual inductance has its maximum value but, at $\theta = 180°$, the lines of flux produced by one current link the other winding "from the other side"; in other words, the sign of M is reversed. Figure 7-16 is a sketch of the three inductances as functions of θ, in accordance with the above reasoning. Example 7-3 uses plausible values and illustrates how the torque varies as a function of the angle θ.

Another way to look at the device under study is to consider the stator and the rotor as two electromagnets, and to call the surfaces at which the flux enters

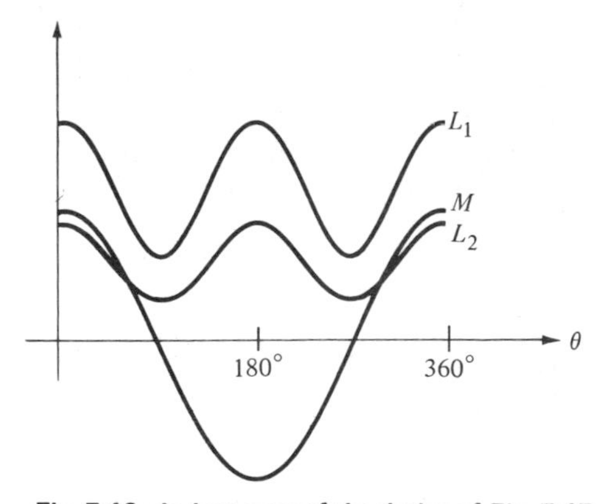

Fig. 7-16. Inductances of the device of Fig. 7-15.

and leaves them, their *poles*. Specifically, the surface at which a flux leaves a core is its *north pole*, the other its *south pole*. In both parts of Fig. 7-15, the stator poles are marked by italic capital letters (*N* and *S*) and the rotor poles by italic lower case letters (*n* and *s*). This point of view is sufficient to show that the rotor is in a state of equilibrium when $\theta = 0$, and that the forces of magnetic attraction and repulsion tend to turn it into this position, which agrees with the more specific results obtained through Eq. 7-38.

Another design, used for many machines, is obtained by making the rotor *cylindrical*, as shown in Fig. 7-17. This requires that the active rotor conductors be placed in *slots* cut into the surface of the rotor core; otherwise, the air gap would have to be unreasonably large to be able to accommodate them. The slots also have the advantage that the conductors can be better secured against centrifugal forces by (nonmagnetic) wedges.

The two broken lines in Fig. 7-17 indicate the stator flux. Note that the magnetic circuit for the stator flux remains the same no matter what the angle θ may be at any moment. Thus, the stator inductance L_1 is a constant, not a function of θ. On the other hand, the rotor flux depends on the angle much as it did in the device of Fig. 7-15 and for the same reason. So there is one reluctance term in the torque equation, not two. The mutual inductance changes about the same way as it did in the previous case.

The poles are again marked in the figure, indicating that, in the position shown, the torque is in the clockwise direction and remains so until the rotor has travelled about 120° to reach a position of equilibrium. In contrast to the poles of the cylindrical rotor, those of the stator are called *salient* poles. In the model of Fig. 7-15, both the stator and the rotor have salient poles.

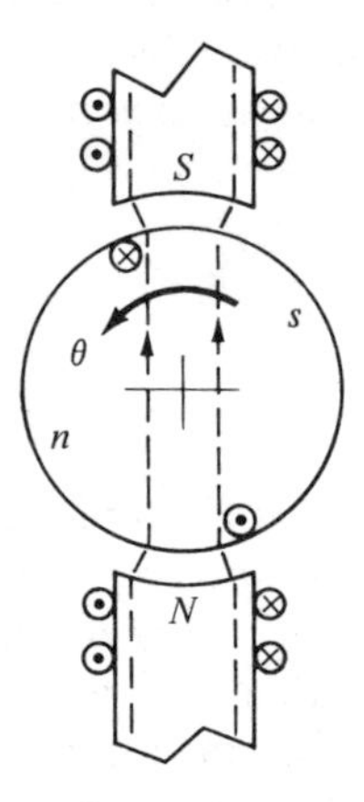

Fig. 7-17. Device with salient-pole stator and cylindrical rotor.

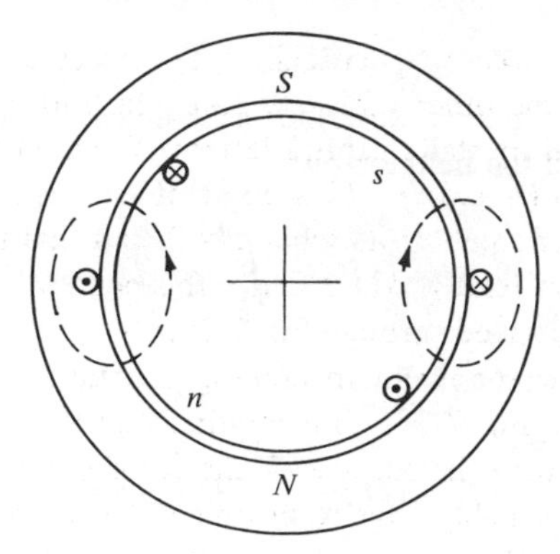

Fig. 7-18. Device with cylindrical stator and rotor.

The final sketch of this section, Fig. 7-18, represents a device with a cylindrical stator and a cylindrical rotor; in other words, with *nonsalient* poles on both stator and rotor. The poles are again marked by the letters N, S, n, and s. The device is drawn with the rotor in the same position as in the previous figure, but, in contrast to it, this time the entire magnetic circuit is shown. Note that the active conductors of both stator and rotor are placed in slots (indicated by being drawn just inside the core surface).

The two broken lines are typical lines of stator flux. It is significant that all lines of flux (except those of leakage fluxes) must cross the air gap twice. To what extent they link the other winding depends on the rotor position. Thus, the mutual inductance M is again a function of the angle θ. The two self-inductances, however, are constants, and the first two terms of the torque equation, Eq. 7-38, are, therefore, both zero.

Devices with salient-pole rotors and cylindrical stators will also be encountered in later chapters. As the reader might expect, they have constant rotor inductances and position-dependent stator inductances.

7-9 EXAMPLES

Example 7-1 (Section 7-4)

The relay mechanism in Fig. 7-4 has a saturation curve that is approximated by $\phi = (M/x)\sqrt{\mathfrak{F}}$ or $\mathfrak{F} = (\phi x/M)^2$, where ϕ is in webers, $\mathfrak{F}$ is in ampere-turns, x is in meters, and $M = 9 \times 10^{-8}$. (a) Find the stored energy as a function of ϕ and x. (b) Find the coenergy as a function of $\mathfrak{F}$ and x. (c) Find the derivatives of energy and coenergy with respect to the gap length x. (d) Find the mechanical force for $x = 0.01$ m and $\phi = 0.0006$ weber.

Solution

(a) The given saturation curve describes a nonlinear relation between ϕ and $\mathfrak{F}$. Use Eq. 7-9 to find the field energy

$$W_m = \int_0^{\phi} \mathfrak{F}\, d\phi = (x/M)^2 \int_0^{\phi} \phi^2\, d\phi = (x/M)^2(\phi^3/3)$$

(b) The coenergy is found as follows

$$W_m' = \int_0^{F} \phi\, d\mathfrak{F} = (M/x) \int_0^{F} \mathfrak{F}^{1/2}\, d\mathfrak{F} = (M/x)(2/3)\, \mathfrak{F}^{3/2}$$

Notice that for this nonlinear problem $W_m' \neq W_m$. Refer to Fig. 7-8. Substitute the function for $\mathfrak{F}$ into the result for coenergy to show

$$W_m' = (M/x)(2/3)(\phi x/M)^3 = (x/M)^2(2/3)\,\phi^3 = 2W_m$$

(c) Differentiate the foregoing answers with respect to x

$$dW_m/dx = (2x/M^2)(\phi^3/3)$$

$$dW_m'/dx = -(M/x)^2(2/3)\,\mathfrak{F}^{3/2}$$

Substitute the equation for $\mathfrak{F}$ into the derivative of coenergy to show $dW_m'/dx = -(M/x)^2(2/3)(\phi x/M)^3 = -(x/M^2)(2/3)\phi^3 = -dW_m/dx$
This result is in agreement with Eqs. 7-12 and 7-17.

(d) Substitute the given values into Eq. 7-12 to find the force.

$$F = -\left(\frac{\partial W_m}{\partial x}\right) = -(x/M^2)(2/3)\,\phi^3 = -178 \text{ newtons}$$

The force on the movable member is exerted in the direction that tends to decrease the air gap.

Example 7-2 (Section 7-5)

The relay mechanism in Fig. 7-4 has reluctance as a function of the gap length x as follows:

$$\mathfrak{R}(x) = 9 \times 10^8(0.003 + x) \text{ mks units}$$

where x is in meters. The coil has 1620 turns and 55 Ω resistance. The external voltage source is 110 v, d-c. (a) Find the energy stored in the magnetic field when the relay is OPEN ($x = 0.006$ m). (b) Find the energy stored in the magnetic field when the relay is CLOSED ($x = 0.001$ m). (c) Find the work done if the relay is allowed to close SLOWLY from $x = 0.006$ m to $x = 0.001$ m. (d) Find the work done if the relay is allowed to close FAST from $x = 0.006$ m to $x = 0.001$ m. (e) Find the coil current as a function of time during the electrical transient that follows the FAST closing of part (d).

Solution

(a) For $x = 0.006$ m, we can find the following. The current in the coil is

$$I_1 = V/R = 110/55 = 2 \text{ amp}$$

The magnetomotive force is

$$\mathfrak{F}_1 = NI_1 = 1620 \times 2 = 3240 \text{ At}$$

The reluctance of the magnetic circuit is

$$\mathfrak{R}_1 = 9 \times 10^8(0.003 + 0.006) = 8.1 \times 10^6$$

The flux across the air gap is

$$\phi_1 = \mathfrak{F}_1/\mathfrak{R}_1 = 3240/(8.1 \times 10^6) = 0.0004 \text{ weber}$$

For a fixed value of x, the relation between ϕ and $\mathfrak{F}$ is a straight line. This is illustrated in Fig. E-7-2; The energy stored in the magnetic field is found from Eq. 7-18.

$$W_1 = \tfrac{1}{2}\,\mathfrak{F}_1\phi_1 = \text{area } 0ac0 = 0.648 \text{ joules}$$

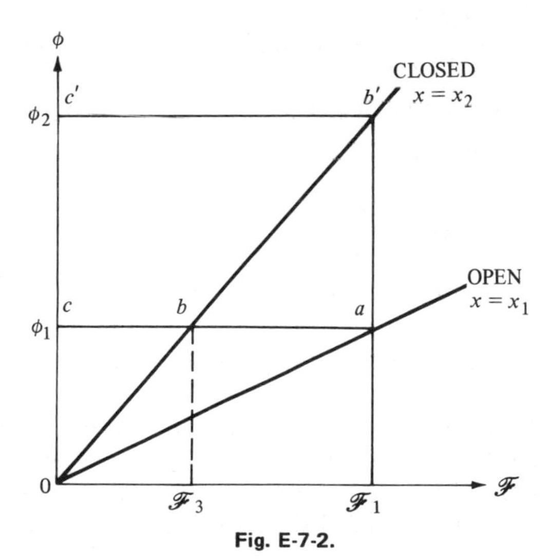

Fig. E-7-2.

(b) For $x = 0.001$ m, we can find the following. The reluctance of the magnetic circuit is

$$\mathcal{R}_2 = 9 \times 10^8(0.003 + 0.001) = 3.6 \times 10^6$$

The flux across the air gap is

$$\phi_2 = \mathcal{F}_1/\mathcal{R}_2 = 3240/(3.6 \times 10^6) = 0.0009 \text{ weber}$$

In the steady state, the current and mmf have the same values as in (a). The energy stored in the magnetic field is found from Eq. 7-18.

$$W_2 = \tfrac{1}{2}\mathcal{F}_1\phi_2 = \text{area } 0b'c'0 = 1.458 \text{ joules}$$

(c) For SLOW closure, the locus of ϕ versus $\mathcal{F}$ is along line $a - b'$. The work done is given by

$$W_{\text{mech}} = \text{area } 0ab'0 = \tfrac{1}{2}\mathcal{F}_1(\phi_2 - \phi_1) = 0.810 \text{ joules}$$

(d) For FAST closure, the locus of ϕ versus $\mathcal{F}$ is along line $a - b$. At point b, the reluctance is $\mathcal{R}_2$, but the flux is still at ϕ_1. The magnetomotive force is

$$\mathcal{F}_3 = \mathcal{R}_2\phi_1 = 3.6 \times 10^6 \times 0.0004 = 1440 \text{ At}$$

The work done is given by

$$W_{\text{mech}} = \text{area } 0ab0 = \tfrac{1}{2}\phi_1(\mathcal{F}_1 - \mathcal{F}_3) = 0.360 \text{ joule}$$

(e) After a FAST closure and during the electrical transient, the ϕ versus $\mathcal{F}$ locus is along line $b - b'$. Choose $t = 0$ as the instant of closing. The current in the coil immediately after closing is

$$i(0+) = \mathcal{F}_3/N = 1440/1620 = 0.89 \text{ amp}$$

The self-inductance of the coil is given by

$$L_2 = N^2\phi/\mathcal{F} = N^2/\mathcal{R}_2 = (1620)^2/(3.6 \times 10^6) = 0.73 \text{ henry}$$

The time constant for the exponential transient is

$$\tau = L_2/R = 0.73/55 = 1/75 = 0.0133 \text{ sec}$$

The current function is

$$i(t) = I_1 - [I_1 - i(0+)]e^{-t/\tau} = 2 - 1.11\, e^{-75t} \qquad \text{for } t > 0$$

Example 7-3 (Section 7-8)

For the doubly excited system in Fig. E-7-3, the inductances are approximated as follows: $L_1 = 11 + 3 \cos 2\theta$, $L_2 = 7 + 2 \cos 2\theta$, $M = 11 \cos \theta$ henrys. The

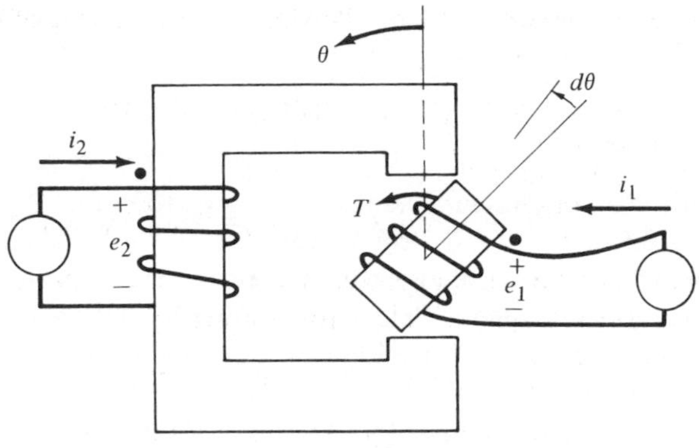

Fig. E-7-3.

coils are energized with direct currents. $I_1 = 0.7$ amp. $I_2 = 0.8$ amp. (a) Find the torque as a function of θ. (b) Find the energy stored in the system as a function of θ.

Solution

(a) Use Eq. 7-38 to find the torque. Find the derivatives of the inductances with respect to θ.

$$dL_1/d\theta = -6 \sin 2\theta, \quad dL_2/d\theta = -4 \sin 2\theta, \quad dM/d\theta = -11 \sin \theta$$

$$\begin{aligned} T &= \tfrac{1}{2} i_1^2 \, dL_1/d\theta + \tfrac{1}{2} i_2^2 \, dL_2/d\theta + i_1 i_2 \, dM/d\theta \\ &= \tfrac{1}{2}(0.7)^2(-6 \sin 2\theta) + \tfrac{1}{2}(0.8)^2(-4 \sin 2\theta) + (0.7)(0.8)(-11 \sin \theta) \\ &= -2.75 \sin 2\theta - 6.16 \sin \theta \text{ newton-meters} \end{aligned}$$

For the position shown in Fig. E-7-3, $\theta = -50°$. The value of torque is $T = +7.43$ nm. This torque acts counterclockwise on the rotor. If this rotor is allowed to turn, it will move to the position where $\theta = 0°$ and where the torque is zero. (The torque is also zero at $\theta = 180°$, but if the position is away from 180°, the direction of the torque will tend to move the rotor toward 0°.) The sign of the mutual term is sensitive to the polarity of the currents. The reluctance torque terms are independent of the polarity of the currents.

(b) The stored energy is given by

$$\begin{aligned} W_m &= \tfrac{1}{2} L_1 i_1^2 + \tfrac{1}{2} L_2 i_2^2 + M i_1 i_2 \\ &= \tfrac{1}{2}(11 + 3 \cos 2\theta)(0.7)^2 + \tfrac{1}{2}(7 + 2 \cos 2\theta)(0.8)^2 + (11 \cos \theta)(0.7)(0.8) \\ &= 4.935 + 1.375 \cos 2\theta + 6.16 \cos \theta \end{aligned}$$

We can see that maximum energy is stored for $\theta = 0°$. Energy stored has positive value for any position of the rotor.

Example 7-4 (Section 7-8)

A machine with two coils has inductances as follows: (on rotor) $L_1 = 0.1$ henry, (on stator) $L_2 = 0.5$ henry, $M = 0.2 \cos\theta$ henry, where θ is the angle of the rotor coil axis displaced counterclockwise with respect to the stator coil axis. Coil 1 (on rotor) is short circuited. Coil 2 (on stator) is energized from a 60-Hz sinusoidal voltage source of 110 v. Resistances of the coils may be neglected. Assume the circuit operates in sinusoidal steady state. θ is set at 30°. (a) Find an expression for the instantaneous torque on the rotor. (b) Find the value of the average torque on the rotor. (c) Determine the direction of this torque.

Solution

(a) In order to find the developed torque, we must first solve an electric circuit problem to determine the values of the currents in the coils. The equivalent circuit is shown in Fig. E-7-4a. Write Kirchhoff's voltage equations

$$\mathbf{V} = +j\omega L_2 \mathbf{I}_2 + j\omega M \mathbf{I}_1$$

$$0 = +j\omega M \mathbf{I}_2 + j\omega L_1 \mathbf{I}_1$$

These yield $\mathbf{I}_1 = -(M/L_1)\mathbf{I}_2$, $\mathbf{I}_2 = \mathbf{V}/j\omega(L_2 - M^2/L_1)$
From the given data, $M = 0.2 \cos 30° = 0.1732$ henry
Choose V for the reference. $\mathbf{V} = 110\underline{/0°}$
Then the currents are $\mathbf{I}_2 = -j\,1.46$ and $\mathbf{I}_1 = +j\,2.53$.
The current functions are

$$i_2(t) = 1.46\sqrt{2} \sin 377\,t, \quad i_1(t) = -2.53\sqrt{2} \sin 377\,t$$

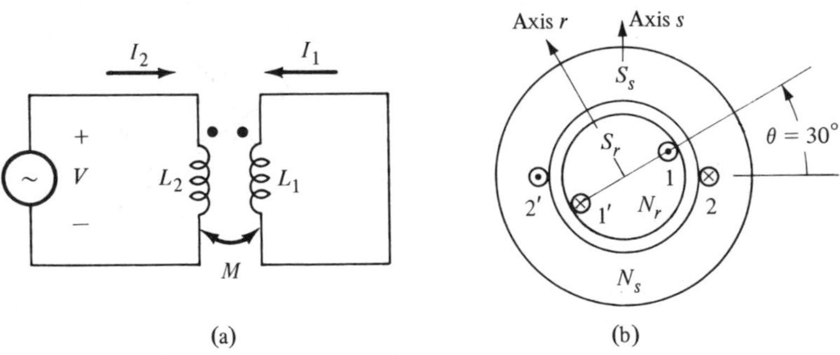

Fig. E-7-4.

Use Eq. 7-38 for the torque

$$dL_1/d\theta = 0, \quad dL_2/d\theta = 0, \quad dM/d\theta = -0.2 \sin\theta$$

$$T = i_1 i_2 dM/d\theta = (2.06 \sin 377t)(-3.58 \sin 377t)(-0.2 \sin 30^\circ)$$

$$= 0.74 \sin^2 377t = 0.74 \left[\tfrac{1}{2} - \tfrac{1}{2} \cos 2(377)\, t\right]$$

$$= 0.37 - 0.37 \cos 754t \text{ nm}$$

(b) Since the average value of the cosine function is zero, we can conclude $T_{ave} = 0.37$ nm.

(c) Figure E-7-4b illustrates the conditions of this problem. At an instant of time when the stator current produces a magnetic field polarity, N_s, S_s, as shown, the rotor current has the direction shown. The rotor field results in a repelling action that makes the torque on the rotor to be counterclockwise.

Example 7-5 (Section 7-8)

A simple device is depicted in Fig. E-7-5a. The ends of the conductors that are in front of the page are labelled a, a', b, b'. The labeled current directions are for positive values of current. Let the rotor turn counterclockwise with constant angular velocity of ω rad, sec. (a) Let $i_a = +10$ amp (d-c) and $i_b = 0$. Determine the voltage polarity in coil $b - b'$ for the instant shown with $\theta = 60^\circ$. (b) Let $i_a = +10$ amp (d-c) and $i_b = +10$ amp (d-c). Determine the direction of the torque on the rotor for the position shown with $\theta = 60^\circ$. (c) For the conditions of part (b), determine whether motor action or generator action is taking place.

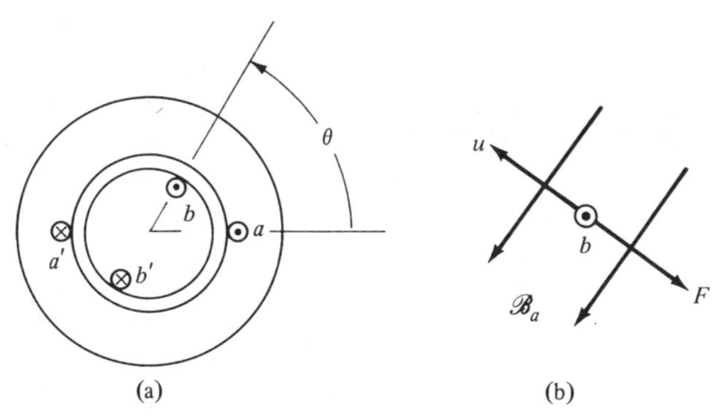

Fig. E-7-5.

Solution

(a) Determine the direction of the magnetic field produced by i_a. Since i_a has positive value, it enters the page in conductor a' on the left and comes out of the page in conductor a on the right. This causes lines of flux in the vicinity of conductor b to be oriented from the stator across the air gap into the rotor. This is illustrated in Fig. E-7-5b. Use the right hand. Thumb points in the direction of u. Forefinger points in the direction of $\mathcal{B}_a$ Middle finger points up. We conclude that b is positive in front of the page. Accordingly, $v_{bb'} > 0$.

(b) Since current i_b has positive value, it comes out of the page in conductor b. The direction of flux across the air gap is the same as in part (a). This is illustrated in Fig. E-7-5b. Use the right hand. Thumb points in the direction of current i_b. Forefinger points in the direction of flux. Middle finger points toward the right at the angle depicted. We conclude that the torque on the rotor is clockwise.

(c) The developed torque on the rotor is clockwise, and this opposes the direction of rotation, which is counterclockwise. A prime mover is required to supply counterclockwise torque to maintain this rotation. Mechanical energy is delivered into this machine and converted into electrical energy. This condition is for generator action.

7-10 PROBLEMS

7-1. The relay mechanism in Fig. 7-4 has a saturation curve that is approximated by $\mathcal{F} = (x/K)\,\phi^{3/2}$ or $\phi = (K\mathcal{F}/x)^{2/3}$ where ϕ is in webers, $\mathcal{F}$ is in ampere-turns, x is in meters, and $K = 2 \times 10^{-10}$. (a) Find the stored energy as a function of ϕ and x. (b) Find the coenergy as a function of $\mathcal{F}$ and x. (c) Find the derivatives of energy and coenergy with respect to the gap length x. (d) Find the mechanical force for $x = 0.006$ m and $\phi = 0.0016$ weber.

7-2. The relay mechanism in Fig. 7-4 has reluctance given by $\mathcal{R}(x) = 7 \times 10^8(0.002 + x)$ mks units, where x is the length of the variable gap in meters. The coil has 980 turns and 30 Ω resistance. The external voltage source is 120 v d-c. (a) Find the energy stored in the magnetic field when the relay is OPEN ($x = 0.005$ m). (b) Find the energy stored in the magnetic field when the relay is CLOSED ($x = 0.002$ m). (c) Find the work done if the relay is allowed to close SLOWLY from $x = 0.005$ m to $x = 0.002$ m. (d) Find the work done if the relay closes FAST from $x = 0.005$ m to $x = 0.002$ m. (e) Find the coil current during the electrical transient after the FAST closing in part (d). (f) Find the work that must be done to pull the relay open FAST from $x = 0.002$ m to $x = 0.005$ m.

7-3. A simple relay-type electromagnet has the self-inductance of its coil given by: $L = 1/(2 + 1000x)$ henry, where x is the length of the variable gap in meters. The coil is energized from a d-c voltage source. When the mechanism closes quickly (nearly instantaneously) from $x = 0.005$ m to $x = 0.002$ m, the amount of mechanical work done is 0.6 joule. (a) Find the amount of energy stored in the magnetic field before closing (i.e., for $x = 0.005$ m). (b) Find the amount of energy stored in the magnetic field a long time after closing (i.e., for $x = 0.002$ m and after any electrical transients). Hint: Construct a ϕ versus $\mathcal{F}$ diagram to establish ratios of values.

7-4. A steel electromagnet is used to support a solid hunk of steel weighing 2000 lb as shown in Fig. P-7-4. A force of 8900 newtons is required to support this weight. The cross-section area of the magnet core (part 1) is 0.01 m^2. The coil has 700 turns. Assume both air gaps are 0.0015 m long. Neglect the reluctance of the steel parts. Neglect fringing in the air gaps. Find the minimum current that can keep the weight from falling.

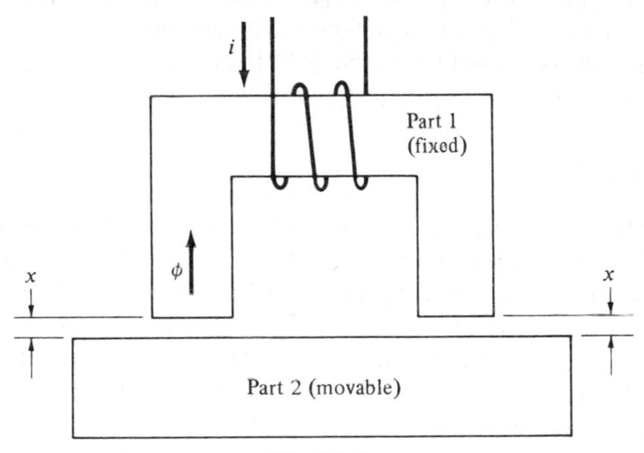

Fig. P-7-4.

7-5. In Fig. P-7-5, the plunger can move only in the vertical direction. Mechanical stops fix the limits of travel so that $0.1 \text{ cm} < x < 1.0 \text{ cm}$. Neglect the reluctance of the iron parts. Neglect leakage and fringing in the air gaps. The coil has 456 turns. In steady state, the current in the coil is 3 amp. (a) Find the reluctance of the magnetic circuit as a function of the variable gap length x. (b) Find the work done if the plunger closes FAST from $x = 1.0$ cm to $x = 0.1$ cm. (c) Find the electric energy delivered into the coil (in excess of copper loss) during the electrical transient that follows the FAST closing.

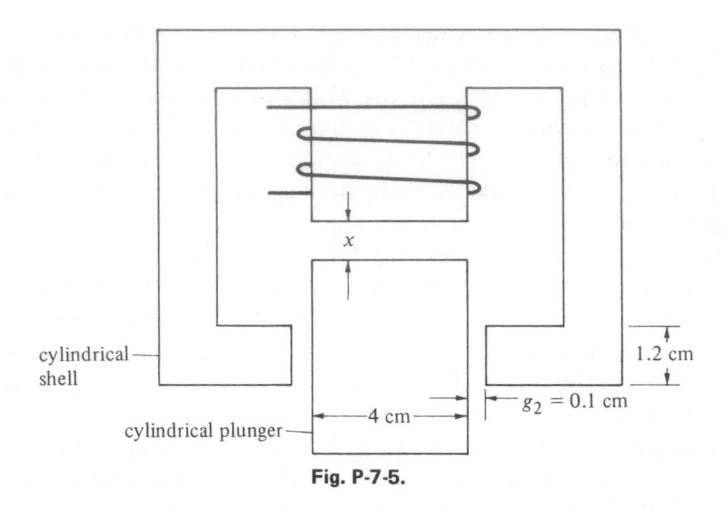

Fig. P-7-5.

7-6. For the plunger magnet in Problem 7-5, solve the following: (a) Find the self-inductance of the coil as a function of the variable gap length x. (b) If the plunger is allowed to move SLOWLY from $x = 1$ cm to $x = 0.1$ cm, find the mechanical work done. (c) For the conditions in part (b), find the amount of electrical energy that is supplied by the electrical source (in excess of copper loss).

7-7. For the system in Fig. E-7-3, the inductances are as approximated: $L_1 = 11 + 3 \cos 2\theta$, $L_2 = 5 + 2 \cos 2\theta$, $M = 7 \cos \theta$ henry. The coils are energized with direct currents. $I_1 = 1$ amp. $I_2 = 2$ amp. (a) Find the torque as a function of θ. (b) Find the torque for $\theta = -30°$. (c) Find the torque for $\theta = +45°$.

7-8. For the doubly excited device in Problem 7-7, let the current be changed to $I_1 = 1$ amp and $I_2 = -2$ amp. (a) Find the torque as a function of θ. (b) Find the torque for $\theta = 150°$. What is its direction? (c) Find the torque for $\theta = 210°$. What is its direction? (d) Find the rotor positions for zero torque. (e) Find the energy stored in the system as a function of θ. (f) Find the rotor position at which stored energy has maximum value.

7-9. A doubly excited machine has inductances in henrys given by: $L_1 = 2 - \cos 2\alpha$, $L_2 = 3 - \cos 2\alpha$, $M = 3 \sin \alpha$, where α is a mechanical angle measured counterclockwise from the given reference position. The current in coil 1 (on rotor) is constant at 12 amp. The current in coil 2 (on stator) is constant at 16 amp. With the rotor at the position for $\alpha = 60°$, find the magnitude and direction of the torque on the rotor.

7-10. A machine with two coils has inductances in henrys as follows: (on rotor) $L_1 = 0.5$, (on stator) $L_2 = 0.8$, $M = 0.6 \cos \theta$ where θ is the angle of the rotor coil axis displaced clockwise with respect to the stator coil axis. Coil 2 (on stator) is short circuited. Coil 1 (on rotor) is energized with a sinusoidal current. $i_1(t) = 2\sqrt{2} \cos 377t$. Resistance of the coils may be neglected. Assume the circuit operates in sinusoidal steady state. $\theta = 45°$. (a) Find an expression for the instantaneous torque on the rotor. (b) Find the value of the average torque on the rotor. (c) Determine the direction of this torque.

7-11. Solve Problem 7-10, but with $\theta = 135°$.

7-12. The machine described in Problem 7-10 is operated with the same excitation and with coil 2 shorted. (a) Find the instantaneous torque as a function of rotor position θ and time t. (b) Find the rotor positions for zero torque. (c) Determine the zero positions toward which the torque tends to turn the rotor.

7-13. A simple device is depicted in Fig. E-7-5a. The labeled current directions are for positive values of current. Let the rotor turn clockwise with constant angular velocity of ω rad/sec. (a) Let $i_a = +10$ amp (d-c) and $i_b = 0$. Determine the voltage polarity in coil $b - b'$ for the instant shown with $\theta = 60°$. (b) Let $i_a = +10$ amp (d-c) and $i_b = +10$ amp (d-c). Determine the direction of the torque on the rotor for the position shown with $\theta = 60°$. (c) For the conditions of part (b), determine whether motor action or generator action is taking place.

7-14. For the machine in Problem 7-13, let the rotation be clockwise. (a) Let $i_b = +10$ amp (d-c) and $i_a = 0$. Determine the voltage polarity in coil $a - a'$ for the instant shown with $\theta = 60°$. (b) Let $i_b = +10$ amp (d-c) and $i_a = -10$ amp (d-c). Determine the direction of the torque on the rotor for the position shown with $\theta = 60°$. (c) For the conditions of part (b), determine whether motor action or generator action is taking place.

8
Distributed Windings

8-1 FLUX DISTRIBUTION

In the discussion of transformers and also of salient-pole stators or rotors, reference was made frequently to *coils* formed from individual *turns* wound around cores by having them connected in series. Indeed, without this possibility of using coils of many turns, no practical electromagnetic devices for power conversion could be built at all.

In the examples just mentioned, all turns of a coil were wound around the same core, forming what is also called a *concentrated winding.* It is characteristic of such a winding that practically all of its flux links all of its turns. By contrast, the active conductors of a cylindrical stator or rotor are placed in slots; not just one pair of slots as in Fig. 7-17 and 7-18, which would make poor use of the material, but rather slots all over the surface like the rotor depicted in Fig. 8.1. A winding whose active conductors are thus distributed over the surface of a stator or rotor is called a *distributed winding.*

Figure 8-2 is a schematic representation of the same rotor with the conductors symbolically represented by little circles drawn under the rotor surface. It is not necessary, in such a diagram, to draw one such little circle for each active conductor. In the case of Fig. 8-2, each circle represents all the conductors in a slot. Together with the same number of conductors in the opposite slot, they form a coil. So this rotor winding consists of as many coils as there are pairs of slots.

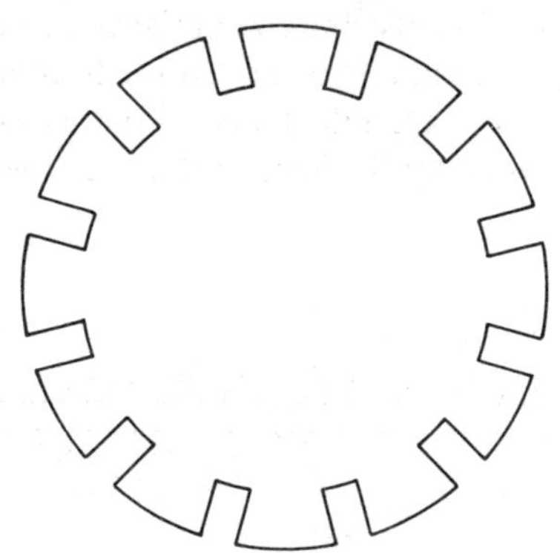

Fig. 8-1. Cylindrical rotor.

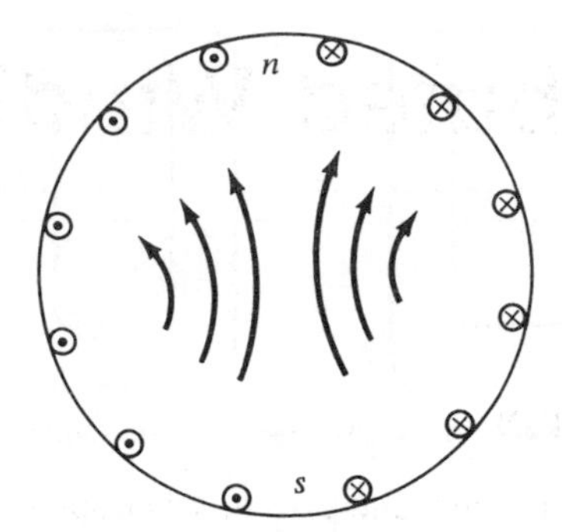

Fig. 8-2. Magnetic field of cylindrical rotor.

Also shown in Fig. 8-2 are the flux configuration and the poles of this rotor. With dots and crosses indicating the current directions for each coil, the flux axis must be vertical and upward. But not all the lines of flux link all the coils, which is typical for distributed windings.

Actually, the flux configuration depends on the entire magnetic circuit, and thereby on the stator. At this point, only the simplest case of a *uniform air gap* is being considered, i.e., both stator and rotor are assumed to be cylindrical. The term *uniform air gap* also implies that the effect of the slots themselves on the flux can be neglected, which means that the slots must be sufficiently small and numerous.

The study of distributed windings is greatly facilitated if the cylindrical surfaces of the stator and the rotor are "developed", i.e., rolled out to appear in diagrams as straight lines. Figure 8-3 shows the same rotor as Fig. 8-2, but developed and facing the stator across a uniform air gap. Figure 8-4 shows the same kind of diagram for the elementary case of a rotor with only a single coil, with some typical lines of flux sketched and the poles marked.

From such a diagram, the flux distribution in the air gap can be approximately obtained if the further assumption is made that the ferromagnetic material used is ideal ($\mu \rightarrow \infty$), which reduces the reluctance of the magnetic circuit to the reluctance of the air gaps crossed by the lines of flux. (Remember the same condi-

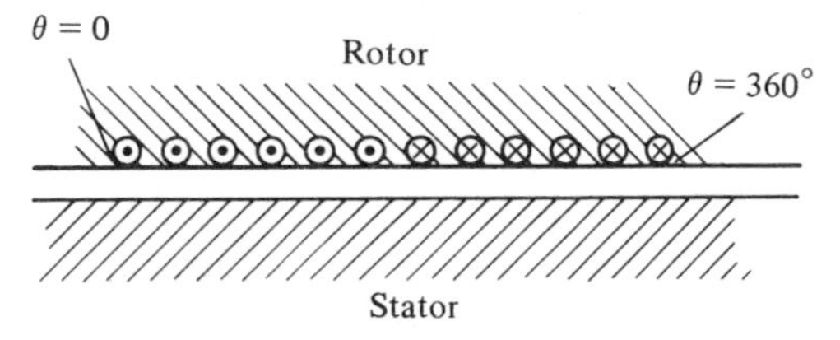

Fig. 8-3. Developed surfaces.

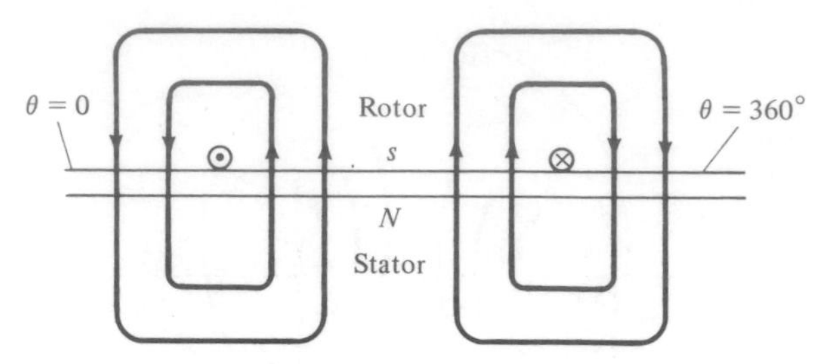

Fig. 8-4. Single rotor coil, flux lines.

tion of infinite permeability was imposed on the core material of a transformer to make that transformer ideal.) This assumption also makes the magnetic circuit *linear*, and it is sometimes reasonably close to the truth as long as the flux density stays far enough below its saturation value.

Using any line of flux as a path of integration, Ampère's law can now be expressed as

$$N_c i = \oint H\,dl \approx 2Hg \tag{8-1}$$

where g is the radial length of the air gap, and N_c the number of turns of the coil (or the number of conductors in each of the two slots). Incidentally, the evaluation of the integral in Eq. 8-1 assumes not only that the core reluctance be negligible, but also that the field in the air gap is uniform. Actually, the lines of flux are in the *radial* direction, and the flux density at the rotor surface is slightly higher than it is at the stator surface. With a reasonably short air gap, this distinction is rather unimportant. At any rate, the values of H and B may always be understood as average values (i.e., taken in the middle of the air gap.)

Equation 8-1 is solved for H, and the result multiplied by the permeability of air, making the flux density in the air gap

$$B \approx \frac{\mu_o N_c\, i}{2g} \tag{8-2}$$

Thus, for the single coil and assuming a uniform air gap and ideal core material, the flux density has the same magnitude at every point of the air gap, and it reverses its sign where the conductors are located. The *flux distribution B* versus θ is a *rectangular wave* as shown in Fig. 8-5. When more realistic cases, with more than one coil, are to be taken up, and when the *resultant* flux distribution of both stator and rotor currents is to be found, this can be done, based on the linear nature of the assumptions made, by *superposition.*

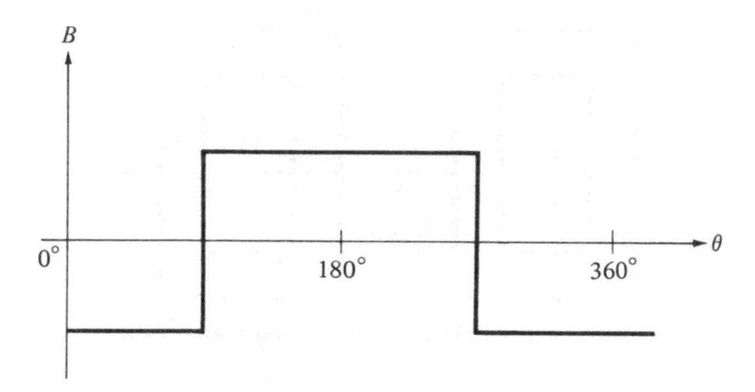

Fig. 8-5. Approximate flux distribution for single rotor coil.

8-2 THE MMF WAVE CONCEPT

To analyze the operation of a given motor or generator, it is necessary to know the flux distribution $B(\theta)$, i.e., the flux density in the air gap as a function of the space angle θ. To obtain this function, there is the method suggested in the previous section, although it is based on assumptions that are never quite satisfied and, in many cases, are not even good approximations. There are also sophisticated methods of flux mapping by digital computation, based on the concept of the magnetic vector potential, but even these methods are not absolutely rigorous (for one thing, they are two-dimensional and thereby neglect "end effects"), and they call for big expenditures of preparation and computing time to produce results valid for just one machine.

Much of this can be avoided, at least for the purpose of gaining an understanding of the subject, by introducing another function of θ, called the *mmf wave* and defined as

$$\mathcal{F}(\theta) = \frac{1}{2} \oint H(\theta)\, dl \tag{8-3)*}$$

where $H(\theta)$ is the magnetic field intensity at some point (described in terms of the angle θ) of the air gap, and where the path of integration is the line of flux that crosses the air gap at that point. Obtaining this function requires no assumption about the shape of the air gap nor is there any need to disregard the reluctance of the core. Even better, the mmf wave is a linear function of the currents, so that *superposition* is a correct method of obtaining a resultant mmf wave from its components. As for the factor $\frac{1}{2}$ in Eq. 8-3, it simply assigns one-half of the loop integral to each of the two poles.

*For fractional pitch windings (see section 8-6) this definition must be modified, but methods of finding and using the mmf wave remain valid.

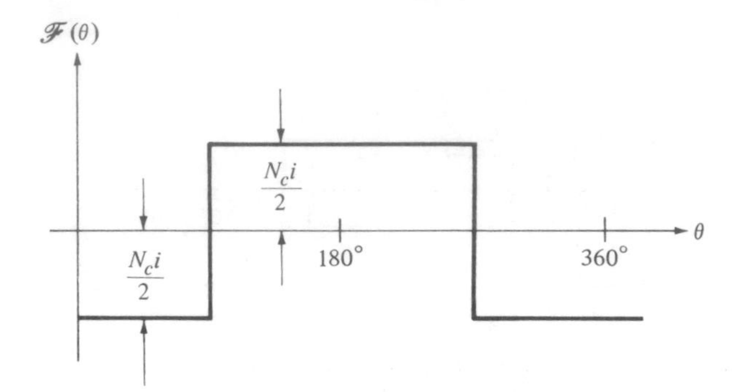

Fig. 8-6. Mmf wave for single coil.

Starting again with the single coil of Fig. 8-4, and using Eqs. 8-1 (without its approximate expression) and 8-3, the magnitude of the mmf wave is

$$\mathfrak{F}(\theta) = \tfrac{1}{2} N_c i \tag{8-4}$$

The corresponding graph of Fig. 8-6 is identical with that of Fig. 8-5 except for a scale change, but the point is that the mmf wave, in contrast to the flux distribution, is not an approximation based on various assumptions. The only inaccuracy still present in the graph of Fig. 8-6 is that it is drawn as if the conductors were of infinitesimal size. This was done only as a matter of convenience, not of necessity; the actual size of the conductors could be taken into consideration by "tilting" the vertical lines to give them finite slopes whereby the rectangular wave would be changed into a trapezoidal one. The benefits obtained from this correction would not justify the complication, and we shall stick to the vertical lines in all such diagrams.

Incidentally, positive values were assigned in this graph to those parts of the wave for which the lines of flux are directed from the stator to the rotor. This is, of course, an entirely arbitrary choice. The next step to be taken is from a single coil to a winding consisting of several coils. Figure 8-7 depicts a developed rotor with *three coils*, and it shows poles and some typical lines of flux. It is

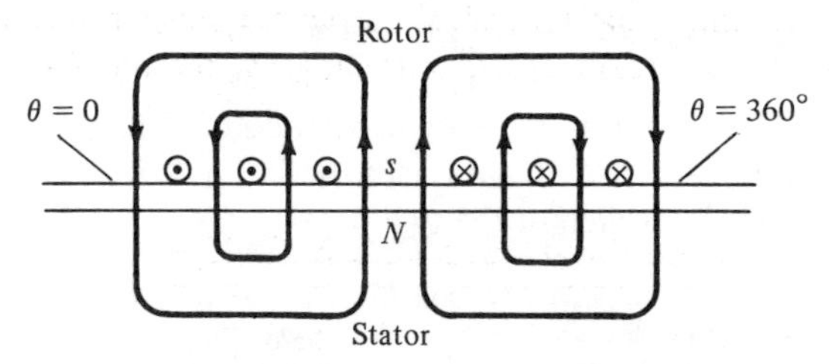

Fig. 8-7. Three rotor coils and flux lines.

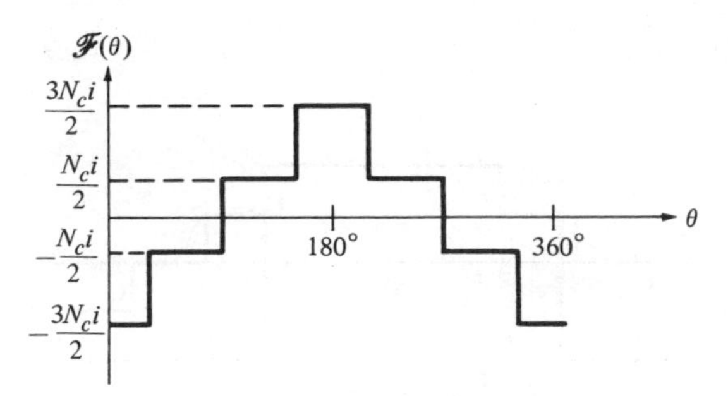

Fig. 8-8. Mmf wave for three coils.

only a sketch based on an assumed uniform air gap, but the mmf wave for this rotor, shown in Fig. 8-8, is valid regardless of the shape of the air gap.

There are several ways to obtain diagrams like that of Fig. 8-8, for any distributed winding. One method, already mentioned, is superposition, i.e., adding the ordinates of diagrams drawn for each coil (like Fig. 8-6, but displaced horizontally against each other). Another way is first to draw a diagram like Fig. 8-7, including typical lines of flux. For instance, in that figure, it is clear that every line crossing the air gap between $\theta = 0$ and $\theta = 30°$, links $3N_ci$ At, so that, for $0 < \theta < 30°$, $\mathcal{F}(\theta) = 3N_ci/2$, etc. Finally, by either one of these two methods it soon becomes clear that the mmf wave consists of steps, each one having the magnitude N_ci, at that value of θ at which the conductors are located. So the wave may be drawn as a sequence of such steps, starting at any arbitrary value for $\mathcal{F}(0)$ and putting the horizontal axis at the end of the operation into such a position that the wave has an average value of zero.

Many windings are more finely distributed, i.e., in more than six slots. The mmf wave then becomes a train of many small steps, always half of them positive and the others negative. For instance, Figs. 8-9 and 8-10 are drawn for a winding of six coils (twelve slots, at $\theta = 15°, 45°, 75°$, etc.)

The process can easily by continued, with the result that the mmf wave gets to be closer and closer to a *triangular* wave (the broken line in Fig. 8-10), while the current distribution approaches more and more that of a uniform *current sheet.* In thus going to the limit, the individual currents disappear, and in their place, there appears another function of the space angle θ, the *angular current density*

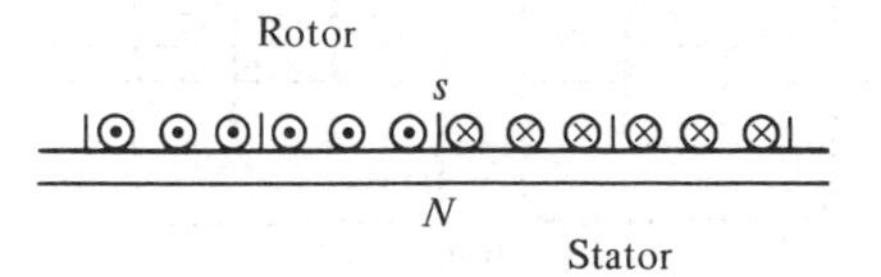

Fig. 8-9. Six-rotor coils.

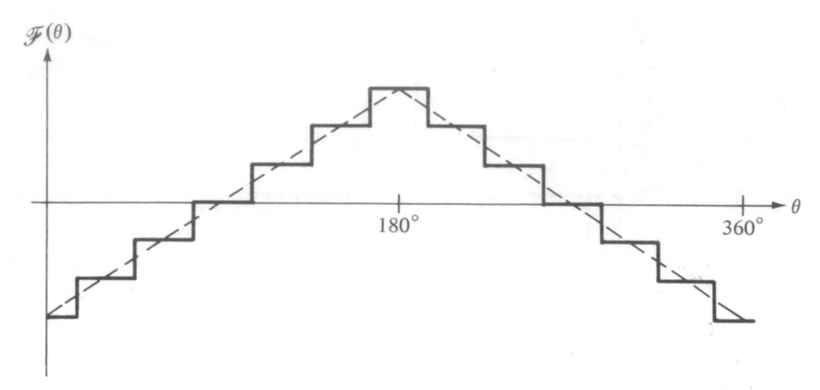

Fig. 8-10. Mmf wave for six coils.

$J(\theta)$, expressed in amperes per radian. For a uniform current sheet, the angular current density is a rectangular wave. Regardless of the shape of $J(\theta)$, the discrete steps of the mmf wave become infinitesimal, with a magnitude $J(\theta)\,dt$, and the mmf wave itself changes from the sum of its steps to the integral

$$\mathcal{F}(\theta) = \int J(\theta)\,d\theta \tag{8-5}$$

where the integration constant has to be chosen so as to give the wave an average value of zero. Equation 8-5 and its inverse

$$J(\theta) = \frac{d\,\mathcal{F}(\theta)}{d\theta} \tag{8-6}$$

remain valid also for a nonuniform current sheet, as well as for discrete currents (in which case the function $J(\theta)$ would be a train of impulse or delta functions).

We have now introduced three different functions of the space angle θ: the flux distribution $B(\theta)$, the mmf wave $\mathcal{F}(\theta)$, and the angular current density $J(\theta)$. They are all *periodic* functions of θ, with the period 2π, because adding 2π (or any multiple of 2π) to any value of θ does not change the values of these functions. Therefore, *Fourier analysis* can be applied to each of these functions; i.e., they can be expressed as series of *sinusoidal* functions of θ, called the *space fundamentals* and *space harmonics.* The reason for doing so is that the operation of addition as well as that of differentiation (and integration) is much easier to perform with sinusoids than with most actual waveshapes. Furthermore, due to the linearity of the operations involved, equations like

$$\mathcal{F}(\theta)_{\text{resultant}} = \mathcal{F}(\theta)_{\text{stator}} + \mathcal{F}(\theta)_{\text{rotor}} \tag{8-7}$$

are valid not only for the waves themselves but also for their space fundamentals and each of their space harmonics.

The reader will see later that it is normally desirable that the flux distributions of a-c motors and generators resemble the sinusoidal waveshape rather than the rectangular one of a concentrated winding or the triangular one of a uniformly distributed winding. Windings can indeed be designed and air gaps shaped to accomplish this purpose.

Most of the following discussions, being mainly preparatory to the study of a-c machines, are based on the assumption of sinusoidal mmf waves. This does not mean that the space harmonics must be neglected. Whenever it is considered worthwhile, their effects can be studied by the same methods as those of the space fundamentals, since they are also sinusoids. Such effects are, however, mostly left for more advanced studies.

8-3 NUMBER OF POLES

All the stators and rotors described so far in this and the previous chapter have one thing in common: they have *two poles* (north and south), whether salient or not. This might seem to be perfectly natural. But actually, the majority of motors and generators are built with several *pairs of poles.* Figure 8-11 illustrates the idea, being a sketch of a four-pole machine with a cylindrical stator and a salient-pole rotor. Only the rotor conductors are shown in the picture, and broken lines are used to indicate typical lines of flux.

For another example, Fig. 8-12 shows a cylindrical rotor with a distributed winding for six poles. (Any even number of poles is possible, with north and

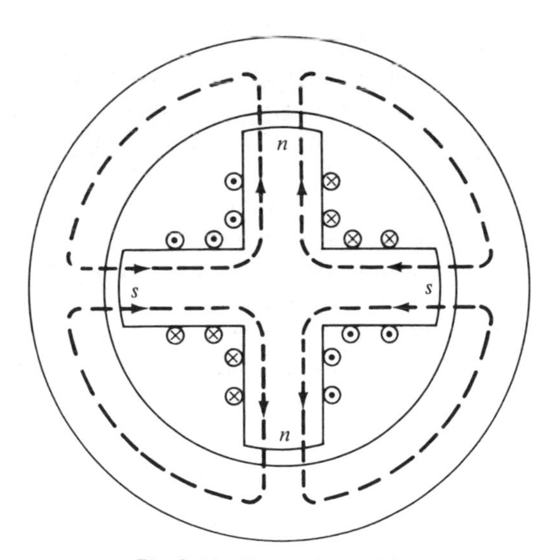

Fig. 8-11. Four-pole machine.

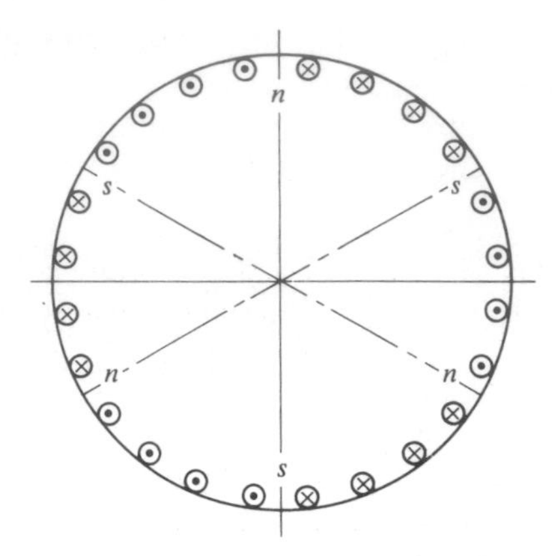

Fig. 8-12. Six-pole rotor.

south poles always alternating.) What produces the poles in the case of a distributed winding is the direction of the currents, indicated in the figure by the usual dots and crosses. Each turn of a coil contains two active conductors with currents in opposite directions. The two conductors forming a turn are located at a distance of one-sixth of the circumference (for this case of six poles) from each other. It follows that the connections in front and back of the paper (i.e., the inactive conductors) can be shortened and material saved thereby if a larger number of poles is used. There are more compelling reasons for the designer to choose a certain number of poles, as will be shown shortly.

Figure 8-13 shows the mmf wave of a winding like that of Fig. 8-11, and also the space fundamental of that wave. There is something redundant about that diagram: one-half of it would be quite sufficient, as long as the number of poles is known (and even that is indicated by the abscissa scale). The shortest angular distance about which these waves are periodic is π rather than 2π, so the space fundamental is a sinusoidal function of 2θ, or generally, for p poles, of $(p/2)\,\theta$. For these and other reasons (which will become clear soon), it has been found convenient to use an angular scale at which the period is always 2π, regardless of the number of poles. This requires the introduction of new units, called *electrical radians* or *electrical degrees*, in contrast to the *mechanical* units (radians or degrees) in which a space angle is conventionally expressed. The relation is

$$\theta_{\text{electrical}} = \frac{p}{2}\,\theta_{\text{mechanical}} \tag{8-8}$$

where p is the number of poles. For instance, Fig. 8-14 shows the space fundamental of a six-pole mmf wave (could be that of the winding of Fig. 8-12) with

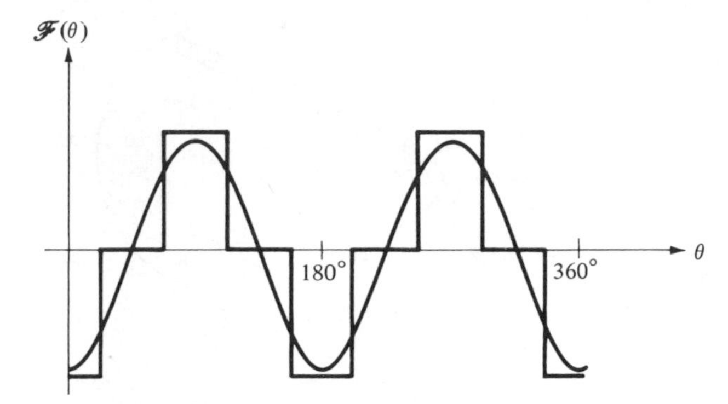

Fig. 8-13. Mmf wave and its space fundamental for rotor of Fig. 8-11.

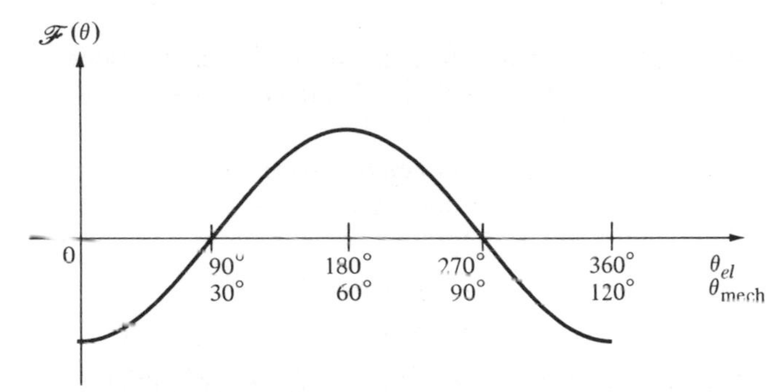

Fig. 8-14. Electrical and mechanical degrees for six poles.

abscissa scales in mechanical and electrical degrees. The angular distance from the axis of one pole to the axis of an adjacent pole is always 180 electrical degrees.

8-4 THE FLUX PER POLE

The mmf waves of the windings on the stator and the rotor are all linear functions of their currents. Therefore, superposition is a correct procedure, and the sum of these waves is the *resultant* mmf wave. This is the one that produces the flux distribution B(θ).

Let it now be assumed that the flux distribution is sinusoidal, or that its space harmonics are not to be considered. With an arbitrary choice of the origin of the θ axis, the flux distribution can then be expressed by the equation

$$B(\theta_{e\ell}) = B_{\text{peak}} \sin \theta_{e\ell} \tag{8-9}$$

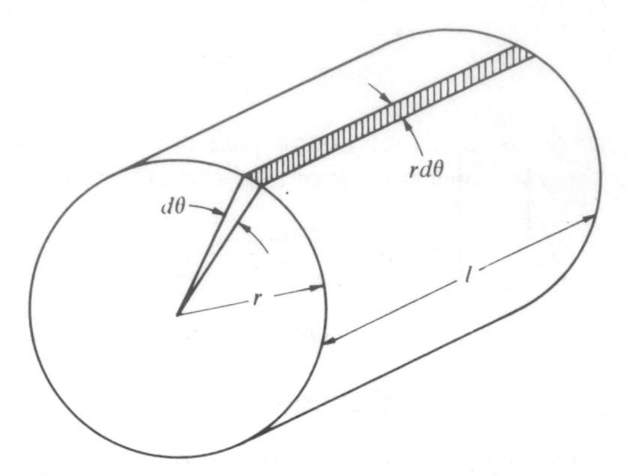

Fig. 8-15. For the calculation of the flux per pole.

where the subscript *peak* is used to indicate the maximum value of a periodic space function, in contrast to *max*, which indicates the maximum value of a periodic time function.

Note that all the lines of flux between $\theta_{e\ell} = 0$ and $\theta_{e\ell} = \pi$ have the same direction. They all belong to the same pole, and their sum is called the *flux per pole*

$$\phi = \int_A B(\theta)\, dA \tag{8-10}$$

where the area element dA can be seen as the shaded area in Fig. 8-15

$$dA = lr\, d\theta_{\text{mech}} = \frac{2lr}{p}\, d\theta_{e\ell} \tag{8-11}$$

Equations 8-9 and 8-11 are now substituted into Eq. 8-10

$$\phi = \int_0^{\pi} B_{\text{peak}} \sin\theta_{e\ell}\, \frac{2lr}{p}\, d\theta_{e\ell} = -\frac{2lr}{p} B_{\text{peak}} \cos\theta_{e\ell} \Big]_0^{\pi} = \frac{4lr}{p} B_{\text{peak}} \tag{8-12}$$

Incidentally, this result could also have been obtained by multiplying the *average* flux density $B_{\text{av}} = (2/\pi)B_{\text{peak}}$ by the cylindrical area per pole $A = lr2\pi/p$.

8-5 MOTIONAL VOLTAGES IN ROTATING MACHINES

There is a voltage induced in every conductor that moves relative to the flux, and it can be expressed by Eq. 7-3: $e = Blu$. However, rotating motion (see Section 7-6) is described in terms of the *angular velocity*

$$\omega_{\text{mech}} = \frac{d\theta_{\text{mech}}}{dt} \tag{8-13}$$

or, sometimes, in terms of the traditional unit of revolutions per minute (rpm). Since there are 2π radians per revolution, the rotating speed in rpm is

$$n = \frac{60}{2\pi} \frac{d\theta_{\text{mech}}}{dt} \tag{8-14}$$

The relation between the velocity u and the angular velocity ω_{mech} is

$$u = \frac{d(r\theta_{\text{mech}})}{dt} = r\omega_{\text{mech}} \tag{8-15}$$

where r is the radius, i.e., the distance from the axis of rotation to the conductor, or, preferably, to a point midway between stator and rotor surfaces. This middle value of r is also that to which all values of $\mathcal{F}(\theta)$ and $B(\theta)$ belong, and so it can be used for all conductors, both on the stator and on the rotor.

There is a choice among several possibilities to be considered: there can be conductors moving past a stationary flux; but there can also be a flux moving past stationary conductors, and there can even be a flux moving past conductors that move at a different speed. These three cases are equivalent to each other because it is the *relative motion* that counts with regard to the induced voltage. Each of the three cases will be encountered when specific types of electric machines are studied in chapters to follow. But at this point, it is sufficient to discuss the first case.

Consider, then, a single conductor moving at a velocity expressed by Eq. 8-15 and a flux distribution in accordance with Eq. 8-9. The voltage induced in this conductor is

$$e = Blu = (B_{\text{peak}} \sin \theta_{e\ell})\, l(r\omega_{\text{mech}}) \qquad 8\text{-}16)$$

It is neither logical nor convenient to have both mechanical and electrical units in the same equation. So the angular velocity is also expressed in electrical units

$$\omega_{e\ell} = \frac{d\theta_{e\ell}}{dt} = \frac{p}{2} \frac{d\theta_{\text{mech}}}{dt} = \frac{p}{2}\, \omega_{\text{mech}} \tag{8-17}$$

whereby Eq. 8-16 is changed into

$$e = \frac{2lr}{p} B_{\text{peak}}\, \omega_{e\ell} \sin \theta_{e\ell} \tag{8-18}$$

A simpler expression results from introducing the flux per pole (from Eq. 8-12)

$$e = \tfrac{1}{2} \phi\, \omega_{e\ell} \sin \theta_{e\ell} \tag{8-19}$$

This is the *instantaneous* value of the motional voltage induced in a conductor when it is located at the angular position $\theta_{e\ell}$. In the steady state, the conductor is moving at a *constant* angular velocity. Let the origin of the time axis be chosen arbitrarily so that the conductor is located at $\theta_{e\ell} = \pi/2$ when $t = 0$. Then its location at the time t is

$$\theta_{e\ell} = \frac{\pi}{2} + \omega_{e\ell} t \tag{8-20}$$

which leads to the following expression for the steady-state voltage as a function of time

$$e = \frac{1}{2}\phi\omega_{e\ell} \sin\left(\frac{\pi}{2} + \omega_{e\ell} t\right) = \frac{1}{2}\phi\omega_{e\ell} \cos \omega_{e\ell} t \tag{8-21}$$

This result applies to a single conductor. A *coil* consists of N_c conductors in one slot and another N_c conductors in a slot located at a distance of π electrical radians from the first slot. At these two locations, the flux density has the same magnitude and the opposite direction. Therefore, the voltages are equal and opposite, e.g., into the paper for one conductor and out of the paper for its counterpart in the other slot. The series connection of these two conductors (in front or back of the paper) forms a turn in which the induced voltage is twice that of the single conductor. The coil is a series connection of N_c turns, which raises its voltage to

$$e_c = N_c \phi \omega_{e\ell} \cos \omega_{e\ell} t \tag{8-22}$$

The fact that this voltage is a sinusoidal time function is entirely due to the assumption that the flux distribution is a sinusoidal function of space. If the space harmonics of the flux distribution are considered, then the voltage wave contains time harmonics; in fact, the waveshape of the voltage induced in a coil (not that of the voltage induced in a distributed winding, however, as will be shown in the next section) as a function of time is the same as that of the flux density as a function of the space angle. The desire to have a sinusoidal flux distribution, mentioned in Section 8-2, is now explained as the desire, basic to power systems engineering, to obtain a sinusoidal steady-state voltage.

Equation 8-22 also shows that the angular velocity of the motion, when expressed in electrical units, is identical with the *radian frequency* $\omega_{e\ell} = 2\pi f$ of the voltage. This is one more reason why electrical units are used for θ and ω.

Another thing seen in Eq. 8-22 is the voltage amplitude

$$E_{\max} = N_c \phi \omega_{e\ell} \tag{8-23}$$

So the rms value of the voltage can be written as

$$E = \frac{E_{\max}}{\sqrt{2}} = \frac{2\pi}{\sqrt{2}} N_c \phi f = 4.44\, N_c \phi f \tag{8-24}$$

The similarity of this equation to the transformer voltage in Eq. 4-25 is no mere coincidence. After all, both are derived from Faraday's induction law. But the flux ϕ in Eq. 8-24 is a constant flux, sinusoidally distributed in space and in relative motion to the coil, whereas ϕ_{max} in Eq. 4-25 is the maximum value of a stationary flux, which varies sinusoidally with time.

One further comment on the voltage equation, Eq. 8-24, refers to the way it was obtained. The derivation was based on the *Blu* formulation of Faraday's law, with an assumed sinusoidal flux distribution. Actually, all active conductors are located in slots, whereas most of the lines of flux must go through the high-permeability teeth between the slots and not through the slots. So it could be argued that only a comparatively low value of flux density, that existing in the slots, should be used in the calculation of the voltage.

This argument has long been refuted, both by observation and by using the other method of calculating the voltage, namely that of the rate of change of flux linkages. The reader is invited to work out problem 8-10 and thereby to satisfy himself that both forms of Faraday's law, *Blu* and $d\lambda/dt$, lead to exactly the same result. In other words, when a coil moves relative to a given flux, the induced voltage is the same whether the conductors of that coil are placed in slots or not.

8-6 WINDING FACTORS

A distributed winding, as introduced at the beginning of this chapter, is a series connection of several coils, located in different pairs of slots. Therefore, the voltage induced in the entire winding is the sum of the coil voltages. In the sinusoidal steady state, the coil voltages all have the same magnitude (they have the same number of turns) and frequency (they move with the same speed relative to the flux) but they are *not in phase* with each other.

Consider two coils, called No. 1 and No. 2, respectively, moving at a constant angular velocity relative to the flux, and let the choice of reference made for Eq. 8-20 be valid for the conductors of coil No. 1:

$$\theta_{1_{e\ell}} = \frac{\pi}{2} + \omega_{e\ell} t \tag{8-25}$$

which results in the voltage of this coil being in the axis of reference:

$$e_1 = N_c \phi \omega_{e\ell} \cos \omega_{e\ell} t \tag{8-26}$$

The slots of coil No. 2 are displaced against those of coil No. 1 by an angle α. Therefore, the location of the conductors of coil No. 2 as a function of time is

$$\theta_{2_{e\ell}} = \frac{\pi}{2} \pm \alpha_{e\ell} + \omega_{e\ell} t \tag{8-27}$$

where the choice of plus or minus depends on whether coil No. 2 is ahead of or behind No. 1 with respect to the direction of motion. The same derivation of the coil voltage as before leads to the result

$$e_2 = N_c \phi \omega_{e\ell} \cos(\omega_{e\ell} t \pm \alpha_{e\ell}) \tag{8-28}$$

Note that the *phase difference* between coil voltages equals the space angle between their slots, provided this angle is expressed in electrical units.

The addition of coil voltages is illustrated by the phasor diagram of Fig. 8-16. This diagram is drawn for three coils but its notation has been chosen to make it valid for c coils (any whole number of them), with the angular distance α between adjacent coils. The coil voltage phasors are $\mathbf{E}_1$, $\mathbf{E}_2$, etc., and their perpendicular bisectants are seen to intersect at a point 0. The distance from this point to the beginning or end of any coil voltage phasor is called b, and the phasor sum of all the coil voltages is called $\mathbf{E}$. The diagram shows that

$$E_1 = 2b \sin \frac{\alpha}{2} \tag{8-29}$$

and, for c coils,

$$E = 2b \sin \frac{c\alpha}{2} \tag{8-30}$$

Clearly, the magnitude of the phasor sum E is less than the algebraic sum cE_1. The ratio of these two values is, therefore, always less than unity, and this ratio is introduced as a correction factor called the *distribution factor* k_d. From the last two equations,

$$k_d = \frac{E}{cE_1} = \frac{\sin \frac{c\alpha}{2}}{c \sin \frac{\alpha}{2}} \tag{8-31}$$

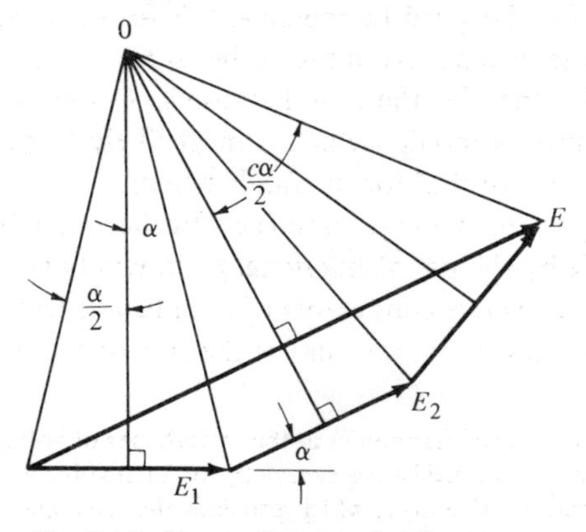

Fig. 8-16. Phasor diagram of coil voltages.

With the aid of this factor, the magnitude of the total steady-state voltage induced in a distributed winding consisting of c coils can be written in terms of the total number of turns of that winding

$$N = cN_c \tag{8-32}$$

The idea is to multiply the coil voltage of Eq. 8-24 by the number of coils c, which gives the algebraic sum, and then to correct for this error by means of the distribution factor:

$$E = 4.44\, Nk_d \phi f \tag{8-33}$$

In principle, slots can be distributed around the entire circumference of a rotor (or stator). For a *single-phase winding*, this means that the space angle $c\alpha$ can approach 180 electrical degrees. In that case, however, the distribution factor becomes rather low. It is not difficult to see why. A diagram like that of Fig. 8-16, but with the angle $c\alpha$ approaching 180 electrical degrees, would show that the voltages induced in the first and last coil of the winding are almost in phase opposition. Thus, together they contribute very little to the total voltage, at any rate not enough to justify their cost. Experience has lead to designs with slots covering only about two-thirds of the circumference as the most economical compromises. For *three-phase windings*, however, the whole problem cannot arise because, as the reader will see a little later, the angle $c\alpha$ cannot exceed 60 electrical degrees anyway.

Another aspect of the distribution factor refers to the waveshape of the induced voltage. The waveshape of the voltage induced in the entire winding is better (i.e., more nearly sinusoidal) than that of the coil voltage. To understand this, consider, for instance, the *third* harmonic of a coil voltage.* It is caused by the third harmonic of the flux distribution. This is a wave with three times as many peaks (poles) as the fundamental. So the angle α in electrical degrees is three times as much for the third harmonic as it is for the fundamental. For instance, two slots whose distance from each other is 60° for the fundamental, are 180° apart from each other for the third harmonic. Consequently, the distribution factor for the third harmonic (which is most likely to be the largest of all harmonics) is much less than that for the fundamental.

A further cleansing of this voltage wave (i.e., further reduction of its harmonics content) is possible by the use of *fractional-pitch* windings. In such windings, the distance between the active conductors of each turn is less than 180 electrical degrees. The voltage induced in any turn is the sum of the two conductor volt-

*Incidentally, there are no *even* harmonics in the waveshapes of voltages induced in rotating electric machines, due to the half-wave symmetry of such voltages. This, in turn, is the consequence of the symmetry of design, which provides the same magnetic circuit for each pole, north and south alike.

ages. For a full-pitch turn,

$$\mathbf{E}_{\text{turn}} = E_{\text{cond}} \underline{/0^\circ} - E_{\text{cond}} \underline{/180^\circ} = 2E_{\text{cond}} \tag{8-34}$$

For a pitch of $\beta < 180^\circ$, the turn voltage can be determined in a straightforward way, without recourse to geometry:

$$\mathbf{E}_{\text{turn}} = E_{\text{cond}} \underline{/0^\circ} - E_{\text{cond}} \underline{/\beta} = E_{\text{cond}} (1 - \cos\beta - j\sin\beta) \tag{8-35}$$

with a magnitude

$$E_{\text{turn}} = E_{\text{cond}} \sqrt{(1 - \cos\beta)^2 + (\sin\beta)^2} = E_{\text{cond}} \sqrt{2 - 2\cos\beta} \tag{8-36}$$

Using the trigonometric identity

$$1 - \cos 2x = 2\sin^2 x$$

the magnitude of the turn voltage becomes

$$E_{\text{turn}} = E_{\text{cond}} \sqrt{4\sin^2 \beta/2} = 2E_{\text{cond}} \sin\beta/2 \tag{8-37}$$

This leads to the definition and calculation of the *pitch factor*

$$k_p = \frac{E_{\text{coil}}}{2N_c E_{\text{cond}}} = \frac{E_{\text{turn}}}{2E_{\text{cond}}} = \sin\beta/2 \tag{8-38}$$

For reasonable values of β, this is only a few percent less than unity. But when it comes to the harmonics in the voltage wave, the angle β is multiplied by the order of the harmonic, and the pitch factors for the harmonics most prominent in the flux distribution are much smaller. Thus, the use of a fractional-pitch winding results in an induced voltage of slightly reduced magnitude but improved waveshape. Since such windings also offer some saving of material, due to the shorter end connections (i.e., the inactive conductors), they are used quite extensively. Examples 8-1 and 8-2 are typical illustrations of distribution factors and pitch factors for fundamentals and harmonics.

Distribution factor and pitch factor can be combined. Their product is called the *winding factor*

$$k_w = k_d k_p \tag{8-39}$$

and Eq. 8-33 is then modified to make it valid for any winding, full- or fractional-pitch:

$$E = 4.44\, N k_w \phi f \tag{8-40}$$

This equation is basic for all a-c generators and motors. The product of the turns number N and the winding factor is sometimes referred to as the effective number of turns.

8-7 THE ELECTROMAGNETIC TORQUE

The reader remembers that an electromagnetic force is exerted on a current-carrying conductor in a magnetic field. For currents flowing in the *axial* direction (i.e., parallel to the axis of rotation) in a field whose direction is *radial*, the direction of the force is *tangential*, and Eq. 7-6 ($F = Bli$) is applicable. For rotating motion, the pertinent quantity is the torque $T = Fr = Blir$, where the meaning of r (radius) is the same as explained in the context of Eq. 8-15.

The following derivation is for the case of one or more *distributed windings* on the rotor. The total torque exerted on the rotor is the sum

$$T = \sum (Blir) = lr \sum (Bi) \tag{8-41}$$

To allow for any kind of current distribution, use will be made of the concept of *angular current density* $J(\theta)$, introduced in Section 8-2. This method deals with infinitesinal current elements $J(\theta)\, d\theta$ instead of discrete currents, and it replaces the sum by the integral

$$T = lr \int_0^{p\pi} B(\theta)\, J(\theta)\, d\theta \tag{8-42}$$

It is understood that the space angle θ is expressed in electrical units, so that the space fundamentals of $B(\theta)$ and $J(\theta)$ become sinusoidal functions of θ, not of $(p/2)\theta$. That is why the upper limit of the integral is $p\pi$, which includes in the integration all conductors around the circumference of the rotor. The subscript $\mathcal{el}$ will be considered unnecessary and will be omitted for the remainder of this chapter.

Analytical expressions for $B(\theta)$ and $J(\theta)$ depend on the origin of the θ axis. The choice made for Eq. 8-9 will be used again, so

$$B(0) = B_{\text{peak}} \sin \theta \tag{8-43}$$

The other space function, $J(\theta)$, has its own peak value, and that is presumably not located at the same angular position as that of $B(\theta)$. It will be found useful later on to write

$$J(\theta) = J_{\text{peak}} \cos (\theta - \delta) \tag{8-44}$$

These two functions are substituted in the torque equation

$$T = lrB_{\text{peak}} J_{\text{peak}} \int_0^{p\pi} \sin \theta \cos (\theta - \delta)\, d\theta \tag{8-45}$$

Using the trigonometric conversion

$$\sin x \cos y = \tfrac{1}{2} [\sin (x + y) + \sin (x - y)] \tag{8-46}$$

the torque becomes

$$T = \frac{1}{2}\, lrB_{\text{peak}}J_{\text{peak}} \int_0^{p\pi} [\sin(2\theta - \delta) + \sin\delta]\, d\theta \tag{8-47}$$

Since p is a whole number, the first term becomes zero after the substitution of the integration limits. The remaining term is the result

$$T = \frac{\pi}{2}\, plrB_{\text{peak}}J_{\text{peak}} \sin\delta \tag{8-48}$$

It is also possible to introduce the *flux per pole* from Eq. 8-12 and thereby obtain an alternate form of the torque equation

$$T = \frac{\pi}{8}\, p^2 \phi J_{\text{peak}} \sin\delta \tag{8-49}$$

In either form, the torque equation shows, as might be expected, how the electromagnetic torque results from the interaction between the electric current and the magnetic flux. In addition, there is an all-important factor $\sin\delta$. What is the significance of the angle δ?

To answer this question, we first refer to Eq. 8-5 and thereby introduce the mmf wave into this discussion

$$\mathcal{F}(\theta) = \int J(\theta)\, d\theta = J_{\text{peak}} \sin(\theta - \delta) \tag{8-50}$$

Comparison between Eqs. 8-43 and 8-50 shows that δ is the angular distance between the space fundamentals of $\mathcal{F}(\theta)$ and $B(\theta)$. So there must be an angular difference between these two waves, or else there is no torque!

There are two possible reasons why such an angle may exist. It must be realized that $\mathcal{F}(\theta)$ is the mmf wave produced only by the *rotor* currents (because it was derived from $J(\theta)$, which was defined in this section as the angular density of the rotor currents), whereas $B(\theta)$ is the distribution of the actual flux, which, in general, is produced by rotor and stator currents *together.* In the absence of stator currents, the angle δ can exist only if the stator has *salient poles.*

Figure 8-17 illustrates this case. Diagram (a) shows the developed surfaces, with (for simplicity's sake) only one coil on the rotor. Diagram (b) is a sketch of the mmf wave and the flux distribution. The mmf wave has the rectangular shape familiar from Section 8-2 (see Fig. 8-6). The flux distribution, drawn as a broken line, would be similar to the mmf wave except for the effect of the salient poles. Between the stator poles, the air gap is so much larger than at the pole surfaces that it reduces the flux density to a small fraction of what it is at the pole surfaces.

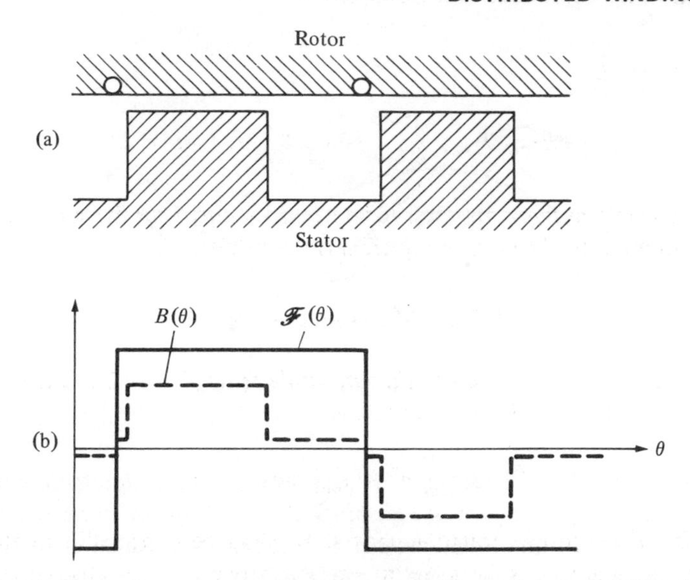

Fig. 8-17. (a) single coil on rotor, no conductors on salient-pole stator; (b) Mmf wave and flux distribution.

Now visualize (no need to calculate) the space fundamentals of the two waves. That of $\mathcal{F}(\theta)$ has its peaks midway between the two coil sides whereas that of $B(\theta)$ has its peaks further to the left (at 90° and 270°). So there is an angle and, thus, a torque. This is the ***reluctance torque*** encountered in Chapter 7, but here it is demonstrated without recourse to linear approximations, and, needless to say, not limited to the single coil of Fig. 8-17. This torque can also be viewed as the result of the attraction between the poles of the rotor flux and the poles of induced magnetism on the stator. The *direction* of the torque is such that it tends to line up these poles against each other, to turn the rotor toward the rest position in which $\delta = 0$ and, thus, $T = 0$.

The other possible reason for the existence of a torque angle δ is that there are currents on both the stator and the rotor, and that the space fundamentals of their mmf waves have their peak values at different angular positions. Figure 8-18 illustrates such a case. If there are no salient poles, the space fundamental of the flux distribution (not shown in the picture) will be in the same position as the *resultant* mmf wave; these two waves will differ only by a scale factor. Thus, the torque angle is the angular difference between the rotor mmf and the resultant mmf, as indicated in the diagram. If the two component mmf waves are in the same position, then the resultant mmf wave and the flux distribution are in the same position, too, and there is no angle δ, and, thus, no torque.

So the torque in the case illustrated by Fig. 8-18 can again be viewed as the result of the attraction between stator and rotor poles, as their tendency to "line

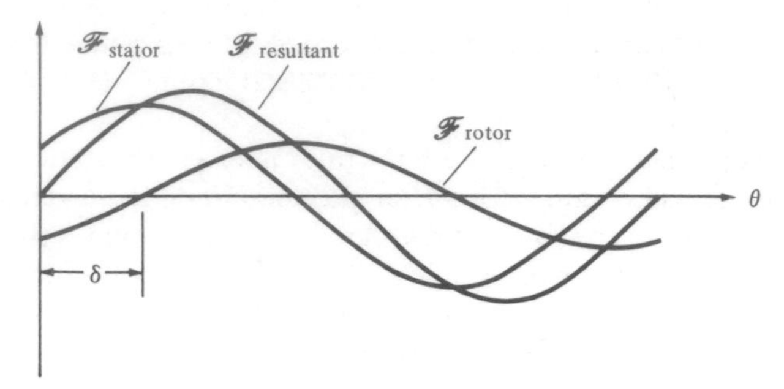

Fig. 8-18. Addition of mmf waves.

up" against each other. It can also be seen as the result of *interaction* of stator and rotor currents, or, as it was done in Section 7-8, on the basis of linear approximations, in terms of the mutual inductance as a function of the rotor position. Finally, a torque angle δ can also exist as a consequence of both causes (salient poles on the stator, and currents on both stator and rotor) acting simultaneously.

The derivation and discussion of the torque equation in this section was all done in terms of the *rotor* mmf wave and the flux. But what would have been the difference if the word *rotor* had consistently been replaced by the word *stator?* The result (Eqs. 8-48 and 8-49) would have been the same, but it would have represented the magnitude of the torque exerted on the *stator*, the angle δ being the space angle between the fundamentals of the flux distribution and the *stator* mmf wave. The stator cannot move but it exerts, by the principle of action and reaction, an equal and opposite torque on the rotor. In other words, the torque on the rotor can be expressed either in terms of the rotor mmf and its angular difference against the flux, or in terms of the stator mmf and *its* angular difference against the flux.

It must also be mentioned that the use of the *Bli* force equation on which all the calculation in this section is based, remains correct in spite of the fact that the conductors are located in slots where the flux density is only a small fraction of the value assigned to $B(\theta)$ in the air gap. The situation is similar to that of the use of the *Blu* form of Faraday's induction law (see the last two paragraphs of Section 8-5). For the electromagnetic torque, specifically, it has repeatedly been established that calculations based on *Bli* lead to the correct result, but that the torque so calculated is mostly exerted on the rotor *core,* not the conductors.

It is well worthwhile to take another look at the torque equations, and especially on the factor sin δ. (That angle δ, incidentally, is often referred to as the torque angle.) Since the torque depends on that angle that is the space difference between the rotor (or stator) mmf and the flux density, it may well be asked how any machine can ever operate at a constant steady-state torque. Does not the rotor mmf have to move relative to the stator mmf? Does this not continuously change the angle δ?

For one thing, in a d-c machine, both the rotor mmf and the stator mmf waves are stationary, as will be shown in a later chapter. Next, in polyphase a-c machines, both the stator and rotor currents produce rotating mmf waves, and that will be the subject of the next section. That leaves single-phase a-c machines, and in those, there is indeed no way for the torque and power to be other than pulsating, as the reader may remember from Section 6-2.

8-8 THE ROTATING FIELD

It is a property of *polyphase windings* that their currents, when they are balanced, produce a rotating mmf wave. The term *polyphase* is used because this statement is valid for any number of phases. Nevertheless, the following analysis is done only for the most important case, that of three phases.

A three-phase winding consists of three single-phase windings in which a symmetrical system of voltages is to be induced. Therefore, these three windings must have equal numbers of turns and equal winding factors (so the voltages will have equal magnitudes), they must move at constant angular velocity relative to a magnetic flux (so the voltages will have the same frequency), and they must be located 120 electrical degrees apart from each other (so the voltages will be 120° out of phase with one another). It is easy enough to see that such windings can also be placed on a stator and have their voltages induced by a rotating magnetic field produced by rotor currents. What is not obvious at all is that balanced currents in stationary windings themselves can also produce a rotating magnetic field.

In Fig. 8-19a, the location of the three windings on the stator is indicated. The symbols a and a' belong to conductors forming a full-pitch coil of phase a; similarly, b and b' represent phase b, etc. The windings may be presumed to be distributed, with the conductors drawn in the diagram located at the center of "phase belts" covering arcs of 60° each. The dots and crosses indicate assumed positive directions, like the familiar current arrows in circuit diagrams. All angles are in electrical units, so the diagram represents the fraction $2/p$ of the circumference.

Using the current in phase a as the axis of reference, and assuming the phase sequence to be $a - b - c$, the currents in the three windings can be written as

$$i_a = I_{\max} \cos \omega t$$

$$i_b = I_{\max} \cos (\omega t - 120°)$$

$$i_c = I_{\max} \cos (\omega t - 240°) \qquad (8\text{-}51)$$

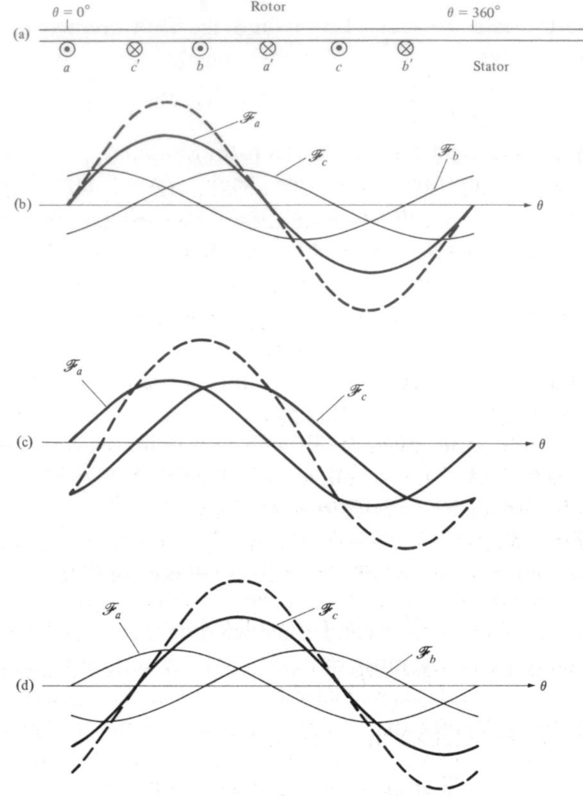

Fig. 8-19. (a) three-phase stator winding; (b) mmf wave at $t = 0$; (c) mmf wave at $t = \dfrac{\pi}{6\omega}$; (d) mmf wave at $t = \dfrac{\pi}{3\omega}$.

The magnitude of an mmf wave depends on the current that produces it. Therefore, an mmf wave produced by a time-varying current is a function of both the space angle θ and the time t, to be denoted $\mathfrak{F}(\theta, t)$. Specifically, at the instant $t = 0$, when the current i_a has its positive maximum value, the space fundamental of its mmf wave is

$$\mathfrak{F}_a(\theta, 0) = \mathfrak{F}_{\text{peak}} \sin \theta \tag{8-52}$$

because the location of the winding is such that the positive peak value of its mmf is located at $\theta = 90°$. This wave is drawn in Fig. 8-19b, together with the space fundamentals of the mmf waves of the other two phase currents for the same instant. These curves are obtained by shifting $\mathfrak{F}_a(\theta, 0)$ to the right by 120° and 240°, respectively, and multiplying its values by $(-\frac{1}{2})$ because, at $t = 0$, both

i_b and i_c have the value $(-\frac{1}{2} I_{max})$. Finally, Fig. 8-19b shows the *resultant* mmf wave for that instant

$$\mathcal{F}(\theta, 0) = \mathcal{F}_a(\theta, 0) + \mathcal{F}_b(\theta, 0) + \mathcal{F}_c(\theta, 0) \tag{8-53}$$

The procedure is repeated in Fig. 8-19c for the instant one-twelfth of a period later, i.e., when $\omega t = 30°$, or $t = \pi/6\omega$. At that instant, $i_a = (\sqrt{3}/2)I_{max}$, $i_b = 0$, and $i_c = (-\sqrt{3}/2)I_{max}$, and the diagram shows the three (actually two, since one of them is zero) component waves and the resultant wave for that instant. Again, after another one-twelfth of a period has passed, at $t = \pi/3\omega$, the current values are $i_a = \frac{1}{2} I_{max}$, $i_b = \frac{1}{2} I_{max}$, and $i_c = -I_{max}$, and Fig. 8-19d shows the corresponding mmf waves, components and resultant.

The procedure can be repeated for any other instant (the reader should try it for himself, e.g., for $t = \pi/2\omega$), leading to the observation that the resultant mmf wave always has the same magnitude, and that it moves steadily to the right, covering the space angle 2π in a full period. That this observation is indeed correct, will now be proved by an *analytical* method.

Return to Eq. 8-52, which is the mmf wave $\mathcal{F}_a$ at the instant $t = 0$, when i_a has its positive maximum value. At all times, $\mathcal{F}_a$ is proportional to i_a. Thus

$$\mathcal{F}_a(\theta, t) = \mathcal{F}_{peak} \sin \theta \cos \omega t \tag{8-54}$$

It is worthwhile to stop at this point, and to discuss this equation in some detail:

(a) For any specific value of time, say at $t = t^*$,

$$\mathcal{F}_a(\theta, t^*) = \mathcal{F}_{peak} \cos \omega t^* \sin \theta \tag{8-55}$$

This is an instantaneous picture, a *snapshot* (like the curves in Fig. 8-19). It is seen to be a sinusoidal function of θ whose peak value depends on the choice of the instant t^* but whose position remains the same at all times.

(b) For any specific value of the angle, say for $\theta = \theta^*$

$$\mathcal{F}_a(\theta^*, t) = \mathcal{F}_{peak} \sin \theta^* \cos \omega t \tag{8-56}$$

This is what a *stationary observer*, placed at the location θ^*, would record. It is a sinusoidal time function whose amplitude depends on the chosen location of the observer, but whose phase angle is the same at any location.

A function of time and space represented by an equation like Eq. 8-54 is known as a *standing wave.* It is characterized by the fact that its peak values always remain at the same location; in the case under study, at $\theta = 90°$ and $\theta = 270°$.

The other two currents produce similar standing mmf waves, with the same peak values, but displaced in space. The peaks of $\mathcal{F}_b(\theta, t)$ are located at $\theta = 210°$ and $\theta = 30°$, displaced by $120°$ against those of $\mathcal{F}_a(\theta, t)$. (Note that $30°$ has been written in lieu of $390°$.) The peaks of $\mathcal{F}_c(\theta, t)$ are located another

120° to the right, at $\theta = 330°$ and $\theta = 150°$. The equations of these two waves are

$$\mathcal{F}_b(\theta, t) = \mathcal{F}_{\text{peak}} \sin(\theta - 120°) \cos(\omega t - 120°) \tag{8-57}$$

and

$$\mathcal{F}_c(\theta, t) = \mathcal{F}_{\text{peak}} \sin(\theta - 240°) \cos(\omega t - 240°) \tag{8-58}$$

The resultant mmf wave is thus

$$\mathcal{F}(\theta, t) = \mathcal{F}_{\text{peak}} [\sin\theta \cos\omega t + \sin(\theta - 120°)\cos(\omega t - 120°) + \sin(\theta - 240°)\cos(\omega t - 240°)] \tag{8-59}$$

With the aid of the trigonometric identy

$$\sin x \cos y = \tfrac{1}{2}[\sin(x + y) + \sin(x - y)] \tag{8-60}$$

this can be put into the form

$$\mathcal{F}(\theta, t) = \tfrac{1}{2}\mathcal{F}_{\text{peak}} [\sin(\theta + \omega t) + \sin(\theta - \omega t) + \sin(\theta + \omega t - 240°) + \sin(\theta - \omega t) + \sin(\theta + \omega t - 480°) + \sin(\theta - \omega t)] \tag{8-61}$$

Of the six terms inside the bracket, the three written on the left side form a symmetrical set of sinuosoids whose sum is zero. (The situation is similar to that encountered in the derivation of three-phase power in Section 6-4.) That leaves the other three terms, and so the final result is

$$\mathcal{F}(\theta, t) = \tfrac{3}{2}\mathcal{F}_{\text{peak}} \sin(\theta - \omega t) \tag{8-62}$$

This, again, deserves a closer look:

(a) The *snapshot*, at the instant $t = t^*$:

$$\mathcal{F}(\theta, t^*) = \tfrac{3}{2}\mathcal{F}_{\text{peak}} \sin(\theta - \omega t^*) \tag{8-63}$$

This is a sinusoidal function of θ whose position changes if the chosen instant t^* is changed, but whose peak value always remains the same. (Note how different the properties of the snapshot of the standing wave are.) Some such snapshots are drawn in Fig. 8-19.

(b) The *stationary observer* at the position $\theta = \theta^*$:

$$\mathcal{F}(\theta^*, t) = \tfrac{3}{2}\mathcal{F}_{\text{peak}} \sin(\theta^* - \omega t) \tag{8-64}$$

which can also be written

$$\mathcal{F}(\theta^*, t) = -\tfrac{3}{2}\mathcal{F}_{\text{peak}} \sin(\omega t - \theta^*) \tag{8-65}$$

This is a sinusoidal time function whose phase angle depends on the position of the observer, but whose amplitude is the same regardless of that location. (Again, note the contrast to the standing wave.)

(c) It is also instructive to check what a *moving observer*, traveling at the constant angular velocity ω, would record. The location of such as observer is

$$\theta = \theta_0 + \omega t \tag{8-66}$$

and so his observation is

$$\mathcal{F}(\theta_0 + \omega t, t) = \tfrac{3}{2} \mathcal{F}_{\text{peak}} \sin \theta_0 \tag{8-67}$$

which is a constant! It depends only on the initial location θ_0 of this observer who remains always in the same position relative to the wave.

Many readers have known before that Eq. 8-62 describes a *traveling* wave. Standing waves and traveling waves occur in many fields of science and engineering. Where such waves are encountered in the study of electric motors and generators, the space variable is not distance but angle; the waves travel in circles, they are *rotating* waves (with the exception of the "linear" motors mentioned at at the end of Section 7-6).

Some further observations can be made either from Eq. 8-62 or from the curves of Fig. 8-19:

(a) This rotating field happens to move in the positive *direction* (from left to right on the diagrams). It would travel in the opposite direction if either the phase sequence of the three currents or the space sequence of the three windings were reversed.

(b) The *speed* of rotation is very significant. Since the field travels the angular distance of 2π electrical radians during each period, its angular velocity is ω electrical radians per second. It is said to rotate at *synchronous speed.* It is a benefit of the use of electrical units that the same symbol stands for both the radian frequency of the currents and the angular velocity of the field.

(c) The *location* of the peak of the traveling wave at the instant when the current of one phase has its positive maximum value is the same as the location of the peak of the standing wave of that phase.

(d) Last, but not least: the *peak value.* In contrast to the other properties of the rotating field, this one depends on the *number of phases.* For three phases, Eq. 8-62 as well as the diagrams of Fig. 8-19 show that the peak value of the resultant mmf wave is $\frac{3}{2}$ times the maximum peak value of each of the standing phase mmf waves. (The peak value of a standing wave is time-varying. *Maximum peak value* means the peak value at the instant when it is largest.)

In the case of *two phases*, which will also be of some interest in later chapters, the phase currents are 90° out of phase with each other, and their windings must be placed 90 electrical degrees apart from each other in space. Both the graphical approach and the analytical derivation are much simpler then for three phases, and the reader is advised to try them both. The result is that the peak value of the resulting rotating mmf wave equals the maximum peak value of the standing mmf wave of each phase.

For *more than three* phases, diagrams like those of Fig. 8-19 become more and more crowded and difficult to draw, but the analytical procedure works well. For m phases ($m \geqslant 3$), there is a phase difference of $2\pi/m$ between adjacent phase currents, and the same angle in electrical units between adjacent phase windings. The equations of the standing mmf waves for each phase look like Eqs. 8-54, 8-57, and 8-58 except for the angles. There are m such equations, and they can be added and the sum written in the form of Eq. 8-61 i.e., with a bracket containing $2m$ terms arranged in m lines. The terms on the left side again add up to zero, and the others are all equal, which leads to the result

$$\mathfrak{F}(\theta, t) = \frac{m}{2} \mathfrak{F}_{\text{peak}} \sin(\theta - \omega t) \tag{8-68}$$

which is valid for two, three, or any other number of phases.

Having completed this section, the reader is now ready to study the principles and operating characteristics of actual generators and motors.

8-9 EXAMPLES

Example 8-1 (Section 8-6)

A three-phase machine has nine slots per pole. A fractional-pitch winding (with $\beta/180^\circ = 7/9$) is distributed in three adjacent slots for one phase. Find the distribution factor and the pitch factor for this winding.

Solution

The location of the coil sides is shown in Fig. E-8-1. One of the coils has its sides in slot 1 and in slot 8. Since nine slots correspond to 180°, the angle between adjacent slots is

$$\alpha = 180^\circ/9 = 20^\circ$$

The number of adjacent slots used for one phase is $c = 3$. The distribution factor is found from Eq. 8-31.

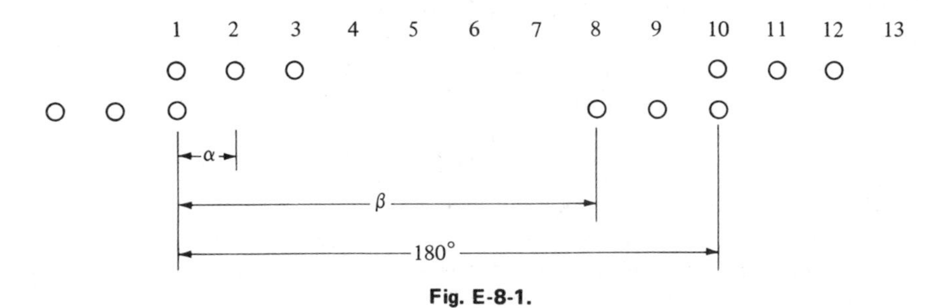

Fig. E-8-1.

$$k_d = \frac{\sin (c\alpha/2)}{c \sin (\alpha/2)} = \frac{\sin (3 \times 20^\circ/2)}{3 \sin (20^\circ/2)} = 0.960$$

The pitch of the coil is $\beta = (\frac{7}{9})(180^\circ) = 140^\circ$. The pitch factor is found from Eq. 8-38

$$k_p = \sin \beta/2 = \sin 70^\circ = 0.940$$

Example 8-2 (Section 8-6)

For the winding in Example 8-1, find the distribution factor and the pitch factor for the fifth harmonic.

Solution

For the harmonic of the order h, the angles α and β are replaced by $h\alpha$ and $h\beta$, respectively. Thus, in our example, the angle to be used in place of α is $5 \times 20^\circ = 100^\circ$, and that in place of β is $5 \times 140^\circ = 700^\circ$.

The distribution factor for the fifth harmonic is

$$k_{d5} = \frac{\sin (3 \times 100^\circ/2)}{3 \sin (100^\circ/2)} = 0.218$$

The pitch factor for the fifth harmonic is

$$k_{p5} = \sin (700^\circ/2) = -0.174$$

Note that both factors are much smaller than for the fundamental

Example 8-3 (Section 8-5)

The machine in Fig. E-8-3 has a cylindrical rotor with radius of 0.11 m and length (into the page) of 0.26 m. The distance across the air gap is 0.003 m. The reluc-

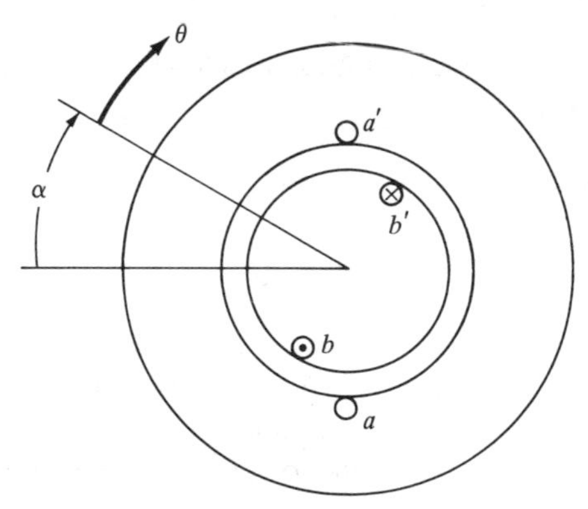

Fig. E-8-3.

tance of the steel parts may be neglected. The stator coil, $a - a'$, has 80 turns. The rotor coil, $b - b'$, has 200 turns. Assume the rotor mmf produces a sinusoidally distributed flux density in the air gap. The rotor turns clockwise at a speed of 3600 rpm (i.e., $\alpha = 377t$). Let the rotor current, i_b, be 15 amp (d-c) with the polarity shown. Find the voltage function in the open-circuited stator coil. Find the rms value of the voltage.

Solution

The mmf in the air gap has the rectangular wave shape of Fig. 8-6. The peak mmf of the sinusoid for the fundamental component is

$$\mathcal{F}_{\text{peak}} = (4/\pi)\,(N_c i/2)$$

The corresponding peak value of the flux density is found by using the magnetic circuit properties of the air gap

$$B_{\text{peak}} = \left(\frac{\mu_0}{g}\right) \mathcal{F}_{\text{peak}} = \left(\frac{4\pi \times 10^{-7}}{0.003}\right)\left(\frac{4}{\pi}\right)\left(\frac{200 \times 15}{2}\right) = 0.8 \text{ weber/m}^2$$

$$\phi = \frac{4lr}{p} B_{\text{peak}} = \frac{4(0.26)(0.11)}{2}\,(0.8) = 0.0458 \text{ weber}$$

The flux linkages of the stator coil are given by

$$\lambda = N_a \phi \cos\alpha = 80 \times 0.0458 \cos 377t = 3.66 \cos 377t$$

The instantaneous value of voltage in winding a is

$$e_{a'a} = d\lambda/dt = -1380 \sin 377t$$

$$e_{aa'} = 1380 \sin 377t \text{ v}$$

The rms value of this voltage is

$$E_{\text{rms}} = E_{\text{max}}/\sqrt{2} = 976 \text{ v}$$

We could also find the rms value of voltage by using Eq. 8-24

$$E = 4.44\, N_c \phi f = 4.44 \times 80 \times 0.0458 \times 60 = 976 \text{ v}$$

Example 8-4 (Section 8-7)

A 60-Hz, six-pole, a-c motor has a voltage of 550 v (rms) generated in its armature winding. The winding has 52 effective turns. When this motor develops 15 kw of power, the amplitude of the sinusoidal field mmf is 460 At. (a) Find the value of the resultant air-gap flux per pole. (b) Find the angle between the flux wave and the mmf wave.

Solution

(a) Use Eq. 8-24 to find the flux per pole

$$\phi = E/(4.44\, N_c f) = 550/(4.44 \times 52 \times 60) = 0.0397 \text{ weber}$$

(b) Use Eq. 8-17 to find the mechanical speed

$$\omega_{\text{mech}} = (2/p)\, \omega_{e\ell} = (2/6)(2\pi\, 60) = 125.7 \text{ rad/sec}$$

Find the mechanical torque from power and speed

$$T = P/\omega_{\text{mech}} = 15{,}000/125.7 = 119 \text{ nm}$$

Use Eq. 8-50 to observe that

$$J_{\text{peak}} = \mathcal{F}_{\text{peak}} = 460 \text{ amp/rad}$$

Use Eq. 8-49 to find the torque angle

$$\sin \delta = T/\left(\frac{\pi}{8} p^2\, \phi\, J_{\text{peak}}\right) = 119/\left(\frac{\pi}{8} \times 6^2 \times 0.0397 \times 460\right) = 0.46$$

$$\delta = \sin^{-1}(0.46) = 27.4°$$

As this machine is a motor, Fig. 8-18 could represent these conditions provided the waves are moving toward the left. The rotor mmf lags by 27.4 electrical degrees behind the resultant mmf.

Example 8-5 (Section 8-8)

Figure E-8-5 shows a two-pole, three-phase machine with a cylindrical rotor. The labeled current directions indicate the directions for positive values of current. The current functions are: $i_a = I_m \cos \omega t$, $i_b = I_m \cos(\omega t - 120°)$, $i_c = I_m \cos(\omega t - 240°)$. Find the magnitude and the location of the resultant mmf for $\omega t = 0$ and for $\omega t = \pi/6$.

Solution

As the mmf produced by each winding is a sinusoidal space function, an interpretation based on phasor arithmetic can be used. Draw lines for the axes of the three windings. When a current is positive, the corresponding phase mmf is oriented in the axis of that phase. Construct three vectors on each axis respectively. Find the three currents and mmfs for $\omega t = 0$

$$i_a(0) = I_m \cos(0) = +I_m \qquad \mathcal{F}_a = +\mathcal{F}_{\text{peak}}$$

$$i_b(0) = I_m \cos(-120°) = -\tfrac{1}{2} I_m \qquad \mathcal{F}_b = -\tfrac{1}{2}\mathcal{F}_{\text{peak}}$$

$$i_c(0) = I_m \cos(-240°) = -\tfrac{1}{2} I_m \qquad \mathcal{F}_c = -\tfrac{1}{2}\mathcal{F}_{\text{peak}}$$

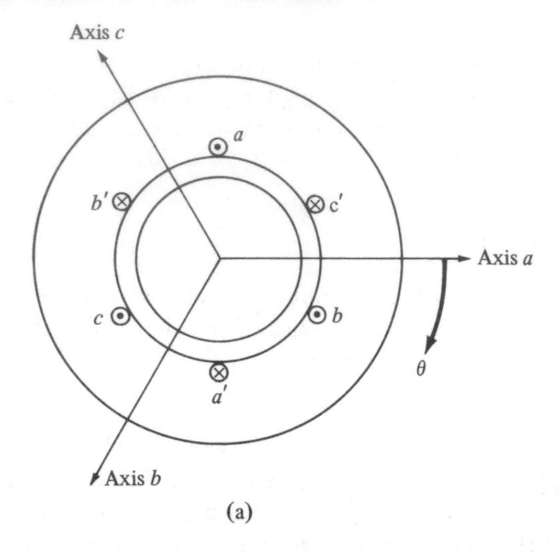

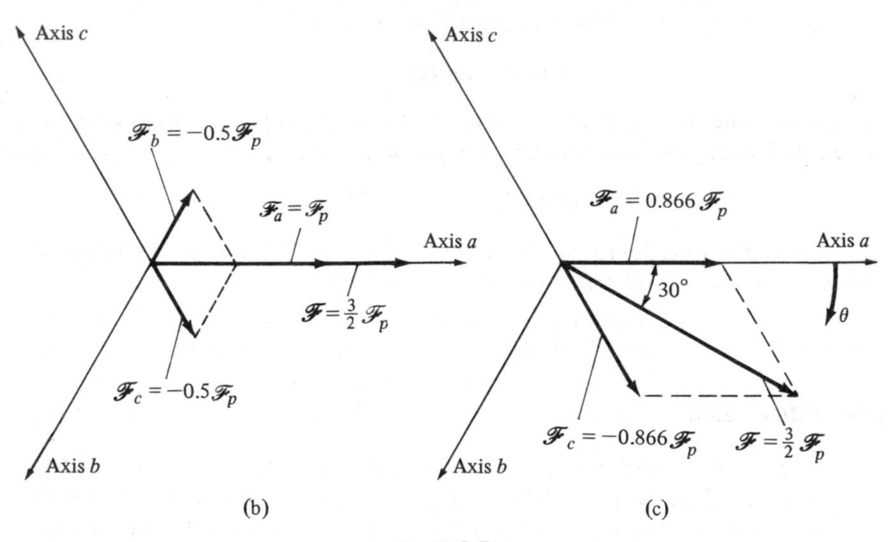

Fig. E-8-5.

Draw $\mathfrak{F}_a$, $\mathfrak{F}_b$, $\mathfrak{F}_c$ on the respective axes. See Fig. E-8-5b. Also refer to Fig. 8-19b. This shows the magnitude is $(\frac{3}{2})\ \mathfrak{F}_p$ and the peak is located at axis a.

Next, for $\omega t = \pi/6$, proceed in similar manner. Find the three currents and mmfs

$$i_a(\pi/6) = I_m \cos(30°) = +0.866\, I_m \qquad \mathfrak{F}_a = +0.866\, \mathfrak{F}_p$$

$$i_b(\pi/6) = I_m \cos(30° - 120°) = 0 \qquad \mathfrak{F}_b = 0$$

$$i_c(\pi/6) = I_m \cos(30° - 240°) = -\ 0.866\, I_m \qquad \mathfrak{F}_c = -0.866\, \mathfrak{F}_p$$

Draw $\mathcal{F}_a$, $\mathcal{F}_b$, $\mathcal{F}_c$ on the respective axes. See Fig. E-8-5c. Also see Fig. 8-19c. This shows the magnitude is $(\frac{3}{2})\,\mathcal{F}_p$ and the peak is displaced by 30° from axis a toward axis b. This direction of movement is in agreement with the phase sequence, $a - b - c$.

Example 8-6 (Section 8-8)

The machine in Fig. E-8-5a has an mmf in the air gap given by $\mathcal{F}(\theta, t) = (\frac{3}{2})\,\mathcal{F}_p \cos(\theta - \omega t - 40°)$. Find the symmetrical three-phase currents that could produce this wave. Use I_m in your answer.

Solution

Since $(\theta - \omega t)$ appears in the equation, we know the wave rotates clockwise and the phase sequence is $a - b - c$. At $t = 0$, the peak of the wave is located at $\theta = 40°$. The peak of the wave is at axis a for $\omega t = -40°$. This is the instant of time that current i_a has its positive maximum. We can write it as

$$i_a(t) = I_m \cos(\omega t + 40°)$$

The peak of the wave is at axis b for $\omega t = +80°$. This is the instant of time that current i_b has its positive maximum. We can write it as

$$i_b(t) = I_m \cos(\omega t - 80°)$$

The peak of the wave is at axis c for $\omega t = +200°$. This is the instant of time that current i_c has its positive maximum. We can write it as

$$i_c(t) = I_m \cos(\omega t - 200°)$$

8-10 PROBLEMS

8-1. A three-phase machine has twelve slots per pole. One phase of a fractional pitch winding (with $\beta/180° = 11/12$) is distributed in four adjacent slots. Find the distribution factor and the pitch factor for (a) the fundamental component, (b) the fifth harmonic, and (c) the seventh harmonic.

8-2. A three-phase machine has fifteen slots per pole. One phase of a fractional pitch winding (with $\beta/180° = 13/15$) is distributed in five adjacent slots. Find the distribution factor and the pitch factor for (a) the fundamental component, (b) the fifth harmonic, and (c) the seventh harmonic.

8-3. A cylindrical rotor machine in Fig. E-8-3 has a radius of 0.17 m and length (into the page) of 0.37 m. The distance across the air gap is 0.0032 m. The reluctance of the steel parts may be neglected. The stator coil, $a - a'$, has 47 turns. The rotor coil, $b - b'$, has 210 turns. Assume the rotor mmf produces a sinusoidally distributed flux density in the air gap. The rotor turns clockwise at a speed of 3000 rpm (i.e., $\alpha = 314t$). The rotor current is $i_b = +16$ amp (d-c). Find the voltage function in the stator coil ($v_{aa'} = ?$).

8-4. The three-phase generator in Fig. P-8-4 has flux per pole of 0.08 weber. Each stator coil has six turns. The rotor is driven clockwise at a speed of 3600 rpm. Choose zero time as the instant when the north pole is at conductor a, as shown. (a) Find the frequency in Hertz. (b) Find the voltage function of each winding. (c) Find the phase sequence. (d) The three stator windings are connected in wye, with a, b, and c tied together. Find all line-to-line voltage functions.

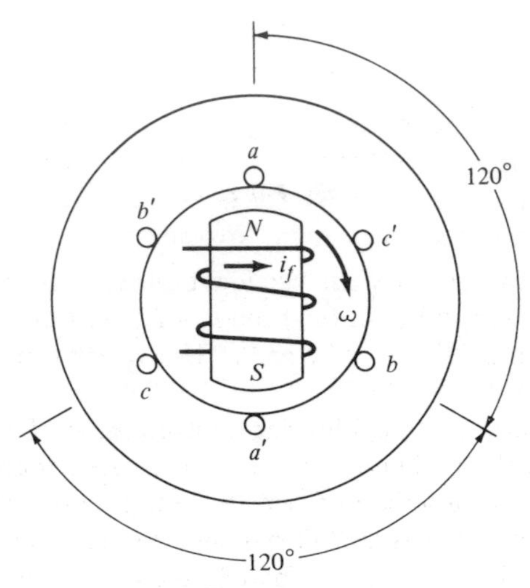

Fig. P-8-4.

8-5. A 50-Hz, eight-pole, a-c motor has a voltage of 760 v (rms) generated in its armature winding. The winding has 56 effective turns. When this motor develops 32 kw of power, the angle between the sinusoidal flux wave and the sinusoidal mmf wave is 33 electrical degrees. (a) Find the value of the resultant air gap flux per pole. (b) Find the amplitude of the mmf wave.

8-6. Solve Example 8-5, for $\omega t = \pi/2$.

8-7. For the machine in Fig. E-8-5a, let the currents be: $i_c = I_m \cos(\omega t)$, $i_b = I_m \cos(\omega t - 120°)$, and $i_a = I_m \cos(\omega t - 240°)$. Find the resultant mmf as a function of θ and t.

8-8. The machine in Fig. E-8-5a has an mmf in the air gap given by: $\mathcal{F}(\theta, t) = (\frac{3}{2})\ \mathcal{F}_p \sin(\theta - \omega t + 60°)$. Find the symmetrical three-phase currents that could produce this wave. Use I_m in your answer.

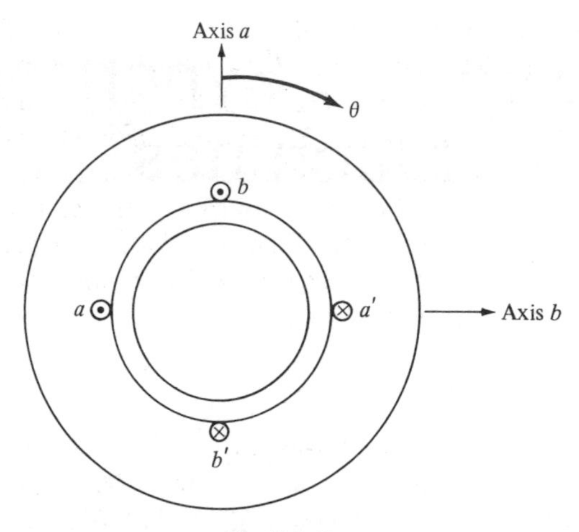

Fig. P-8-9.

8-9. The machine in Fig. P-8-9 has two identical stator coils located 90° apart. The currents are $i_a = I_m \cos(\omega t)$ and $i_b = I_m \cos(\omega t - 90°)$. Find the resultant mmf as a function of θ and t. Use $\mathcal{F}_p$ as the maximum peak mmf of one phase.

8-10. A four-pole rotor has the following dimensions: axial length 0.6 m, radius 0.4 m. The flux distribution at the rotor surface has the space fundamental $B(\theta) = 1.25 \sin\theta$ weber/m^2, where the angle θ is expressed in electrical units. The rotor rotates at 900 rpm and, for each pair of poles, it carries a full-pitch coil of 50 turns. The two coils are connected in series. (a) Find the total flux linkages of the rotor circuit as a function of the rotor position (expressed by the electrical angle θ in such a way that, at the position $\theta = 0$, each coil links the entire flux per pole). (b) From the result of part (a), find the voltage induced in the rotor circuit as a function of the rotor position. (c) Find the velocity of each rotor conductor in m/sec. (Assume the conductors are at the rotor surface.) (d) Find the voltage induced in the rotor circuit by means of the *Blu* equation.

9

Three-Phase Synchronous Machines

9-1 FIELD EXCITATION

Among the various types of electromagnetic energy-converters to be studied in this book, the synchronous machine is the one that embodies the idea of *relative motion* between electric conductors and a magnetic field in its purest form. This machine also has the distinction of being practically the only type of *generator* for major power systems, in addition to being widely used as a *motor*.

The model device depicted in Fig. 7-13 may be used as a starting point. It shows how an alternating voltage is induced in an *armature* rotating in a stationary magnetic field of constant magnitude. To produce such a field, there is the *field winding* carrying a constant *field current* (the popular abbreviation d-c, standing for *direct current* is applicable), which is, in general, supplied from a separate *auxiliary* source.

The need for a constant current to provide the *excitation* of the magnetic circuit is characteristic of the synchronous machine, whether it is operating as a generator or as a motor. The auxiliary source has the function of supplying the magnetic field energy, but, once a steady state is reached in the field circuit, any further expenditure of energy in this circuit merely covers the loss caused by its resistance. There is no difficulty in designing the field winding in such a way that its steady-state power consumption is only a small fraction of the power rating of the machine itself.

The field circuit can be energized from any d-c source, such as a battery, a d-c generator, or rectified a-c. For instance, many power plants have prime movers that drive both their synchronous generators and their auxiliary d-c generators (called *exciters*) on the same shaft. Another possibility, steadily gaining in importance (particularly for the field supply of synchronous motors), is the use of *rectifiers* converting the available a-c supply into d-c. Both d-c generators and rectifiers lend themselves well to adjustment of their output, which is very useful for control of operation of synchronous machines, as will be seen in later chapters.

Another look at the model of Fig. 7-13 shows the need for some connection between the rotating armature coil and the stationary outside circuit (containing

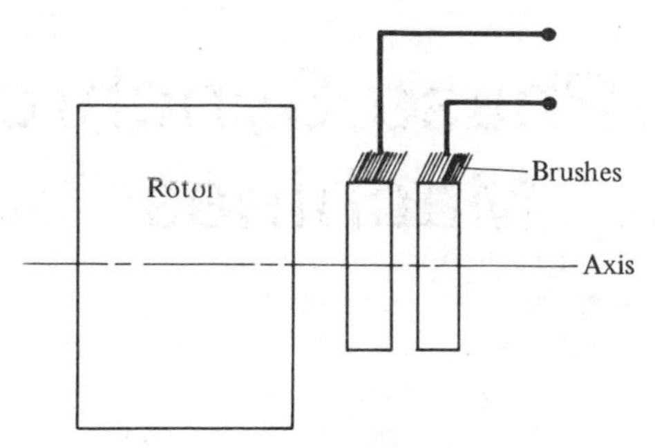

Fig. 9-1. Slip rings.

the load in the case of a generator, the source in the case of a motor). Such a *sliding contact* can be obtained by means of *slip rings*, a common enough construction element of several types of rotating electric machines. Figure 9-1 is a sketch intended to describe the idea. The slip rings have blank metallic cylindrical surfaces that are connected to the beginning and end of the rotor winding (or windings); they are attached to the shaft so that they rotate with the rotor (in fact, they may be considered to be part of the rotor). The *brushes* are stationary pieces of conducting material (either metal or carbon), pressed against the slip rings by means of springs in order to ensure a good contact. Inevitably, the contact is never perfect (a voltage drop on the order of magnitude of 1 v is attributed to the *contact resistance*), and there is also some power lost due to the *friction* between each brush and its ring.

The need for an auxiliary source and slip rings is clearly a drawback of the synchronous machine. Is there no way to avoid them? There is, but at a cost that makes it worthwhile only for exceptional cases. The part of the magnetic circuit that carries the field winding can be replaced by a *permanent magnet.* But flux densities obtainable from electromagnets are much higher than from permanent magnets. Thus, the presence of a field circuit raises the possible rating of a machine of given size considerably.

Another possibility, again important only for special cases, has to be mentioned in this context. The synchronous machine is capable of operating as a singly excited device, on the principle presented in Section 7-4. This subject will be discussed, together with the effect of salient poles, in a later chapter.

9-2 STATOR AND ROTOR

Since it is the *relative* motion between armature and field that matters, the functions of stator and rotor can be interchanged. If the design of a synchronous machine were based on the model of Fig. 7-13, the armature winding would be on the rotor, and the field winding on the stator. In such a machine, the magnetic field would be stationary.

Actually, synchronous machines are practically always built the other way: the armature winding on the stator, the field winding on the rotor, the magnetic field rotating. There are good reasons for this choice. It must be kept in mind that the armature circuit is the main circuit, the one at whose terminals electric power is delivered to the load or obtained from the source, the one in which the voltage is induced; briefly, then, the one in which the power conversion takes place. The field circuit, no matter how vitally important, remains an *auxiliary* circuit, whose voltage is likely to be a small fraction of that of the armature circuit.

It would be unnatural, against all good engineering judgment, to have the conductors of the main circuit located on the rotor, where they would be subject to centrifugal forces and vibration, and where it would be harder to find room for them and their insulation.

Another point to be considered is that of the slip rings. They are the connection between whichever winding is located on the rotor and its outside circuit. If there must be such a sliding connection, it is preferable to have it in the auxiliary rather than in the main circuit. Also, whereas a three-phase armature on the rotor would require at least three slip rings, the field winding on the rotor needs only two. It is even possible to avoid the use of slip rings entirely: this requires the use of an auxiliary a-c generator (*exciter*) with a rotating armature and of a rotating set of rectifiers, all on the same shaft (see Section 11-10).

So the machine to be studied in this and the next few chapters has, in its usual form, a rotor carrying a winding that is *excited* from some d-c source. Figure 9-2 shows an example of such a rotor, this one with two salient poles. In this diagram, the poles are drawn somewhat more realistically than in previous sketches (e.g., Fig. 8-11). The magnetic circuit is widened at the rotor poles, in order to reduce the reluctance of the air gap. The *pole shoes* thereby obtained are also useful in helping to keep the field winding (cross-hatched in the figure) in its place.

When the rotor rotates, its mmf wave rotates with it, without changing its shape. So there is a travelling *rotor mmf wave* whose space fundamental can be described by the equation

$$\mathfrak{F}_r(\theta, t) = \mathfrak{F}_{r\,\text{peak}} \cos(\theta - \omega t) \tag{9-1}$$

where the angle θ is counted in the direction of rotation, with the origin assigned to the location of the pole axis at the time $t = 0$. The symbol ω stands for the angular velocity of the rotor. In the case of more than two poles, the equation remains correct provided both the angle θ and its rate of change ω are expressed in electrical units, so that the distance between adjacent pole axes is always counted as $180°$, or π radians.

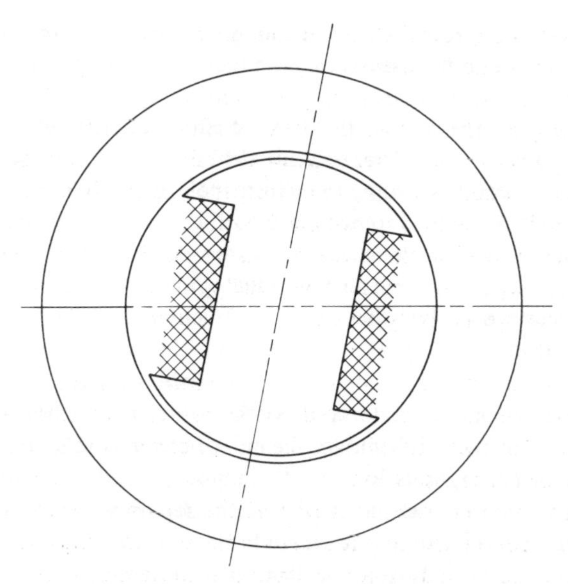

Fig. 9-2. Machine with salient-pole rotor.

When the stator conductors carry currents, they produce a *stator mmf wave.* As the reader knows, currents in a single-phase armature winding produce a standing mmf wave. But this can never lead to a constant steady-state torque. It was shown in Section 8-7 that such a torque requires a constant space angle between the flux distribution and the mmf wave of either the stator or the rotor. In other words, it requires that the two component mmf waves be stationary with respect to each other and thereby to the resultant mmf wave and to the flux distribution. So it is only the *rotating* mmf wave of *polyphase* armature currents that interacts with the rotor mmf wave to produce a constant torque.

It is true that single-phase synchronous machines have their place in the world of power-converting devices. They have no constant torque, no uniform flow of power. Their torques are pulsating, and they work because these torques can have nonzero average values. Their analysis is more difficult than that of polyphase machines, and will be left to a much later chapter. This and the following chapters will be devoted to the study of *three-phase* machines (other numbers of phases being much less important and no different in the methods of treatment).

The picture in our minds' eyes should be, therefore, a picture of two rotating mmf waves (stator and rotor) that combine to form a *resultant* mmf wave. This is the one that produces the rotating flux whose value may be substituted into the voltage equation, Eq. 8-40, or the torque equation, Eq. 8-49.

The magnitude of the flux produced by a given mmf depends entirely on the reluctance of the magnetic circuit. To produce a constant flux, the resultant mmf wave must, therefore, always encounter the same reluctance. This eliminates any thoughts about possible salient poles on the *stator:* the stator must be cylindrical.

For the *rotor*, on the other hand, there is a choice: it may either be cylindrical or have salient poles, because the rotating field is, in the steady state, motionless relative to the rotor. Basically, the salient-pole rotor is the preferred design because it contributes a reluctance term to the total torque (see Eq. 7-38 and Section 7-8), and also because salient poles are helpful for the dissipation of heat by convection. The only reason for using cylindrical rotors arises when the centrifugal forces would be too strong for salient poles. That is the case with large, high-speed machines; practically, it means the case of generators driven by steam turbines.

The fact that the steady-state flux is moving relative to the stator but not relative to the rotor has a bearing on the choice of the core material. The stator must be made of *laminations* in order to reduce the eddy-current loss (see Section 4-4). But the rotor may be made of *solid* ferromagnetic material, which is much less expensive than laminations. Only the portion of the rotor core closest to the air gap is frequently made of laminations, because of the harmonics of the air gap flux distribution, which are due mainly to the slots and teeth of the stator.

9-3 SYNCHRONOUS SPEED

The most characteristic property of the synchronous machine, the one to which it owes its name, is the rigid relationship between the rotating speed of the machine and the frequency of the voltages and currents in the armature.

Figure 9-3 is a developed view of a three-phase stator, similar to that of Fig. 8-19a except that the single coils are replaced by distributed windings, occupying the indicated spaces. The diagram covers 360 electrical degrees, or the fraction $2/p$ of the stator circumference.

To induce voltages of the phase sequence $a - b - c$ in these windings, the flux must travel from left to right, so that the voltage induced in phase b lags that

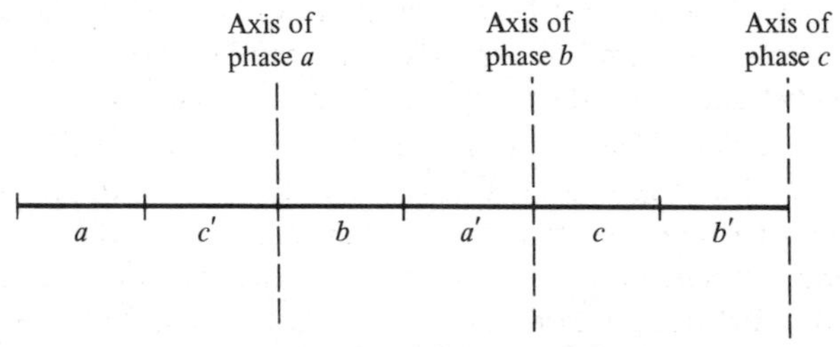

Fig. 9-3. Three-phase armature.

induced in phase a by 120°, etc. The time it takes the flux to travel 360 electrical degrees equals one period of induced voltage. The same relation exists between the stator currents and the direction and speed of their mmf wave. As the reader remembers from Section 8-8, balanced currents of phase sequence $a - b - c$ in a stator winding arranged like that of Fig. 9-3 produce an mmf wave traveling from left to right, and of such a speed that it covers 360 electrical degrees during one current period.

In other words, the flux distribution and the stator mmf both travel in the same direction and at the same speed, and so, consequently, does the resultant mmf wave and the rotor mmf wave. In terms of the frequency f (in Hertz), this speed, known as the *synchronous speed*, is

$$\omega_s = 2\pi f \text{ electrical radians per second} \tag{9-2}$$

or

$$\omega_s = \frac{2}{p} 2\pi f = \frac{4}{p} \pi f \text{ mechanical radians per second} \tag{9-3}$$

or, using the conventional unit of revolutions per minute,

$$n_s = \frac{60}{2\pi} \frac{2}{p} 2\pi f = \frac{120 f}{p} \text{ rpm} \tag{9-4}$$

Since this is the speed of the rotor mmf, it is also the speed of the rotor itself. Synchronous machines (in the steady state) *must* run at synchronous speed, in contrast to induction machines (to be studied in later chapters), which operate at other speeds (and, therefore, are sometimes called asynchronous machines).

The term synchronous speed is meaningful only in relation to the frequency of the armature voltages and currents. For the frequency of 60 Hz used in the power systems of the US, Eq. 9-4 leads to the following statements:

A two-pole synchronous machine runs at 3600 rpm.
A four-pole synchronous machine runs at 1800 rpm.
A six-pole synchronous machine runs at 1200 rpm, etc.

So the number of poles must be chosen in accordance with the desired speed. Essentially, electric motors and generators are the less expensive the faster they run. It could be asked why, then, any synchronous machines should be built with more than two poles. The answer lies in the *other* machine that is attached to the same shaft. For a generator, this is the prime mover; for a motor, it is the load. It might be suggested that these other machines could run on their own shafts, driving the generator or being driven by the motor over a set of gears (or other drives). But in that case, the cost of these gears, their power losses and their maintenance problems would usually outweight the advantages of such a scheme. Generators, in particular, are practically always mounted on the same shaft with their prime movers. Typically, steam turbines are fast-running machines for which 3600 and 1800 rpm are the only suitable speeds for 60 Hz. Various types of waterwheels, on the other hand, require lower speeds and, therefore, generators with larger numbers of poles.

9-4 THE ADDITION OF MMF WAVES

The key to the study of synchronous machines, to understand how they respond to different operating conditions, lies in the combination of the two component mmf waves. The following analysis is based on the assumption of sinusoidal wave shapes, which means only that it is restricted to the space fundamentals of the mmf waves. A study of the effects of space harmonics would be based on the same principles and techniques.

We are considering, then, two sinusoidal waves traveling in the same direction and at the same speed, but with different peak values and space angles. They can be described by the equations

$$\mathfrak{F}_a(\theta, t) = A \cos(\theta - \theta_a - \omega t) \tag{9-5}$$

for the *armature* (stator) mmf wave, and

$$\mathfrak{F}_f(\theta, t) = F \cos(\theta - \theta_f - \omega t) \tag{9-6}$$

for the *field* (rotor) mmf wave. The new symbols A and F represent the peak values of the two waves, and θ_a and θ_f are the locations of these peaks at the instant $t = 0$.

There are several ways to show that the *resultant* mmf wave has the form

$$\mathfrak{F}_{\text{res}}(\theta, t) = R \cos(\theta - \theta_{\text{res}} - \omega t) \tag{9-7}$$

and to find the values of R and θ_{res}, the peak value of the resultant mmf wave and its initial position. Mathematically, the most elegant method begins with a substitution according to Euler's equation: $\cos\alpha = \mathcal{R}e\, e^{j\alpha}$. A different procedure is chosen here, because it makes use of the representation of sinusoidal time functions by *phasors*, a technique well known to the reader (and recapitulated in Section 4-1). In addition, this method makes it possible to use phasors representing mmf waves in conjunction with phasors representing voltages and currents.

Mmf waves are functions of both time and space, whereas phasors can represent only functions of time. Therefore, recourse is taken once again to the idea of a *stationary observer*, previously introduced in Section 8-8. For such an observer, stationed at $\theta = \theta^*$, a rotating wave appears as a sinusoidal time function. For instance, the armature mmf wave, as registered by this observer, is

$$\mathfrak{F}_a(\theta^*, t) = A \cos(\theta^* - \theta_a - \omega t) = A \cos[\omega t + (\theta_a - \theta^*)] \tag{9-8}$$

which can be represented by the phasor

$$\mathbf{A} = A \underline{/\theta_a - \theta^*} \tag{9-9}$$

Results obtained this way are not restricted by the chosen location θ^*. If the observer is placed somewhere else, the magnitude of the phasor $\mathbf{A}$ remains the same, and its phase angle changes by the same amount as the location of the observer. The choice of θ^* is merely a matter of convenience.

Figure 9-4a shows the arrangement of a three-phase armature winding on the stator, repeated from Fig. 9-3. In addition, a few conductors of phase a are drawn, with indications of the assumed positive direction. Phase a is chosen (arbitrarily) to be the *reference phase*, meaning that, whenever a *phase current* or a *phase voltage* is mentioned, it is understood to be the one in phase *a.*

For this armature, Fig. 9-4b gives a snapshot of the armature mmf wave at that instant at which the phase current (i_a) has its positive maximum value. Thus the peak of $\mathcal{F}_a$ appears in the axis of phase a, and its being positive means that the corresponding lines of flux are pointing upward (referring to Fig. 9-4a), or out

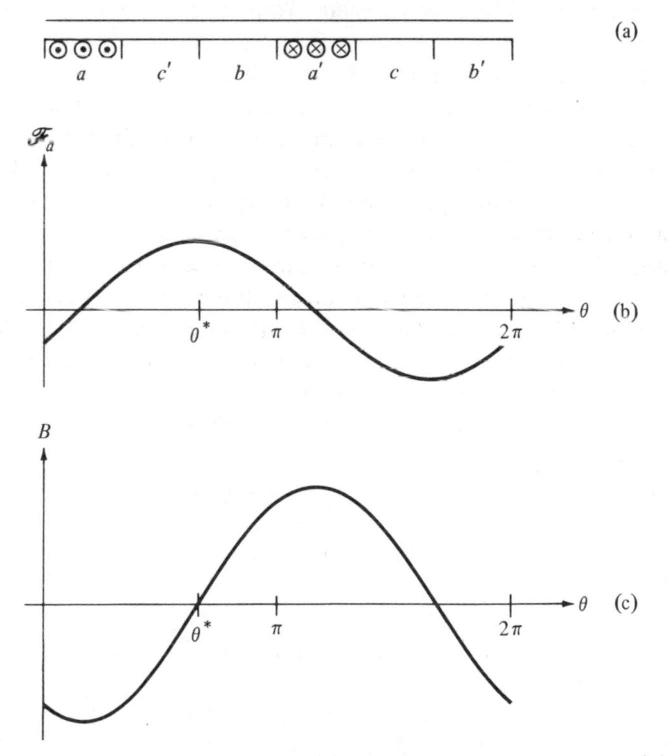

Fig. 9-4. Location of stationary observer: (a) positive direction for phase a; (b) armature mmf wave at instant when $i_a = I_{max}$; (c) flux distribution at instant when $e = E_{max}$.

of the stator. Now, if the stationary observer is placed in the axis of phase a, as indicated in Fig. 9-4b, then, at the instant of this figure, the observer registers the positive maximum value of **A**. In other words, due to this choice of observer location, the phasors representing *the armature mmf and the armature current are in phase*, in addition to their magnitudes being proportional to each other.

To establish a further phase relation, Fig. 9-4c shows the flux distribution at the instant when the induced voltage in phase a has its positive maximum value. The reader should check the polarity of the flux, using the right-hand rule given in the footnote of Section 7-1. It must be remembered, however, that the direction of the thumb represents the motion of the conductors relative to the field, which, in this case, is to the *left* because the conductors are stationary and the field is traveling to the right. So the observer at his location θ^* must have registered a positive maximum value of flux density a quarter-period before the instant of the diagram. In other words, using phasors in accordance with this observer location, *the flux distribution leads the induced voltage by 90°*. Again, the magnitudes of these two quantities are directly proportional to each other. (Eq. 8-40 states that the voltage is proportional to the flux, which, in turn, by Eq. 8-12 is proportional to the peak value of the flux distribution.)

Next comes the relation between the flux distribution and the resultant mmf wave. It was already mentioned in Section 9-2 that this relation depends only on the reluctance of the magnetic circuit. But while this reluctance is constant in the steady state, it may change from one operating condition to another, in case the rotor is built with salient poles. This presents a complication to the analysis. The present study will, therefore, be limited to machines with cylindrical rotors, leaving the effects of salient poles to a later chapter. The idea is that results obtained for cylindrical-rotor machines are valid qualitatively for all synchronous machines, even if they are not accurate quantitatively for machines with salient-pole rotors. Thus the flux distribution and the resultant mmf wave will now be considered to be in phase with each other. Their relation is described by a *saturation curve* B_{peak} versus R.

Finally, the induced voltage mentioned above differs from the *terminal voltage* of the armature in the same way in which the corresponding quantities of a transformer differ from each other. The armature winding is not ideal and, thus, has the properties of *resistance* and *leakage inductance* (*leakage* referring to all those lines of flux that link stator conductors but do not cross the air gap). So each phase of the armature winding can be represented by the equivalent circuit of Fig. 9-5. For the sinusoidal steady state, the phasor equation

$$\mathbf{E} = \mathbf{V} + R_a\mathbf{I}_a + jX_l\mathbf{I}_a \tag{9-10}$$

expresses Kirchhoff's voltage law for the circuit of Fig. 9-5. Of the symbols used in this equation and in the diagram, R_a stands for armature resistance, X_l for

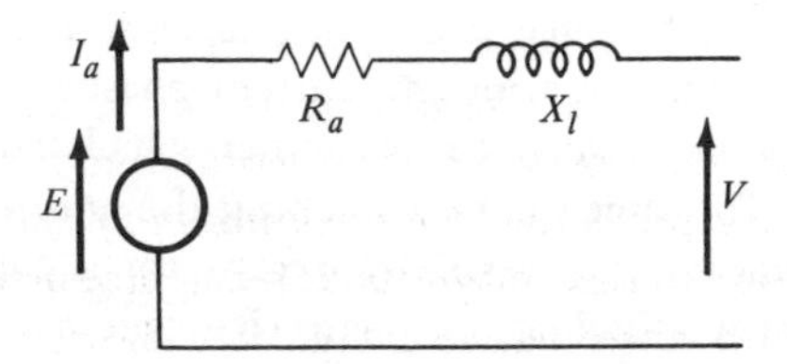

Fig. 9-5. Armature circuit per phase.

leakage reactance, and I_a for armature current. (The reader may also, if he prefers, interpret the subscript a as indicating the reference phase a).

The various statements and relations developed in this section will all be used and combined in the following one.

9-5 OPERATING CONDITIONS

The reader is referred to Fig. 7-3b, in which generator and motor operation are illustrated for a device whose induced voltage has a given direction. The device acts as a generator when the current flows through it in the direction of the voltage, i.e., from the lower to the higher potential, and as a motor if the current direction is reversed.

When voltage and current are *sinusoidal*, their relative directions may change during every cycle. What counts is the condition that prevails during the greater part of every period. The equation for *average power* in the sinusoidal steady state, $P = VI \cos \theta$, applied to the circuit of Fig. 9-5, makes it clear that the machine delivers power to a load (acts as a generator) when the angle θ, the phase difference between V and I, is in the first or fourth quadrant, making the power factor $\cos \theta$ positive.* With such an angle, voltage and current are in the same direction for the greater part of every period, which explains why the average power is positive. It also follows that the machine is receiving electric power at the terminals when the angle between V and I (as defined in Fig. 9-5) is in the second or third quadrant.

The synchronous machine can operate either as a motor or as a generator. Its operating condition is completely determined by the values of the three quantities V, I, and θ. For every given operating condition, the relations obtained in the previous section can be combined to give a picture of what is going on.

The procedure begins with the choice of V (arbitrary but plausible) as the axis of reference, whereby the two phasors **V** and **I** are defined. They are substituted

*It must be mentioned that the Greek letter θ was used since Section 7-6 as the symbol for a space angle. From here on, however, it reverts to its previous meaning, familiar from circuit analysis.

into Eq. 9-10, resulting in the phasor **E**. The phasor **A** representing the armature mmf wave is in phase with the armature current phasor $\mathbf{I}_a$, and proportional to it. The factor of proportionality can be either calculated from design data or obtained from tests. The same can be said about the *saturation curve* E versus R (identical with the curve B_{peak} versus R except for a change of the ordinate scale). This curve permits finding the phasor **R** (representing the resultant mmf wave) that leads the voltage phasor **E** by 90°, and thereby the phasor **F** representing the field mmf wave, since

$$\mathbf{F} = \mathbf{R} - \mathbf{A} \tag{9-11}$$

The significance of this result lies in the fact that **F** is the *field excitation* that is required if the machine is to operate as specified (i.e., with the given values of V, I, and θ). The magnitude F is proportional to the field current I_f, and the angle δ between **F** and **R** is the familiar *torque angle* appearing in Eq. 8-49.

Figures 9-6 and 9-7 illustrate the procedure, without any numerical values. They represent two arbitrary examples, one for generator and the other for motor operation (determined by the quadrant of the current phasor $\mathbf{I}_a$, since the voltage phasor **V** is in the axis of reference). In both diagrams, the magnitudes R_aI_a and X_lI_a are somewhat exaggerated, i.e., drawn larger in relation to V than they would be in most practical cases. The numerical calculations in Examples 9-1 and 9-2 are based on operating conditions similar to those described in the two diagrams and on realistic values.

The reader can see, both in the figures and in the calculated examples, that the torque angle δ must be reversed if the machine is to change between motor and generator operation. This is only natural since the electromagnetic torque in a motor is in the direction of rotation, whereas in a generator, it opposes the

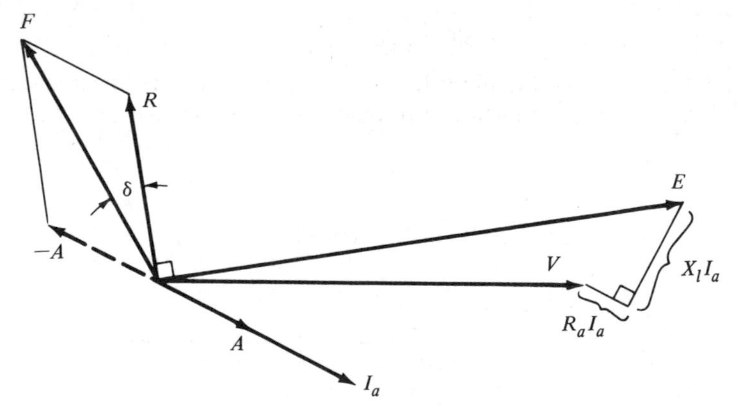

Fig. 9-6. Generator operation.

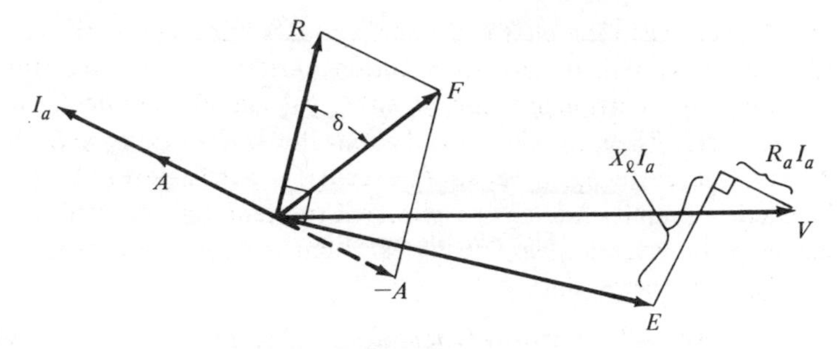

Fig. 9-7. Motor operation.

motion. It is also interesting to study how the magnitude relations among the mmf waves change with changing operating conditions. This will be left for later sections, however, after a much simpler, even though less rigorous, method of analysis has been developed.

9-6 LINEAR ANALYSIS

If fluxes were proportional to the mmfs that produce them: in other words, if magnetic circuits were linear, their study could in many cases be considerably simplified by the use of *superposition.* In particular, this can be said about synchronous machines. The contrast between linear and nonlinear analysis can be illustrated by a graphic tabulation of the pertinent relations.

In Fig. 9-8, the arrows refer to the relations established in Section 9-4 and used in Section 9-5. The heart of the procedure lies in the addition of the two component mmf waves, each of which is proportional to its current or currents. At the end of the sequence, there is the induced voltage that is proportional to the flux and lags it by 90°. The nonlinear relation is the one between the resultant mmf wave R and the flux; it is the familiar saturation curve.

Figure 9-9 shows how the procedure can be changed in the linear case, when the saturation curve is a straight line through the origin. In that case, each of the two mmf waves A and F is thought of as producing its own flux, and each such

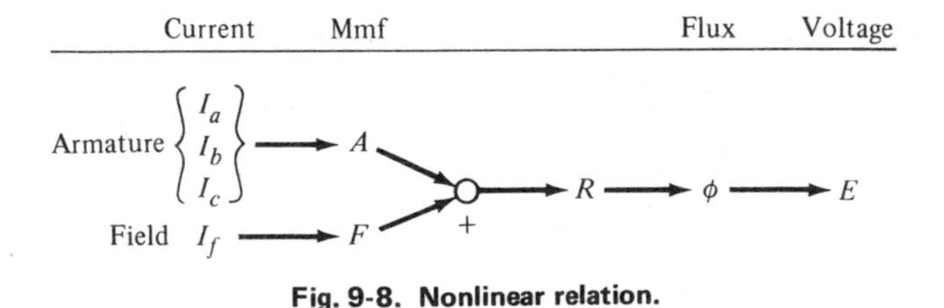

Fig. 9-8. Nonlinear relation.

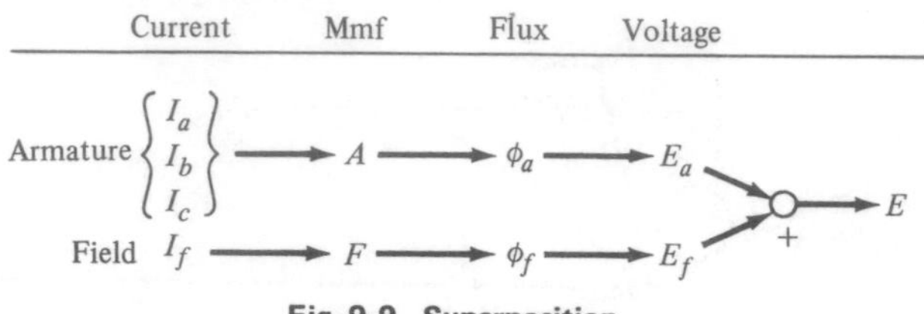

Fig. 9-9. Superposition.

flux its own voltage. All the arrows from left to right, from currents to voltages, indicate proportional relations, and each component voltage lags its mmf wave by 90°. Specifically, a phasor relation can be established between the voltage E_a and the armature current (in the reference phase):

$$\mathbf{E}_a = -jX_a\mathbf{I}_a \tag{9-12}$$

because the symbol X (reactance) stands for the ratio of a voltage over a current when the two are 90° out of phase. What's more, the use of this symbol is not only formally justified but also physically meaningful because E_a is a voltage of self-induction, a voltage induced in each phase of the armature winding due to the flux produced by the currents in that same winding.

The effect of the flux produced by armature currents is generally called *armature reaction.* Accordingly, the reactance X_a is given the awkward name of *armature reaction reactance*, a name that surely could not have survived long in engineering practice if it were not for the fact that this reactance by itself does not appear in the final results.

Now the phasors representing the two component voltages are added, in accordance with the principle of superposition (the sum of the linear effects of several causes equals the effect of the sum of the causes):

$$\mathbf{E} = \mathbf{E}_f - jX_a\mathbf{I}_a \tag{9-13}$$

The purpose of the whole procedure becomes clearer when Eq. 9-10 is substituted into Eq. 9-13, whereby the terminal voltage is brought into the picture, for instance in the form

$$\mathbf{E}_f = \mathbf{V} + R_a\mathbf{I}_a + jX_l\mathbf{I}_a + jX_a\mathbf{I}_a \tag{9-14}$$

This equation suggests that there are two reactances in series. This is illustrated by Fig. 9-10, an equivalent circuit for one phase of the armature winding from which Eqs. 9-10, 9-12, and 9-13 can be read. In contrast to Fig. 9-5, this diagram displays a new source E_f, called the *excitation voltage*, since it is the voltage that would be induced in the absence of armature currents by the field current alone.

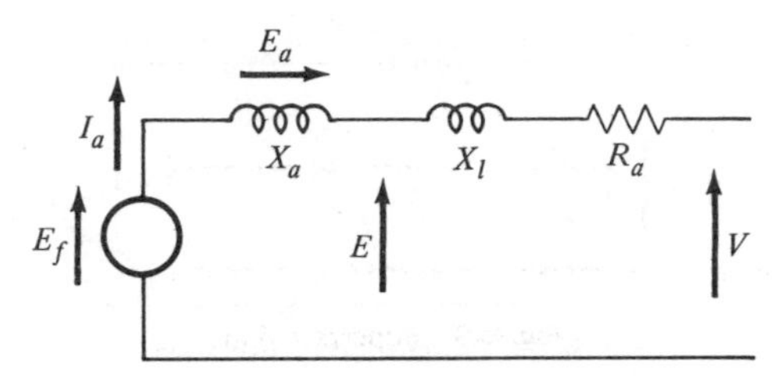

Fig. 9-10. Equivalent circuit.

As to the two reactances, it is instructive to compare them. The leakage reactance is to be considered as an inevitable imperfection of the armature winding, just like the leakage reactances of transformer windings. It is also (at least, as a close approximation) a linear circuit element because the path of leakage flux is largely through air. Armature reaction, on the other hand, is not an imperfection, but rather an essential feature of the operation of synchronous machines. The flux in the air gap is produced by the combined action of the stator and rotor currents, just as the flux in a transformer is produced by the combined action of the primary and secondary currents. But the representation of armature reaction by a reactance is an artifice made possible only by ignoring the nonlinearity of the magnetic circuit. This casts a shadow of uncertainty over the value to be given to this reactance. Still, whatever numerical value may be chosen for it, it is sure to be much larger than the leakage reactance, and the two reactances can be added in any case, as Eq. 9-14 and Fig. 9-10 indicate. Thus

$$X_l + X_a = X_s \tag{9-15}$$

which is called the *synchronous reactance*, because it is *the* reactance that appears in the most frequently used method of analysis of the synchronous machine. (This eliminates the *armature reaction reactance* from further consideration, as promised.) With the introduction of the synchronous reactance, Eq. 9-14 is reduced to

$$\mathbf{E}_f = \mathbf{V} + R_a\mathbf{I}_a + jX_s\mathbf{I}_a \tag{9-16}$$

and the equivalent circuit to that of Fig. 9-11.

A striking feature of this new equivalent circuit is that the induced voltage E has been obliterated. What is left is a description of the synchronous machine by an ideal source E_f (dependent on the field current I_f, but independent of any voltage or current in the armature current, which was not true of the source E in Fig. 9.5) and a complex impedance $R_a + jX_s$; in other words, a Thévenin type of equivalent circuit.

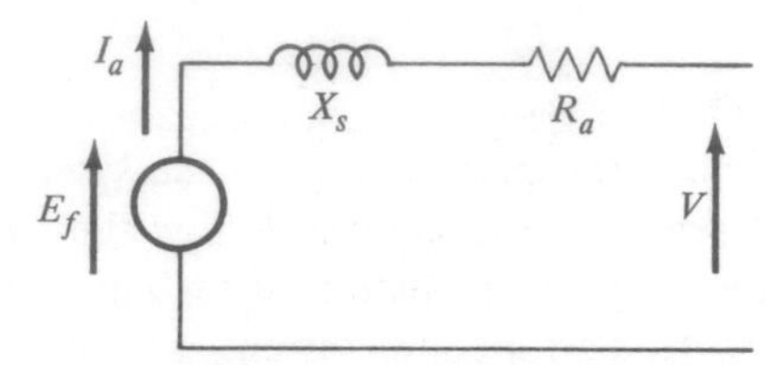

Fig. 9-11. Synchronous reactance.

It must not be forgotten that this simple representation was made possible by a linear approximation. No matter how the value of X_s is chosen, it will never fit all operating conditions with more than some limited accuracy. For this reason, a good case can be made for disregarding the armature resistance. This means a considerable gain of simplicity, and a comparatively slight further loss of accuracy, since this resistance (an imperfection responsible for power losses) must be much smaller than the synchronous reactance. Most of the rest of this chapter and of the next two chapters will be based on the simplified equivalent circuit of Fig. 9-12, and on the corresponding equation

$$\mathbf{E}_f = \mathbf{V} + jX_s\mathbf{I}_a \tag{9-17}$$

Results thus obtained will be qualitatively significant even though quantitatively suspect. Only when power losses and efficiency of the synchronous machine are to be studied, must the armature resistance be considered.

9-7 OPEN- AND SHORT-CIRCUIT CHARACTERISTICS

Two methods to analyze the operation of the synchronous machine have now been developed. The one based on the *addition of mmf waves* is more rigorous but also more complicated. The introduction of the *synchronous reactance* sacrifices accuracy, but makes everything much easier (which makes it practically indispensable for the study of entire power systems for which synchronous machines are only components). Whichever of the two methods is used, some

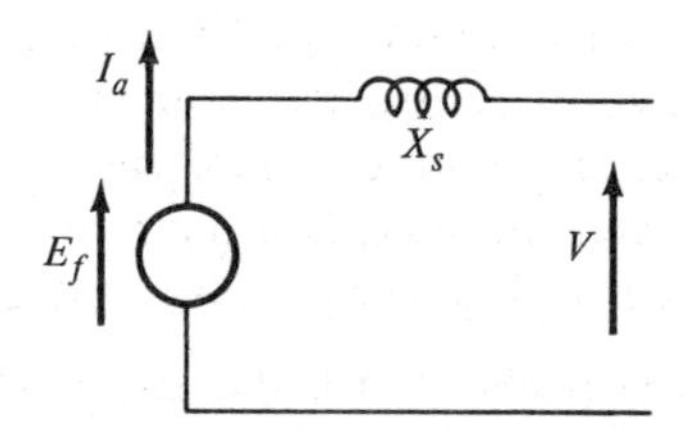

Fig. 9-12. Simplified equivalent circuit.

properties of the specific machine to be studied must be known in numerical form. How are they obtained?

Since machines are built for specific purposes, designers must be able to determine their numerical properties, to anticipate how they will operate. For that purpose, designers are using a combination of techniques, many of them based on approximations, trial-and-error computations, and previous experience. When a machine is built, the final proof that it meets its specifications must be by *testing.* Thus, it is necessary, for the manufacturer as well as for the purchaser of the machine, to determine the numerical properties of the machine experimentally. The situation is the same as that encountered in the study of transformers (Section 5-9).

There are tests and methods to evaluate them for the purpose of obtaining the numerical data needed for the analysis based on the addition of mmf waves. For those, however, the reader will be referred to some of the older texts.* In this book, we shall concentrate on tests that serve to find numerical values of synchronous reactance.

The basic idea is suggested by a look at the simple (Thévenin equivalent) circuit of Fig. 9-12. Clearly, the synchronous reactance can be found as the ratio of *open-circuit voltage* over *short-circuit current*

$$X_s = \frac{V_{oc}}{I_{sc}} \tag{9-18}$$

where both V_{oc} and I_{sc} must be phase quantities (not line quantities). But there is a complication: both V_{oc} and I_{sc} are dependent on the field excitation, and so is their ratio.

In an *open-circuit test*, the machine is driven at its intended synchronous speed (regardless of whether it is meant to be used as a generator or as a motor), supplied with an adjustable field current from an auxiliary source, but left open-circuited at the armature terminals. Under these circumstances, there is no armature current, therefore no stator mmf, and the flux is entirely due to the field current. Also, there is no voltage drop across the armature resistance and the leakage reactance. In other words, $E_f = E = V_{oc}$, which can be measured at the armature terminals.

By adjusting the field current to several different values, points of a *saturation curve* are obtained because its abscissa, the resultant mmf wave, equals the rotor mmf wave (in the absence of a stator current), which is proportional to the field current, while the ordinate, the flux, is proportional to the induced voltage. This form of saturation curve, V_{oc} versus I_f, is called the *open-circuit character-*

*For instance, M. Liwschitz-Garik and C. Whipple, *Electric Machinery*, D. Van Nostrand Co., 1946, or A. F. Puchstein, T. C. Lloyd, and A. G. Conrad, *Alternating Current Machines*, third edition, John Wiley & Sons, 1954.

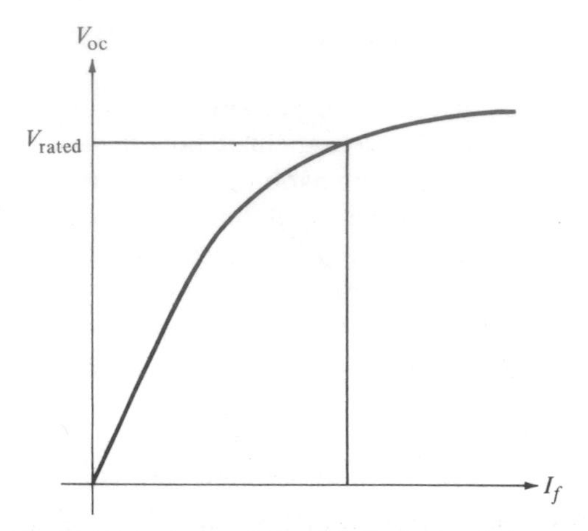

Fig. 9-13. Open-circuit characteristic.

istic (OCC) of the machine. Figure 9-13 depicts the typical shape of such a curve, and shows how rated voltage can be expected to be somewhat above the knee, in the interest of proper utilization of the core material.

A *short-circuit test* is conducted in a similar fashion. Again, the machine is driven at its synchronous speed and excited from an auxiliary source. The field current is varied but, this time, the armature windings are short-circuited, and the currents at their terminals are measured.

The *short-circuit characteristic* (SCC) thus obtained (i.e., the curve I_{sc} versus I_f) turns out to be a straight line over any reasonable range of values. In other words, the magnetic circuit is not saturated in this test. It is not difficult to show why this is so, by pointing at Eq. 9-10, with V set equal to zero. In this equation, the voltage drops across the imperfections R_a and X_l cannot be more than small fractions of the rated voltage, even if the armature current is raised far beyond its rated value. Thus, in the short-circuit test, the induced voltage E is only a small fraction of rated value whereas, in any normal operation, E and V are at the same order of magnitude. Consequently, in the short-circuit test, the flux is much smaller than under operating conditions, and thus it is unsaturated.

In Fig. 9-14, both the open- and the short-circuit characteristic are drawn in the same system of axes, but there are two independent ordinate scales. According to Eq. 9-18, the synchronous reactance is the ratio of the ordinates of these two curves. The diagram serves to dramatize the uncertainty about the numerical value to be attributed to this "constant." If it is taken at a value of field current at which the magnetic circuit is not saturated in the open-circuit test, it is called the *unsaturated* synchronous reactance. For instance

$$X_{s_{unsat}} = \frac{ca}{ba} \tag{9-19}$$

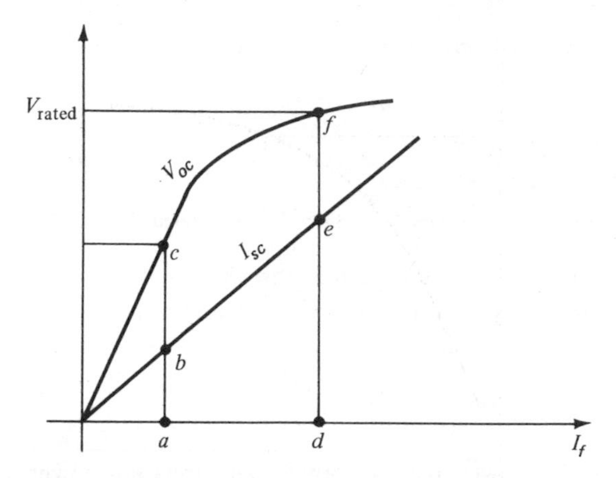

Fig. 9-14. Unsaturated and saturated synchronous reactance.

It is clearly greater than a similar ratio taken at a value of field current at which the open-circuit characteristic is no longer a straight line. For instance,

$$X_{s_{\text{sat}}} = \frac{fd}{ed} \tag{9-20}$$

Experience has shown that a useful value of this *saturated* synchronous reactance, one that can be substituted into Eq. 9-16 or Eq. 9-17 leading to fairly good approximations, is obtained by choosing point f as the point where the ordinate of the open-circuit characteristic equals rated voltage.

9-8 NORMALIZED QUANTITIES

For numerical calculations, the use of normalized (per unit) values, first encountered in Section 5-9 in the study of power transformers, can be meaningful and advantageous for synchronous machines also. As the reader recalls, a quantity is normalized by being divided by a base quantity. Rated values are useful choices for base quantities, as they were for transformers.

For three-phase machines, there is the additional benefit that the distinction between phase and line quantities can be made irrelevant. This result is obtained by the choice of different bases for phase and line quantities. For a *wye connection*, where $V_L = \sqrt{3}\,V_P$, the bases are

$$V_{P\,\text{base}} = V_{P\,\text{rated}} \tag{9-21}$$

and

$$V_{L\,\text{base}} = V_{L\,\text{rated}} = \sqrt{3}\,V_{P\,\text{rated}} \tag{9-22}$$

with the pleasing result that normalized phase and line voltages equal each other.

Similarly, in a balanced *delta connection*, where $I_L = \sqrt{3} I_P$,*

$$I_{P\,\text{base}} = I_{P\,\text{rated}} \tag{9-23}$$

and

$$I_{L\,\text{base}} = I_{L\,\text{rated}} = \sqrt{3} I_{P\,\text{rated}} \tag{9-24}$$

Resistances, reactances, and impedances (as well as their reciprocals) can only appear in "per phase" equations. Thus, the impedance base must be

$$Z_{\text{base}} = \frac{V_{P\,\text{rated}}}{I_{P\,\text{rated}}} \tag{9-25}$$

Finally, the base for power, reactive power, and apparent power is the apparent power at the armature terminals (generator output, or motor input) under rated conditions:

$$P_{a\,\text{base}} = 3 V_{P\,\text{rated}} I_{P\,\text{rated}} = \sqrt{3} V_{L\,\text{rated}} I_{L\,\text{rated}} \tag{9-26}$$

Consequently, the factors 3 or $\sqrt{3}$ never appear in equations thus normalized. For instance, the output power of a generator operating at rated conditions equals the power factor. An example will illustrate the whole idea.

9-9 EXAMPLES

Example 9-1 (Section 9-5)

A cylindrical-rotor, three-phase, two-pole, 3600-rpm synchronous machine has a rating as a generator of 3000 kva, 6600 v, and 60 Hz. Each phase has resistance and leakage reactance of 0.06 Ω and 0.72 Ω, respectively. The stator mmf produced by 262 amp is 12,400 At/pole. Generated voltage per phase, for different excitations is given by:

E	3000	3620	3720	3810	3900	3980	4200	v
R	10,000	15,000	16,000	17,000	18,000	19,000	24,000	At/pole

This machine is operated as a generator with rated terminal voltage and rated load current with power factor of 0.9 lagging. Find the generated voltage and the excitation required in the rotor field.

*All current symbols in this section refer to armature currents. The subscript a has been omitted as unnecessary in this context.

Solution

Rated voltage is $6600/\sqrt{3} = 3810$ v per phase. Rated current is 3,000,000/(3 X 3810) = 262 amp. Use the equivalent circuit of Fig. 9-5. Use the terminal voltage as reference.

$$\mathbf{V} = 3810 \underline{/0^\circ} = 3810 + j\,0 \text{ v}$$

$$\mathbf{I}_a = 262 \underline{/-25.8^\circ} \text{ amp}$$

$$R_a\mathbf{I}_a = (0.06 \underline{/0^\circ})(262 \underline{/-25.8^\circ}) = 15.7 \underline{/-25.8^\circ} = 14.1 - j\,6.8 \text{ v}$$

$$jX_l\mathbf{I}_a = (0.72 \underline{/90^\circ})(262 \underline{/-25.8^\circ}) = 189 \underline{/64.2^\circ} = 82.4 + j\,170 \text{ v}$$

Use Eq. 9-10 to find **E**

$$\mathbf{E} = \mathbf{V} + R_a\mathbf{I}_a + jX_l\mathbf{I}_a = 3906 + j\,163 = 3909 \underline{/2.39^\circ} \text{ v}$$

From the saturation curve, find R

$$\mathbf{R} = 18{,}100 \underline{/90^\circ + 2.39^\circ} = 18{,}100 \underline{/92.4^\circ} = -755 + j\,18{,}080 \text{ At/pole}$$

$$\mathbf{A} = 12{,}400 \underline{/-25.8^\circ} = 11{,}160 - j\,5410 \text{ At/pole}$$

Use Eq. 9-11 to find **F**

$$\mathbf{F} = \mathbf{R} - \mathbf{A} = -11{,}915 + j\,23{,}490 = 26{,}340 \underline{/117^\circ} \text{ At/pole}$$

The quantities in this problem are illustrated by Fig. 9-6. As this is a generator, notice that **F** leads **R** by 24.6°. This method of problem solving is called the *General Method.*

Example 9-2 (Section 9-5)

The machine of Example 9-1 is operated as a synchronous motor with rated terminal voltage and rated current. The field current is set to make the power factor 0.9 lagging. Find the generated voltage and the excitation required in the rotor field.

Solution

Use the equivalent circuit of Fig. 9-5. Recall that the current direction shown is for generator action. Motor performance, with lagging power factor, will have the current phasor located in the second quadrant.

$$\mathbf{I}_a = 262 \underline{/154.2^\circ} \text{ amp}$$

$$R_a\mathbf{I}_a = (0.06 \underline{/0^\circ})(262 \underline{/154.2^\circ}) = 15.7 \underline{/154.2^\circ} = -14.1 + j\,68 \text{ v}$$

$$jX_l\mathbf{I}_a = (0.72\ \underline{/90^\circ})(262\ \underline{/154.2^\circ}) = 189\ \underline{/244.2^\circ} = -82.4 - j\,170\text{ v}$$

$$\mathbf{V} = 3810\ \underline{/0^\circ} = 3810 + j\,0\text{ v}$$

Use Eq. 9-10 to find **E**

$$\mathbf{E} = \mathbf{V} + R_a\mathbf{I}_a + jX_l\mathbf{I}_a = 3713 - j\,163 = 3717\ \underline{/-2.5^\circ}\text{ v}$$

From the saturation curve, find R

$$\mathbf{R} = 15{,}970\ \underline{/90^\circ - 2.5^\circ} = 15{,}970\ \underline{/87.5^\circ} = 697 + j\,15{,}950\text{ At/pole}$$

$$\mathbf{A} = 12{,}400\ \underline{/154.2^\circ} = -11{,}160 + j\,5400\text{ At/pole}$$

Use Eq. 9-11 to find **F**

$$\mathbf{F} = \mathbf{R} - \mathbf{A} = 11{,}860 + j\,10{,}550 = 15{,}870\ \underline{/41.6^\circ}\text{ At/pole}$$

The quantities in this problem are illustrated by Fig. 9-7. As this is a motor, notice that **F** lags **R** by 45.9°. Also notice that the motor, with lagging current, is underexcited, while the generator in Example 9-1 was overexcited.

Example 9-3 (Section 9-7)

The machine in Example 9-1 is operated in a short-circuit test with F equal to 17,000 At/pole. The short-circuit current is 337 amp. Find the saturated synchronous reactance.

Solution

Use Fig. 9-14 and Eq. 9-20

$$X_{s\,\text{sat}} = \frac{fd}{ed} = \frac{3810\text{ v}}{337\text{ amp}} = 11.3\ \Omega$$

Notice that the synchronous reactance is approximately 15 times as large as the leakage reactance for this machine. Also notice that the short-circuit current is larger than rated current for this machine. The ratio 337 amp/262 amp = 1.29 is called the *short-circuit ratio.*

Example 9-4 (Section 9-8)

For the machine in Examples 9-1 and 9-3, (a) Find the line voltage, current and power in per-unit. (b) Find the voltages (per phase) in Example 9-1 in per-unit. (c) Find the resistance and reactances in per-unit.

Solution

(a)

$V_{L\,\text{base}} = 6600$ v $\qquad V_L = 6600$ v

$I_{L\,\text{base}} = 262$ amp $\qquad I_L = 262$ amp

$P_{a\,\text{base}} = 3000$ kva $\qquad$ P.F. = 0.9 lag

$V_L = V_L/V_{L\,\text{base}} = 1$ pu $\qquad P = 2700$ kw

$I_L = I_L/I_{L\,\text{base}} = 1$ pu

$P = P/P_{a\,\text{base}} = 2700/3000 = 0.9$ pu

(b)

$V_{p\,\text{base}} = 3810$ v

$V = V/V_{p\,\text{base}} = 3810\text{ v}/3810\text{ v} = 1$ pu

$E = E/V_{p\,\text{base}} = 3909\text{ v}/3810\text{ v} = 1.026$ pu

$R_aI_a = R_aI_a/V_{p\,\text{base}} = 15.7\text{ v}/3810\text{ v} = 0.004$ pu

$X_lI_a = X_lI_a/V_{p\,\text{base}} = 189\text{ v}/3810\text{ v} = 0.05$ pu

(c)

$Z_{\text{base}} = V_{p\,\text{rated}}/I_{p\,\text{rated}} = 3810\text{ v}/262\text{ amp} = 14.5\ \Omega$

$R_a = R_a/Z_{\text{base}} = 0.06\ \Omega/14.5\ \Omega = 0.004$ pu

$X_l = X_l/Z_{\text{base}} = 0.72\ \Omega/14.5\ \Omega = 0.05$ pu

$X_s = X_s/Z_{\text{base}} = 11.3\ \Omega/14.5\ \Omega = 0.78$ pu

Compare R_a and X_l with R_aI_a and X_lI_a in part (b). Observe that $1/X_{s\,\text{sat}} = 1/0.78 = 1.29$ = short-circuit ratio. That is, the short-circuit ratio is the reciprocal of the saturated synchronous reactance expressed in per-unit.

9-10 PROBLEMS

9-1. The machine in Example 9-1 is operated as a generator with rated terminal voltage and rated load current with power factor of 0.8 lagging. Find the generated voltage and the excitation required in the rotor field.

9-2. The machine in Example 9-1 is operated as a synchronous motor with rated terminal voltage and rated current. The field current is set to make the power factor be 0.8 lagging. Find the generated voltage and the excitation required in the rotor field.

9-3. A 60-Hz, two-pole synchronous motor has its shaft directly coupled to the shaft of a fourteen-pole a-c generator. Find the frequency of the voltage produced in the generator.

9-4. Energy is to be transferred from a 60-Hz system to a 50-Hz system through a motor-generator set. Find the minimum number of poles that can be used for the synchronous motor and for the a-c generator, respectively. Find the shaft speed for this set.

9-5. A 50-kva, three-phase, Y-connected, 220-v (line to line), six-pole, 60-Hz, synchronous machine has resistance and leakage reactance (per phase) of 0.03 Ω and 0.087 Ω, respectively. No-load test data are:

OCC	E(per phase)	116	124	130	135 v
	I_f	1.75	2.0	2.25	2.5 amp
SCC	I_a	121	138	155	173 amp

Find the saturated synchronous reactance.

9-6. A 900-kva, 60-Hz, 1039 v (line to line), 500 amp, three-phase, Y-connected synchronous machine has per-phase parameters of 0.02 Ω resistance and 0.08 Ω leakage reactance. The OCC is given by

$$E = \frac{832\ I_f}{5 + I_f} \quad \text{or} \quad I_f = \frac{5\ E}{832 - E}$$

where E is the no-load voltage per phase and I_f is the d-c field current. The SCC is given by

$$I_{a(\mathrm{SC})} = 50.2\ I_f$$

where $I_{a(\mathrm{SC})}$ is the short-circuit armature current per phase and I_f is the d-c field current. Find the saturated synchronous reactance.

9-7. The machine in Problem 9-5 is operated with I_a = 104.8 amp. Find I_a, $R_a I_a$, and $X_l I_a$ in per-unit. Also find R_a, X_l, and X_s in per-unit.

9-8. The machine in Problem 9-6 is operated with I_a = 350 amp. Find I_a, $R_a I_a$, and $X_l I_a$ in per-unit. Also find R_a, X_l, and X_s in per-unit.

9-9. A three-phase, Y-connected, cylindrical-rotor synchronous machine is driven at rated speed. On the open-circuit characteristic, the rated voltage point is given by 400 v per phase for 8 amp of field current. One point on the short-circuit characteristic is given by an armature current of 360 amp for 6 amp of field current. Find the saturated synchronous reactance for this machine.

10
Synchronous Motors

10-1 THE INFINITE BUS

If a synchronous machine is to operate as a *motor*, it must receive electric power at its armature terminals. Since this study concentrates on three-phase machines, we shall visualize three armature terminals connected to a three-phase power system. This system, in turn, is energized by one or more generators whose function it is to supply the entire load, including "our" motor, with the electric power it requires, at substantially constant values of voltage and frequency.

In the next chapter, when the operation of synchronous generators will be studied, the reader will get a good look at the problems of how to deal with changes of magnitude and frequency of the voltage at various points of the power system. In studying motor operation, however, it is best to assume that these problems are successfully solved, even to the extent that the input voltages are purely sinusoidal, and of constant magnitude and frequency.

A frequently used term for such a supply is *infinite bus*. To a reader unfamiliar with this expression, it should first be pointed out that *bus* is an abbreviation of the Latin *omnibus*, which means *for all* (thus the familiar use of the same word for a public transportation vehicle). In the professional idiom of the electric power engineer, a bus (or, more elaborately, a *bus bar*) is a system of large conductors by which, in a power plant or substation, all circuits carrying incoming and outgoing power, including generators, transformers, transmission lines, etc, are connected in parallel.

An actual bus is always *finite* in the sense that all currents entering and leaving the bus must be finite, so that adding another generator or another load to those already connected to the bus changes the total current, and thereby indirectly the voltage (due to the voltage drop across the impedance of every generator and/or transmission line supplying the bus). An *infinite bus* is one whose total curent is so large that adding another generator or load does not make any difference. It is comparable to a lake or pond that is so large that taking some buckets of water out of it or pouring them into it cannot change its level.

In terms of circuit analysis, an infinite bus is the equivalent of an ideal voltage source connected directly to the armature terminals. The terminal voltage V thereby becomes the natural choice for the axis of reference. (It was already so used in Figs. 9-6 and 9-7.)

10-2 STEADY-STATE OPERATION

The following steady-state analysis of the synchronous motor is based on the linear approximation introduced in Section 9-6, thus on the use of the synchronous reactance concept. The armature resistance will be disregarded, together with the effects of saturation, of salient poles, and of the space harmonics of the flux distribution. This leaves an armature circuit per-phase reduced to that of Fig. 9-11. It is reproduced, with some modifications, in Fig. 10-1.

Some of the changes are of minor importance: assumed positive voltage polarities are indicated by plus and minus signs instead of the arrows used previously, and the terminals are drawn on the left-hand side, to suggest a power flow from left to right. What is significant is the reversal of the current arrow, with respect to the voltage. It requires that Eq. 9-17, representing Kirchhoff's law, be rewritten:

$$\mathbf{E}_f = \mathbf{V} - jX_s\mathbf{I}_a \tag{10-1}$$

With the current arrow of Fig. 10-1, the motor may be more easily compared to other circuits receiving electric power, like resistances or impedances. The machine so represented operates as a motor if the phase angle between voltage and current is in the first or fourth quadrant. In this case, the power factor (the cosine of this phase angle) is positive, and so the electric input power of the motor is positive. (By contrast, if the current arrow of Fig. 9-11 were used, motor operation would be characterized by negative power at the armature terminals.)

With the armature connected to an infinite bus, and the terminal voltage V chosen as the axis of reference, the armature *current* phasor of a motor may be anywhere in the first or fourth quadrant, between 90° and -90°. Just where it lies within these limits, depends on the two independent variables of motor operation: the mechanical load and the field excitation.

To illustrate these relations, Fig. 10-2 shows phasor diagrams representing three possible cases: the armature current in phase with the terminal voltage (θ = 0), leading it, and lagging it. The three diagrams are the graphic equivalent

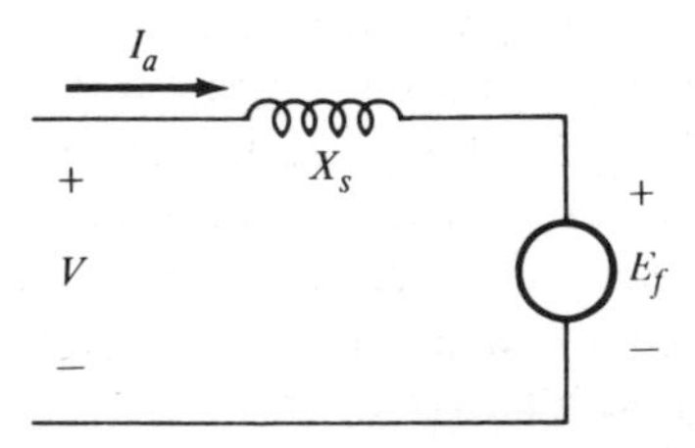

Fig. 10-1. Current direction choice for motor operation.

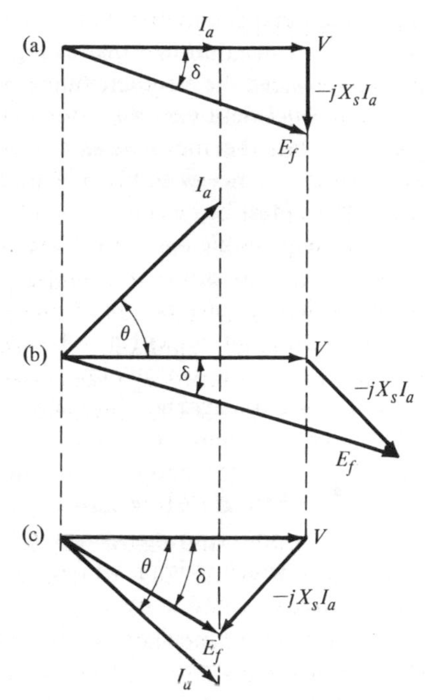

Fig. 10-2. Phasor diagrams: (a) current in phase with voltage; (b) current leading voltage, (c) current lagging voltage.

of Eq. 10-1. They are drawn for the same amount of electric power since the real component of I_a has the same value in each case. Also, the reactive power in the cases (b) and (c) is the same since the imaginary component of I_a is the same in these two cases.

The most striking difference between the three diagrams is that of the magnitudes of the *excitation voltage* E_f. The reader recalls that E_f is the voltage that would be induced in the armature by the flux due to the field current alone, i.e., in the absence of any armature current. Therefore, the magnitude of E_f depends entirely on that of the *field current* I_f. Since the whole concept of synchronous reactance is based on an assumption of constant saturation, the relation between these two variables must be assumed to be linear, with a suitably chosen factor of proportionality (See the last paragraph in Section 9-7).

So it can be learned from the three diagrams of Fig. 10-2 that the phase angle θ, and thereby the power factor cos θ, depend on the *field excitation*. (In a qualitative sense, this term *field excitation* can be understood to mean either E_f or I_f, since increasing one of them also means increasing the other, and vice versa.) For a load requiring a certain amount of power, the power factor can be adjusted by varying the field excitation.

Let it, for instance, be desired that the motor, while driving a given load, draw current at unity power factor. A diagram like that of Fig. 10-2a, or the equivalent analytical procedure, determines the field excitation required for that purpose. This is known as the *normal* field excitation for that load. If the actual field excitation is more than that, the motor is said to be *overexcited*, and it must draw leading current in accordance with Fig. 10-2b. Similarly, the current drawn by an *underexcited* motor must be lagging.

These statements can be made plausible by reference to well-known properties of passive circuits. A lagging current is one that supplies an inductive circuit, one that helps to store magnetic field energy. If an underexcited synchronous motor is thought of as one whose field current is not large enough to store the required field energy, then the armature current must be capable of making up for that deficiency; thus, it must be lagging. Logically, an overexcited motor needs the opposite kind of reactive current, i.e., a leading one.

It is often desirable to operate a synchronous motor with normal field excitation, because at unity power factor the armature current has the smallest possible magnitude for a given amount of power, thus causing the lowest possible copper loss. But there are also cases in which it is preferable to operate the motor at other power factors. For instance, in an industrial plant whose total electric load is predominantly inductive, an overexcited synchronous motor would help to raise the overall power factor of the plant. It will also be seen later on that overexcitation tends to improve the stability of a synchronous motor. At any rate, it is characteristic of the synchronous motor (in contrast to the induction motor) that its power factor can be arbitrarily adjusted.

10-3 THE TORQUE ANGLE

The reader is invited to take another look at the three phasor diagrams of Fig. 10-2. It might strike him that the phasor representing the excitation voltage E_f always appears in the fourth quadrant. It is not hard to see that this is no accident. In a motor, the armature current I_a must be in the first or fourth quadrant; thus, it must have a positive real component, i.e., one pointing to the right. Therefore, the phasor $(-jX_s\mathbf{I}_a)$, which lags I_a by 90° must have a negative imaginary component, one pointing downward.

So the field excitation voltage E_f must *lag* the terminal voltage V, if the machine is to act as a motor. By contrast, if it is to be a generator, E_f has to *lead* V, as will be confirmed in the next chapter. In other words, the angle between these two voltages, called δ in Fig. 10-2, has to be reversed for a change between motor and generator operation, a change that calls for a reversal of the electromagnetic torque.

It must be explained that this angle δ is related to, but not identical with the angle given the same symbol in Chapters 8 and 9. There, it was the angle between the rotor mmf wave and the flux distribution. Using the reasoning given in Section 9-6, and Figs. 9-8 and 9-9, it can be seen that the same angle appears between the voltages E_f and E, since E_f lags the rotor mmf wave F by 90°, whereas E lags the resultant mmf wave R by 90°. But when the equivalent circuit was simplified by the merger of two reactances into one (see Figs. 9-10 and 9-11), the voltage E disappeared. In its place, there is the terminal voltage V, which differs from E only by the effects of the imperfections of the armature winding.

From here on, the symbol δ will be used consistently for the phase difference between E_f and V, as it is in Fig. 10-2. It will now be shown that this new angle δ deserves being called a torque angle (as did the old one). This will be done by means of a derivation that begins by equating the imaginary parts of both sides of the complex equation, Eq. 10-1.

$$\mathcal{I}_m \mathbf{E}_f = \mathcal{I}_m \mathbf{V} - \mathcal{I}_m (jX_s\mathbf{I}_a) \tag{10-2}$$

Since $\mathbf{E}_f = E_f \underline{/-\delta}$, $\mathbf{V} = V\underline{/0}$, and $(-jX_s\mathbf{I}_a) = X_s\mathbf{I}_a \underline{/-\theta - 90°}$, Eq. 10-2 becomes

$$-E_f \sin \delta = -X_s I_a \cos \theta \tag{10-3}$$

which may be confirmed by a look at the geometry of the phasor diagrams of Fig. 10-2. Now solve for the real part of I_a

$$I_a \cos \theta = \frac{E_f}{X_s} \sin \delta \tag{10-4}$$

and substitute this into the general expression for power in a balanced three-phase circuit, applied to the motor armature (and remembering that V and I_a are phase quantities)

$$P = 3VI_a \cos \theta \tag{10-5}$$

resulting in a new expression for power

$$P = \frac{3VE_f}{X_s} \sin \delta \tag{10-6}$$

Since the armature resistance is being neglected, there are no armature copper losses to consider, and so P is the amount of power that is converted into mechanical form

$$P = T\omega_{s_m} \tag{10-7}$$

where ω_{s_m} is the synchronous angular velocity in mechanical radians per second. Thus, the electromagnetic torque is

$$T = \frac{1}{\omega_{s_m}} \frac{3VE_f}{X_s} \sin \delta \tag{10-8}$$

This result puts the spotlight on the angle δ. For one thing, it confirms that a change of its sign means a change between motor and generator operation. Furthermore, within the limits of motor operation, the equation indicates that the angle δ must change with the load (i.e., increase when the load increases, etc.) In particular, when E_f and V are in phase, the power and the torque are zero. This can be confirmed from a phasor diagram describing this condition. If both phasors $\mathbf{E}_f$ and $\mathbf{V}$ are in the axis of reference, then the phasor of their difference $(-jX_s\mathbf{I}_a)$ must also be horizontal, and, thus, the current phasor must be vertical, either upward (leading) for overexcitation or downward for underexcitation but in any case vertical, so that the power factor $\cos\theta$ is zero.

On the other hand, the sine of an angle cannot be more than unity. So Eq. 10-8 says that, for a motor supplied from an infinite bus (which means that V, ω_s, and X_s are all constant), for a given field excitation (E_f is constant), there is a *maximum torque*. It is

$$T_{\max} = \frac{1}{\omega_{s_m}} \frac{3VE_f}{X_s} \tag{10-9}$$

and it occurs when the torque angle reaches 90°.

10-4 δ AS A SPACE ANGLE

In addition to being defined as a phase difference (which is a normalized *time* interval), the torque angle δ also has a physical significance as a *space* angle, as will be explained now.

Figure 10-3 represents a developed armature (stator) surface, with the location of the conductors of phase a (the reference phase) indicated. In addition, each of the two diagrams displays a broken line as a schematic description of the rotor poles (assumed to be salient in order to be more easily visualized), and a sinusoid representing the field (rotor) mmf wave F, which is the one responsible for the excitation voltage E_f. Both diagrams are to be understood as snapshots taken at the instant *when the terminal voltage has its maximum value*. The rotor and all waves are assumed to move from left to right.

In Fig. 10-3a, the rotor poles are exactly opposite the conductors of phase a. Thus, the voltage E_f must have its maximum value at this instant. So the two voltages E_f and V have their maximum values at the same instant; thus, they are in phase with each other, and the electromagnetic torque is zero. This diagram depicts an *idealized no-load* condition, because it shows what would occur if an ideal motor (one with zero power losses) were to run at no-load, i.e., idling, without driving anything. An actual motor does have power losses (to be studied later), and even when it is idling, it still needs a certain electromagnetic torque, although one that is much smaller than at full load.

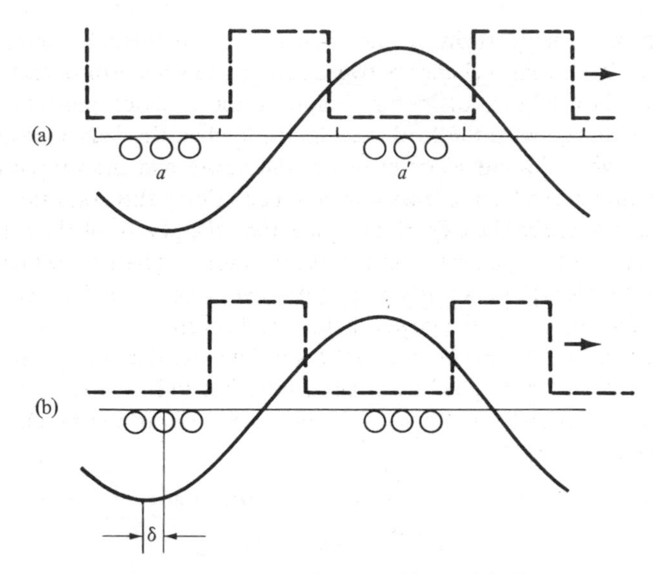

Fig. 10-3. Two snapshots at the instant when $v = V_{max}$. (a) E_f in phase with V; (b) E_f lagging V.

Now consider the case illustrated in Fig. 10-3b. This time, at the instant when $v = V_{max}$, the rotor poles (and the rotor mmf wave) have not yet reached the position facing the armature conductors, so the excitation voltage has not yet reached its maximum value; in other words, it lags the terminal voltage. So this diagram depicts a case of a motor operating under load.

Since the rotor travels 2π electrical radians during every cycle, the phase angle by which E_f lags V equals the space angle by which the *rotor positions* in the two diagrams differ from each other, expressed in electrical radians or degrees. Therefore, the torque angle δ can be defined as the space angle (in electric units) by which the rotor position at any instant differs from what it would be at the same instant in the ideal no-load case. It is so marked in Fig. 10-3b.

10-5 TRANSIENT CONDITIONS

In any steady-state condition, the rotor runs at synchronous speed, thereby maintaining a constant value of the torque angle δ. Consider, for instance, a motor running at no-load (idling), corresponding approximately to the case of Fig. 10-3a where $\delta = 0$. Now let the load torque be a step function of time, rising from zero to T_1 at the instant t_1. According to Eq. 10-8, this calls for a new value of δ, say δ_1. When a steady state is reached again, the rotor has "fallen back" by the same angle δ_1, and its position may be that of Fig. 10-3b.

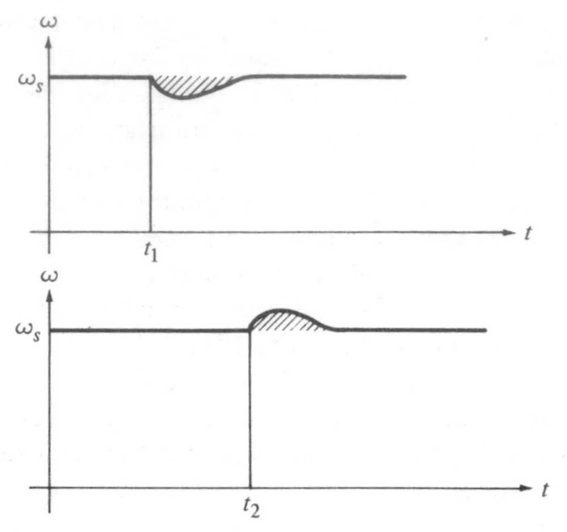

Fig. 10-4. Transient conditions. (a) motor load increasing at $t = t_1$; (b) motor load decreasing at $t = t_2$.

To bring about that change of δ, there must have been a time interval during which the rotor moved a little more slowly, i.e., at slightly less than synchronous speed.

Figure 10-4a describes what could happen during this *transient* interval in terms of angular velocity. (The actual wave shape will be discussed in the next two sections.) The diagram is somewhat exaggerated: the moment of inertia of an actual rotor and its load would cause the relative range of the speed change to be much less than shown. Figure 10-4b is a similar picture describing the case when the load torque is decreased at the instant t_2.

So the actual speed of a synchronous motor is *not always* synchronous speed, only in the steady state. In general, the angular velocity is

$$\omega = \omega_s - \frac{d\delta}{dt} \tag{10-10}$$

where ω and ω_s must be expressed in *electric* radians per second, because the angle δ is always expressed in electric units. The minus sign in Eq. 10-10 assumes (arbitrarily) that the angle δ is considered positive for motor operation (i.e., when E_f lags V).

By integrating Eq. 10-10 from some instant before the beginning of the transient interval to some instant after the end of this interval, it is seen that the shaded areas in the diagrams of Fig. 10-4 represent the amount by which the torque angle δ changes during the transient interval.

The reader can now begin to understand (after perhaps having wondered for some time) how a synchronous motor maintains its synchronous speed in the steady state. It happens "automatically," and the key lies in the torque angle. Suppose, for instance, the motor "wanted" to run a little more slowly. Equation 10-10 indicates that this would bring about an increase of δ, and Eq. 10-8 confirms that this would increase the electromagnetic torque, which in turn would accelerate the motor back to its synchronous speed. Similarly, a momentary increase in speed would result in a temporary decrease of δ, and thereby of the torque, which would restore synchronous speed.

This kind of game does actually occur whenever the load torque changes. A steady state constitutes a condition of ***equilibrium*** in which the electromagnetic torque equals the opposing load torque. (This ignores the torque caused by the "rotational" losses, which are to be discussed later. At this point, this torque may be either disregarded for the sake of greater simplicity, or it may be thought to be already included in the load torque.) A change of load torque upsets this equilibrium and produces a resulting torque that must change the speed of the motor:

$$T - T_{\text{load}} = J \frac{d\omega_m}{dt} \tag{10-11}$$

where J is the *moment of inertia*, i.e., the quantity that takes the place of the mass when Newton's law of motion is applied to rotating motion. In this equation, the angular velocity must be expressed in mechanical units, which is the meaning of the subscript m.

So an increased load torque does indeed slow down the motor temporarily, in agreement with Fig. 10-4a, until the angle δ is sufficiently increased to restore the torque equilibrium and the synchronous speed. Similarly, the motor responds to a decreasing load torque by a temporary increase in speed and, thereby, a reduction of the angle δ, again leading to a new state of equilibrium.

Comparable chains of events are brought about by changes of field excitation. Suppose, for instance, a synchronous motor is driving its load in the steady state, when an operator chooses to raise the field excitation (presumably in order to change the power factor). By Eq. 10-8, this must increase the electromagnetic torque, which temporarily increases the speed. The reader should be able to visualize (and confirm from Eq. 10-10) that this reduces the torque angle and thereby the torque until equilibrium is restored. In the new steady state, the angle δ is smaller than it was before the change, which checks against any pair of diagrams in Fig. 10-2.

10-6 ANALYTICAL APPROACH

To find out more about what happens in a transient interval, some substitutions will be made in Eq. 10-11. The mechanical angular velocity is obtained by changing Eq. 10-10 from electrical to mechanical units for a machine with p poles:

$$\omega_m = \omega_{s_m} - \frac{2}{p}\frac{d\delta}{dt} \tag{10-12}$$

Its derivative with respect to time is

$$\frac{d\omega_m}{dt} = -\frac{2}{p}\frac{d^2\delta}{dt^2} \tag{10-13}$$

For the electromagnetic torque, Eq. 10-8 is abbreviated to

$$T = K \sin \delta \tag{10-14}$$

where K is a constant for a given motor supplied from an infinite bus and operating at constant field excitation.* With these substitutions, Eq. 10-11 may be rearranged to read

$$\frac{2J}{p}\frac{d^2\delta}{dt^2} + K \sin \delta = T_{\text{load}} \tag{10-15}$$

This *differential equation* displays the fact that, in a synchronous machine, both the inertial torque and the electromagnetic torque are functions of the same variable, the all-important angle δ. To obtain an analytical solution in general terms, i.e., without specific numerical values and without digital or analog computation, this equation will now be *linearized*, which limits its validity to small values of δ for which $\sin \delta \approx \delta$. So

$$\frac{2J}{p}\frac{d^2\delta}{dt^2} + K\delta \approx T_{\text{load}} \tag{10-16}$$

which is a linear second-order differential equation of a type very familiar to most readers from the study of many mechanical as well as electrical phenomena (pendulum, spring, LC circuit, etc.). Its solution, obtained by either classical (time domain) or transform methods, may be written as

$$\delta = \frac{T_{\text{load}}}{K} + \left[\delta(0+) - \frac{T_{\text{load}}}{K}\right] \cos \omega_0 t + \frac{1}{\omega_0}\frac{d\delta}{dt}(0+) \sin \omega_0 t \tag{10-17}$$

This means that the angle δ performs harmonic *oscillations* at the radian frequency

$$\omega_0 = \sqrt{\frac{pK}{2J}} \tag{10-18}$$

*Strictly speaking, this constant is valid only at synchronous speed. The justification of its use in the context of speed variations lies in the fact that such variations normally amount to only small fractions of synchronous speed.

and, like the variables in the mechanical and electrical analogies mentioned above, it never settles down to its steady-state value

$$\delta_{ss} = \frac{T_{load}}{K} \tag{10-19}$$

at which the electromagnetic torque would equal the load torque.

In order to illustrate this result graphically, consider the special case of a load torque rising suddenly, at the time $t = 0$, from zero (ideal no-load) to T_{load}. In this case, the initial value of the angle δ is zero, and so is the initial value of its derivative, as can be seen from Eq. 10-12, with $\omega_m(0+) = \omega_{s_m}$. Thus, for this case

$$\delta = \frac{T_{load}}{K} - \frac{T_{load}}{K} \cos \omega_0 t \tag{10-20}$$

and the mechanical angular velocity, obtained from Eq. 10-12, is

$$\omega_m = \omega_{s_m} - \frac{2\omega_0 T_{load}}{pK} \sin \omega_0 t \tag{10-21}$$

Both results are illustrated in Fig. 10-5.

The reader might like to follow the course of events, as it is shown by this diagram. At the initial instant, the motor runs at synchronous speed, and the angle δ is zero. From there on, there is a load torque that opposes the motion and, at first, slows down the motor. The angle δ increases and reaches the value called for by the load torque after a quarter-period. (This refers to the period of the oscillations, $\tau = 2\pi/\omega_0$, where ω_0 is given in Eq. 10-18, and has nothing to do with the period of the alternating currents and voltages in the armature circuit.) But at this instant the speed is less than synchronous, and during the next quarter-period, while the speed returns to its synchronous value, the angle δ reaches twice its steady-state value. And so it goes on, without any chance ever to reach a steady state.

The linear approximation of Eq. 10-16 limits the validity of Eqs. 10-17 to 10-21 and of Fig. 10-5 to fairly small values of the torque angle. When δ gets to be too

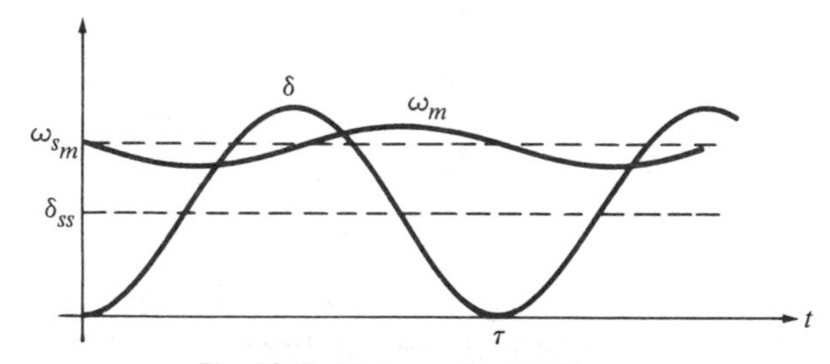

Fig. 10-5. Undamped oscillations.

large for that, the response waveforms are no longer sinusoidal, but the principles remain the same. In particular, if changes of δ do not cause any power consumption, these changes will continue "forever" as ***undamped*** oscillations.

Equation 10-15 is a second-order differential equation, indicating that the system it describes contains two independent energy-storing elements. One of them is easily enough identified as the ***inertia*** of the rotating masses that causes the storage of kinetic energy. The other may be called the ***stiffness*** of the machine, in analogy to the elastic property of a spring, which "wants" to stay in its rest position and resists being forced away from it, just as the synchronous motor runs with an angle $\delta = 0$ unless it is forced to behave otherwise. Its electromagnetic torque increases with increasing values of δ (Eq. 10-14), just as the elastic force of a spring is proportional to its stress.

10-7 DAMPING

Undamped oscillations are obtained as the solutions of differential equations like 10-15 or 10-16, but they hardly ever occur in nature. This is so because these equations fail to make any provision for the presence of energy-consuming elements in the otherwise oscillatory system. Actually, as the reader knows, a pendulum has some friction, an LC circuit has some resistance. As a result, oscillations in such systems are either damped or completely suppressed (overdamped).

Figure 10-6 shows a typical underdamped response in terms of the angular velocity and the torque angle of a synchronous motor whose load torque is a step function of time, as it was in the previous diagram. The variables oscillate with steadily diminishing amplitudes, reaching a final steady state asymptotically, i.e., in theory after infinite time but practically after only a few time constants. (The time constant of an exponentially damped sinusoid is the time at which its envelope is the fraction $1/e \approx 37$ percent of its initial value.) The degree of damping may be expressed by the damping ratio, a constant that relates the period of the undamped oscillations to the time constant.

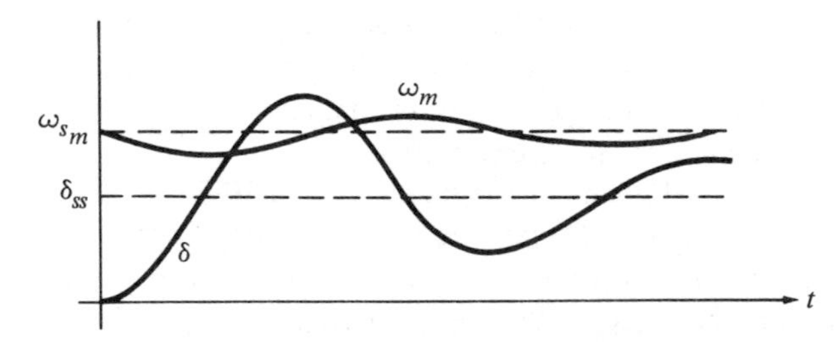

Fig. 10-6. Damped oscillations.

Such an oscillatory transient behavior (in contrast to an overdamped one) of a synchronous machine is sometimes called ***hunting***, and it can have serious disadvantages. For one thing, it can lead to a loss of synchronism, as will be seen in the next section. It also raises the possibility of resonance: if the load torque of the motor (or the driving torque of the prime mover, in the case of a generator) is time-varying, it might have a periodic component whose period happens to coincide with that of the transient oscillation of the synchronous machine, leading to increasing rather than decaying amplitudes.

What helps to damp the natural oscillations of a synchronous machine? The answer is that any closed current path on the rotor serves to produce a *damping torque*. At synchronous speed, there is no voltage induced in such a path because there is no relative motion between the rotor and the magnetic field. But in a transient condition, the mmf wave produced by the stator currents continues to rotate at synchronous speed while the rotor itself runs at a slightly different speed. No calculation is required to show that the result of this relative motion must be some damping. A voltage induced in a conducting path causes a flow of current and thereby the consumption of power in the resistance of that path, and the reader knows that any power consumption caused by oscillations tends to damp them.

Any closed current path—the term includes the field winding (even though that is closed through its source and possibly other external elements), and it includes eddy-current paths in the rotor, which exist to some extent even in a laminated core. So every synchronous machine has some damping. But most synchronous motors (and many generators) are equipped with additional short-circuited windings on the rotor, in order to obtain sufficient damping. They are called *damper windings* or, by their French name, *amortisseur windings*, and they are essentially identical with a certain type of rotor windings (known as *squirrel cages*) much used for induction motors. They do not interfere with the steady-state operation of synchronous machines, because they carry currents and produce torques only at speeds other than synchronous. In fact, it will be shown in the study of induction machines that such a torque is roughly proportional to the "slip speed" $\omega_{s_m} - \omega_m$, at least for speeds not too far away from synchronism. Consequently, as Eq. 10-12 indicates, a damping torque is represented by a first-order term (i.e., a term containing the first derivative of δ) to be added to the left side of Eq. 10-15 (or 10-16), in analogy to the way a resistance term is added to the equation of an LC circuit.

On the other hand, the torque expressed by Eq. 10-8 (or 10-14) is effective only at synchronous speed. To see what happens to this torque at other speeds, consider a rotor rotating at a constant but not synchronous speed. This condition can be viewed as a case of the angle δ changing at a uniform rate. So this torque would have to alternate with the sine of the angle, and its average value would be zero. This is what is meant by the frequently heard statement that a synchronous motor *cannot* run at any other than synchronous speed.

To return to the damper winding of a synchronous motor: it also serves another purpose, apart from that of suppressing hunting. It enables the motor to *start*.

Consider a synchronous motor at *standstill*, with its armature disconnected from its source (or infinite bus). Now let that connection be made, by closing the appropriate switch. Since the motor speed at that moment is not synchronous (but zero), the motor could not start up at all if it were not for its damping torque, which is mainly produced by the damper winding. Due to that torque, the motor does accelerate until it reaches a speed close enough to synchronism so that sin δ stays positive long enough for the motor to be "pulled into" synchronism. Further details about starting will be discussed in the context of induction motors.

10-8 STABILITY

The discussion of both the steady state and the transient operation of the synchronous motor has focused on the importance of the torque angle δ, but up to now the possibility that this angle might exceed 90° has not yet been considered. The diagram of Fig. 10-7 represents Eq. 10-14, and it has been drawn in order to raise and illustrate the pertinent question: for a given value of the load torque, say T_1, could the motor operate in the steady state at the angle δ_2, as well as at δ_1?

This question can be answered with the aid of the reasoning used previously in Section 10-5. For instance, assuming that steady-state operation at δ_2 were possible, consider the effects of a slight increase of the load torque. As the reader

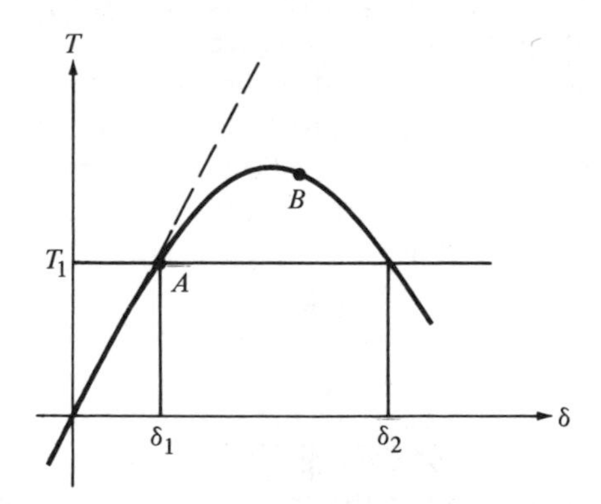

Fig. 10-7. Torque versus torque angle.

knows, this produces an instantaneous decrease of speed, which raises the angle δ. But this time, the increase of the torque angle makes the electromagnetic torque smaller, not larger. As a result, the imbalance of the torques gets worse, not better, and the motor speed goes down further. Instead of reaching synchronous speed again, the motor "loses synchronism," or "falls out of step," and comes to a full stop.

This situation is a typical case of an ***unstable equilibrium***. The simplest and most familiar example of this phenomenon is observed by anybody who tries to balance an object, for instance a pencil, on one point (see Fig. 10-8). The position shown in this diagram is one of equilibrium because the weight of the pencil and the opposing force cancel each other. But the slightest horizontal force upsets this equilibrium, because, after the motion it causes, the two vertical forces form a couple that tends to turn the pencil farther away from its position of equilibrium. The pencil falls down.

As a further illustration of the instability of the motor running at the angle δ_2, let the load torque decrease rather than increase. The resulting chain of events brings the motor to its *stable* condition of equilibrium, at the angle δ_1. The point is that in no case can operation at δ_2 be maintained in the face of even the most minute change of load (or of field excitation).

So a synchronous motor can operate in a stable steady state only at a torque angle of less than 90°. The maximum torque expressed in Eq. 10-9 is the ***stability limit*** of the motor. It follows that the stability problem of the synchronous machine is a consequence of the ***nonlinearity*** of the relation between the torque and the torque angle. If the linearized relation of Eq. 10-16 were generally valid, not only for small values of δ, (see the broken line in Fig. 10-7), there would be no way the motor could lose its stability. (For the benefit of readers familiar with classical feedback control theory, which is so much concerned with the question of stability, it may be added that an analysis of the linearized equation, Eq. 10-16, with or without a first-order term for damping added, will confirm this statement.)

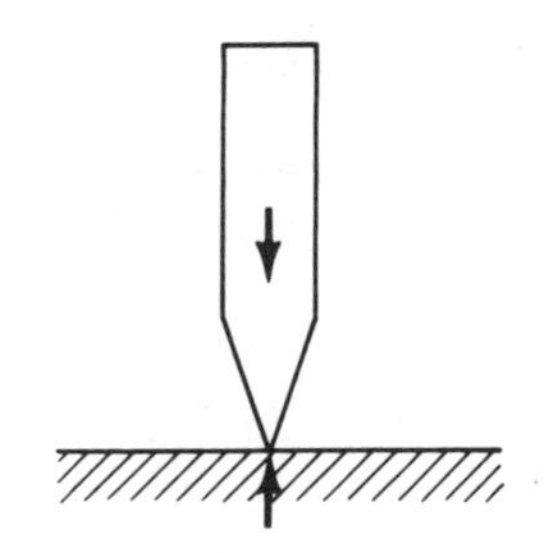

Fig. 10-8. Unstable equilibrium.

It has now been shown that a synchronous motor may operate at any value of δ between zero and 90°, in a condition known as ***steady-state stability***, or static stability. But there are cases in which the motor is unable to reach such a condition, namely when this condition would have to be reached as a consequence of a load change that is too big or too sudden. Figures 10-5 and 10-6 show how the angle δ tends to "overshoot the mark," in response to a step function load torque, to an extent depending on the damping. (Even though these figures are drawn for the linearized case of Eq. 10-16, the general shape of the curves remains valid in any case, except in the overdamped case.) It can happen that the steady state cannot be reached because of this overshoot. Such a case is referred to as a lack of ***transient stability***.

To examine this possibility, refer to Fig. 10-7. Assume again that the load torque rises suddenly from zero to T_1, at the time $t = 0$. The motor slows down instantaneously and thereby permits the angle δ to rise from zero to δ_1. But at the instant when this value is reached, (point A in Fig. 10-7), the motor speed is less than synchronous, and so the motor cannot yet settle down to its new steady state. Instead, the angle δ keeps increasing, which makes $T > T_1$ to accelerate the rotor back to its synchronous speed. While this goes on, δ increases until synchronous speed is reached. If this occurs at a value of δ, like that of point B of the diagram, then the resulting torque, still positive, accelerates the rotor further, above synchronous speed, which decreases δ toward δ_1. Oscillations continue, with decaying amplitudes due to damping. So the final steady state can be reached, even though δ becomes temporarily more than 90°. But if synchronous speed is not reached at point B, and if the angle δ increases above δ_2 before this speed is reached, then the motor cannot regain synchronism and steady state; it falls *out of step* and slows down to a standstill.

If the field excitation is raised, the range of torque values for steady-state operation is increased and the transient stability is improved because the margin of safety for maintaining synchronism is increased. This can be seen from Eq. 10-8, which indicates that more field excitation (a higher value of E_f) lifts the entire curve of Fig. 10-7 upward and thereby increases the maximum torque.

10-9 ANALYTICAL STUDY OF TRANSIENT STABILITY

Let a synchronous motor operate in the steady state, with its torque angle at some value called δ_0 (which need not be zero). Then let the load torque be increased suddenly, at the instant t_0, to a new value requiring a new torque angle called δ_1. Assume that the response is neither overdamped nor unstable. The torque angle reaches its final value δ_1 for the first time at the instant t_1, and then overshoots it, reaching its maximum value δ_2 at the time t_2.

These events are sketched in Fig. 10-9. The problem is to find the maximum

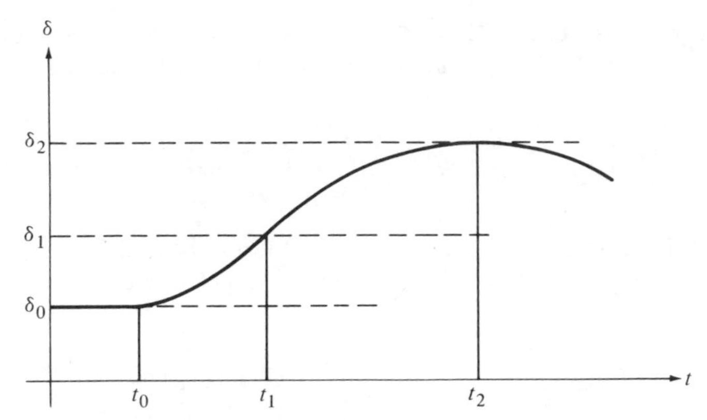

Fig. 10-9. Torque angle versus time.

torque angle δ_2 because transient stability requires that, at this maximum value δ_2, the electromagnetic torque be larger than the load torque.

The solution is based on the *kinetic energy*

$$W_k = \frac{J\omega_m^2}{2} \tag{10-22}$$

of the rotating masses. Note that the angular velocity is synchronous at the instant t_0, and that it has the same value again at the instant t_2 (from Eq. 10-12). Therefore, the kinetic energy has the same value at these two instants, or

$$W_k(t_2) - W_k(t_0) = 0 \tag{10-23}$$

The change of kinetic energy during this time interval (from t_0 to t_2) can be expressed as the integral of its infinitesimal increments. For this purpose, let Eq. 10-22 be differentiated with respect to time (by means of the "chain rule"):

$$\frac{dW_k}{dt} = \frac{J}{2}(2\omega_m)\frac{d\omega_m}{dt} = J\omega_m\frac{d\omega_m}{dt} \tag{10-24}$$

Thus Eq. 10-23 can be rewritten in the form

$$\int_{t_0}^{t_2} \frac{dW_k}{dt}\,dt = \int_{t_0}^{t_2} J\omega_m \frac{d\omega_m}{dt}\,dt = 0 \tag{10-25}$$

and, by substituting Eq. 10-12,

$$\int_{t_0}^{t_2} J\left(\omega_{s_m} - \frac{2}{p}\frac{d\delta}{dt}\right)\frac{d\omega_m}{dt}\,dt = 0 \tag{10-26}$$

which breaks up the integral into a sum of two terms. In the second term, the *accelerating torque* $T - T_{\text{load}}$ is introduced from Eq. 10-11, leading to

$$J\omega_{s_m} \int_{t_0}^{t_2} \frac{d\omega_m}{dt}\, dt - \frac{2}{p} \int_{t_0}^{t_2} (T - T_{\text{load}}) \frac{d\delta}{dt}\, dt = 0 \qquad (10\text{-}27)$$

In each of the two terms, the integration variable may be changed, together with the corresponding integration limits. So the equation becomes

$$J\omega_{s_m} \int_{\omega_{m_0}}^{\omega_{m_2}} d\omega_m - \frac{2}{p} \int_{\delta_0}^{\delta_2} (T - T_{\text{load}})\, d\delta = 0 \qquad (10\text{-}28)$$

The first term is zero because ω_m has the same (synchronous) value at the two instants t_0 and t_2, leaving the significant result

$$\int_{\delta_0}^{\delta_2} (T - T_{\text{load}})\, d\delta = 0 \qquad (10\text{-}29)$$

Figure 10-10 illustrates this equation. The integral is represented by the difference of the two shaded areas. To satisfy Eq. 10-29, the *two areas must be equal* to each other, which determines the angle δ_2. In the case shown in the diagram, the electromagnetic torque is larger than the load torque when the maximum angle δ_2 is reached, which means that the machine maintains synchronism. Thus the T versus δ diagram provides a clear and elegant method of determining whether or not a certain step load increase throws the motor out of synchronism.*

The beauty of Eq. 10-29 lies in the fact that it remains valid regardless of what the relation between T and δ is. (For instance, it will be seen in Chapter 12 that the simple Eq. 10-14 has to be modified for machines with salient poles.) It is even valid for *damped* oscillations, provided the damping torque is included in T_{load}. But in that case the integral cannot be interpreted as an area in a diagram. What can be done is to use the equal areas criterion without the damping term, and to conclude that, if that criterion indicates stability, the damping can only be beneficial, by increasing the margin of stability.

*This equal areas method has been known for a long time, but it has been wrongly presented in several textbooks. The point is that the area in the T versus δ diagram is *not* the change in kinetic energy.

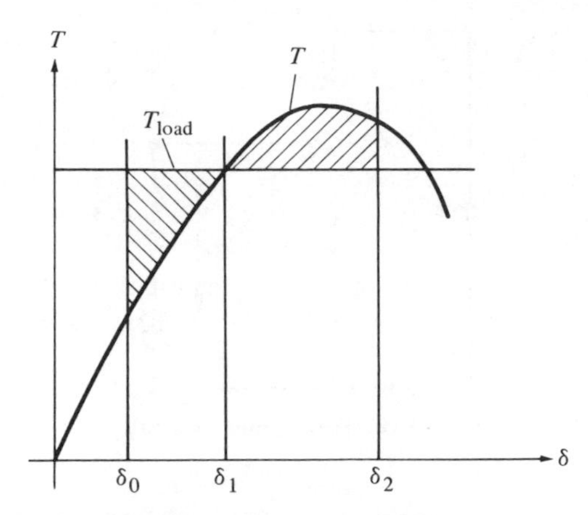

Fig. 10-10. Transient stability.

10-10 METHODS OF CONTROLLING FIELD CURRENT

In the past, *rheostats* have been used to adjust the magnitude of d-c field currents. Electric motors and generators have long lives, so many rheostats are still in use. Their big disadvantage is the loss of energy due to heating in the rheostat.

Figure 10-11 shows a rheostat connected in series with a field winding. The position of the moving arm sets the value of resistance R_1 and permits adjustment of the field current between a minimum of $V/(R_1 + R_f)$ and a maximum of V/R_f. This circuit is used for the field winding of motors where it is desired to keep the field current above some minimum value.

Figure 10-12 shows a rheostat used as a resistance voltage divider to give an adjustable voltage across the field winding. This permits the field current to be

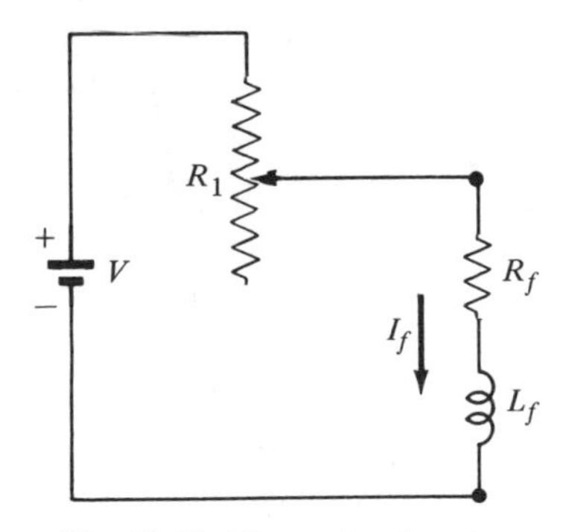

Fig. 10-11. Two-point rheostat.

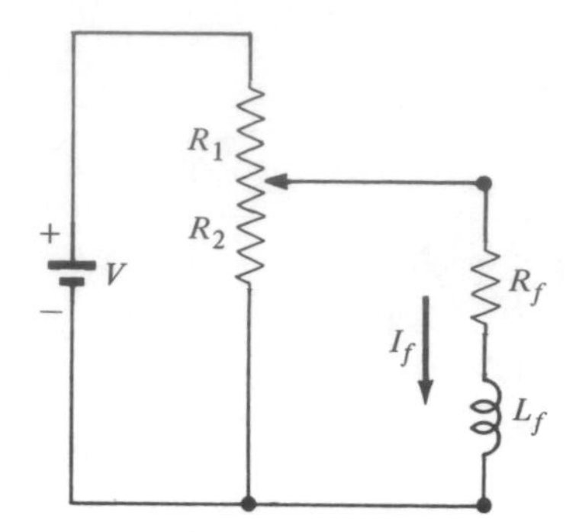

Fig. 10-12. Three-point rheostat.

adjusted between zero and a maximum of V/R_f. This circuit is used for the field winding of generators where the ability to set the field current to zero is important enough to accept the heat loss in both R_1 and R_2.

New installations usually use *diodes* and *thyristors* to supply and control the direct current needed in field windings of electric machines. This also eliminates the need for a d-c source, a need that previously often constituted a drawback of synchronous machines. A solid-state diode consists of one PN junction. A typical voltage-current characteristic is shown in Fig. 10-13. Solving problems for circuits containing diodes is simplified by assuming that the diode functions as a switch that is closed when conditions permit forward current, and open when conditions permit reverse voltage. For such an ideal diode, one of two cases must hold at all times: (1) If $v = 0$, then $i \geqslant 0$. This is the condition for forward current. We say the diode is conducting. (2) If $i = 0$, then $v \leqslant 0$. This is the condition for reverse voltage, and we say the diode is blocking.

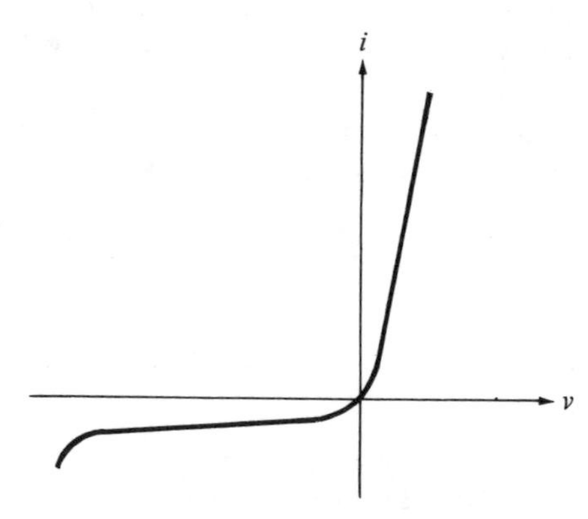

Fig. 10-13. Voltage-current characteristic of a diode.

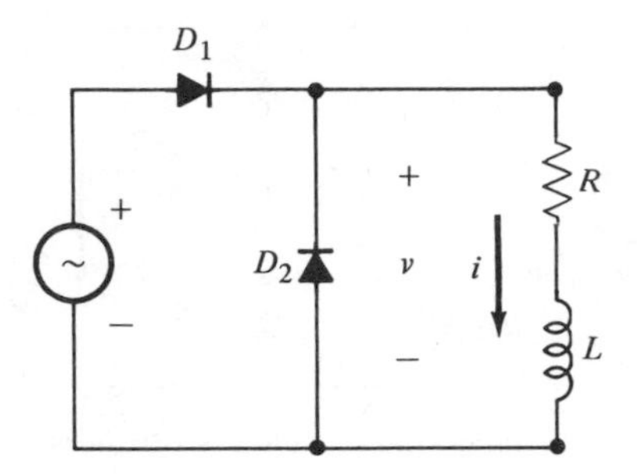

Fig. 10-14. Half-wave diode rectifier circuit.

Figure 10-14 shows a single-phase, *half-wave* diode rectifier supplying a field circuit. Clearly, both diodes cannot conduct simultaneously. When the source voltage is positive, $D2$ is blocking, $D1$ is conducting, and a positive half-cycle of voltage is impressed on the terminals of the field winding. When the source voltage is negative, $D1$ is blocking, $D2$ is conducting, and the voltage across the terminals of the field winding is zero. As $D2$ will always provide a path for current, if no other path is available, this is called a freewheeling diode. This prevents excessive voltage that could appear if the energy stored in the self-inductance had to be dissipated quickly by interrupting the current. The function of voltage across the terminals of the field winding is shown in Fig. 10-15. The average value of this function is V_{max}/π. The d-c component of the field current is $V_{max}/\pi R$. The field voltage has sizeable harmonics. The inductance of the field circuit performs as a low pass filter that allows the d-c component of current to flow and nearly eliminates the harmonics. Note that the source current is a rectangular pulse of current during a positive half-cycle of voltage and that it is zero during a negative half-cycle of voltage.

Figure 10-16 shows a single-phase, *full-wave* diode rectifier supplying a field circuit. Clearly $D1$ and $D3$ cannot conduct simultaneously, and similarly for $D2$ and $D4$. When the source voltage is positive, $D1$ and $D2$ conduct. The positive half-cycle of voltage is impressed on the field circuit. When the source voltage is negative, $D3$ and $D4$ conduct. The source is then connected to the field

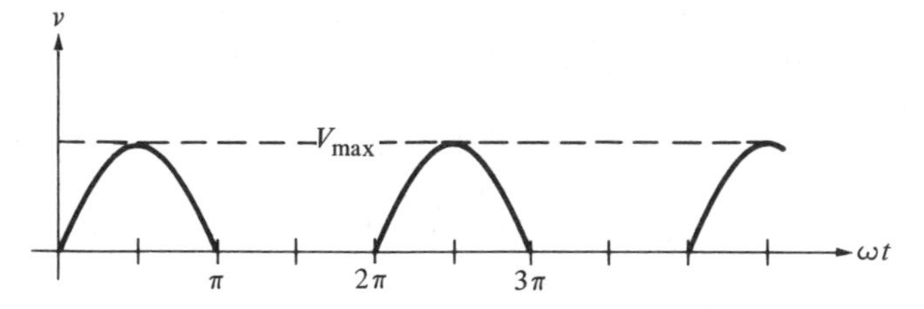

Fig. 10-15. Load voltage of half-wave rectifier.

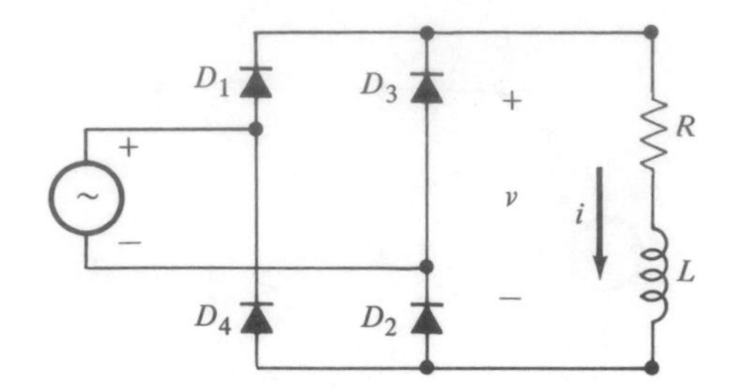

Fig. 10-16. Single-phase, full-wave diode rectifier.

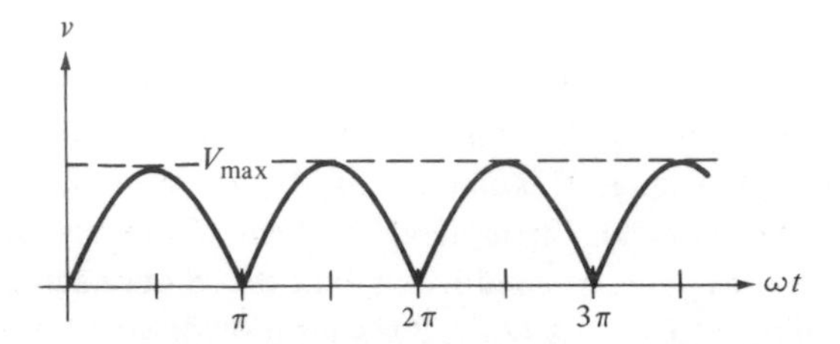

Fig. 10-17. Load voltage of full-wave rectifier.

circuit so as to again impress a positive half-cycle of voltage. The function of voltage across the terminals of the field winding is shown in Figure 10-17. The average value of this voltage function is $2V_{max}/\pi$. The d-c component of the field current is $2V_{max}/\pi R$. One advantage of the full-wave circuit over the half-wave circuit is the delivery of more power to the field winding. Another advantage is that the current through the source is alternating rectangular pulses. The field current has extremely small harmonics. The source current has sizeable harmonics, but it has no d-c component.

The *thyristor* is a semiconductor device containing three internal PN junctions in series. It has three terminals. Two of these, the anode and the cathode, may be thought of as the terminals of a switch. The third one, called the *gate*, permits the control of whether the switch is open or closed. The v - i characteristic of a thyristor is shown in Fig. 10-18. With no gate current, the thyristor behaves like an open switch. During this OFF-STATE, it may absorb either forward or reverse voltage with only small current. With gate current applied, the thyristor will perform essentially the same as a diode. We call this the ON-STATE. If the thyristor has forward voltage while OFF, the transition to ON can be effected by applying gate current. If forward current is flowing in the ON-STATE, the gate loses its control. The thyristor will be turned off only when the circuit conditions establish reverse voltage between the cathode and the anode. After the current is zero, control by the gate is reestablished. It is helpful to assume ideal-

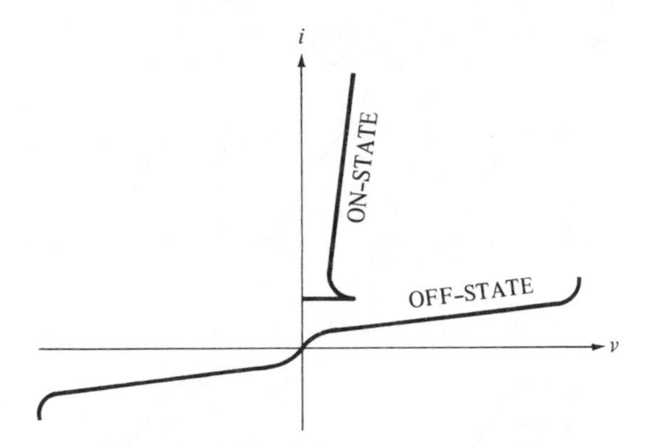

Fig. 10-18. Voltage-current characteristic of thyristor.

ized properties for a thyristor. In the OFF-STATE, it is an open switch. In the ON-STATE, it is an ideal diode.

Figure 10-19 shows a two-pulse converter circuit for energizing a field winding. Diodes $D1$ and $D2$ can play the role of a freewheeling diode, so the voltage across the field can never be negative. As a positive half-cycle of the source voltage begins, thyristor T_1 can be OFF. Let a firing pulse be applied to the gate at $\omega t_1 = \alpha$. We call α the firing angle. $T1$ goes into the ON-STATE, and current flows through $T1$ and $D1$. The remaining portion of a half-wave of voltage is impressed on the field winding. From π to $\pi + \alpha$, it is freewheeling with thyristor $T2$ in OFF-STATE. At $\omega t_2 = \pi + \alpha$, a firing pulse is applied to $T2$. $T2$ goes into the ON-STATE, and current flows through $D2$ and $T2$. Figure 10-20 shows the voltage function across the terminals of the field winding. As the firing angle α is varied from zero to π, the average value of field voltage varies from $2V_{max}/\pi$ to zero, respectively. The average of the voltage function is given by $V_{ave} = (1 + \cos \alpha) V_{max}/\pi$. The source current consists of alternating rectangular pulses. The fundamental component of the source current lags behind the source voltage with a phase angle equal to $\alpha/2$.

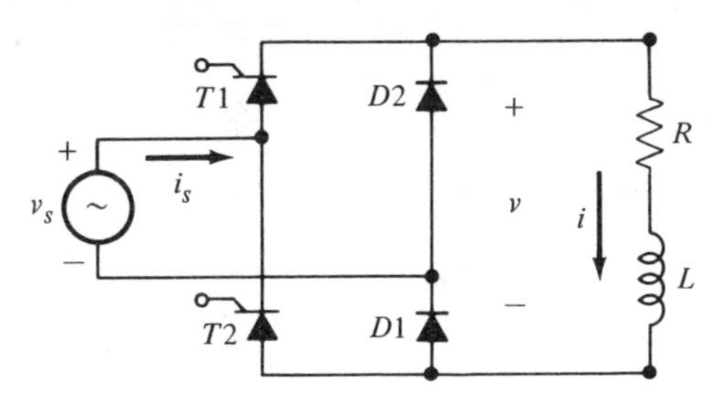

Fig. 10-19. Two-pulse converter.

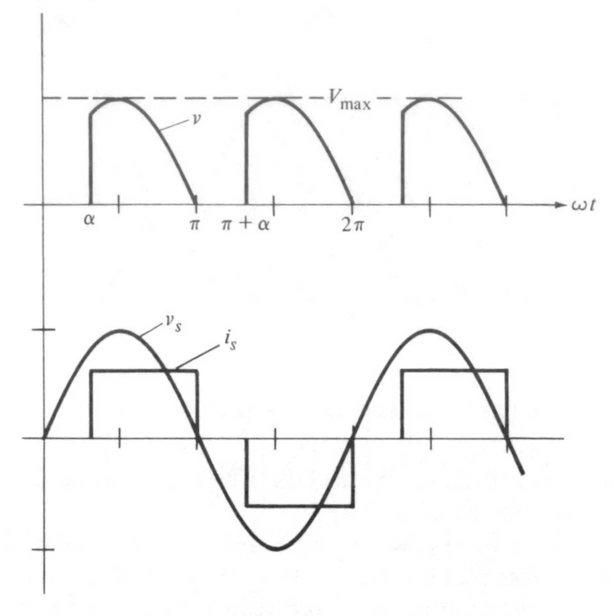

Fig. 10-20. Waveforms of a two-pulse converter.

For high power, a three-phase bridge converter is used. The circuit shown in Fig. 10-21 has an inductive load. The load current is assumed to be constant d-c, as the inductance effectively filters out all a-c components. The voltage source is symmetrical three-phase with *a-b-c* phase sequence. Figure 10-22 shows some of the waveforms associated with this circuit. Observe that the three line-to-line voltages are shown, and the negative of these three functions is also shown. Let t' denote $(\omega t)(180°/\pi)$. This can be considered a normalized time, namely time divided by $T/360$, where T is the period. The idea is that

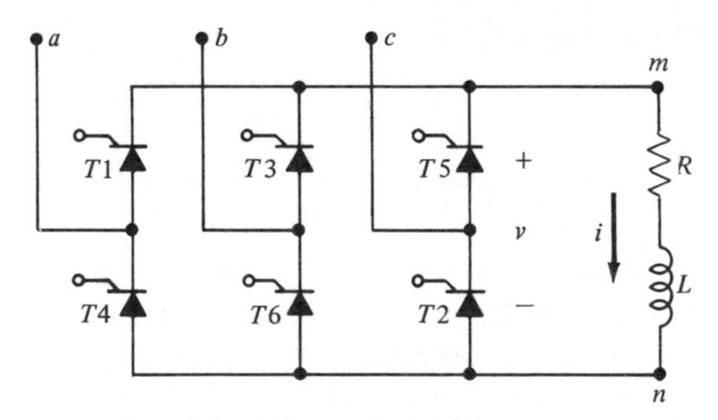

Fig. 10-21. Three-phase bridge converter.

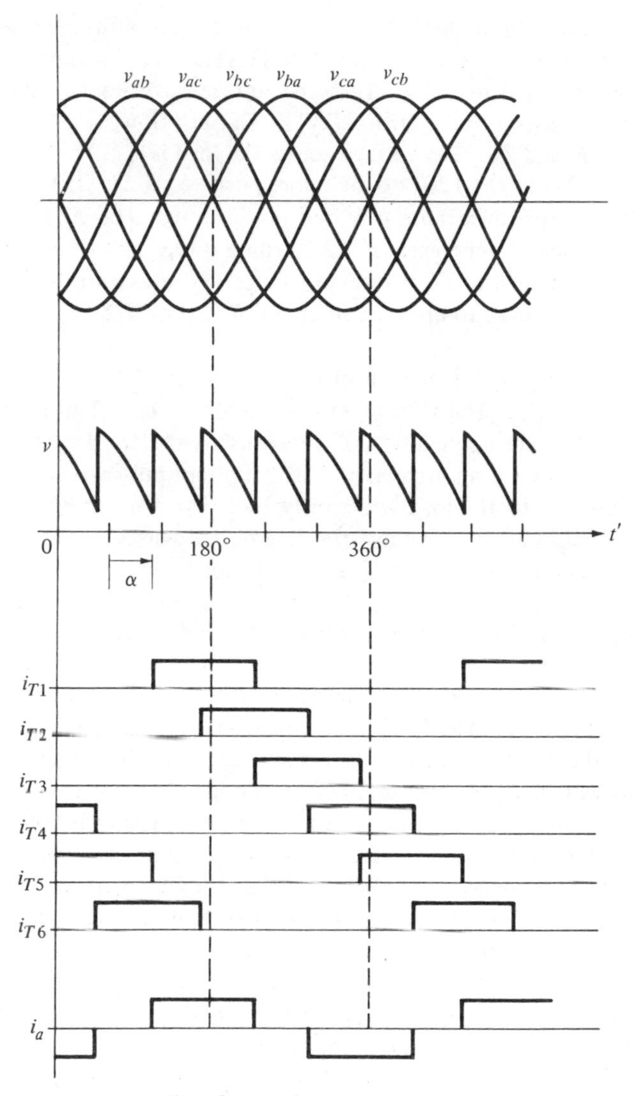

Fig. 10-22. Waveforms of three-phase bridge converter.

only two thyristors conduct simultaneously, thus connecting the load to two of the source terminals. Proper symmetrical control of the thyristors will result in six identical pulses of load voltage during one cycle of the source.

First, assume *T*5 and *T*6 are conducting as the 60° point is approached. Let *T*1 be OFF. After t' passes 60°, v_{ac} is positive and is the forward voltage across

$T1$, because c is tied to m through $T5$. At $t' = 60° + \alpha$, a firing pulse is applied to the gate of $T1$, so it goes into the ON-STATE. $T6$ continues in the ON-STATE. Now, a is tied to m by $T1$, so the voltage v_{ac} is impressed on $T5$ as reverse voltage. $T5$ goes into the OFF-STATE. Current flows from line a to line b through $T1, R, L$, and $T6$. The voltage across the load is v_{ab}.

Next, observe that at $t' = 120°$ v_{bc} becomes positive. $T2$ is OFF, and v_{bc} is the forward voltage across it because b is tied to n through $T6$. At $t' = 120° + \alpha$, a firing pulse is applied to the gate of $T2$, turning it ON. Now, c is tied to n, so the voltage v_{bc} is applied to $T6$ as reverse voltage. $T6$ goes into the OFF-STATE. Current flows from line a to line c through $T1, R, L$, and $T2$. The voltage across the load is v_{ac}.

Next, at $t' = 180°$, v_{ba} becomes positive. At $t' = 180° + \alpha$, $T3$ is turned ON, and $T1$ is turned OFF. The voltage across the load is v_{bc}. This process continues. Firing pulses are applied at $60°$ intervals to the thyristors in sequence. Each thyristor is ON for an interval of $120°$. Voltage pulses of $60°$ duration are successively applied to the load while only two thyristors are ON at any time. These voltage pulses are segments of the line-to-line voltages. The value of the firing angle α determines the wave shape of the load voltage. For α greater than $60°$, the load voltage v can have momentary negative values. The self-inductance forces current to flow through the two thyristors that are ON, even if the voltage v is negative. Figure 10-22 is drawn for $\alpha = 45°$. The average value of load voltage is $V_{ave} = (3V_{max}/\pi) \cos \alpha$. The average value of load voltages is adjustable between zero (for $\alpha = 90°$) and $3V_{max}/\pi$ (for $\alpha = 0°$).

For $\alpha = 0°$, the performance of this circuit duplicates that of a bridge rectifier using six diodes instead of thyristors. Using diodes, the transfer of current from one diode to another occurs at $60°$, $120°$, $180°$, etc. when the voltage function would produce forward voltage across one of the diodes. Using thyristors enables the firing to be delayed by the angle α and, thus, to effect the adjustment of the load voltage. The angle α has been chosen to be measured from the point in the cycle when the transition would have occured if diodes were used.

Only one line current is shown in Fig. 10-22. The current flowing from the source into terminal a is $i_a = i_{T1} - i_{T4}$. This current consists of alternating rectangular pulses. The fundamental component of the current i_a lags behind the line voltage v_{ab} by the angle $30° + \alpha$. This current lags behind the phase voltage v_{aN} by the angle α.

Actual circuits will not quite match ideal cases in detail. The transfer of current from one thyristor to another cannot occur instantaneously. The transition time can be a small portion of a pulse duration, so the wave shapes are close to ideal. Diodes and thyristors do not have zero forward voltage when conducting. Accordingly, actual voltage across the load will be slightly less than ideal. There is heat produced in the thyristors or diodes, so proper heat sinks must be provided to carry the heat away fast enough to maintain safe temperatures in the solid-state devices.

10-11 EXAMPLES

Example 10-1 (Section 10-3)

A 75-HP, three-phase, six-pole, 60-Hz, Y-connected, cylindrical-rotor synchronous motor has synchronous reactance of 9.6 Ω per phase. Its rated terminal voltage is 500 v per phase. (a) Find the value of excitation voltage that makes maximum torque to be 120 percent of rated torque. (b) The machine is operated with the excitation voltage set as in part (a). For rated load torque, find the armature current, the power factor, and the torque angle.

Solution

(a) Synchronous speed is

$$\omega_s = 4\pi f/p = 4\pi(60)/6 = 40\pi \text{ rad/sec}$$

$$P_{\text{rated}} = 75 \times 746 = 56{,}000 \text{ w}$$

$$T_{\text{rated}} = 56{,}000/40\pi = 446 \text{ nm}$$

The excitation voltage can be found from Eq. 10-9

$$T_{\max} = 3(500)\,(E_f)/40\pi(9.6) = 1.2(446)$$

$$E_f = 430 \text{ v}$$

(b) The torque angle can be found from Eq. 10-6

$$P = \frac{3(500)\,(430)}{9.6} \sin\delta = 56{,}000$$

$$\delta = 56.5^\circ$$

For the equivalent circuit of Fig. 10-1, Kirchhoff's voltage equation is given by Eq. 10-1. Use the terminal voltage as the reference

$$\mathbf{V} = 500\,\underline{/0^\circ}$$

For motor operation, the excitation voltage lags behind the terminal voltage by angle δ

$$\mathbf{E}_f = 430\,\underline{/-56.5^\circ}$$

Solve Eq. 10-1 for the armature current

$$\mathbf{I}_a = \frac{\mathbf{V} - \mathbf{E}_f}{jX_s} = \frac{500 \angle 0^\circ - 430 \angle -56.5^\circ}{j9.6} = 53.3 \angle -45.3^\circ$$

Thus, $I_a = 53.3$ amp. The power factor is P.F. = cos 45.3° = 0.7 lagging.

Example 10-2 (Section 10-7)

A 2400-HP, 5500-v (line to line), four-pole, 50-Hz synchronous motor has synchronous reactance of 11.3 Ω per phase. The moment of inertia of the motor and the connected load is 677 kgm². The damping is 728 nm/rad/sec. The excitation voltage is the same magnitude as the terminal voltage. (a) Find the undamped angular frequency and the damping ratio. (b) Let the load be periodic with $T_L(t) = T_1 \cos(7.11\, t)$ nm. Find the value of T_1 that will limit δ_{max} to 1 radian.

Solution

(a) With a damping term, Eq. 10-15 has the following form

$$\frac{2J}{p}\frac{d^2\delta}{dt^2} + K_D \frac{d\delta}{dt} + K \sin\delta = T_L$$

Use Eq. 10-8 to find K

$$T = \frac{1}{50\pi}\,\frac{3(3180)(3180)}{11.3}\sin\delta = 17{,}100 \sin\delta \text{ nm}$$

$$K = 17{,}100 \text{ nm/rad}$$

The linearized approximation and the characteristic equation are

$$\frac{2(677)}{4}\frac{d^2\delta}{dt^2} + 728\frac{d\delta}{dt} + 17{,}100\,\delta = 0$$

$$s^2 + 2.15\,s + 50.5 = 0$$

The general form of this equation is

$$s^2 + 2\zeta\omega_0 s + \omega_0^2 = 0$$

from which we find the undamped natural frequency

$$\omega_0 = \sqrt{50.5} = 7.11 \text{ rad/sec}$$

We also find the damping ratio ζ

$$\zeta = 2.15/2(7.11) = 0.15$$

(b) The frequency of the load pulsations is the same as the undamped natural frequency of this system. We must find the sinusoidal steady-state solution of the linearized equation

$$338 \frac{d^2\delta}{dt^2} + 728 \frac{d\delta}{dt} + 17{,}100\delta = T_1 \cos(7.11t)$$

Use $\omega = 7.11$. The solution is obtained from

$$\delta(t) = \mathcal{R}e \, [T_1 e^{j\omega t}/(338(j\omega)^2 + 728(j\omega) + 17{,}100)]$$
$$= \mathcal{R}e \, [(T_1/j5200)e^{j\omega t}] = -(T_1/5200) \sin(7.11t)$$

If δ is not to exceed 1 rad, then $T_1 = 5200$ nm. Observe that a constant torque of 5200 nm would only require δ to be 0.304 rad, while the pulsating load causes the swings to be approximately three times as great.

10-12 PROBLEMS

10-1. A three-phase, Y-connected, cylindrical-rotor synchronous motor has synchronous reactance of 7 Ω per phase. One point on the open-circuit characteristic is given by 400 v per phase for a field current of 3.33 amp. This motor is operated with terminal voltage of 400 v per phase. The armature current is 50 amp with power factor of 0.85 leading. Find the excitation voltage and the corresponding field current.

10-2. A three-phase, Y-connected, cylindrical-rotor synchronous motor has synchronous reactance of 1.9 Ω. It is operated with terminal voltage of 254 v per phase. The field current is set to give an excitation voltage of 380 v per phase. The load is enough to make the torque angle be 30 electrical degrees. Find the armature current and the power factor.

10-3. A three-phase, Y-connected cylindrical-rotor synchronous motor has synchronous reactance of 4 Ω per phase. It is operated with terminal voltage of 254 v per phase and armature current of 40 amp with power factor of 0.8 leading. Find the excitation voltage, E_f, and the torque angle, δ, for this operating condition.

10-4. A three-phase, 440-v (line to line), 60-Hz, Y-connected, cylindrical-rotor synchronous motor has synchronous reactance of 2.6 Ω per phase. With the motor running at no load, the field is overexcited and set to make the line current to be 19.5 amp. With the same field setting, load is then added to become 100 HP. Find the torque angle δ for this operating condition.

10-5. A three-phase, cylindrical-rotor synchronous motor operates with terminal voltage of 127 v per phase, and excitation voltage of 159 v per phase.

The armature current is 130 amp at unity power factor. Find the synchronous reactance for this machine.

10-6. A three-phase, cylindrical-rotor synchronous motor has a synchronous reactance of 0.8 per unit. Energy is supplied from an infinite bus, with $V_{bus} = 1$ per unit, through a transmission line whose inductive reactance is 0.2 per unit per phase. The load connected to the motor requires 70 percent of rated motor torque. The excitation voltage is 1.35 per unit. Find the line current and the terminal voltage of the motor.

10-7. A three-phase synchronous motor has a synchronous reactance of 0.7 per unit. Rated voltage on the open-circuit characteristic is obtained by field current of 0.82 per unit. This motor is to be operated at rated voltage, rated kva, and 0.8 leading power factor. Find the per-unit field current for this condition.

10-8. A three-phase, cylindrical-rotor synchronous motor has six poles and the frequency is 60 Hz. When the load power is changed from no-load to a load of 80 kw, the rotor position shifts 12 mechanical degrees from its no-load position. Find the maximum power that this motor can develop (i.e., the pull-out power). Also find the maximum torque.

10-9. A three-phase, Y-connected, cylindrical-rotor synchronous motor has synchronous reactance of 3 Ω per phase. The terminal voltage is 254 v per phase. When operating with a power of 43 kw, the field current is adjusted to make the torque angle be 30 electrical degrees. For the same field current, find the maximum power that can be developed. Also find the rms magnitude of the armature current at pull-out.

10-10. A 2400-HP, 5500-v (line to line), four-pole, 50-Hz, synchronous motor has synchronous reactance of 14 Ω per phase. The moment of inertia of the motor and the load is 460 kgm^2. The damping is 364 nm/rad/sec. (a) Find the undamped angular frequency and the damping ratio. (b) Let the load be periodic with $T_L(t) = T_1 \cos(10.9\, t)$ nm. Find the value of T_1 to limit δ_{max} to 1 radian.

10-11. The machine in Problem 10-10, with excitation voltage of 5000 v per phase, is operating at no-load. Find the value of load torque T_1 that can be applied as a step function and assure that the machine can maintain synchronism.

10-12. The machine in Problem 10-10 is operating with load torque at one-half of rated value. Find the minimum setting for the excitation voltage, E_f, to be sure that an increase from one-half of rated load to full load (i.e., a step function of one-half of rated torque) does not cause the machine to lose synchronism.

10-13. For the series field rheostat circuit in Fig. 10-11, the d-c voltage source is 250 v and the field resistance is 100 Ω. Find the power in the field resistance and the power in the rheostat for (a) $R_1 = 100$ Ω, (b) $R_1 = 200$ Ω.

10-14. For the three-point rheostat circuit shown in Fig. 10-12, the d-c voltage source is 250 v and the field resistance is 100 Ω. The total resistance in the rheostat $(R_1 + R_2)$ is 200 Ω. Find the power in the field resistance, the power in R_2, and the power in R_1 for (a) $R_1 = 58.6$ Ω, (b) $R_1 = 100$ Ω, (c) $R_1 = 200$ Ω.

10-15. For the full-wave rectifier circuit shown in Fig. 10-16, the source voltage is 325 sin $377t$. The load resistance is 130 Ω. Find the rms value of forward current through diode $D1$. Find the value of peak reverse voltage across diode $D1$.

10-16. For the two-pulse converter shown in Fig. 10-19, the source voltage is 325 sin $377t$. The load resistance is 130 Ω. The firing angle is 60°. Find the rms value of the forward current and the peak value of reverse voltage for (a) thyristor $T1$, (b) diode $D1$.

10-17. For the three-phase bridge converter shown in Fig. 10-21, $v_{ab}(t) = 325$ sin $377t$. The load resistance is 130 Ω. The firing angle α is 45°. Find the rms value of forward current through thyristor $T1$. Find the peak reverse voltage across thyristor $T1$.

10-18. For the converter in Problem 10-17, the firing angle is changed to 75°. Find the load current. Sketch the load voltage waveform.

11 Synchronous Generators

11-1 THE LOAD

As the reader knows, the terms *synchronous motor* and *synchronous generator* refer not to two different types of machines but rather to two different modes of operation of the same machine. They are discussed in two separate chapters because many of their problems of operation are different. Nevertheless, much of what was learned in the previous chapter (on motors) is also valid for generators, and some of the contents of this chapter are equally applicable to motors.

Since a generator is basically a *voltage source* (although not an ideal one), its load can be expressed in terms of its output *current*. For the sinusoidal steady state, this means the current magnitude and phase angle, or the combination of these two pieces of information, i.e., the current *phasor*. It is also possible to express the load in terms of (average) power and reactive power, or to combine these two values to form the complex power. The relations between these various load quantities, for balanced three-phase systems, are given in Section 6-4.

In addition, the term *load*, as applied to a generator, may also mean the various *devices* that draw current from the generator and consume power supplied by it. One such device is the familiar linear *resistor*, used for a wide variety of practical purposes, e.g., as a heater, a toaster, an incandescent lamp, etc. The current it draws is directly proportional to the magnitude (and independent of the frequency) of the voltage supplied to it. Other types of loads are equivalent to impedances; they draw currents whose magnitudes and phase angles depend on both the magnitude and the frequency of the voltage. A major part of the load of most generators consists of electric motors. The quantitative contribution of a motor to the total load depends again on the magnitude and frequency of the voltage, but also on that motor's own mechanical load.

Each individual load device is designed to operate at its rated voltage and frequency, and its operation is likely to be adversely affected by changes in these two quantities. For instance, an incandescent lamp operating at too low a voltage gives less light than intended, whereas at too high a voltage it is brighter but has a shorter lifetime. This fact alone is sufficient to require generators to be operated at substantially constant *magnitudes* of output voltage.

The requirement to maintain a constant *frequency* is much more stringent. This is so largely because the load of a generator may include electric clocks, which are small synchronous motors, running at strictly synchronous speed. To guarantee that such clocks show the right time, generators must not only maintain their rated frequency within very small tolerances, but must also compensate for their departures from that frequency. In other words, both the frequency and its time integral are constantly checked and corrected, usually by automatic devices.

The control mechanisms for the magnitude and the frequency of the output voltage of a synchronous generator are essentially separate. Adjusting the frequency means adjusting the speed with which the generator is driven. It requires manipulating whatever affects the speed of the prime mover: the steam valve of a turbine, the water gate of a waterwheel, the throttle of an internal combustion engine, etc. Once the frequency is right, the way to adjust the voltage is by changing the field excitation, using the control devices (rheostat or thyristor) discussed in Section 10-10.

11-2 STEADY-STATE OPERATION

For this analysis, the same equivalent circuit will be used as in the previous chapter for motor operation, except for the assumed positive direction of the armature current. The diagram of Fig. 11-1 has a current arrow that makes the *output* power positive when the current phasor I_a is in the first or fourth quadrant, with the terminal voltage V in the axis of reference. A comparison with Fig. 10-1 also shows that the two diagrams are drawn in such a way as to suggest a power flow from left to right. Finally, the reader is reminded that the diagram represents one phase of a three-phase armature circuit.

Kirchhoff's voltage law applied to the circuit of Fig. 11-1 reads

$$\mathbf{E}_f = \mathbf{V} + jX_s\mathbf{I}_a \tag{11-1}$$

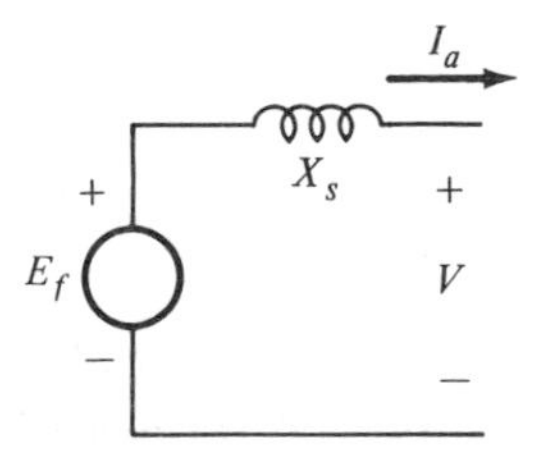

Fig. 11-1. Current direction choice for generator operation.

which is identical with Eq. 9-17 (but different from Eq. 10-1, which refers to Fig. 10-1). It is graphically represented in Fig. 11-2, which is the counterpart, for generator operation, of Fig. 10-2. For these diagrams, the terminal voltage V is assumed to be of constant (rated) magnitude, and it is used as the axis of reference. The phase angle of the current depends entirely on the load; the three phasor diagrams shown represent the cases of loads with unity, lagging, and leading power factors, respectively. The power is assumed to be the same in all three cases.

The following observations can be made from the diagrams:

(a) The phasor of the field excitation voltage for any case of generator operation *leads* the terminal voltage phasor, in contrast to motor operation. In other words, the torque angle δ is reversed, as might have been expected. (The reader should ascertain that this is *not* the consequence of the choice of current arrow direction. If that arrow is reversed, the plus sign in Eq. 11-1 must be replaced by a minus sign.)

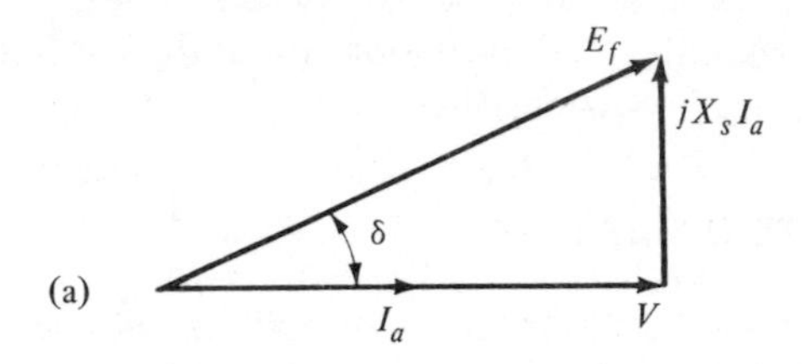

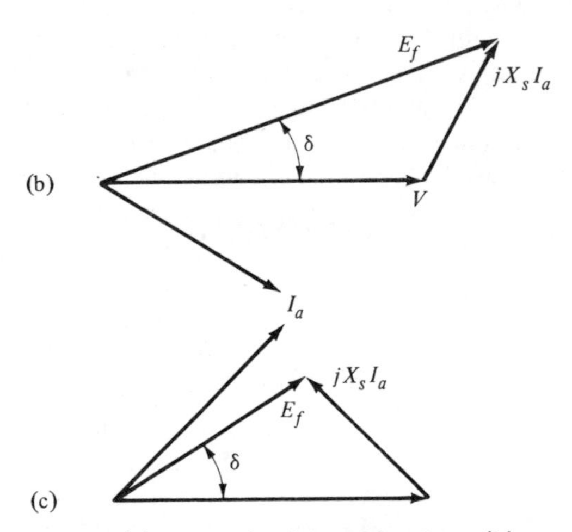

Fig. 11-2. Phasor diagrams: (a) current in phase with voltage; (b) current lagging voltage; (c) current leading voltage.

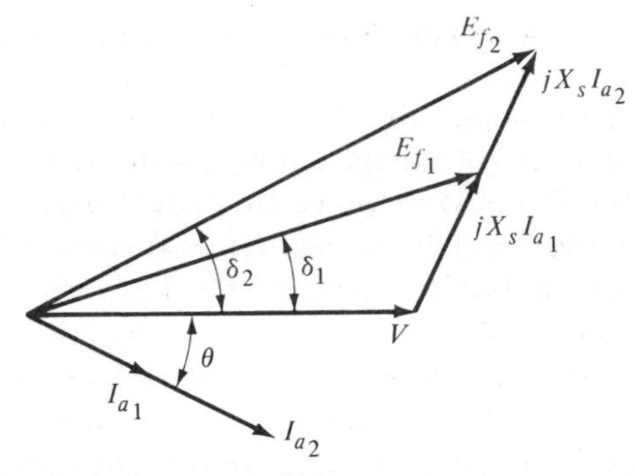

Fig. 11-3. Comparison of two load conditions (same power factor).

(b) Calling the amount of field excitation for the unity-power factor load (Fig. 11-1a) normal, the other two diagrams show that a *lagging* (inductive) load requires *overexcitation*, whereas a *leading* (capacitive) load calls for *under-excitation*. These rules are just the opposite of the ones for motor operation. It is plausible to think of a lagging load as one that needs the higher field excitation because it has to store magnetic field energy, and the rule for a leading load then follows logically.

It must be pointed out that the amount of field excitation needed depends not only on the phase angle, but also on the magnitude of the current drawn by the load. As an arbitrarily chosen example, Fig. 11-3 illustrates the case of a lagging load again (as in Fig. 11-2b), but for two different values of I_a, both at the same angle θ. Clearly, an increased load current calls for an increase in field excitation, if the assumption of a constant terminal voltage is to remain valid.

11-3 VOLTAGE REGULATION

What happens if the field excitation of a generator is not adjusted with changing load? The answer is that the terminal voltage cannot be expected to remain constant. Out of the infinite variety of possible cases, this study will be limited to the most significant ones: a change from *full-load* (rated load) to *no-load* (open-circuit), for any given power factor.

Begin with rated operating conditions: $V = V_{\text{rated}}$, and $I = I_{\text{rated}} = P_{a\ \text{rated}} / 3V_{\text{rated}}$, in terms of phase (not line) values of voltage and current. Next, the field excitation voltage is obtained from the phasor equation Eq. 11-1, either analytically, or graphically as in Fig. 11-2. (Note that this result depends on the power factor of the load.) The problem is to determine what happens to the terminal voltage, if the load is now disconnected, i.e., if the generator is made to operate at no-load but without any further adjustment of the field current.

It might be thought, and a look at Fig. 11-1 might seem to confirm, that E_f is the no-load voltage and the answer to the problem. Actually, however, the value thus obtained would be quite far from correct, in most cases. The point is that the use of synchronous reactance is based on the assumption of *linearity*. Therefore, the field excitation voltage E_f obtained from Eq. 11-1 must be considered as a fictitious voltage, namely, the voltage that would be induced in the armature by the field current alone, in the absence of an armature current, if the *saturation* were unchanged. In the actual open-circuit condition, when the armature current is zero, the saturation may be quite different.

The situation is illustrated by Fig. 11-4 in which the open-circuit characteristic is redrawn from Fig. 9-13. What the concept of synchronous reactance does is to replace that curve by a straight line through the origin, to represent the function E_f versus I_f for the degree of saturation that corresponds to the point of intersection. (Saturation is related to the reluctance of the magnetic circuit. For instance, at point c of the diagram, there is more saturation, and the reluctance is larger, than at point a or b). The line $0c$ is the relation between E_f and I_f for the saturation of point c, etc.

Practical experience has shown and confirmed that reasonably good results are obtained by choosing that straight line that intersects the open-circuit characteristic at the point whose ordinate is rated voltage. Having determined the value of E_f for a given full-load condition (from Eq. 11-1), the correspond-

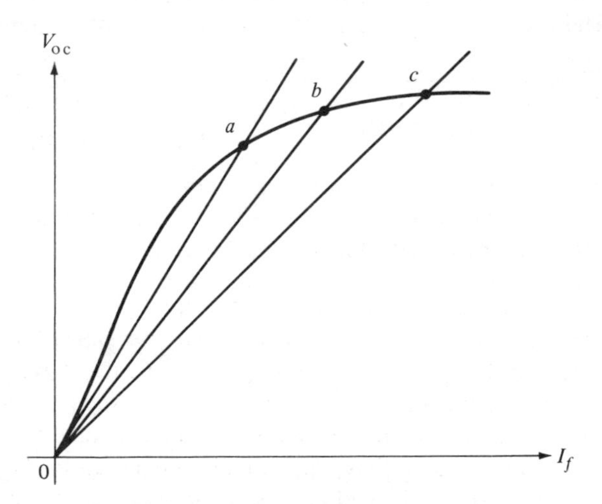

Fig. 11-4. Saturated synchronous reactances.

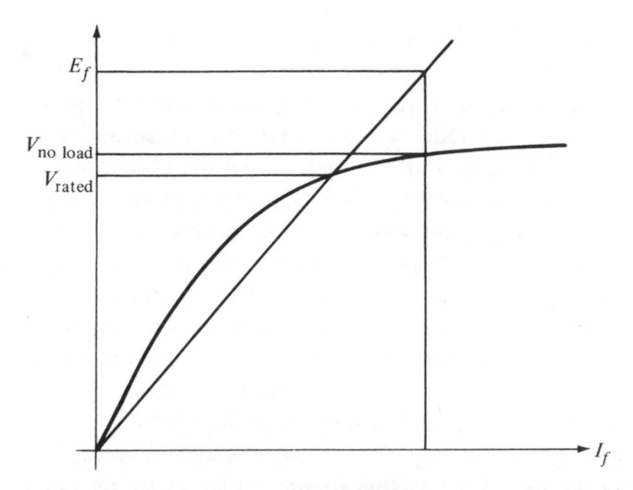

Fig. 11-5. Regulation obtained by saturated synchronous reactance method.

ing value of the field current is then found on that line. (See Fig. 11-5.) The point is that, even though E_f is a fictitious quantity, the value of I_f is real and measurable; thus, the magnitude of the open-circuit voltage is related to that of the field current by the open-circuit characteristic, as shown in Fig. 11-5.

Finally, the term *voltage regulation*, or simply *regulation*, is defined as it was for power transformers (Section 5-6) as the normalized difference between the full-load and no-load values of the terminal voltage

$$\epsilon = \frac{V_{\text{NL}} - V_{\text{FL}}}{V_{\text{FL}}} = \frac{V_{\text{open-circuit}} - V_{\text{rated}}}{V_{\text{rated}}} \tag{11-2}$$

The whole procedure, known as the saturated synchronous reactance method, is illustrated by Example 11-1. It should be noted that different values of regulation are obtained for different load power factors, just as with transformers.

11-4 POWER SYSTEM OPERATION

The discussion of generator operation in this chapter up to this point is based on a seemingly natural assumption, namely that the generator under study has to supply its own load. This leads to the requirement that every load change has to be met by an adjustment of the prime mover torque (in order to maintain the frequency) and of the field excitation (in order to maintain the voltage). Actually, this assumption describes a case that is only occasionally encountered. The normal situation is one in which several, or even many, generators work together to supply the entire load of a power system, or of several interconnected power systems.

In that case, each generator delivers a portion of the total power (and of the total reactive power) demanded at any instant by the entire load of the system or systems. The *distribution* of the load among the generators may be planned (and constantly readjusted) to satisfy requirements of reliability and economy. For instance, there should always be enough generators in operation ready to meet any foreseeable increase of the load. (Such generators, not delivering any power but capable of doing so at short notice, are referred to as *spinning reserve*.) Also, it is desirable that every generator be operating at that load condition at which it is most efficient, and differences in operating cost between the various generators should be taken into consideration. For instance, generators driven by waterwheels do not consume any fuel; thus, their operating expense is mainly depreciation of the initial cost and interest. To a large extent, the same can be said of generators driven by steam turbines whose power is obtained from nuclear fission. On the other hand, when fossil fuels (oil, coal, or gas) are the prime sources of energy, then every kilowatt hour generated has to be paid for in terms of fuel consumption.

This was just a quick and superficial glimpse of the problems of operating (to say nothing of planning and designing) an electric power system. These problems are beyond the scope of this book. What will be discussed here is merely how to operate an individual generator to satisfy the demands made on it by the power system operators.

In general, it is hardly feasible and certainly not desirable to adjust each generator in the system simultaneously in response to every load change. Instead, the generators may be divided into two groups: one or more of them are to be continuously adjusted to maintain the magnitude and frequency of the voltage; all the others are thereby, in effect, connected to an *infinite bus* (see Section 10-1), as are all the loads, including the synchronous motors studied in the previous chapter.

11-5 GENERATOR AND INFINITE BUS

The operation of a generator connected to an infinite bus is significantly different from that of a single generator, as discussed in the first three sections of this chapter. Since the magnitude and frequency of the terminal voltage are fixed quantities, they are not affected by any manipulation of that generator or its prime mover. So the question may be asked: what happens when the generator field current or the prime mover torque are adjusted?

The first point to be kept in mind is that, since the frequency is fixed, the steady-state *speed* of the generator and the prime mover cannot change either.

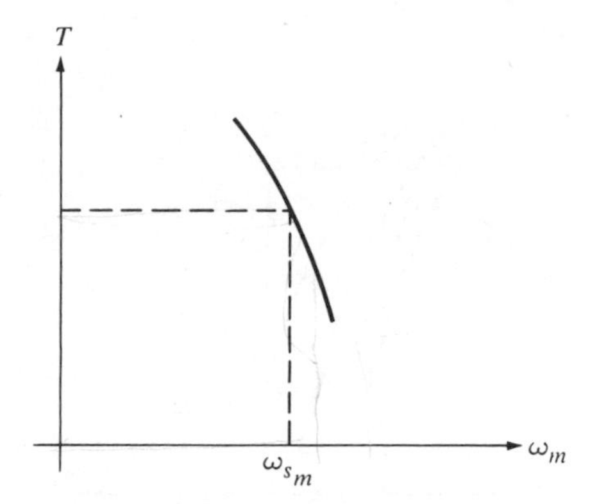

Fig. 11-6. Prime mover characteristic.

It is strictly held to its synchronous value defined in Section 9-3. Any instantaneous departure from that speed changes the electromagnetic torque so as to accelerate or decelerate the rotor back to synchronous speed, just as it does in motor operation.

The next point to be understood is that the prime mover has a definite relationship between its speed and its torque. Figure 11-6 shows what the graph of such a relationship, called the *speed-torque characteristic* of the prime mover, may look like. The negative slope of the curve is both natural (to carry a higher load, a machine must slow down, just as a human being would) and necessary in the interest of stability (as will be explained later). Similar characteristics, incidentally, will be encountered and discussed at greater length in the study of induction motors and d-c motors. At this point, however, we are concerned with devices of mechanical engineering, like turbines or internal combustion engines.

This characteristic determines the torque produced by the prime mover at synchronous speed, and thereby the output power of the prime mover, which is the input power of the generator. Deducting the power losses of the generator from its input leads to a practically constant amount of power contributed to the electric power system by the generator.

There is only one way to change that power: the prime mover's speed-torque characteristic itself must be changed. This can be done by adjusting whatever it is that controls the power input to the prime mover. For instance, to increase the power, one must open the steam valve, or "step on the gas," etc. This shifts the curve to the right, as sketched in Fig. 11-7, and thereby raises the torque at synchronous speed, from *a* to *b* to *c*.

In the case of such a change, there is a *transient* interval between the initial and the final steady state. For instance, when the prime mover input is raised,

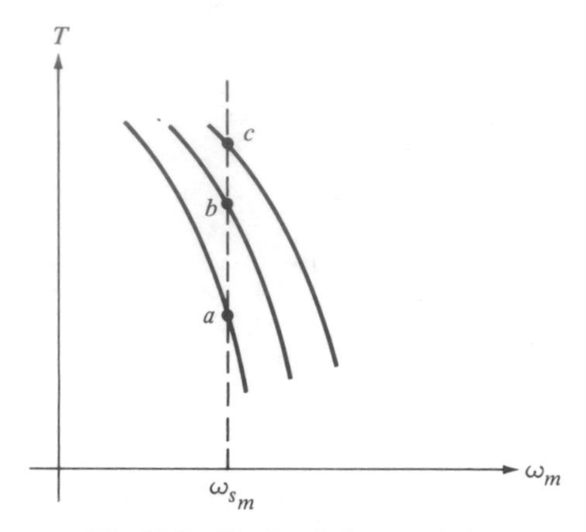

Fig. 11-7. Family of characteristics.

the increased torque must at first accelerate the rotor and thereby advance it relative to its previous steady-state position. In other words, the angle δ is increased, and so is the electromagnetic torque, according to Eq. 10-8, which is valid for motor as well as for generator operation. There is a difference, however: the angle δ and the torque are both reversed; now the torque opposes the motion, as it must for generator operation. So the rotor eventually returns to its synchronous speed, probably with some oscillations similar to those drawn in Fig. 10-6. In the new steady state, the torque angle is increased. Thus, the final result of the shift of the prime mover characteristic is that the generator contributes more power to the infinite bus. Figure 11-8 shows two phasor diagrams illustrating the conditions before and after the change. In this example, the generator was originally operating at normal field excitation

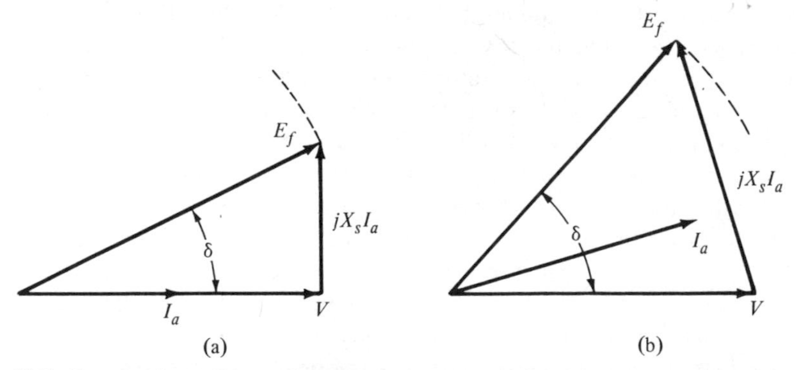

Fig. 11-8. Two load conditions, same field excitation: (a) original steady state; (b) final steady state after increase of prime mover torque.

(unity power factor) for its load. In the final steady state, it is somewhat underexcited (delivering leading current) for the increased load, since its field excitation voltage has the same magnitude as before.

The other possible adjustment of a generator connected to an infinite bus is that of the *field excitation*. Here it must be emphasized that such an adjustment cannot change the power, which is fixed by the generator's synchronous speed and the prime mover's speed-torque characteristic. What actually happens can be seen from phasor diagrams like those of Fig. 11-2. For example, let the generator be normally excited (as in Fig. 11-2a), and then let the field excitation be increased. This produces a momentary increase in the electromagnetic torque. Since this torque opposes the motion, it tends to retard the rotor and thereby to reduce the angle δ, leading toward a new steady-state condition like that of Fig. 11-2b, in which the generator, now overexcited, delivers a lagging armature current. In other words, the generator now has an output of lagging reactive power.

Similarly, a decrease in field excitation forces the generator to deliver leading reactive power, as seen in Fig. 11-2c. Generally speaking, then, any adjustment of the field excitation changes the *kilovars*, whereas the prime mover characteristic must be adjusted to change the *kilowatts*.

11-6 SYNCHRONOUS REACTORS

There is a special condition to be studied, for its theoretical and its practical significance. The prime mover (of a generator connected to an infinite bus) can be adjusted to make the power output of the generator equal to zero. In such a condition, the two voltage phasors $\mathbf{E}_f$ and $\mathbf{V}$ must be in phase with each other (so that their difference, the power angle δ, is zero), but they may well have different *magnitudes*. In that case, there is a nonzero armature current, obtained from Eq. 11-1 as

$$\mathbf{I}_a = \frac{\mathbf{E}_f - \mathbf{V}}{jX_s} \tag{11-3}$$

This current phasor is at a right angle against that of the terminal voltage, as can be seen from Eq. 11-3 or from the fact that the power factor $\cos\theta$ must be zero. The current either leads or lags the terminal voltage by 90°, depending on whether E_f is smaller or larger than V; in other words, whether the machine is under- or overexcited.

With the phase angle $\theta = \pm 90^\circ$, the synchronous machine is in a borderline condition between generator and motor operation, unless the power losses in the machine are considered. Strictly speaking, if the phase angle θ is to be *exactly* 90°, the synchronous machine must be driven by a prime mover whose function it is to supply the power losses. In that case, the synchronous machine should be considered as a generator with a zero power output. Actually, the same purpose can be served much more economically by omitting the prime mover and operating the synchronous machine as a motor at no-load. Since its power losses are covered from the electric power system, its power factor can only be close to zero, like that of an impedance whose real part is much smaller than its imaginary part, i.e., its reactance. This is why a synchronous machine in this mode of operation is usually referred to as a *synchronous reactor*.

There are several reasons why synchronous reactors are used in power systems. The most frequent purpose is the improvement of the power factor. In many power systems, a substantial amount of inductive (lagging) reactive power is needed, due to the presence of many electromagnetic devices like induction motors and transformers, and to the inductances of overhead transmission lines. An overexcited synchronous reactor in parallel to the load can be viewed either as a motor drawing capacitive (leading) reactive power to compensate for the inductive one of the load, or as a generator supplying the load with the inductive reactive power it needs.

Such a *power factor correction* is illustrated in Fig. 11-9 for a power system consisting of a generator, transmission line, and load only (no infinite bus in this case). The main property of the transmission line is its inductive reactance (to which may be added that of transformers at either or both ends of the line). The phasor diagram of Fig. 11-9a shows voltage and current relations without any power factor correction, whereas in Fig. 11-9b, an overexcited synchronous reactor has been added. The voltage at the load terminals is assumed to be the same in both cases. The following conclusions can be drawn:

(a) The total current (drawn by the combination of load plus reactor) is smaller than the load current alone. This can be confirmed from the magnitude relation

$$I_p = \frac{P}{3V_p \cos\theta} \tag{11-4}$$

which is written here in terms of phase (not line) quantities, for a balanced three-phase circuit. As a result, the generator, transmission line, and transformers may have lower ratings and, thus, be less expensive than they would be without power factor correction, or else the power losses in these components are substantially reduced.

(b) The terminal voltage of the generator (V_{gen} in the diagram) differs less from V_{load} when the reactor is added. In the power system of Fig. 11-9, this means that the generator needs less voltage adjustment with changing load, leaving this function largely to the reactor. In the more likely case that the

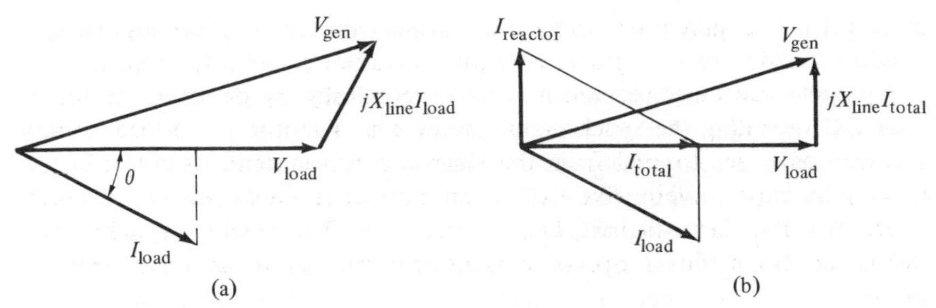

Fig. 11-9. Power factor correction: (a) system without reactor, lagging load current; (b) with reactor.

generator is supplying additional loads and should operate at a constant terminal voltage, voltage fluctuations at the terminals of "our" load can be minimized by adjusting the reactor. More generally, then, synchronous reactors can be used also to adjust the voltages at various points of power systems.

It must be mentioned that (static) capacitors can be, and are being, used for the same purposes in power systems. But synchronous reactors lend themselves well to smooth automatic control methods, since it is only the field current that has to be adjusted, and no switching in a power circuit is required. In addition, synchronous reactors can also be operated with underexcitation, to be equivalent to inductors, something that capacitors cannot do. The need for such an operation can arise at times of low power demand, particularly in systems in which transmission and distribution circuits consist largely of underground cables, because such cables have inherently more capacitance and less inductance than overhead lines.

11-7 POWER LOSSES AND EFFICIENCY

The subject of this section is basically valid for motors as well as for generators. It is being discussed in *this* chapter because most generators have higher power ratings than most motors, so that the economic aspects of generator operation may be considered as even more significant than those of motor operation.

The nature of power losses in a synchronous machine is to a large extent the same as what was encountered in the study of transformers. The main difference is that the motion of a rotating machine is responsible for additional losses that do not occur in a transformer. Such losses are caused by the inevitable *friction* between the rotor shaft and the bearings that support the rotor. There is some additional friction between slip rings and brushes. Furthermore, there is *windage* produced by the motion of the rotor. Windage, however, is not only a cause of power being lost, i.e., converted into heat, it is also a desirable factor in getting the heat dissipated by convection (see Section 1-6).

It should be noted that friction and windage losses do not at all depend on the load condition of the machine. They are functions of the speed only; thus, for most synchronous machines, they are constant. The same thing can be said, at least as an approximation, of the *core losses* (hysteresis and eddy-current losses). These losses depend on the speed and the magnitude of the rotating magnetic flux. For a constant terminal voltage, this flux does not vary with the load except as a result of imperfections (resistance and leakage reactance) of the armature windings.

What these two groups of losses have in common is not only their independence of the load condition, but also the fact that they occur as a consequence of the rotation of the rotor and the flux. For this reason, they are often lumped together under the name *rotational losses* (with the symbol P_{rot}). They account for the difference between the input torque of a generator (or the output torque of a motor) on the one hand, and the electromagnetic torque on the other. It is the electromagnetic torque T whose product by the mechanical angular velocity ω_m is the power converted from mechanical into electrical form (or vice versa).

In addition, there are the so-called copper losses, i.e., the power losses in the resistances of the stator and rotor windings. Here, the difference from the transformer is that, in the synchronous machine, only the *armature* (stator) windings carry load currents, whereas, in the transformer, both the primary and the secondary windings carry currents that depend on the load. Consequently, in the synchronous machine, only the copper losses in the stator windings constitute *load losses*, i.e., power losses that are proportional to the square of the load current.

By contrast, the current in the *field* (rotor) winding is arbitrarily adjusted to some desired value, either for the purpose of obtaining a certain value of terminal voltage, or (in the case of an infinite bus) to obtain the desired amount of reactive power. In any case, the field current does not vary by itself when the load changes. Thus, the power loss in the field winding may be considered constant.

Figure 11-10 is a graphical description of the various power conversions that occur in a synchronous generator operating in the steady state. The main direction of *power flow* is from left to right. Note that the generator has two inputs, a mechanical one (from the prime mover) and an electric one (from the source of the field circuit), and that the latter is entirely consumed in the field circuit resistance. A similar power flow diagram can be drawn for a synchronous motor, showing two electric inputs, the various losses, and the mechanical output.

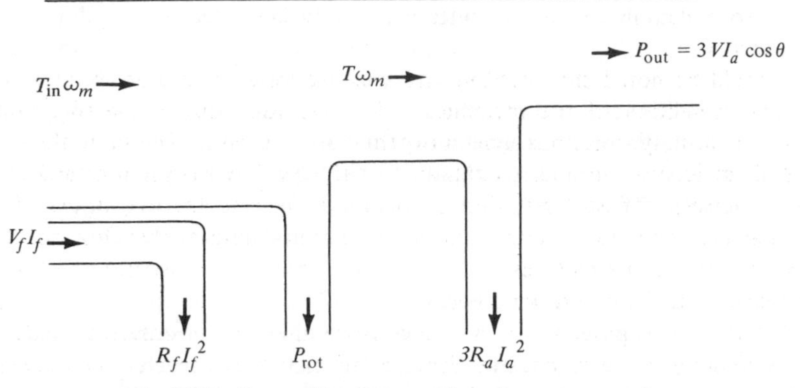

Fig. 11-10. Power flow diagram of synchronous generator.

Based on the division of the power losses into constant losses and load losses, the shape of the efficiency-versus-load curve and the location of its maximum can be found in the same way as for transformers. The reader is referred to Eq. 5-56 and to Fig. 5-11. For any constant power factor of the load, the *maximum efficiency* occurs at that load at which the load losses equal the constant losses. In the case of the synchronous generator, this statement can be derived by differentiating the efficiency

$$\eta = \frac{3VI_a \cos\theta}{3VI_a \cos\theta + P_{rot} + R_f I_f^2 + 3R_a I_a^2} \tag{11-5}$$

with respect to I_a (holding all other parameters constant), and setting the derivative equal to zero, with the result

$$P_{rot} + R_f I_f^2 = 3R_a I_a^2 \tag{11-6}$$

11-8 STABILITY

The subjects of steady-state stability and transient stability were first encountered in the context of motor operation, where they can arise from changes in the mechanical load. Analogous problems in the operation of a synchronous generator connected to an infinite bus can arise from changes in the speed-torque characteristic of the prime mover. The prime mover torque must not be raised above the maximum electromagnetic torque of the generator (Eq. 10-9), and there are definite limits for sudden changes of the prime mover torque.

Generally speaking, a synchronous machine loses its synchronism if the power angle δ exceeds its steady-state limit of 90° and is unable to return below that value. A generator, in such a condition, would be driven by its prime mover at continuously rising speeds, and some automatic safety device would have to curtail the power input to the prime mover drastically to avoid serious damage. Such circumstances, however, are most unlikely to occur in any power plant. For one thing, the prime mover has its own torque limitations, which are presumably intended to match those of the generator. And as far as problems of transient stability are concerned, it should be realized that major changes in the prime mover torque probably cannot be produced so suddenly that they would resemble step functions.

Another hypothetical case of instability could arise if the speed-torque characteristic of the prime mover had a positive slope. Consider, for instance, the possibility of a momentary decrease, no matter how slight, of the prime mover torque. It would bring about an instantaneous decrease in speed which, with a positive-slope characteristic, would further reduce the prime mover torque, leading toward a complete loss of power, at least in the case of a generator supplying its own load.

None of this should be taken to imply that stability problems in the operation of generators exist only in theory. On the contrary; current issues of professional publications are filled with papers on problems of power system stability. But such problems involve much more than the properties of generators and their prime movers. Among other things, such studies have to consider the presence of the numerous and intricate automatic control devices that, as the reader knows, serve to maintain the magnitude and frequency of the voltage in the face of constantly and unpredictably changing loads. In addition, disturbances to be considered do not have to come from the prime mover. In fact, they mostly originate in the electric network to which the various generators are connected, their interplay with each other and with switching apparatus, transmission and distribution circuits, and loads. Such studies require a much more extensive background in subjects like feedback control theory than is expected of the reader of this book, and must, therefore, be considered to be beyond its scope.

11-9 ELECTRIC TRANSIENT PHENOMENA

The transient phenomena discussed in Chapter 10, and again referred to in the foregoing section, are *mechanical* in nature. The time functions involved are space angles and angular velocities, and the energy-storing elements are inertia and *stiffness*, all mechanical quantities. Yet the reader knows very well that inductances are also energy-storing elements and that they are responsible for transient phenomena in electric circuits.

It is possible to separate such *electrical* transients from mechanical ones whenever the time constants (or periods) that characterize them are sufficiently far apart from each other. Fortunately, this is the case for many electric motors and generators. Consider the case when mechanical transients are much slower than electric ones. Then, in the first short time interval following a disturbance (no matter whether this is mechanical or electrical in nature), the electric currents change very quickly before the speed of the machine can undergo any substantial change. By the time that the mechanical transient phenomena reach significant proportions, the electric ones have already decayed to very small, possibly negligible magnitudes, a condition sometimes referred to as a quasi-steady state. In other cases, separating electric from mechanical transients does not lead to trustworthy numerical results, but has still some utility for the physical insight it gives.

The need to study electrical transients in synchronous machines arises mostly for generators that may be subjected to sudden major changes in the electric circuit connected to their armature terminals. The most drastic changes of this kind are not due to loads being switched on or off, but rather to accidental short circuits or breaks in transmission conductors. The classical case is a three-phase short circuit in the vicinity of the armature terminals of a generator.

Unfortunately, the analysis of such a case is far more complicated than it might appear at a first glance to a reader familiar with circuits consisting of inductances and resistances. Simply short circuiting the terminals in the equivalent circuit of Fig. 11-1 will not do. The picture of a rotating stator mmf wave at rest relative to the field circuit, a picture which is basic to the entire method of analysis of synchronous machines used up to this point, does not do justice to electric transient phenomena. The first reason for this is that any changes of armature currents (other than the balanced sinusoidal ones of the steady state) induce voltages by transformer action in the *field circuit* and thereby cause the field current to change.

For purely qualitative purposes, a useful picture can be obtained by thinking of the field winding as the secondary of a transformer whose primary is the armature winding. In the steady state, when no voltages are induced in this secondary, it is as if this transformer were *open circuited*. For this condition,

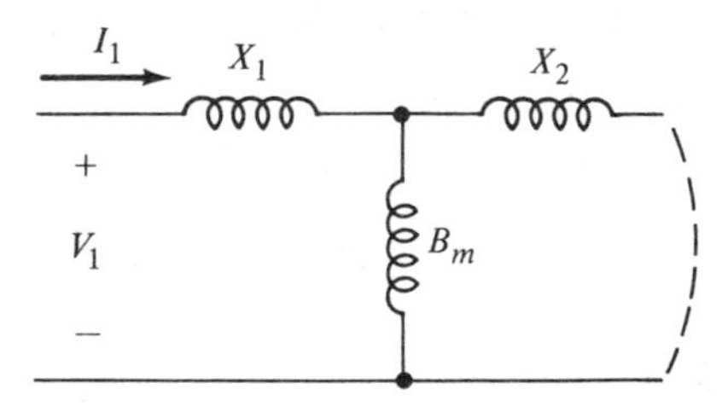

Fig. 11-11. Simplified transformer equivalent circuit.

its primary may be described by the synchronous reactance X_s. On the other hand, under transient conditions (e.g., after a sudden disturbance), the field circuit resembles much more nearly a *short-circuited* transformer secondary. For this case, a new equivalent reactance describing the armature winding (the primary) may be introduced. It is known as the *transient reactance* X_s', and it is much smaller than X_s.

The fact that the driving-point reactance of a transformer is much smaller in the short-circuit condition than in the open-circuit condition can be confirmed by a look at an equivalent circuit. For instance, the diagram drawn as Fig. 11-11 is the same as that of Fig. 5-7, except that the energy-consuming elements are neglected, and the ideal transformer is omitted because it is irrelevant when only a comparison of open-circuit and short-circuit conditions is to be studied. A short-circuit connection is drawn as a broken line.

This picture is complicated by the fact that most synchronous machines are equipped with *damper windings* (see Section 10-7). It has been found that a damper winding tends to reduce further the equivalent reactance of the armature winding to a new value called the *subtransient reactance* with the symbol X_s''.

An oscillograph of an armature current of a generator undergoing such a condition shows a transient component that actually decays at a rate according to two distinguishable time constants. The very rapid decay at the first short time interval following the disturbance corresponds to the smaller time constant τ'', whereas the subsequent slower decay is determined by a larger time constant τ'. It is logical to associate the lower reactance X_s'' to the lower time constant τ'', since the time constant of an RL circuit is L/R. So the whole transient phenomenon can be divided into a "subtransient interval," attributed mainly to the effect of the damper winding, and a "transient interval," in which the field winding is the determining factor.

For an analytical approach to such problems, aiming at a determination of the various currents as functions of time, it must, first of all, be realized that the three armature currents have different initial values at the instant of the disturbance. Consequently, there is no *per-phase* description; each current must be found as a separate time function. Even if there is no damper winding, equations must be set up for four circuits (three armature, one field), each with resistance and self-inductance and each coupled to all the others by mutual inductances. Worst of all, the mutual inductances between any one armature circuit and the field circuit change periodically as the field winding rotates relative to the armature windings. If there is a damper winding (which is the more frequent case), there are additional equations, additional terms in each equation, and other complications.

This system of equations can be made manageable by means of a transformation, the principles of which can be understood after the concept of direct- and quadrature-axis quantities has been introduced. For this, the reader is referred to the next chapter.

11-10 BRUSHLESS EXCITATION FOR A LARGE ALTERNATOR

A synchronous generator as large as 500,000 kva could require 1500 kw to supply the d-c field on the rotor. If slip rings are used, the electric energy lost in the brush contact is several kilowatts. The slip rings and brushes are mechanical devices that wear, so they require attention and maintenance. A scheme for supplying and controlling the d-c field of a large alternator is shown in Fig. 11-12. The idea is to mount a rectifier on the shaft, so d-c can be supplied to the field of the alternator by a direct connection. The input energy for the rectifier is obtained from the a-c armature windings on the

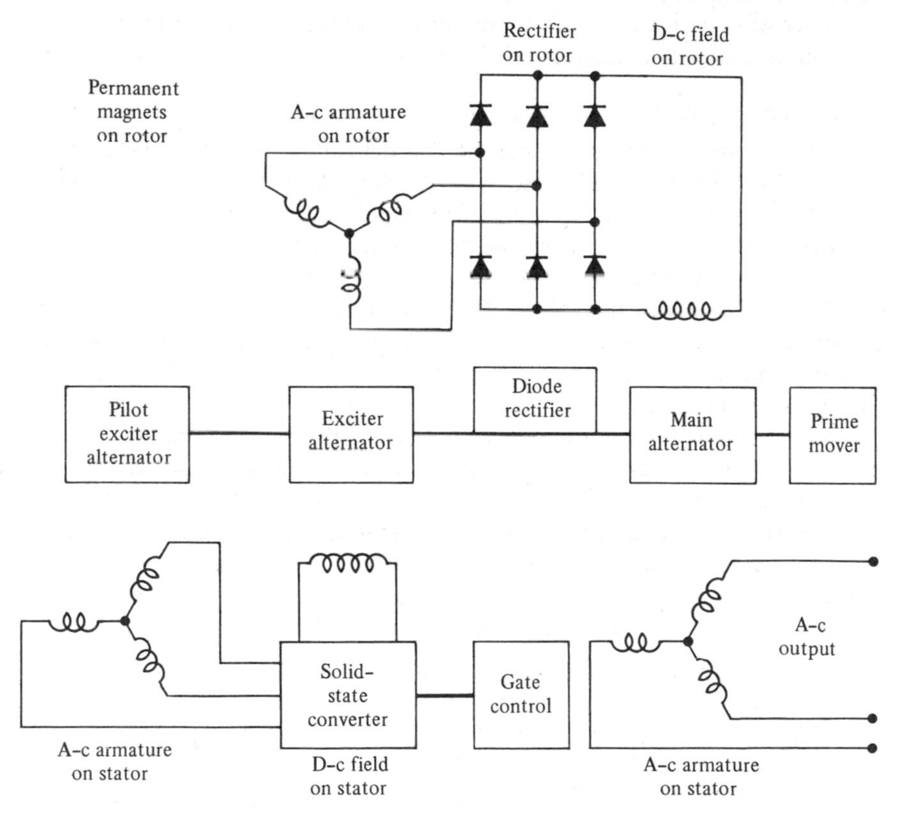

Fig. 11-12. Schematic of brushless excitation for a large alternator.

rotor of the exciter. This exciter is an example of an alternator with the a-c windings on the rotor and with the d-c field on the stator. The d-c field of the exciter is controlled by a solid state converter. One example is the thyristor bridge described in Section 10-10. An increase in the exciter field current would result in an increase in the voltage generated in the exciter armature. This increased voltage would cause an increase in the d-c field current of the main alternator. It is desirable to operate without depending on some external source of electricity, so a third generator, called a pilot exciter, with a permanent magnet rotor is used to supply the energy through the converter to the exciter field. A 2000 kva. exciter could require a pilot exciter of 15 kva. The rotors of the main generator, the exciter, and the pilot exciter are all mounted on the same shaft. So, the prime mover, in driving the shaft, must supply energy to all three machines.

11-11 EXAMPLES

Example 11-1 (Section 11-3)

The alternator in Example 9-1 has synchronous reactance of 11.3 Ω per phase. It is operated as a generator with rated terminal voltage and rated load current with power factor of 0.9 lagging. Use the synchronous reactance method to find the excitation voltage, the excitation required in the rotor field, and the voltage regulation.

Solution

The armature resistance is neglected, so we use the equivalent circuit in Fig. 11-1. Use the terminal voltage as the reference

$$\mathbf{V} = 3810 \underline{/0^\circ} = 3810 + j\,0 \text{ v}$$

$$\mathbf{I}_a = 262 \underline{/-25.8^\circ} \text{ amp}$$

$$jX_s\mathbf{I}_a = (11.3 \underline{/90^\circ})(262 \underline{/-25.8^\circ}) = 2960 \underline{/64.2^\circ} = 1290 + j\,2660 \text{ v}$$

Use Eq. 11-1 to find the excitation voltage

$$\mathbf{E}_f = \mathbf{V} + jX_s\mathbf{I}_a = 5100 + j\,2660 = 5750 \underline{/27.5^\circ} \text{ v}$$

Figure 11-2b illustrates these conditions. The excitation is determined as shown in Fig. 11-5. The excitation for rated voltage of 3810 volts at no-load is 17,000 At/pole. The excitation for E_f is (17,000 × 5750)/3810 = 25,700 At/pole.

Compare this with the answer to Example 9-1. From the open-circuit characteristic, find the no-load voltage that would be generated for $F = 25{,}700$

At/pole

$$V_{NL} = 4270 \text{ v}$$

The voltage regulation is found using Eq. 11-2

$$\epsilon = (V_{NL} - V_{FL})/V_{FL} = (4270 - 3810)/3810 = 0.12$$

Example 11-2 (Section 11-5)

A synchronous generator supplies power to a large system through a transmission line. The receiving system is equivalent to an infinite bus whose voltage is 1 per unit. The synchronous impedance of the generator is 1 per unit. The reactance of the transmission line is 0.4 per unit, based on the generator rating. Resistances are negligible. The power output is adjusted to be 1 per unit. The excitation of the generator is adjusted so that the power factor at the infinite bus is unity. (a) Find the voltage and the power factor at the generator terminals. (b) Find the excitation voltage of the generator. (c) With the excitation of part (b), find the maximum power that could be delivered. Find the terminal voltage for this condition.

Solution

(a) The equivalent circuit of the generator and the transmission line is shown in Fig. E-11-2a. Use V_B as the reference

$$\mathbf{V}_B = 1 \underline{/0^\circ} = 1 + j\,0 \text{ pu}$$

$$\mathbf{I}_a = 1 \underline{/0^\circ} = 1 + j\,0 \text{ pu}$$

The terminal voltage of the generator is the input voltage to the transmission line, and can be found from Kirchhoff's voltage equation

$$\mathbf{V} = \mathbf{V}_B + jX_T\,\mathbf{I}_a = (1 + j\,0) + j\,(0.4)\,(1 + j\,0) = 1 + j\,0.4 = 1.08 \underline{/21.8^\circ} \text{ pu}$$

The power factor is P.F. = cos 21.8° = 0.93 lagging. The phasor diagram is shown in Fig. E-11-2(b).

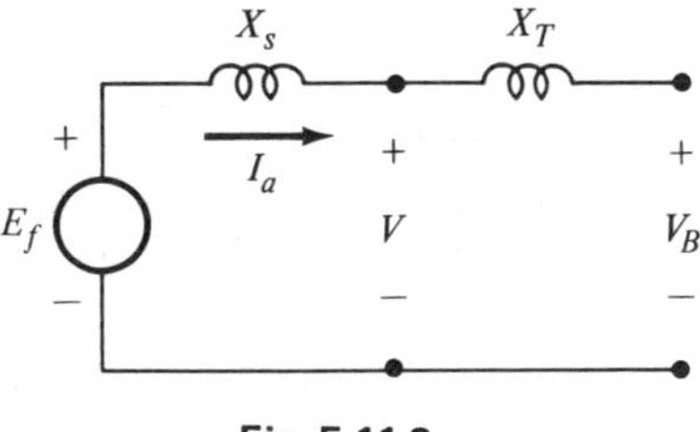

Fig. E-11-2a.

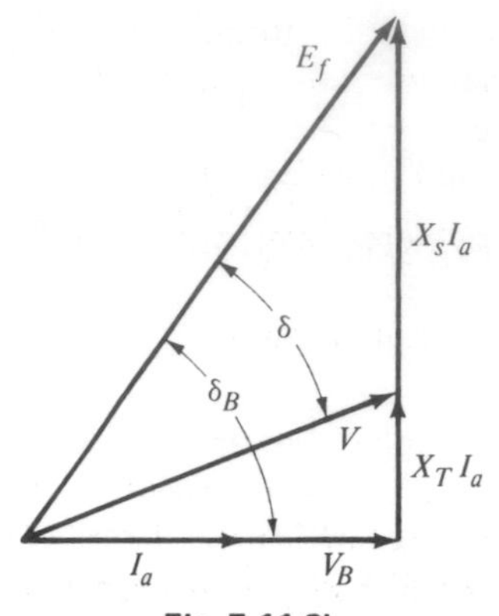

Fig. E-11-2b.

(b) The excitation voltage of the generator can be found using Kirchhoff's voltage equation for the generator

$$\mathbf{E}_f = \mathbf{V} + jX_s\mathbf{I}_a = (1 + j\,0.4) + j\,(1)\,(1 + j\,0) = 1 + j\,1.4 = 1.72\;\underline{/54.5^\circ}\;\text{pu}$$

(c) The power formula in Eq. 10-6 can be modified to apply to this system. Use bus voltage in place of the generator terminal voltage. Use total reactance in place of X_s alone. Use angle δ_B in place of δ. Maximum power occurs for $\delta_B = 90^\circ$. We can solve for one phase

$$P_{\text{max}} = \frac{V_B E_f}{X_s + X_T} = \frac{(1)\,(1.72)}{1.4} = 1.23\;\text{pu}$$

using rated phase power as the basis.

$$\mathbf{E}_f \text{ leads } \mathbf{V}_B \text{ by the angle } \delta_B = 90^\circ$$

$$\mathbf{E}_f = 1.72\;\underline{/90^\circ} = 0 + j\,1.72\;\text{pu}$$

Find the current from Kirchhoff's voltage equation for the system

$$\mathbf{I}_a = \frac{\mathbf{E}_f - \mathbf{V}_B}{j(X_s + X_T)} = \frac{(0 + j\,1.72) - (1 + j\,0)}{j\,(1.4)}$$

$$= 1.42\;\underline{/30.2^\circ} = 1.23 + j\,0.71\;\text{pu}$$

The terminal voltage can be found from

$$\mathbf{V} = \mathbf{V}_B + jX_T\mathbf{I}_a = (1 + j\,0) + j\,(0.4)\,(1.23 + j\,0.71)$$

$$= 0.72 + j\,0.49 = 0.87\;\underline{/34.6^\circ}\;\text{pu}$$

For the pull-out condition, the torque angle δ for the generator is $90° - 34.6° = 55.4°$. We can use this as a check on the computation and to check the power formula

$$P = \frac{V E_f}{X_s} \sin \delta = \frac{(0.87)(1.72)}{(1)} \sin 55.4° = 1.23 \text{ pu}$$

which agrees with the previous answer.

11-12 PROBLEMS

11-1. A two-pole, 60-Hz, Y-connected, three-phase synchronous generator is rated 5000 kva, 2400 v per phase, 694 amp per phase. Data for the open-circuit characteristic are as follows:

E	2110	2570	2860	2990	3080	3140	v per phase
I_f	60	80	100	120	140	160	amp

The synchronous reactance is 4.05 Ω per phase. This machine is operated with rated terminal voltage, with rated load with unity power factor. Find the excitation voltage, the field current, and the voltage regulation.

11-2. For the machine in Problem 11-1, find the voltage regulation for power factor of 0.8 lagging.

11-3. For the machine in Problem 11-1, find the voltage regulation for a power factor of 0.8 leading.

11-4. A 60-Hz, three-phase synchronous generator is rated 1000 v per phase and 200 amp per phase. The synchronous reactance is 3.89 Ω per phase. The open-circuit characteristic is approximated by the following:

$$E = \frac{2400\, I_f}{9 + I_f} \qquad \text{or} \qquad I_f = \frac{9\, E}{2400 - E}$$

where E is the no-load volts per phase and I_f is the field current in amperes. For operation with rated terminal voltage and rated current with power factor of 0.85 lagging, find the excitation voltage, the field current, and the voltage regulation.

11-5. For the machine in Problem 11-4, find the voltage regulation for power factor of 0.85 leading.

11-6. The machine in Problem 11-4 is operated with a field current of 10.4 amp. The Y-connected load impedance is $4 + j\,3$ Ω per phase. Find the load current and the terminal voltage.

11-7. The machine in Problem 11-4 is operated with a field current of 9.6 amp. The Y-connected load impedance is $5.2 + j\,2.8\ \Omega$ per phase. Find the load current and the terminal voltage.

11-8. A three-phase, Y-connected alternator has a cylindrical rotor. With an open circuit at the stator terminals, a d-c field current of 20 amp makes the terminal voltage to be 254 v per phase. Next, a Y-connected, pure resistance load with 1.1 Ω per phase is connected to the stator terminals. Now, it takes 26 amp of d-c field current to make the terminal voltage to be 254 v per phase. Find the synchronous reactance for this machine.

11-9. A three-phase alternator has synchronous reactance of 1 Ω per phase. The open-circuit characteristic is approximated by

$$E = \frac{800\,I_f}{4 + I_f} \qquad \text{or} \qquad I_f = \frac{4\,E}{800 - E}$$

where E is the open circuit volts per phase and I_f is the d-c field current in amperes. The terminals of this machine are tied to an infinite bus of 500 v per phase. The prime mover is adjusted to deliver 600 kw. The field current is adjusted to make the armature current to be 470 amp with lagging power factor. Find the field current.

11-10. The machine in Problem 11-9 has its field current adjusted to make the torque angle to be 31 electrical degrees. Find the field current and the armature current.

11-11. The machine in Problem 11-9 has its field current adjusted to make the power factor be 0.9 leading. Find the field current and the armature current.

11-12. The machine in Problem 11-9 is operated as a synchronous reactor by having its prime mover adjusted to deliver zero power, while the terminals are still connected to the infinite bus with 500 v per phase. Find the corresponding armature current for the following values of field current: 2, 4, 6, 6.67, 8, 10, and 12 amp. Draw a graph of the rms magnitude of armature current versus the field current.

11-13. A three-phase, Y-connected synchronous generator is operated as a synchronous reactor with its terminals tied to an infinite bus of 254 v per phase. With the field current set at 1.6 amp, the armature current is 32 amp. With the field current set at 1.15 amp, the armature current is approximately zero. (a) Find the synchronous reactance. (b) Find the armature current for a field current of 0.93 amp.

11-14. Two four-pole, three-phase alternators operate in parallel to supply a common load of 105 kw and 70 kvar. (lagging). The speed-versus-load characteristics of the prime movers can be approximated by straight lines. Machine A runs at 1830 rpm at no-load and at 1770 rpm for 60 kw. Machine B runs at 1815 rpm at no-load and at 1785 rpm at 60 kw. (a) Find the speed at which the system runs. (b) Find the average power delivered by each generator. (c) The field currents are adjusted so that the terminal voltage is rated value and so that the power factor of each machine is the same as the load power factor. Find the reactive power of each machine.

11-15. Two four-pole, three-phase alternators operate in parallel to supply a common load of 75 kw and 40 kvar (lagging). The speed-versus-load characteristic of the prime movers can be approximated by straight lines. Machine A runs at 1815 rpm at no-load and at 1785 rpm at 60 kw. Machine B runs at 1800 rpm at no-load and at 1785 rpm at 60 kw. (a) Find the speed at which this system runs. (b) Find the average power delivered by each generator. (c) The field currents are adjusted so that the terminal voltage is rated value and so that the armature currents of both machines have equal value. Find the power factor of each machine.

11-16. The machine in Problem 11-1 has armature resistance of 0.022 Ω per phase. The field winding resistance is 0.81 Ω. The rotational losses amount to 32 kw. The machine is operated at rated load with power factor of 0.8 lagging (as in Problem 11-2). Find the efficiency.

11-17. A three-phase, Y-connected synchronous generator is operated with its armature terminals connected to an infinite bus having rated voltage. Its synchronous reactance is 1 per unit. The excitation voltage is 1.6 per unit. Find the torque angle for operation with rated current. Find the power factor.

11-18. The machine of Problem 11-17 delivers energy through a transmission line to an infinite bus having rated voltage. The inductive reactance per phase of the transmission line is 0.4 per unit, based on the generator rating. The excitation voltage is 1.6 per unit. The armature current is rated value. Find the terminal voltage of the generator. Find the power factor at the terminals of the generator.

12

Synchronous Machines with Salient Poles

12-1 DEVICES AND MODELS

The study of engineering devices begins almost invariably with *simplifying assumptions.* After this method has led to results, the effects of the assumptions upon the results can be investigated, both in theory and empirically. In some cases, such investigations reveal no significant difference between the simplified *model* and the actual device. In other cases, the simplifications may have led to results so worthless that a new beginning has to be made, with fewer, or different, such assumptions. Many more cases are found to lie between these extremes.

In describing the *synchronous machine* and its operation in the last several chapters, a number of simplifying assumptions were made. In particular, the introduction of the concept of synchronous reactance had to be based on an assumption of *linearity*, or, in more specific terms, constant saturation. This assumption holds fairly well over a wide range of operating conditions, especially when the machine is connected to an infinite bus, or otherwise as long as its terminal voltage is held to substantially constant values, which is normally the case. The one exception encountered in these pages, the open-circuit operation of a generator, illustrates what difficulties may be expected to arise when the assumption is no longer valid.

The chosen model of the synchronous machine is also based on the assumption that all mmf waves and flux distributions are *sinusoidal*; in other words, that only their space fundamentals need to be considered. This is a different case inasmuch as the methods used for the analysis of the model, regardless of how closely the model approximates the device, can always be expanded to take the space harmonics into consideration. Since this adds but little to a basic understanding of the synchronous machine, it is left to more advanced studies.

Another simplification was made in Section 9-4 where the effect of *salient poles* was put aside. The consideration of this effect does add something significant to the analysis of machines equipped with salient poles. Furthermore, it leads, rather surprisingly, to new methods useful even in the study of machines with cylindrical rotors.

12-2 DIRECT AXIS AND QUADRATURE AXIS

A few diagrams may be helpful to clarify the issue. Figure 12-1 is a sketch of a synchronous machine whose rotor has two salient poles. Only one stator coil is drawn, just to indicate the axis of the stationary mmf wave of the reference phase (a) of the armature current. The positive direction of rotation is marked (counterclockwise), and so are the positive direction of the field current and the corresponding polarities (*n* for north and *s* for south) of the rotor poles. Finally, the choice of positive armature circuit direction (indicated by cross and dot) is such that, at the instant of the diagram, the excitation voltage induced in phase *a* is positive if the field current and the direction of rotation are positive. (Check this with the directional rule stated in Section 7-1, remembering that it refers to the direction of motion of the armature conductor relative to the field, which is opposite to the rotor motion.)

This diagram is, in effect, reproduced in Fig. 12-2a where it is developed for greater convenience. Then, in Fig. 12-2b, there is a snapshot of the space fundamental of the *rotor mmf wave*, for the instant of the two previous diagrams. As the rotor moves, this wave always moves with it, with its peaks always facing the pole centers. The axis of this wave is called the *direct axis*, and it is marked d - d in both Figs. 12-1 and 12-2a.

By contrast, the location of the axis of the rotating *stator mmf wave* depends on the phase angle of the armature current. Suppose, for instance, that this current is in phase with the excitation voltage. In this case, at the instant depicted by the diagrams, both the excitation voltage and the armature current have positive maximum values, and the stator mmf wave is located 90 electrical

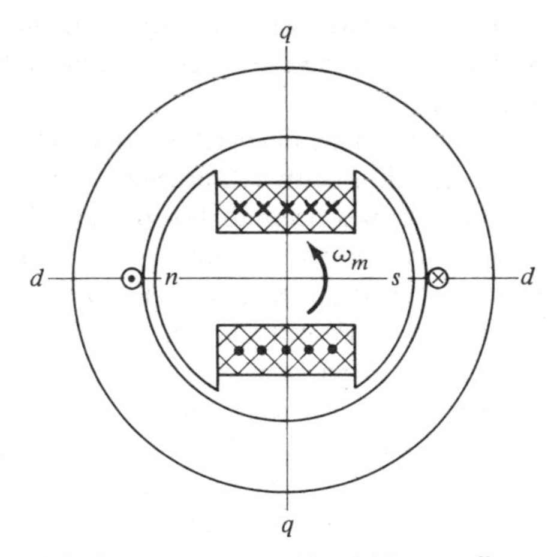

Fig. 12-1. Synchronous machine with two salient poles.

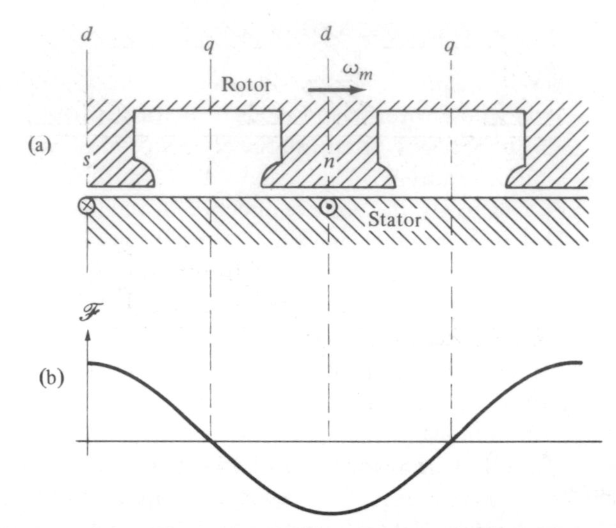

Fig. 12-2. Developed view: (a) stator and rotor; (b) rotor mmf wave.

degrees to the left (in terms of Fig. 12-2b) of the rotor mmf wave. This is true not only for the stationary mmf wave of the current in the reference phase, but also for the rotating mmf wave produced by the three armature currents together, in accordance with paragraph (c) (the third of four observations about rotating armature mmf waves) near the end of Section 8-8. So the axis of the stator mmf wave is midway between two adjacent poles, in a location known as the *quadrature axis* because it is in space quadrature with (i.e., 90 electrical degrees away from) the direct axis. See the q - q markings in Figs. 12-1 and 12-2a. Both the direct and the quadrature axis are thus defined in relation to the rotor, and they both rotate at all times at the same speed as the rotor.

It is also possible for the axis of the stator mmf wave to coincide with the direct axis. This occurs when the armature current is 90° out of phase with the excitation voltage. In this case, the armature conductors of Fig. 12-2 again have positive maximum values of excitation voltage, but the armature current is zero at that instant. Thus, the rotating armature mmf wave is a quarter of a wave length (i.e., 90 electrical degrees) away from where it was in the previous case, which puts its axis right into the direct axis.

To recapitulate: the axis of the armature mmf wave coincides with the quadrature axis when I_a is in phase with E_f, and it coincides with the direct axis when I_a is 90° out of phase with E_f. These are two special cases; with any other phase relationship between I_a and E_f, the axis of the armature mmf wave is somewhere in between the direct and the quadrature axis.

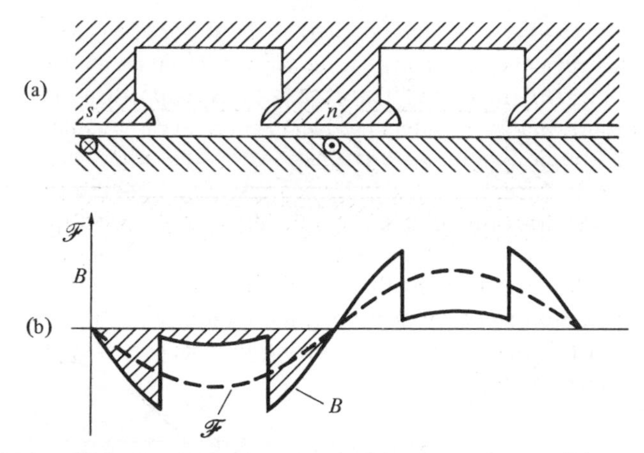

Fig. 12-3. Stator mmf wave in quadrature axis: (a) stator and rotor; (b) stator mmf wave and flux distribution.

12-3 SYNCHRONOUS REACTANCES FOR SALIENT-POLE MACHINES

The relation between an mmf wave and the flux it produces depends on the magnetic circuit. In the case of a cylindrical rotor, an mmf wave of given magnitude produces the same amount of flux no matter whether it is in the direct or in the quadrature axis, or anywhere in between, because it faces the same magnetic circuit everywhere. This is not so if the rotor has salient poles.

Figures 12-3 and 12-4 represent the two extreme cases discussed in the previ-

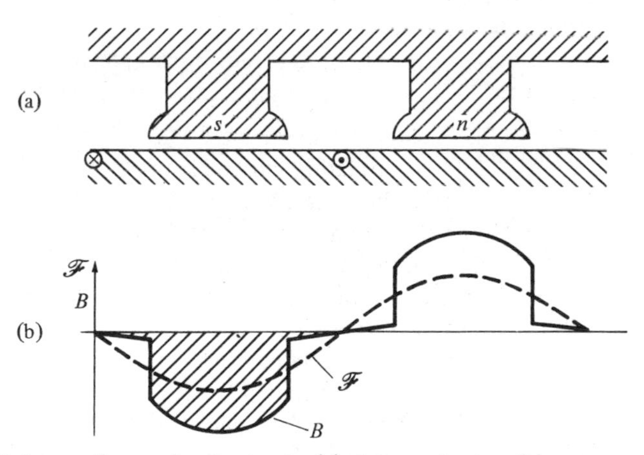

Fig. 12-4. Stator mmf wave in direct axis: (a) stator and rotor; (b) stator mmf wave and flux distribution.

ous section. Both figures contain sketches of the stator and rotor similar to Fig. 12-2a, but this time they are meant to be snapshots taken at the instant when the *armature current* (in the reference phase) has its positive maximum value. If, at that instant, the pole centers are facing the armature conductors, as they do in Fig. 12-3, then I_a and E_f are in phase with each other, and the armature mmf wave is in the quadrature axis. By contrast, Fig. 12-4 represents the other extreme, with the armature mmf wave in the direct axis. In either figure, part b is a snapshot of the armature mmf wave (the broken line) and a reasonable approximation of the corresponding flux distribution.

These curves are obtained by neglecting the reluctance of the iron part of the magnetic circuit (in other words, by assuming an ideal core material). Thus, the mmf wave is taken to be

$$\mathcal{F}(\theta) \approx gH(\theta) = \frac{g}{\mu_0} B(\theta) \tag{12-1}$$

similar to Eq. 8-2 but not restricted to a single coil. This way, the flux distribution is proportional to the mmf wave and inversely proportional to the length g of the air gap. If the curves had somehow been drawn accurately, without any approximation, they would be slightly distorted. They would also display some effect of fringing near the pole edges, but their main features would be the same. The essential observation can be made from the diagrams, without any calculation: the *flux per pole* (which is proportional to the shaded area) is decidedly larger when the axis of the mmf wave coincides with the direct axis than when it coincides with the quadrature axis. In other words: an mmf wave in the direct axis produces more flux than in the quadrature axis.

The reader is now referred to Section 9-6 in which the idea of superposition was applied to the effects of the two rotating mmf waves produced by the armature currents and the field current. Specifically, a voltage E_a induced by the rotating armature flux ϕ_a appears in Fig. 9-9 and is expressed in Eq. 9-12 as proportional to the armature current and lagging it by 90°. If this concept is to be used for a salient-pole machine, the coefficient X_a (called the armature reaction reactance) must be given a different value depending on the location of the armature mmf wave relative to the rotor poles. Considering the two extreme cases depicted in Figs. 12-3 and 12-4: when the armature mmf wave is in the quadrature axis, this reactance has a value X_{aq}, which is smaller than the value X_{ad} that it takes on for an armature mmf wave in the direct axis.

What is to be done if the axis of the armature mmf wave does not coincide with either the direct or the quadrature axis? Is it necessary to use a different value of armature reaction reactance for every angular position? The French engineer A. Blondel was the first to suggest an idea that turned out to be extraordinarily fruitful, even beyond these immediate questions: let the armature mmf wave be decomposed into two *component waves*, one in the direct axis and one in the quadrature axis; as a result, only two values of armature reaction reactance, namely X_{ad} and X_{aq}, have to be used.

The way to accomplish this follows from the results obtained in the previous section. It is the angle between the armature current I_a and the excitation voltage E_f that determines the location of the axis of the armature mmf wave relative to the direct and quadrature axis. Consequently, each of the three armature currents is split into two components, one in phase with the excitation voltage and one in phase quadrature to it. This is illustrated in Fig. 12-5, which is drawn (as an arbitrary example) for an overexcited generator, similar to that of Fig. 10-2b.

Using the symbol ψ for the phase difference between I_a and E_f, the phasor diagram of Fig. 12-5 shows the magnitude relations

$$I_q = I_a \cos \psi \tag{12-2}$$

and

$$I_d = I_a \sin \psi \tag{12-3}$$

The subscripts given to these *component currents* indicate the location of their mmf waves. The current I_q in the reference phase, together with its counterparts in the other two phases, produces a rotating mmf wave called A_q whose axis coincides with the quadrature axis. Similarly, I_d and its two counterparts produce a rotating mmf wave called A_d whose axis coincides with the direct axis.

The sign of I_d (plus or minus) depends on the arbitrary definition of the angle ψ (as the angle by which E_f leads I_a, or vice versa). Also, this same definition determines whether ψ is the sum or the difference of the torque angle δ and the

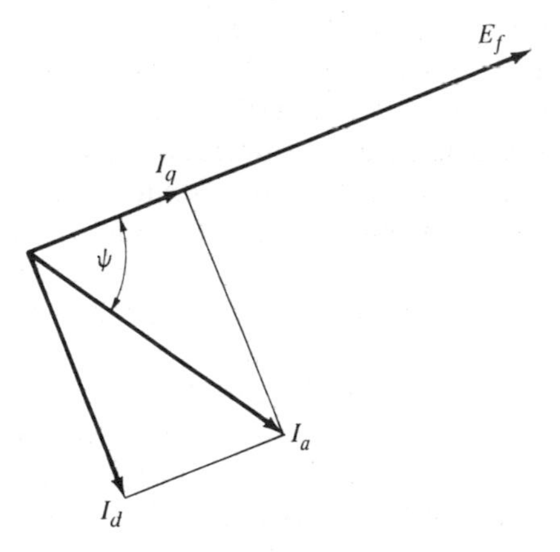

Fig. 12-5. Current components.

power factor angle θ. At any rate, for the phasors representing the component currents,

$$\mathbf{I}_d + \mathbf{I}_q = \mathbf{I}_a \tag{12-4}$$

The synchronous machine may thus be considered to have three, not two, separate mmf waves, all rotating in the same direction and at the same speed. Instead of combining them to a resultant mmf wave to which the corresponding flux would depend on the angular location of its axis, *superposition* is once again called to the rescue. Each mmf wave is thought of as producing its own flux, and each flux as inducing its own voltage. In other words, the thought process that led from Fig. 9-8 to Fig. 9-9 is repeated, this time with three mmf waves instead of two. The result is illustrated in Fig. 12-6. The total induced voltage is the phasor sum

$$\mathbf{E} = \mathbf{E}_f + \mathbf{E}_q + \mathbf{E}_d \tag{12-5}$$

The voltages are related to the component currents in a way similar to that expressed in Eq. 9-12, except that each current is associated to its own reactance. Thus

$$\mathbf{E} = \mathbf{E}_f - jX_{aq}\mathbf{I}_q - jX_{ad}\mathbf{I}_d \tag{12-6}$$

Now the imperfections of the armature windings are introduced, by substituting Eq. 9-10 into Eq. 12-6. (These imperfections, resistance and leakage reactance, are independent of the angle ψ.) Solving for the excitation voltage

$$\mathbf{E}_f = \mathbf{V} + R_a\mathbf{I}_a + jX_l\mathbf{I}_a + jX_{aq}\mathbf{I}_q + jX_{ad}\mathbf{I}_d \tag{12-7}$$

Next, substitute Eq. 12-4 into Eq. 12-7, collect the terms with the same current component, and neglect the resistance term, as was done throughout most of the preceding chapters.

$$\mathbf{E}_f = \mathbf{V} + j(X_l + X_{aq})\mathbf{I}_q + j(X_l + X_{ad})\mathbf{I}_d \tag{12-8}$$

Finally, the leakage reactance may be combined with each of the two armature reaction reactances (similar to Eq. 9-15), resulting in *two synchronous reactances*, one associated with the direct axis and one with the quadrature axis

$$X_l + X_{aq} = X_q \tag{12-9}$$

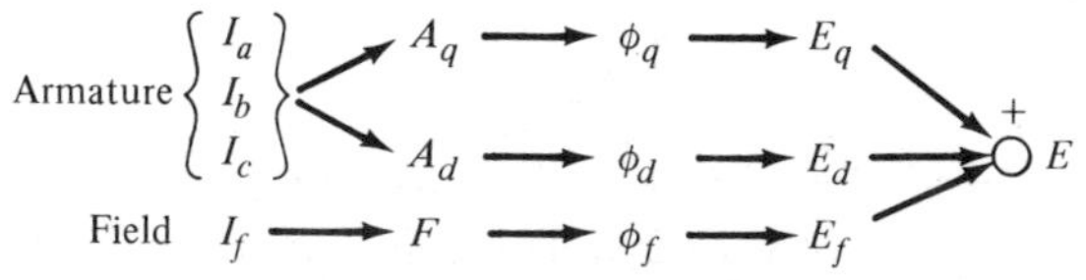

$$\begin{array}{l} \text{Armature}\left\{\begin{matrix} I_a \\ I_b \\ I_c \end{matrix}\right\} \begin{matrix} \nearrow A_q \longrightarrow \phi_q \longrightarrow E_q \searrow \\ \searrow A_d \longrightarrow \phi_d \longrightarrow E_d \rightarrow \overset{+}{\bigcirc}\, E \end{matrix} \\ \text{Field} \quad I_f \longrightarrow F \longrightarrow \phi_f \longrightarrow E_f \nearrow \end{array}$$

Fig. 12-6. Superposition for salient-pole machine.

and

$$X_l + X_{ad} = X_d \tag{12-10}$$

This leads to the final form of the voltage equation for salient-pole machines

$$\mathbf{E}_f = \mathbf{V} + jX_q \mathbf{I}_q + jX_d \mathbf{I}_d \tag{12-11}$$

This whole derivation parallels that of Eq. 9-17 in the ninth chapter, and it is based on the same choice of positive current direction, which is the one used in Chapters 9 and 11 for *generator* operation. If the current arrow is reversed for *motor* operation (as it is in Chapter 10), the equation becomes

$$\mathbf{E}_f = \mathbf{V} - jX_q \mathbf{I}_q - jX_d \mathbf{I}_d \tag{12-12}$$

12-4 PHASOR DIAGRAMS

Equations 12-11 and 12-12, just derived above, are modified, or extended, versions of Eqs. 11-1 and 10-1, respectively. They are modified to account for the effect of salient poles. If these equations (12-11 or 12-12) are applied to cylindrical-rotor machines, for which $X_q = X_d = X_s$, they revert to their simpler original form (with the aid of Eq. 12-4). So it ought to be possible to illustrate these equations, for any operating condition of a salient-pole machine, by phasor diagrams, just as Eq. 10-1 was illustrated in Fig. 10-2, and Eq. 11-1 in Fig. 11-2.

A first attempt to do this runs into a peculiar difficulty. Starting out with a given operating condition, the known variables are the phasors representing the terminal voltage V and the armature current I_a. But the component currents I_q and I_d depend on the phase angle of the excitation voltage E_f, which is yet to be obtained. It is possible to overcome this difficulty by trial and error: begin by assuming a plausible direction for the phasor of E_f, and improve it in a series of iterative drawings or calculations until the result is sufficiently close to the assumption. But there is a simple and elegant way to make such a procedure unnecessary.

Substituting for $\mathbf{I}_q$ (from Eq. 12-4) changes Eq. 12-11 into

$$\mathbf{E}_f = \mathbf{V} + jX_d \mathbf{I}_d + jX_q (\mathbf{I}_a - \mathbf{I}_d) \tag{12-13}$$

Now collect the terms with the same current phasor

$$\mathbf{E}_f = \mathbf{V} + jX_q \mathbf{I}_a + j(X_d - X_q) \mathbf{I}_d \tag{12-14}$$

In this equation, the last term is seen to lead the current component I_d by 90°. But since this component, by its definition, is 90° out of phase with E_f, that last term of Eq. 12-14 is in phase with E_f. Thus, the sum of the first two terms on the right side of the equation must also be in phase with E_f. This sum can be

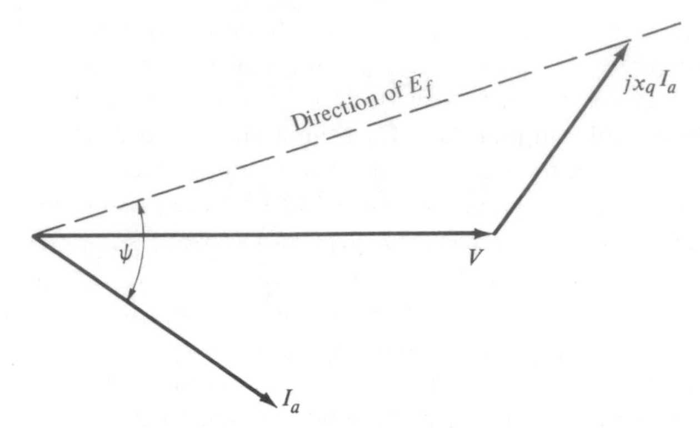

Fig. 12-7. Finding the direction of E_f.

calculated or drawn without prior knowledge of the angle of E_f. Figure 12-7 demonstrates this method, for the (arbitrarily chosen) example of a generator delivering lagging current. Once the direction of E_f, and thereby angle ψ, has been determined, the component currents can be found from Eqs. 12-2 and 12-3, and then the excitation voltage can be found, either from Eq. 12-11 or from Eq. 12-14, either analytically or graphically. See example 12-1.

For a graphical construction of the phasor diagram, it is not even necessary to find the component currents I_q and I_d. Figure 12-8 is drawn for the same machine as Fig. 12-7, and for the same operating condition. The line *ab* represents the phasor $jX_q \mathbf{I}_a$, and its end point *b* determines the direction of E_f and thereby

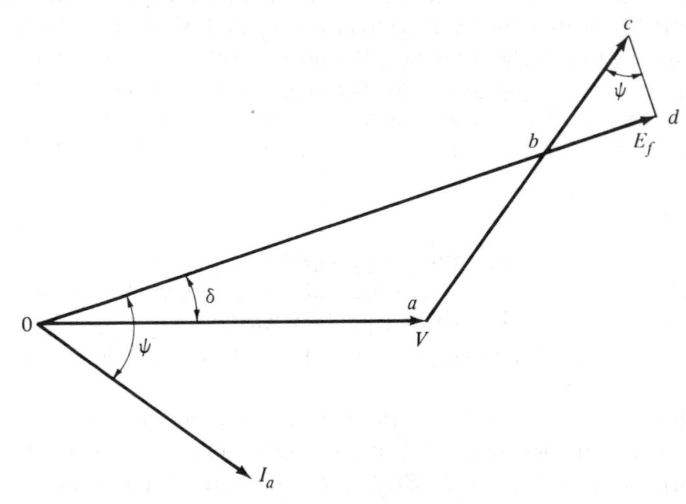

Fig. 12-8. Graphical procedure.

the angles δ and ψ. Now this line is continued to point c, with the distance $ac = X_d I_a$, or $bc = (X_d - X_q) I_a$. Then a line cd is drawn perpendicular to E_f, forming the angle ψ at point c, which makes $bd = (X_d - X_q) I_a \sin \psi = (X_d - X_q) I_d$ (by Eq. 12-3). This completes the diagram, point d being the end point of the phasor of E_f, according to Eq. 12-14.

The reader should repeat this procedure for a generator delivering leading current, and also for motor operation (both leading and lagging). If such diagrams are drawn for the same amount of power (same values of V and $I_a \cos \theta$), they show, for one thing, that the rules for *over- and underexcitation*, as they were found for motors in Chapter 10 and for generators in Chapter 11, are valid for salient-pole machines.

The diagrams also illustrate the effect of salient poles on the magnitude and angle of the field excitation voltage. Let Fig. 12-8 be used as an example. If the machine had a cylindrical rotor rather than salient poles, its synchronous reactance would have the single value $X_s = X_d$, corresponding to a uniform air gap. The excitation voltage, as determined by Eq. 11-1, would appear in the phasor diagram as a straight line connecting the origin to point c. The actual excitation voltage for the salient-pole machine has just slightly less magnitude but a significantly smaller angle. So the salient poles make the machine "stiffer" (the torque angle δ changes less for a given load change). It may also be expected that they improve the steady-state and transient stability. This latter conclusion, however, can be confirmed only after the relation between the electromagnetic torque and the angle δ has been established for salient-pole machines.

12-5 TORQUE AND TORQUE ANGLE

A torque equation will be derived, as was done in Chapter 10 for cylindrical-rotor machines, from the power that is converted from mechanical into electrical form or vice versa. This power differs from the power at the armature terminals only by the power loss in the armature resistance (see Fig. 11-10), which is neglected in this whole chapter as it was in most of the two preceding ones. Thus

$$T = \frac{P}{\omega_{sm}} = \frac{3VI_a \cos \theta}{\omega_{sm}} \tag{12-15}$$

As an aid to the further derivation, the phasor diagram of Figs. 12-7 and 12-8 is redrawn once more in Fig. 12-9, with the component currents I_q and I_d added. As this diagram shows, $\mathbf{I}_q = I_q \underline{/\delta}$ and $\mathbf{I}_d = I_d \underline{/\delta - 90^\circ}$. Therefore, equating the real parts of both sides of the complex equation, Eq. 12-4, leads to

$$I_a \cos \theta = I_q \cos \delta + I_d \sin \delta \tag{12-16}$$

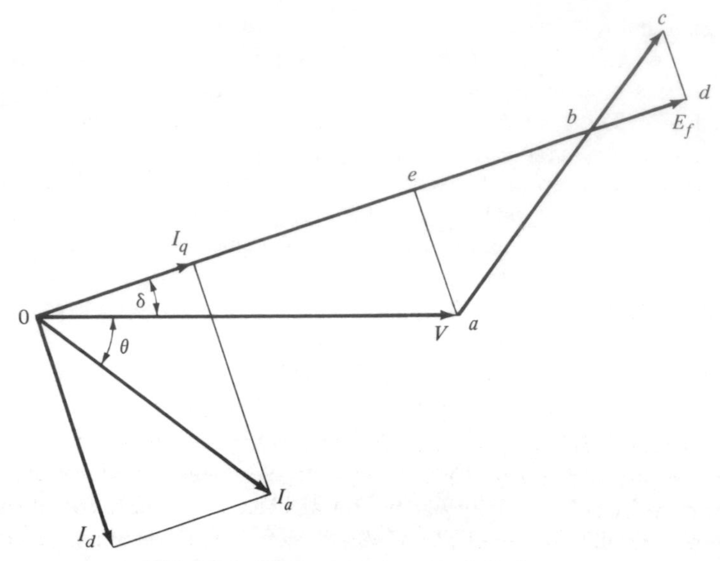

Fig. 12-9. Diagram for torque calculation.

which can be substituted into the previous equation

$$T = \frac{3V}{\omega_{s_m}}(I_q \cos \delta + I_d \sin \delta) \tag{12-17}$$

This substitution introduces the angle δ into the torque equation, but it does so at the cost of also introducing the component currents. To remove them from the equation, another line is drawn in Fig. 12-9, this one from point *a* in the direction perpendicular to that of the phasor of E_f, reaching it at point *e*. The significance of this point lies in the fact that the polygon 0*aed* describes the phasor relation of Eq. 12-11, just as the polygon 0*abd* describes Eq. 12-14. Thus

$$ae = X_q I_q \tag{12-18}$$

and

$$ed = X_d I_d \tag{12-19}$$

Solve these two equations for the component currents, with a look at the triangle 0*ae*, as follows:

$$I_q = \frac{ae}{X_q} = \frac{V \sin \delta}{X_q} \tag{12-20}$$

and

$$I_d = \frac{ed}{X_d} = \frac{E_f - V\cos\delta}{X_d} \tag{12-21}$$

These expressions for I_q and I_d are substituted into Eq. 12-17, with the result

$$T = \frac{3V}{\omega_{s_m}}\left(\frac{V}{X_q}\sin\delta\cos\delta + \frac{E_f - V\cos\delta}{X_d}\sin\delta\right) \tag{12-22}$$

which may be rearranged to read

$$T = \frac{3V^2}{\omega_{s_m}}\left(\frac{1}{X_q} - \frac{1}{X_d}\right)\sin\delta\cos\delta + \frac{3VE_f}{X_d}\sin\delta \tag{12-23}$$

Before this result is discussed, it should be pointed out that its derivation was based on the phasor diagram of Fig. 12-9, which describes a generator delivering lagging current. Could a diagram drawn for a different operating condition lead to a different result? To dispel such doubts, consider the case of a generator delivering leading current (or a motor drawing lagging current) at a phase angle θ exceeding the torque angle in magnitude. In such a case, the current phasor $\mathbf{I}_d$ lies in the second (or third) quadrant; thus, its real part is negative, and the plus signs in Eqs. 12-16 and 12-17 must be changed into minus signs. But in the same cases, point d lies to the left, not the right, of point e; thus, $de = V\cos\delta - E_f$, which is the negative of what was used in Eq. 12-21 The two minus signs cancel each other, and the result, Eq. 12-22 or 12-23, remains unchanged.

In discussing this result, the reader is reminded that the aim of this section was to establish a torque equation similar to Eq. 10-8 but valid for a salient-pole machine also. In that sense, the striking feature of Eq. 12-23 is that it expresses the effect of salient poles in terms of the difference between X_q and X_d in the first term. If this equation is applied to a cylindrical-rotor machine, where $X_q = X_d = X_s$, it reverts back to Eq. 10-8.

It may be further noticed that the first term in Eq. 12-23 does not contain the field excitation at all. So this is a torque that exists even in the absence of any field excitation, due only to the different reluctances in the direct and quadrature axis. It is, therefore, logical to name the two terms of Eq. 12-23 the *reluctance torque* and the *excitation torque*

$$T = T_r + T_f \tag{12-24}$$

The possibility of obtaining torques in singly excited devices, based on differences of reluctances, was already discussed in Chapters 7 and 8. The only application of this idea that produces a torque that is continuous in one direction is the synchronous machine. To study this torque further, its expression in Eq. 12-23 is now changed to

$$T_r = \frac{3V^2}{\omega_{s_m}} \frac{X_d - X_q}{2X_d X_q} \sin 2\delta \tag{12-25}$$

(using the familiar trigonometric identity $\sin 2\delta = 2 \sin \delta \cos \delta$). In this form, the expression shows more clearly that it is the difference between X_d and X_q that is responsible for the production of this torque.

The appearance of the angle 2δ is also significant, and can be understood by examining a diagram similar to that of Fig. 12-3 or Fig. 12-4. In the absence of any field excitation, the salient poles on the rotor are only poles of induced magnetism, without any fixed polarities. Whichever rotor pole happens to face the north pole of the stator flux becomes an induced south pole, and vice versa. The reluctance torque must be a periodic function of 2δ, not of δ, because a motion of the rotor of 180 electric degrees relative to the rotating armature flux would, in the absence of a field current, lead to the same, not the opposite condition.

The presence of a reluctance torque changes the relation between torque and torque angle. In place of the pure sinusoidal relation of Eq. 10-8 and Fig. 10-7, which is valid for the machine with a cylindrical rotor, there is now a sum of two functions of δ, given by Eq. 12-23 and depicted in Fig. 12-10. A look at this diagram confirms that the salient poles enable the machine to change its torque without as great a change of the angle δ as it would need without them. This is the greater "stiffness" of the machine, mentioned at the end of the previous section, where it was deduced from the phasor diagram. Another thing Fig. 12-10

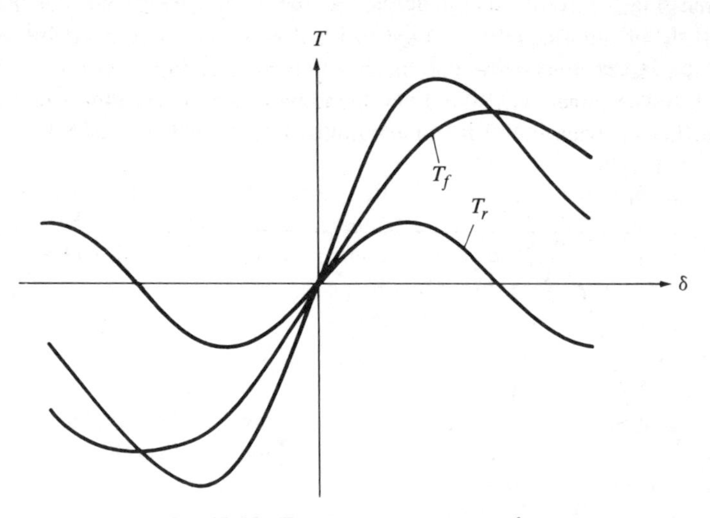

Fig. 12-10. Torque versus torque angle.

makes clear is that, due to the salient poles, the maximum torque (which is the steady-state stability limit) is increased and that it occurs at a torque angle of less than 90°. Incidentally, the figure is drawn for both positive and negative values of the angle δ, and the choice of which side represents motor operation and which generator operation can be made arbitrarily.

12-6 RELUCTANCE MOTORS

A synchronous machine with salient poles does not need any field excitation in order to work. Nevertheless, most synchronous machines are equipped with field windings even though these windings and the connections to their auxiliary sources add considerably to the cost of the machines. The excitation torque is usually a multiple (typically, something like three times) of the reluctance torque, so that a machine without field excitation would have a very low power rating for its size. In addition, synchronous machines without field excitation must inevitably operate at very low power factors.

To verify that last statement, set the field excitation voltage in Eq. 12-12 (that is the equation for *motor* operation) equal to zero. It may then be written

$$\mathbf{V} = jX_q \mathbf{I}_q + jX_d \mathbf{I}_d \tag{12-26}$$

A phasor diagram illustrating this equation is drawn in Fig. 12-11. It must be understood that δ is a meaningful angle even if there is no E_f. For one thing, the existence of δ can be understood by assuming the field excitation voltage to be very small, i.e., small enough to be negligible compared to the other voltages, but not zero so that δ can be defined as usual as the phase angle between V and E_f. Or, preferably, let the angle ψ be defined first as the space angle between the axis of the stator mmf wave and the quadrature axis, and then use the relation $\delta = \theta - \psi$. The large value of θ, corresponding to a very low power factor $\cos\theta$, is by no means exaggerated in the diagram. Rather, it is typical for a synchro-

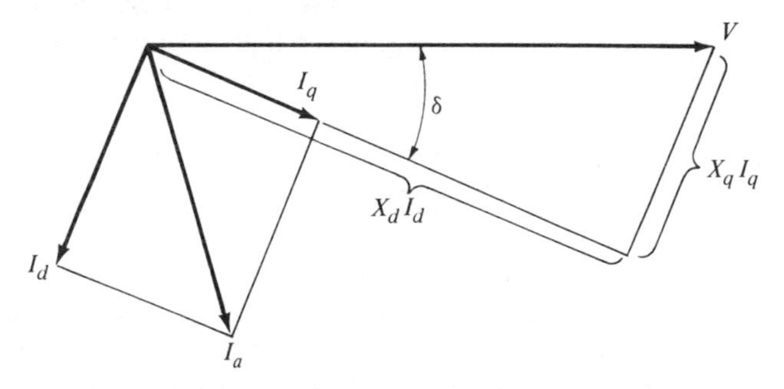

Fig. 12-11. Phasor diagram of reluctance motor.

nous motor without field excitation. Nor should this come as a surprise if it is remembered that an underexcited motor draws lagging current. Just let the field excitation be gradually reduced toward zero (for a given load) and watch the current become more and more lagging.

Apart from such drawbacks, if a synchronous machine is to operate as a generator without field excitation (a *reluctance generator*), there has to be at least one other generator (*with* field excitation) connected to the same load, and this other generator needs additional field excitation in order to deliver the lagging current needed to compensate for the leading current delivered by the reluctance generator. All in all, reluctance generators may be dismissed as having negligible practical importance.

This is not so for the reluctance *motor.* Its typical use is for electric *clocks* or other timing devices. For such purposes, the amount of power needed is so small that a low power factor and a low efficiency are of no practical consequence, whereas simplicity, ruggedness, and low cost are prime considerations.

12-7 INDUCTANCES

It was pointed out at the end of Chapter 11 that a mathematical study of a synchronous machine in an electrical *transient* condition requires that the concept of the rotating magnetic field, useful as it is for any study of steady-state (and quasi-steady-state) operation, be abandoned. In its place, a system of differential equations must be set up, one equation for each circuit, with their self- and mutual inductances and resistances. For the purpose of this introductory discussion, damper windings will be left out of consideration, which still leaves three armature circuits and one field circuit.

The use of inductances always implies an assumption of constant saturation. So the differential equations will be considered as linear, although they do not have constant coefficients because some of the inductances are *time-varying.* Such a possibility was already encountered in Section 7-8, and it will now be investigated for the specific case of the three-phase synchronous machine.

Of all the inductances involved, the simplest is the self-inductance L_f of the field winding. It is constant because the stator is cylindrical and offers the same reluctance to the rotor mmf wave no matter what its position. For the same reason, the self-inductances of the three armature circuits are constant if the rotor is cylindrical. In the case of a salient-pole rotor, however, each of these self-inductances has a maximum value at the instant when the mmf wave of its phase is in the direct axis, and a minimum value when it is in the quadrature axis.

To obtain mathematical expressions for these time-varying inductances, we begin by introducing the symbol φ for the space angle around the stator circumference, in electrical units. (This takes the place of the symbol θ that was used in

Chapters 7 and 8 for this angle.) Let this angle have its zero value at the axis of phase a, and let time be zero at the instant when the direct axis is at $\varphi = 0$. This means that the direct axis is always located at an angular position $\omega_s t$, and, thus, the self-inductance of phase a of the armature winding has the form

$$L_a = L_1 + L_2 \cos 2\omega_s t \tag{12-27}$$

where $L_1 > L_2$, and the higher harmonics are neglected. The self-inductances of the other two phases are similar, but they reach their maximum values one-third of a period later and earlier, respectively. Thus

$$L_b = L_1 + L_2 \cos 2(\omega_s t - 120°) \tag{12-28}$$

and

$$L_c = L_1 + L_2 \cos 2(\omega_s t + 120°) \tag{12-29}$$

The *mutual inductances* between the field winding and each of the armature windings are time-varying, regardless of whether there are salient poles or not. They are zero for whichever phase has its mmf wave in the quadrature axis, and their signs alternate with the polarities of the rotor poles

$$M_{af} = M_1 \cos \omega_s t \tag{12-30}$$

$$M_{bf} = M_1 \cos (\omega_s t - 120°) \tag{12-31}$$

and

$$M_{cf} = M_1 \cos (\omega_s t + 120°) \tag{12-32}$$

Lastly, there are the mutual inductances between phases of the armature winding, and they are a little trickier. Figure 12-12 presents a developed view of the stator surface, showing the location of the conductors of phases a and b, with their positive directions indicated by dots and crosses. It can be seen that these mutual inductances must be assigned negative values because those lines of flux that link both phases a and b link them in opposite directions relative to their dots and crosses. It will also be noticed that these lines are centered around the

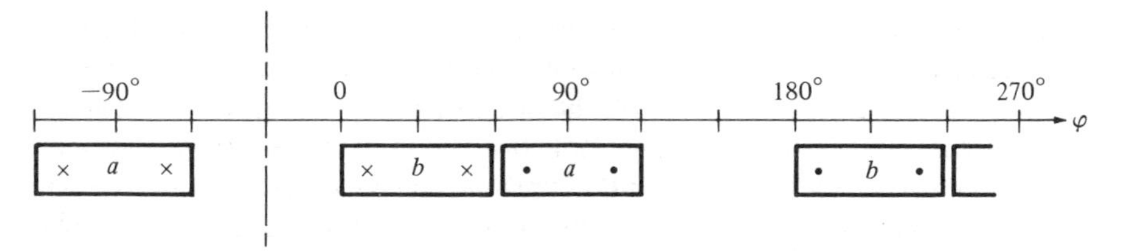

Fig. 12-12. Location of armature conductors, phases *a* and *b*.

axis of symmetry of these two phases. (See the dash-dot line in the figure.) Therefore, if the rotor has salient poles, the mutual inductance M_{ab} has its maximum value at the instant when this axis of symmetry coincides with the direct axis. All of which leads to the expressions

$$M_{ab} = -M_2 - M_3 \cos 2(\omega_s t + 30°) \tag{12-33}$$

$$M_{bc} = -M_2 - M_3 \cos 2(\omega_s t - 90°) \tag{12-34}$$

$$M_{ca} = -M_2 - M_3 \cos 2(\omega_s t + 150°) \tag{12-35}$$

For the case of a cylindrical rotor, simply set $M_3 = 0$. Notice that all time variations due to salient poles are periodic functions of $2\omega_s t$, whereas those that involve the field winding are periodic functions of $\omega_s t$.

12-8 TRANSFORMATION OF VARIABLES

The differential equations mentioned in the previous section are the voltage equations of the four circuits. Their basic form is

$$v = Ri + \frac{d\lambda}{dt} \tag{12-36}$$

where λ represents the flux linkages, and the variables are further identified by subscripts. For instance, for the circuit of phase a, the subscript a is added to the symbols of current, voltage, and flux linkages. Then the inductances are introduced (as they were in Eqs. 7-32 and 7-33, which are a system of two equations for two magnetically coupled circuits), for instance in

$$\lambda_a = L_a i_a + M_{ab} i_b + M_{ca} i_c + M_{af} i_f \tag{12-37}$$

Next, Eq. 12-37 is substituted into Eq. 12-36 (for phase a):

$$v_a = R_a i_a + i_a \frac{dL_a}{dt} + L_a \frac{di_a}{dt} + i_b \frac{dM_{ab}}{dt} + M_{ab} \frac{di_b}{dt}$$

$$+ i_c \frac{dM_{ca}}{dt} + M_{ca} \frac{di_c}{dt} + i_f \frac{dM_{af}}{dt} + M_{af} \frac{di_f}{dt} \tag{12-38}$$

This, in turn, requires substitution of Eqs. 12-27, 12-30, 12-33 and 12-35 for the inductances. The whole procedure is to be repeated for the other three circuits.

In these lengthy equations, the reader can recognize the transformer voltages and the motional voltages. Beyond that, the equations are far too unwieldy for solution by even sophisticated computing equipment. The way to make them manageable is a remarkable transformation that was developed in the late 1920s (long before the advent of digital computers) and for which the names of R. H. Park, R. E. Doherty, and C. A. Nickle are remembered. The idea is related to the salient-pole theory discussed in this chapter, but is is applicable to any synchronous machine, whether it has salient poles or not.

The actual procedure is too lengthy to be given here, and what follows is merely a sketch of the basic principle. The three armature currents i_a, i_b, and i_c are replaced by three fictitious currents with the symbols i_d, i_q and i_0. Of these, i_d and i_q can be viewed as the currents in two fictitious rotating windings centered around the direct and the quadrature axis, respectively. They are given such values that, in a balanced condition, when $i_a + i_b + i_c = 0$, they produce the same flux, at any instant, as the actual phase currents in the armature. The relations between these two currents and their fluxes (i.e., their inductances) are not time-varying, and there is no magnetic coupling between i_q on one hand and i_d and i_f on the other. The third fictitious current i_0 is needed to make the transformation possible when the sum of the three phase currents is not zero.

Similar transformations must be made in replacing the three armature terminal voltages v_a, v_b, and v_c by a set of fictitious voltages named v_d, v_q and v_0. In terms of all these six new variables and two original ones (i_f and v_f), the transformed equations can be written in a form that contains fewer terms, and in which *all coefficients are constant.* In addition, the actual steady-state and transient inductances L_d, L_q, L_d', and L_q' can be obtained from these coefficients. (Also the subtransient inductances, if the damper windings are included in the system of equations.) For the whole procedure, involving the transformation of the variables and the necessary manipulation of the equations, the reader is referred to the advanced literature on that subject.*

12-9 EXAMPLES

Example 12-1 (Section 12-4)

A 1500 kva, Y-connected, 2300-v (line to line), three-phase, salient-pole synchronous generator has reactances $X_d = 1.95$ and $X_q = 1.40\ \Omega$ per phase. All losses may be neglected. Find the excitation voltage for operation at rated kva and power factor of 0.85 lagging.

Solution

Voltage per phase is $2300/\sqrt{3} = 1328$ v. Rated current is $1{,}500{,}000/(3 \times 1328) = 377$ amp. Use the terminal voltage as the reference

$$\mathbf{V} = 1328 \underline{/0^\circ} = 1328 + j\,0 \text{ v}$$

$$\mathbf{I}_a = 377 \underline{/-31.8^\circ} = 320 - j\,199 \text{ amp}$$

*E.g., E. W. Kimbark, *Power System Stability*, vol. 3, John Wiley & Sons, 1956.

We propose to find E_f from Eq. 12-14. Let $\mathbf{E}' = \mathbf{V} + jX_q \mathbf{I}_a$. E' has the same direction as E_f on a phasor diagram.

$$\mathbf{E}' = \mathbf{V} + jX_q \mathbf{I}_a = (1328 + j\,0) + j(1.40)(320 - j\,199)$$
$$= 1607 + j\,448 = 1668 \underline{/15.6^\circ} \text{ v}$$

The phase difference between E' and I_a is angle ψ

$$\psi = 15.6^\circ - (-31.8^\circ) = 47.4^\circ$$

Use Eq. 12-3 to find I_d

$$I_d = I_a \sin \psi = 377 \sin 47.4^\circ = 278 \text{ amp}$$
$$(X_d - X_q) I_d = (1.95 - 1.40)\,278 = 153 \text{ v}$$

Since $\mathbf{E}_f$, $\mathbf{E}'$, and $j(X_d - X_q)\,\mathbf{I}_d$ are in phase, it is only necessary to add magnitudes

$$E_f = E' + (X_d - X_q) I_d = 1668 + 153 = 1821 \text{ v}$$

Example 12-2 (Section 12-4)

The reactances X_d and X_q of a salient-pole synchronous motor are 1.00 and 0.60 per unit, respectively. The armature resistance is negligible. Find the excitation voltage when the motor is operating at rated terminal voltage, rated kva, and with power factor of 0.8 leading.

Solution

The positive direction of armature current for motor operation is chosen as in Fig. 10-1. Use armature current as the reference

$$\mathbf{I}_a = 1 \underline{/0^\circ} = 1 + j\,0 \text{ pu}$$
$$\mathbf{V} = 1 \underline{/-36.9^\circ} = 0.8 - j\,0.6 \text{ pu}$$

Equation 12-12 is written for motor operation. We want to solve for E_f using an equation similar to Eq. 12-14. For a motor it becomes

$$\mathbf{E}_f = \mathbf{V} - jX_q \mathbf{I}_a - j(X_d - X_q)\,\mathbf{I}_d$$

First, we must find the angle of E_f

$$\mathbf{E}' = \mathbf{V} - jX_q \mathbf{I}_a = (0.8 - j\,0.6) - j\,(0.6)\,(1 + j\,0)$$
$$= 0.8 - j\,1.2 = 1.44 \underline{/-56.3^\circ} \text{ pu}$$

The angle between E_f and I_a is $\psi = -56.3^\circ$. Use Eq. 12-3 to find I_d

$$I_d = I_a \sin \psi = 1 \sin(-56.3^\circ) = -0.832 \text{ pu}$$

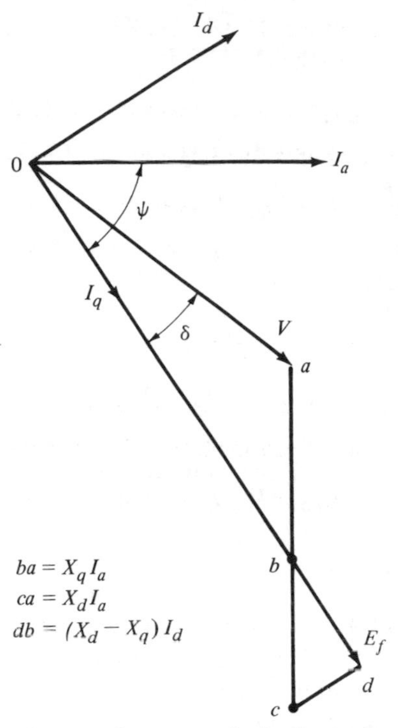

Fig. E-12-2. Phasor diagram of an overexcited salient pole synchronous motor.

Interpret the minus sign to show that $\mathbf{I}_d$ leads $\mathbf{E}_f$ by 90°. Since $\mathbf{E}_f$, $\mathbf{E}'$, and $-j(X_d - X_q)\,\mathbf{I}_d$ are in phase, it is only necessary to combine magnitudes

$$E_f = E' + (X_d - X_q) I_d = 1.44 + 0.33 = 1.77 \text{ pu}$$

Figure E-12-2 is the phasor diagram that illustrates the quantities in this example.

12-10 PROBLEMS

12-1. For the machine in Example 12-1, find the excitation voltage for operation at rated kva and power factor of 0.7 lagging.

12-2. A three-phase salient-pole synchronous generator is operated with terminal voltage of 254 v per phase. The excitation voltage is 410 v per phase. The armature current is 20 amp. The power factor angle is 30 degrees. The torque angle δ is 20 electrical degrees. Find the reactances X_d and X_q for this machine.

12-3. A three-phase salient-pole synchronous generator is connected to an infinite bus of rated voltage and frequency. The reactances are $X_d = 1.2$ per unit and $X_q = 0.7$ per unit. The generator output power to the bus is 0.55 per unit. The torque angle δ is 30 electrical degrees. Find the line current in per unit.

12-4. The salient-pole synchronous motor in Example 12-2 has its field current reduced until the power factor has become 0.6 lagging. Find the excitation voltage for this condition.

12-5. The salient-pole synchronous motor in Example 12-2 has its excitation adjusted to 1.00 per unit. When operated at no-load, the armature current is found to be negligible. With this field current, the mechanical load is increased until the motor loses synchronism. Find the pull-out power in per unit. Find the current in per-unit at pull-out.

12-6. A salient-pole synchronous motor is operating with rated terminal voltage. If the field current is zero, the motor can develop maximum torque of 0.3 per unit. If it runs at no-load and with the field current set to make the excitation voltage be 1.7 per unit, the armature current is 0.8 per unit. Find the reactances X_d and X_q in per-unit.

13

Three-Phase Induction Machines

13-1 HISTORICAL BACKGROUND

It may be difficult to imagine nowadays that in the early years of electric power engineering, there was a serious controversy about the relative merits of direct-current and alternating-current power systems. The advocates of a-c systems, lead by innovators like Tesla,* were opposed by no less a giant of the engineering world than Edison,† who had pioneered the first (d-c) system. Even though the advantages of high-voltage power transmission and the need for transformers were known and appreciated, it took another breakthrough, the invention of the *induction motor* (attributed to Tesla) to turn the tide in favor of a-c.

Induction motors are simpler and cheaper to build and operate than either synchronous motors or d-c motors. It is true that each of these three types of electric motors still has, even today, its field of application where it is preferred to the others; but induction motors are more frequently used by far than others. On the other hand, *generator* operation of induction machines is quite rare, for reasons that will be explained in due course.

This and the following chapter deal only with *three-phase* induction machines, just as the previous four chapters were limited to three-phase synchronous machines. The extension of these studies to single-phase machines will be left for later. In the case of induction machines, their very principles are based on the rotating magnetic field of polyphase currents. So it will require additional preliminary studies to explain even the possibility of single-phase operation.

13-2 PRINCIPLE OF OPERATION

The basic idea behind the induction motor is that the *rotating magnetic field* produced by a set of polyphase currents ought to be capable of producing rotating motion, without the need for a separate source of rotor excitation that characterizes most synchronous motors. That this is indeed the case can be confirmed

*Nikola Tesla, 1856–1943.
†Thomas A. Edison, 1847–1931.

in several ways. The following line of reasoning is based on general principles well known to the reader, and does not require any mathematical expressions.

Let the stator of a rotating device be cylindrical and carry three-phase windings, just like the stator of a synchronous machine. For the rotor, choose the simplest possible design, a cylinder made of a solid material that is both ferromagnetic and conducting (for instance iron). When the stator windings are connected to balanced three-phase voltages, they carry balanced three-phase currents and produce a magnetic field rotating at synchronous speed. The motion of this field relative to the rotor (which may at first be thought to be stationary) produces eddy currents and thereby causes the consumption of power (i.e., its conversion into heat) in the rotor.

The last sentence would remain valid if the actual stator with its three-phase excitation were replaced by a part surrounding the rotor, equipped with a winding carrying a constant current, and rotating at synchronous speed. This modified device would exhibit generator action, because the motion of the rotating outside part (which could hardly be called a stator) would be the cause of the power consumption in the (stationary) rotor. Consequently, the rotor would exert a *torque* opposing the motion of the outside part. Now, the rotor cannot possibly "tell" the difference, and so it responds in the same way to the rotating field of the actual three-phase stator, namely with a torque in the direction opposite to the direction of the field rotation. But since the stator cannot move, it *reacts* by exerting a torque of equal magnitude on the rotor, in the direction of the rotating field.

This reasoning proves, first of all, that a device with a three-phase stator and a solid iron rotor can operate as a motor. The reasoning remains valid for any relative motion between the rotating field and the rotor. Thus, it shows that a torque exists in such a device regardless of whether the rotor is stationary or moving, as long as it does not rotate at synchronous speed (at which the relative speed between field and rotor would be zero). It is typical for the induction motor that it does *not* operate at synchronous speed, a property that accounts for another name given to it, that of *asynchronous motor.* As for its more popular appellation, it is called the induction motor because its rotor currents are due to induction rather than to some separate source of excitation, and it is the interaction between these induced rotor currents and the magnetic field that makes the power conversion possible.

13-3 ROTOR TYPES

The device described in the previous section (cylindrical stator with three-phase windings, solid iron rotor) is not the normal kind of induction motor. The weakness of that design lies in the rotor, which has to combine two functions: it

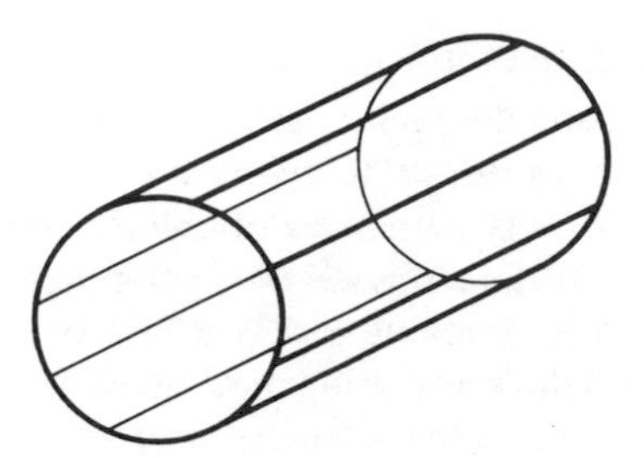

Fig. 13-1. Squirrel cage.

is a part of the magnetic circuit, and it also carries electric currents, both within the same piece of material. It is preferable to separate these two functions, to use separate pieces of different materials for them. In most actual induction motors, the cylindrical rotor core is made of ferromagnetic laminations that minimize the flow of eddy currents, whereas separate prescribed paths, namely conductors embedded in slots, are provided for the electric currents in the rotor.

There are two distinct types of rotor design based on this principle. *Wound rotors* carry three-phase windings similar to those on the stator, i.e., sets of coils consisting of many turns. The other construction, commonly referred to as the *squirrel cage*, has individual conductors, one in each slot, connected to each other by end rings as shown in Fig. 13-1 (they usually have more than the eight conductors of that figure).

It does not require any detailed analysis to see that a squirrel cage is much simpler and cheaper than a wound rotor (for similar ratings). No wonder that the great majority of all induction machines are built with squirrel cage rotors. There are cases, however, when it is desired to insert additional circuit elements into the rotor circuit, for purposes of speed control and starting. Only in such cases does the wound-rotor type become competitive and sometimes preferable.

On the other hand, when it comes to *studying* the operation of induction motors, it is easier to start with the wound-rotor type. So the following theory will be developed for a wound-rotor induction motor, and it will be shown a little later how this theory can be applied to a squirrel-cage motor.

13-4 REMEMBER TRANSFORMERS?

Of all rotating electric machines, the wound-rotor induction motor is most closely related to the transformer. In fact, this motor *can* be used as a transformer, with the stator winding as the primary and the rotor winding as the secondary (or the other way around). For such a use, it is only necessary that the rotor be *blocked* (i.e., kept from running, held stationary), and that the rotor terminals be accessible so that they can be connected to an outside load (or source) circuit. The differences between such a transformer and the device studied in Chapter 5 are worth noting:

(a) This motor would be a *three-phase* transformer since the stator and rotor cores would serve as a magnetic circuit for all the three phases of the stator and rotor windings. But, in contrast to the device mentioned in the last paragraph of Section 6-5, the three stationary mmf waves would combine to form a rotating mmf wave and thereby produce a rotating magnetic flux.

(b) It would be an inferior transformer, with *imperfections* far exceeding those of a conventional power transformer. The main drawback would be the presence of an *air gap* that would inevitably raise the reluctance of the magnetic circuit and thereby the magnetizing susceptance B_m. In addition, having all conductors located in *slots* would result in larger leakage reactances X_1 and X_2.

Nevertheless, the theory of the power transformer is basic to that of the induction motor. To account for the differences just mentioned, it is merely necessary (a) to write all equations and draw all diagrams on a *per-phase* basis, which is the normal procedure for balanced three-phase circuits in the steady state (and also in the quasi-steady state, i.e., for slow mechanical transient conditions), and (b) to limit any use of an *approximate* equivalent circuit to cases in which qualitatively meaningful results rather than numerically accurate ones are sought. (This is true because approximate equivalent circuits of transformers are the more accurate the smaller their imperfections are.)

A transformer in the sinusoidal steady state can be described by the set of equations 5-23, 5-24, 5-31, 5-32, and 5-33, which are embodied in the equivalent circuit of Fig. 5-6. When they are applied to a wound-rotor induction motor with a blocked rotor, the one needed modification is to set $V_2 = 0$, since the rotor winding must be closed in order to carry a current. (Provisions may be made for the insertion of outside elements into the rotor circuit, as was mentioned above, but there is no need to consider that at this point.) So the equivalent circuit diagram is redrawn in Fig. 13-2 with a short-circuited secondary. A sharp-eyed reader might also notice that the second subscript of the admittance Y_{ϕ_1} has been omitted. This was done because there will be no need to transfer this element to the secondary side.

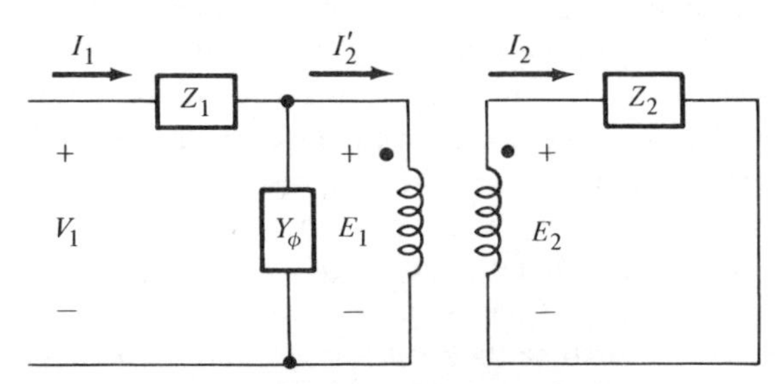

Fig. 13-2. Short-circuited transformer.

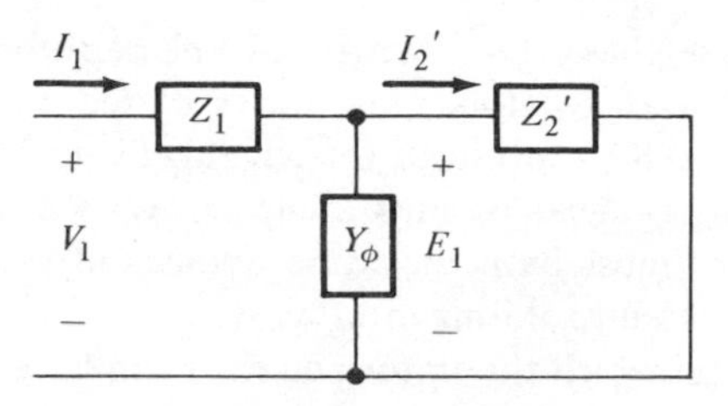

Fig. 13-3. Induction machine at standstill.

The passive elements of this circuit, as the reader remembers, represent the imperfections of the transformer. Specifically, the impedances $\mathbf{Z}_1$ and $\mathbf{Z}_2$ are series combinations of the winding resistances and leakage reactances. The admittance $\mathbf{Y}_\phi$ is a parallel combination of the core loss conductance and the magnetizing susceptance. All voltages and currents are phase (not line) quantities.

Regarding the ideal transformer of the diagram, its ratio of transformation is the *voltage ratio* $a = E_1/E_2$. For the static transformer with its concentrated windings, this also equals the turns ratio N_1/N_2, as can be confirmed by applying Eq. 4-25 to these two voltages that are induced by the same flux. But for a rotating machine with distributed windings, the applicable voltage equation is the one numbered 8-40. Since the stator and rotor windings may well have different winding factors, the voltage ratio is

$$a = \frac{E_1}{E_2} = \frac{N_1 k_{w_1}}{N_2 k_{w_2}} \tag{13-1}$$

The next step is to transfer the secondary impedance to the primary side (a procedure first seen in Section 5-5) where it is given the symbol Z_2'. This makes the voltages on both sides of the ideal transformer equal to zero; thus, it may be omitted as trivial, which leads to the simpler equivalent circuit of Fig. 13-3 without loss of accuracy. This diagram represents one phase of the induction motor *at standstill*, with the actual stator voltage and current appearing at the primary terminals.

This equivalent circuit still needs to be modified to describe a *running* induction motor. But first, it will be explained that this circuit can represent a motor with a *squirrel-cage* rotor, as well as one with a wound rotor. The key to this explanation lies in viewing a squirrel cage as a set of single-turn windings for a large number of phases.

13-5 THE SQUIRREL CAGE AS A POLYPHASE WINDING

It may come as a surprise to some readers that the rotor of an induction motor could be wound for a different number of *phases* than the stator. For instance, there could be a three-phase stator with a two-phase rotor. This is so because they would both produce rotating mmf waves of the same direction and the same speed (see Section 8-8), although the winding factor and the space harmonics might constitute drawbacks of such a choice. On the other hand, the stator and rotor mmf waves must have the same number of *poles*, so that they can combine to form a resultant rotating mmf wave.

Can such a motor (in which the stator and the rotor have different numbers of phases) be represented by the equivalent circuit of Fig. 13-3? The answer is yes, but the method of transferring the rotor impedance Z_2 to the stator side is somewhat different. The ratio of phase voltages expressed in Eq. 13-1 remains valid, but the reciprocal of that ratio cannot be used to refer the rotor current to the stator; in other words, $I_2' \neq I_2/a$. To understand this, recall that the current ratio of an ideal transformer is based on the cancellation of the secondary mmf by the primary mmf. For a static single-phase transformer, this leads to $N_1 I_2' = N_2 I_2$. Polyphase windings, on the other hand, produce rotating mmf waves whose peak values are proportional not only to their currents, but also to their number of phases. Thus, if a stator of m_1 phases and a rotor of m_2 phases constituted an ideal transformer, with the voltage ratio a, their current ratio would be

$$b = \frac{m_2}{m_1} \frac{1}{a} \tag{13-2}$$

So the rotor current referred to the stator is

$$I_2' = bI_2 = \frac{m_2}{m_1} \frac{1}{a} I_2 \tag{13-3}$$

Finally, *impedances* are transferred from the rotor to the stator side by being multiplied by the voltage ratio and divided by the current ratio. Thus

$$Z_2' = \frac{a}{b} Z_2 = \frac{m_1}{m_2} a^2 Z_2 \tag{13-4}$$

Incidentally, Eqs. 13-1, 13-3, and 13-4 are not only valid for magnitudes, but also as complex equations (for voltage and current phasors and complex impedances), since the ratios a and b are real numbers.

To confirm that Eqs. 13-1, 13-3, and 13-4 make Fig. 13-3 correct and meaningful, compare the actual rotor voltage equation

$$\mathbf{E}_2 = \mathbf{Z}_2 \mathbf{I}_2 \tag{13-5}$$

to the one shown by the diagram

$$\mathbf{E}_1 = \mathbf{Z}_2' \mathbf{I}_2' = \left(\frac{a}{b} \mathbf{Z}_2\right) (b\mathbf{I}_2) = a\mathbf{E}_2 \tag{13-6}$$

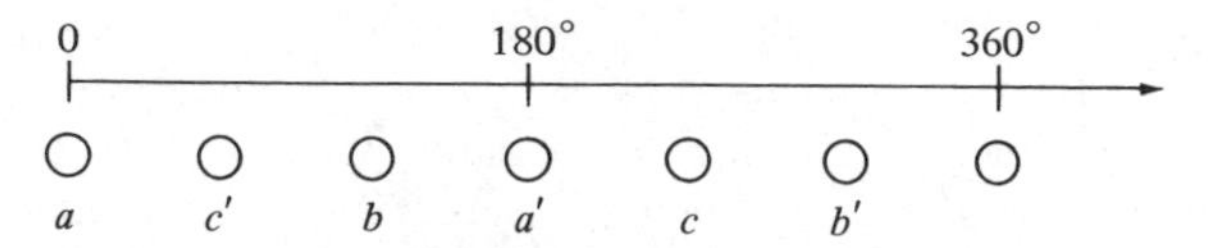

Fig. 13-4. Single coils for three phases.

Furthermore, the power consumption in the actual rotor resistance is correctly shown in the diagram, since the identity

$$P_2 = m_2 I_2^2 \,\mathcal{R}_e(\mathbf{Z}_2) = m_1 I_2'^2 \,\mathcal{R}_e(\mathbf{Z}_2') \tag{13-7}$$

can be readily verified by substitution of Eqs. 13-3 and 13-4.

Next, it will be shown that a squirrel cage can be viewed as a polyphase winding. Consider first a three-phase, two-pole stator facing a rotor having just six equally spaced conductors (bars) connected to each other by metallic end rings. This rotor could be considered as having a three-phase winding, each phase consisting of just one turn (two conductors opposite to each other). The developed sketch of Fig. 13-4 illustrates this rotor and shows the angular distance of 120° between phases.

Similarly, if the rotor had 10 conductors instead of 6, it could be viewed as a five-phase winding with one turn per phase. Generally, with c individual rotor conductors, one per slot, located inside a two-pole stator, the number of rotor phases is

$$m_2 = \frac{c}{2} \tag{13-8}$$

If the stator winding has more than two poles, however, the angular distance between rotor slots must be expressed in electrical units if it is to equal a phase angle. For instance, if Fig. 13-4 is meant to represent 360 electrical degrees in a four-pole machine, then it shows only half the circumference, and there must be 12 rather than 6 rotor conductors. Thus, for a p-pole machine, the number of phases of a c-slot rotor is

$$m_2 = \frac{2}{p}\frac{c}{2} = \frac{c}{p} \tag{13-9}$$

with single turns constituting the windings of each phase.

It must be understood that the number of poles of a machine is entirely determined by the stator, since the stator windings are the ones that are energized from a (three-phase) source. The rotor carries currents only by induction (transformer action), and their phase differences always equal the angular distances of their conductors in electrical units. As an example, a 36-bar squirrel cage inside a 2-pole stator has 18 phases, but the same squirrel cage inside a 4-pole stator has only 9 phases, etc.

Now that the squirrel-cage rotor is viewed as a wound rotor with m_2 phases, its impedance can be transferred to the three-phase stator, in accordance with Eq. 13-4, with $m_1 = 3$ and the voltage ratio a equal to the effective number of stator turns $N_1 k_{w_1}$, since the number of rotor turns is unity. Thus the equivalent circuit of Fig. 13-3 is valid regardless of the number of rotor phases. It should be mentioned that the actual rotor impedance and the ratio by which it is transferred to the stator are of interest mainly to the designer, not the analyst. The study of the motor performance is based on the value of $\mathbf{Z}_2'$, not $\mathbf{Z}_2$.

13-6 INTRODUCING THE SLIP

When the rotor of an induction machine moves, major changes take place. Since it is the *relative motion* between field and rotor that is responsible for the induced voltage in the rotor, this voltage must be expected to change with the rotor speed. For a numerical expression of this speed, there is a choice between the angular velocity ω_m in mechanical radians per second, and the traditional unit of revolutions per minute (rpm). The relation

$$n = \frac{60\,\omega_m}{2\pi} \tag{13-10}$$

is easily established by the facts that there are 60 seconds in a minute, and 2π radians in a revolution. Regardless of units used, the relative speed between field and rotor, also called the *slip speed*, is the difference between the synchronous speed at which the field rotates and the actual rotor speed.

The voltage induced in a coil by the relative motion of a sinusoidally distributed magnetic field was the subject of Section 8-5, where it was shown that both the magnitude and the frequency of this voltage are proportional to the speed of this relative motion. Thus, the following ratios may be formed for the magnitude and the frequency of the voltage E_2 induced in each phase of the rotor winding

$$\frac{E_{2\,\text{run}}}{E_{2\,\text{stand}}} = \frac{\omega_{sm} - \omega_m}{\omega_{sm}} \tag{13-11}$$

and

$$\frac{f_{2\,\text{run}}}{f_{2\,\text{stand}}} = \frac{\omega_{sm} - \omega_m}{\omega_{sm}} \tag{13-12}$$

because at standstill, $\omega_m = 0$.

On the right side of each of these two equations, the numerator is the slip speed. Dividing it by the synchronous speed, which is a constant (i.e., not depending on the operating condition of the motor), means *normalizing* the slip speed with respect to the synchronous speed. Thus, the right side of the last two equations is the normalized slip speed, briefly known as the *slip* of an induction machine.

$$s = \frac{\omega_{sm} - \omega_m}{\omega_{sm}} = \frac{n_s - n}{n_s} \tag{13-13}$$

This definition is illustrated in Fig. 13-5 (by arbitrary choice, in terms of n rather than ω_m). The reader will often encounter the facts that the slip is unity at standstill, and zero at synchronous speed.

The slip will turn out to be a most important quantity, one that characterizes the operating condition of an induction machine in much the same way as the torque angle δ for the synchronous machine. In terms of the slip, Eqs. 13-11 and 13-12 can be rewritten in simpler form:

$$E_{2\,\text{run}} = s\,E_{2\,\text{stand}} \tag{13-14}$$

and

$$f_{2\,\text{run}} = s\,f_{2\,\text{stand}} = s\,f \tag{13-15}$$

where f without subscript is the stator frequency, which is the frequency of the source or power system by which the stator is energized. Equation 13-15 introduces the frequency of the voltages and currents in the rotor circuit, logically named the *slip frequency*.

At this point, it is possible to demonstrate the remarkable and most significant fact that the *rotor mmf wave* always rotates at synchronous speed, regardless of the speed at which the rotor itself rotates. The reader knows from Section 8-8 that polyphase currents produce a rotating mmf wave whose speed is proportional to the frequency of these currents. So, if the stator currents whose frequency is f produce an mmf wave rotating at the speed n_s, the rotor currents whose frequency is sf must produce an mmf wave rotating at the speed sn_s. But that is its speed relative to the rotor. Relative to the stator, the speed of the rotor mmf wave is the sum

$$sn_s + n = \frac{n_s - n}{n_s}\,n_s + n = n_s \tag{13-16}$$

Fig. 13-5. Speed scale and slip scale.

This shows that the two mmf waves, although produced by currents of different frequencies, are always rotating at the same (synchronous) speed, no matter what the speed of the rotor itself, so that these waves can be combined to form a resultant mmf wave, just as in the synchronous machine.

13-7 THE EQUIVALENT CIRCUIT

The diagram of Fig. 13-3 describes a short-circuited transformer that has a constant voltage ratio $E_1/E_2 = a$, and the same frequency on the primary and secondary side. It is perfectly applicable to an induction machine at standstill. By contrast, in the rotor of a *running* induction machine, both the magnitude and the frequency of the induced voltage have been seen to be functions of the speed. In addition, the rotor impedance $\mathbf{Z}_2$ also depends on the speed since its imaginary part is the rotor reactance, or $2\pi f_2$ times the rotor leakage inductance. So it is by no means obvious that a slight modification will suffice to make the equivalent circuit applicable to the running induction machine.

It will be found convenient to define the symbol X_2 to mean the rotor reactance at standstill, i.e., at unity slip. This makes the *actual rotor impedance* per phase (according to Eq. 13-15)

$$\mathbf{Z}_2 = R_2 + jsX_2 \tag{13-17}$$

and the voltage equation for one phase of the actual rotor circuit becomes

$$s\mathbf{E}_{2\,\text{stand}} = (R_2 + jsX_2)\,\mathbf{I}_2 \tag{13-18}$$

The following step might appear to be trivial, but is in fact the key to the equivalent circuit of induction machines. Simply divide both sides of Eq. 13-18 by the slip s:

$$\mathbf{E}_{2\,\text{stand}} = (R_2/s + jX_2)\,\mathbf{I}_2 \tag{13-19}$$

The significance of this equation is that it describes the secondary circuit of a fictitious transformer, one with a constant voltage ratio and with the same frequency of both sides; in other words, a *fictitious stationary rotor* with a constant reactance but with a variable resistance. This fictitious stationary rotor carries the same current as the actual rotating rotor, and, thus, it produces the same mmf wave.

The stator does not "know" whether it faces the actual rotating rotor or the fictitious stationary one. Thus, the latter may be substituted for the former, which makes it possible to transfer the secondary impedance to the primary side (according to Eq. 13-4), leading to an equivalent circuit that differs from that of Fig. 13-3 only by having a variable resistance. Figure 13-6 illustrates this equivalent circuit. The impedance of each phase of the fictitious rotor transferred to

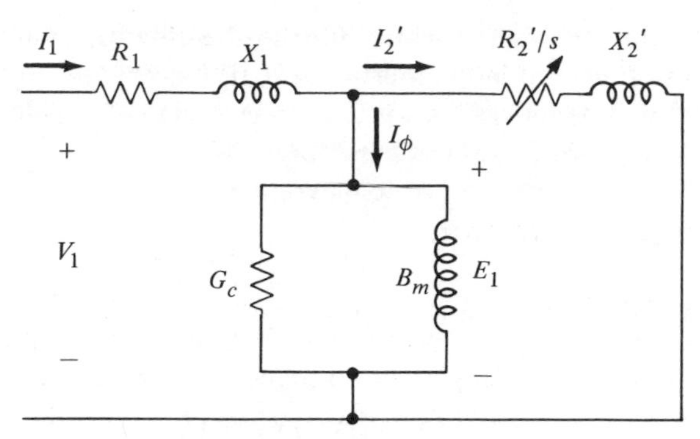

Fig. 13-6. Equivalent circuit.

the stator is marked by primes

$$\mathbf{Z}_2' = R_2'/s + jX_2' \tag{13-20}$$

The arrow drawn through the symbol of the resistance R_2'/s serves to draw attention to the fact that it is a variable, namely a function of the slip. For the sake of consistency, the other complex elements are also broken up into their real and imaginary parts:

$$\mathbf{Z}_1 = R_1 + jX_1 \tag{13-21}$$

and

$$\mathbf{Y}_\phi = G_c + jB_m \tag{13-22}$$

13-8 OPERATING CHARACTERISTICS

The equivalent circuit of Fig. 13-6 will be put to work immediately. Let rated voltage be applied to the stator terminals, and let its phasor $\mathbf{V}_1$ be chosen as the axis of reference. Then all other steady-state quantities can be calculated for any given or assumed or arbitrarily chosen value of slip. In particular, the stator current can be found as

$$\mathbf{I}_1 = \frac{\mathbf{V}_1}{R_1 + jX_1 + \cfrac{1}{G_c + jB_m + \cfrac{1}{R_2'/s + jX_2'}}} \tag{13-23}$$

For practical ways to handle this and some of the following calculations, see Example 13-1. If they are to be repeated for many values of slip, they are, of course, done on the digital computer.

There are only resistances and inductances in the circuit. Therefore, the current phasor must come out in the fourth quadrant (except for negative values of s, a subject that will be looked into in a later section). The cosine of the phase angle θ of $\mathbf{I}_1$ is the power factor of the motor.

Having found $\mathbf{I}_1$, one may proceed to determine

$$\mathbf{E}_1 = \mathbf{V}_1 - (R_1 + jX_1)\,\mathbf{I}_1 \tag{13-24}$$

and the rotor current referred to the stator

$$\mathbf{I}_2' = \frac{\mathbf{E}_1}{R_2'/s + jX_2'} \tag{13-25}$$

which could also have been found by current division.

The most important quantities to be determined from the equivalent circuit are the various amounts of power received, lost, or converted by the induction motor. The input power is

$$P_{\text{in}} = 3V_1 I_1 \cos\theta \tag{13-26}$$

Of the three-energy-consuming elements shown by the equivalent circuit, two clearly represent losses. The power losses in the stator resistances, commonly referred to as *stator copper losses*, are

$$P_{R_1} = 3R_1 I_1^2 \tag{13-27}$$

and the *core losses* are

$$P_c = 3G_c E_1^2 \tag{13-28}$$

The third element, R_2'/s, is fictitious, but the power consumed by it is significant. To understand this, look at this power as what is left if the losses in R_1 and G_c are subtracted from the input power. According to the equivalent circuit

$$P_g = P_{\text{in}} - P_{R_1} - P_c = 3(R_2'/s)\,I_2'^2 \tag{13-29}$$

This quantity is known as the *power across the air gap*, or the *air gap power* for short. (This explains the subscript g, for one thing.) The idea is that the input power is supplied to the motor at its *stator* terminals, and that the losses P_{R_1} and P_c occur in the stator windings and the stator core, respectively. (This is not quite true; part of the core losses occur in the rotor core, but that is a minor part only, as will be seen a little later.) So the remainder, the power P_g, is being transferred to the rotor across the air gap.

What happens to this power in the rotor? The part of it that is consumed in the rotor resistance (the *rotor copper losses*) can be expressed as

$$P_{R_2} = 3R_2' I_2'^2 \tag{13-30}$$

because, as was shown in Section 13-5, the transfer of the rotor impedance to the stator circuit leaves the power unchanged. A comparison of the last two equations indicates that

$$P_{R_2} = sP_g \tag{13-31}$$

The purpose of a motor is to convert electrical into *mechanical power*, not to raise its temperature. If a motor is to have a reasonable efficiency, no group of power losses may amount to more than a small fraction of the input power. Thus, by Eq. 13-29, P_g must be in the same order of magnitude as P_{in}, just a little smaller, and by Eq. 13-31, the slip at any normal operating condition must be a small fraction of unity. In other words, the rated speed of an induction motor must be only a little below synchronous speed. Typical values of slip at rated condition are around 0.05.

What is left of P_g after subtracting the rotor copper losses is the power converted into mechanical form. From Eq. 13-31, it is

$$P_{\text{mech}} = P_g - P_{R_2} = (1 - s)\, P_g \tag{13-32}$$

In this context, it may be worth mentioning that the term *slip* originates from the analogy between the induction motor and a friction clutch or belt drive. The amount of power lost in such a device by friction depends on how much it "slips", i.e., on the difference in speed between the driving part and the driven part, just as the rotor copper losses of an induction motor depend on the slip speed.

The division of the air gap power into rotor copper losses and mechanical power, expressed by the last two equations, can be shown in the equivalent circuit. Let the fictitious resistance R_2'/s be split into two parts

$$R_2'/s = R_2' + (1 - s)\, R_2'/s \tag{13-33}$$

In Fig. 13-7, the resistance symbolizing the mechanical power is drawn on the right side, making the whole equivalent circuit look like that of a transformer with an adjustable load resistance.

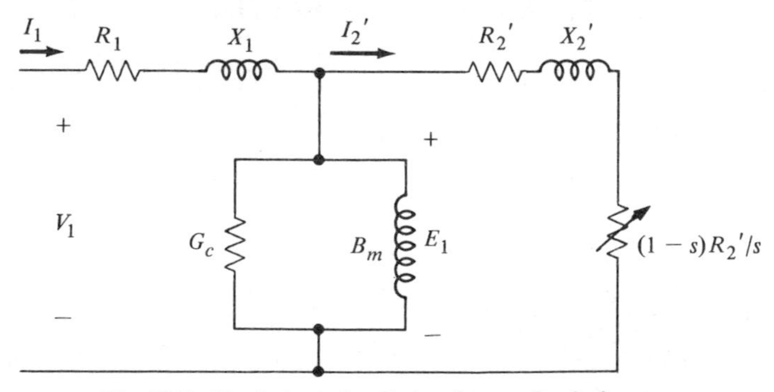

Fig. 13-7. Equivalent circuit showing mechanical power.

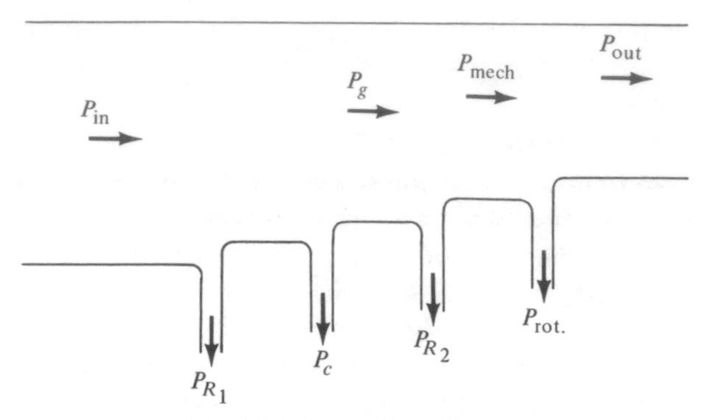

Fig. 13-8. Power flow diagram.

The mechanical power expressed in Eq. 13-32 is not the output power of the induction motor because it still includes the *rotational losses* that are essentially the power losses due to friction and windage. Thus, the *output power* is

$$P_{out} = P_{mech} - P_{rot} \tag{13-34}$$

Being a purely mechanical quantity, the rotational losses are not represented by any element of the equivalent circuit.

To recapitulate all the power relations discussed, the *power flow diagram* of Fig. 13-8 shows the input power, all losses and the output power. A comparison with the power flow diagram of a synchronous machine (drawn for a generator in Fig. 11-10) reveals an interesting difference: the rotational losses of a synchronous machine were defined to include the core losses as well as friction and windage, whereas, for the induction machine, these two groups of losses are separated. There is a reason for this: the rotating mmf wave of the induction machine is entirely the product of polyphase currents and not at all of the motion of the rotor; by contrast, the d-c-excited rotor of a synchronous machine produces no rotating mmf wave (and, thus, no core losses) unless the rotor itself rotates.

At this point, it can also be explained why the core losses of an induction machine occur mainly in its stator. According to Eqs. 4-18 (for the hysteresis loss) and 4-19 (for the eddy-current loss), all core losses depend on the core volume, the frequency, and the flux density. The fact that the volume of the stator greatly exceeds that of the rotor can be seen from a sketch like that of Fig. 7-18. Even more importantly, the rotor frequency at any steady-state operating condition is a small fraction of the stator frequency, a fact that reduces the core losses occurring in the rotor to an insignificant portion of the total core losses.

The *efficiency* of the induction motor may be seen from Fig. 13-8 to be

$$\eta = \frac{P_{out}}{P_{out} + P_c + P_{R_1} + P_{R_2} + P_{rot}} \tag{13-35}$$

For typical numerical calculations, see Example 13-2.

Finally, the *electromagnetic torque* can be found, just as it was in the case of the synchronous machine, from the power converted.

$$T = \frac{P_{mech}}{\omega_m} \tag{13-36}$$

Another expression for this torque is obtained by substitution of $\omega_m = (1 - s)\,\omega_{s_m}$ (from Eq. 13-13, the definition of the slip) in the denominator, and of Eq. 13-32 in the numerator:

$$T = \frac{(1 - s)\,P_g}{(1 - s)\,\omega_{s_m}} = \frac{P_g}{\omega_{s_m}} \tag{13-37}$$

where P_g may be calculated as the power consumed by the fictitious resistance R_2'/s in Fig. 13-6. One advantage of the last torque expression is that only the numerator is variable; the denominator is a constant, independent of the load condition. It also enhances the significance of the air gap power P_g by showing it to be proportional to the electromagnetic torque. In addition, this expression is applicable to the starting condition when the slip is unity and the torque cannot be calculated directly from Eq. 13-36, which becomes an indeterminate form.

Now all variable quantities (stator current, output power, losses, torque, etc.) can be obtained for any chosen value of the slip, the equivalent circuit consisting of nothing but constant elements except for the effect of the slip. Repeating these procedures for many values of slip (presumably using the computer) leads to tabulations or graphs of these quantities as functions of the slip. Also, corresponding values of any two variables may be plotted against each other. Such relationships, as, for instance, efficiency versus output power, or stator current versus torque, are often called characteristics of the motor. The most important of them, the *torque-speed characteristic*, will be discussed in the next section.

13-9 BETWEEN NO-LOAD AND STANDSTILL

The electromagnetic torque accelerates the rotor in the forward direction (unless it is exceeded by an opposing torque). Therefore, in an *idealized no-load* condition (i.e., in the absence of both load and rotational losses), the motor speed increases until it reaches the synchronous value. This is so because, at *synchronous speed*, the electromagnetic torque is zero, a fact that was previously deduced by physical reasoning, and that can be confirmed from the equivalent circuit (Fig. 13-6): synchronous speed means $s = 0$, so that the branch with the resistance R_2'/s becomes an *open circuit*, making the current I_2' and, thus, the mechanical power, as well as the torque, disappear.

Now let a load torque be applied to the shaft. Its effect is to slow down the motor, and if it is strong enough it eventually brings the motor to its other extreme condition, that of *standstill.* With $s = 1$, the "load resistance" of Fig. 13-7 becomes zero, as in a *short-circuited* transformer. Accordingly, the mechanical power is once again zero, as it has to be for zero speed.

Thus, it has been shown that $P_{mech} = 0$ at both extremes of synchronous speed and zero speed. In between, with $0 < s < 1$, this power has positive values, and it must reach a maximum somewhere within this range.

The reader is presumably familiar with the principle of *impedance matching* to obtain maximum power in the circuit of Fig. 13-9. The adjustable impedance $\mathbf{Z}_b$, in series with the constant impedance $\mathbf{Z}_a$, consumes zero power in the two extreme cases of short circuit ($Z_b = 0$) and open circuit ($Z_b \to \infty$). In between, the power consumed by $\mathbf{Z}_b$ has a maximum value if $\mathbf{Z}_b$ is adjusted to the conjugate of $\mathbf{Z}_a$.

Less well known is how $\mathbf{Z}_b$ can be adjusted for maximum power if only the magnitude, not the angle of Z_b is adjustable. For this case, call $\mathbf{Z}_b = Z \underline{/\theta}$ and $\mathbf{Z}_a = R_a + jX_a$, and differentiate

$$P_b = I^2 \,\mathcal{R}e\, \mathbf{Z}_b = \frac{V^2 Z \cos\theta}{(R_a + Z\cos\theta)^2 + (X_a + Z\sin\theta)^2} \tag{13-38}$$

with respect to Z, with V, R_a, X_a, and θ held constant. Then the numerator of the derivative is set equal to zero:

$$0 = V^2 \cos\theta(R_a^2 + 2R_aZ\cos\theta + Z^2\cos^2\theta + X_a^2 + 2X_aZ\sin\theta + Z^2\sin^2\theta)$$
$$- V^2 Z\cos\theta\,[2(R_a + Z\cos\theta)\cos\theta + 2(X_a + Z\sin\theta)\sin\theta] \tag{13-39}$$

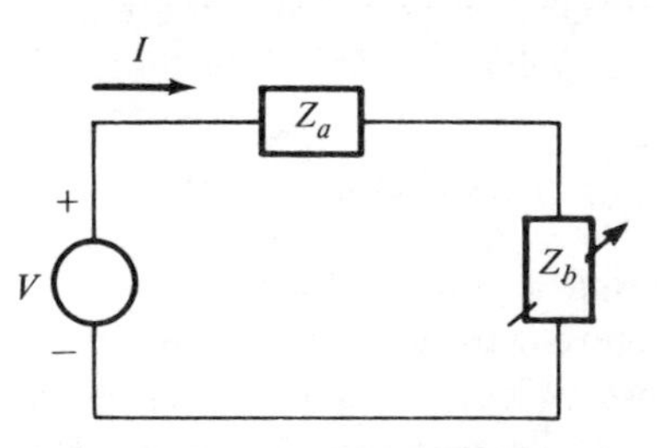

Fig. 13-9. Impedance matching.

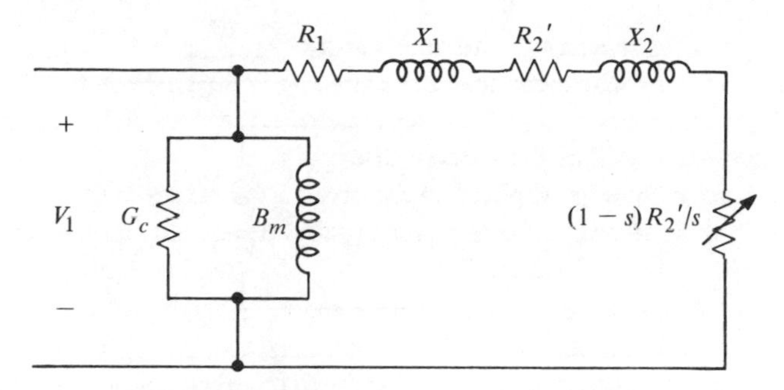

Fig. 13-10. Approximate equivalent circuit.

Several terms cancel out, leaving only

$$0 = V^2 \cos\theta\,(R_a^2 + X_a^2 - Z^2\cos^2\theta - Z^2\sin^2\theta) \tag{13-40}$$

Thus, the answer is

$$Z = \sqrt{R_a^2 + X_a^2} \tag{13-41}$$

In words: if the angle of $\mathbf{Z}_b$ is fixed, then maximum power is obtained by matching the *magnitudes* of the two impedances. For instance, if $\mathbf{Z}_b$ is a pure resistance, then its value should be adjusted to equal the magnitude of the fixed impedance to which it is connected in series.

If this theorem is to be applied to the induction motor, its equivalent circuit must first be modified to take the form of Fig. 13-9. There are two possible schemes.

(a) Use an *approximate* equivalent circuit, with the admittance $\mathbf{Y}_\phi$ shifted to the input terminals. Fig. 13-10 is obtained by performing this slight-of-hand operation on the circuit of Fig. 13-7. It was stated earlier that this results in more substantial inaccuracies than for the transformer. Yet it is acceptable whenever only qualitative results are desired. The circuit of Fig. 13-10 permits the matching of impedance magnitudes. Accordingly, the mechanical power has a maximum value at the speed that makes

$$(1 - s)R_2'/s \approx \sqrt{(R_1 + R_2')^2 + (X_1 + X_2')^2} \tag{13-42}$$

where the "approximately equal" sign is used as a reminder that an approximate equivalent circuit was used.

(b) Without sacrificing accuracy, an all-series circuit can be obtained by replacing the part of the equivalent circuit to the left of terminals a and b (see Fig. 13-11) by its equivalent according to *Thévenin's theorem.* The ideal voltage

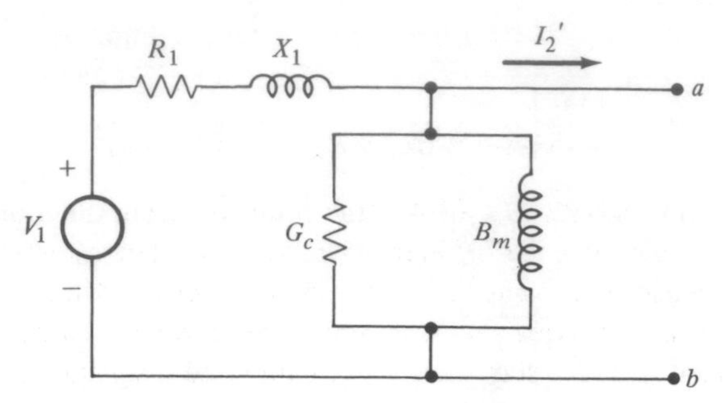

Fig. 13-11. Circuit for application of Thevenin's theorem.

source V_1 represents the infinite bus from which the motor is considered to be supplied.

The result of this operation is the circuit of Fig. 13-12. The new source V_1^* is the open-circuit voltage (obtained by voltage division)

$$\mathbf{V}_1^* = \frac{V_1 \dfrac{1}{G_c + jB_m}}{R_1 + jX_1 + \dfrac{1}{G_c + jB_m}} \tag{13-43}$$

The new impedance $\mathbf{Z}_1^* = R_1^* + jX_1^*$ is the driving-point impedance at the terminals a and b, with the sourŷe V_1 replaced by a short-circuit connection. Thus, it is the parallel combination of the two impedances $\mathbf{Z}_1$ and $1/\mathbf{Y}_\phi$

$$R_1^* + jX_1^* = \frac{(R_1 + jX_1) \dfrac{1}{G_c + jB_m}}{R_1 + jX_1 + \dfrac{1}{G_c + jB_m}} \tag{13-44}$$

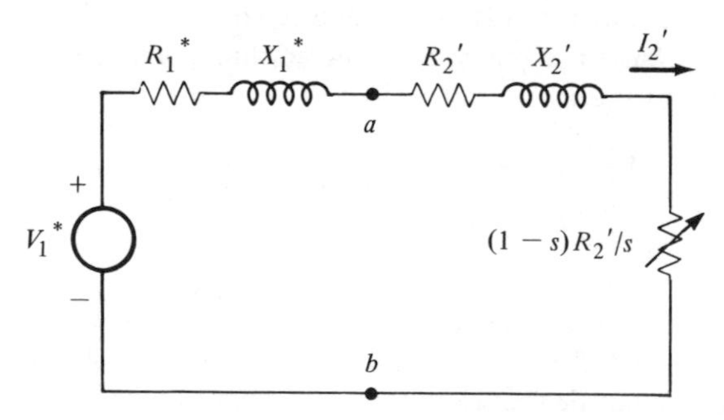

Fig. 13-12. Circuit obtained from Thevenin's theorem.

From the circuit of Fig. 13-12, the condition for maximum mechanical power is obtained as

$$(1 - s)R_2'/s = \sqrt{(R_1^* + R_2')^2 + (X_1^* + X_2')^2} \qquad (13\text{-}45)$$

This result is identical with Eq. 13-42 (the result of approximation) except for the asterisks. To appraise the difference, Eq. 13-44 offers a direct comparison of the two impedances $\mathbf{Z}_1$ and $\mathbf{Z}_1^*$. It must be kept mind that all circuit elements appearing in this equation represent imperfections. Thus, each of them must have a value of much less than 1 pu, with rated quantities chosen as bases of a per-unit system. Now the denominator of $\mathbf{Z}_1^*$ is the sum of $\mathbf{Z}_1$ and $1/\mathbf{Y}_\phi$, two terms of which the first must be much smaller than the second. Consequently, the terms $1/\mathbf{Y}_\phi$ in the numerator and denominator can "almost" be canceled; in fact, if $\mathbf{Z}_1$ were negligible compared to $1/\mathbf{Y}_\phi$, the result would be $\mathbf{Z}_1^* = \mathbf{Z}_1$. So Eq. 13-45 may be looked at as an improved version of Eq. 13-42, one in which the error due to the use of the approximate equivalent circuit is corrected.

The relation between $\mathbf{V}_1^*$ and $\mathbf{V}_1$ (Eq. 13-43) is the same as that between $\mathbf{Z}_1^*$ and $\mathbf{Z}_1$. Again, $\mathbf{V}_1^*$ is the corrected value to be used in the calculation of maximum power. In the remainder of this section, the corrected values will be used consistently, and it will be understood that leaving out the asterisks would be the way to avoid the work of evaluating Eqs. 13-43 and 13-44, and would lead to similar but less accurate results.

The *electromagnetic torque* is obtained, according to Eq. 13-37, as proportional to the air gap power, which is the power consumed by the fictitious resistance R_2'/s. Figure 13-13 is the same circuit as the previous diagram, but redrawn to display this resistance. From it, the condition for *maximum torque* can be found to be

$$R_2'/s = \sqrt{R_1^{*2} + (X_1^* + X_2')^2} \qquad (13\text{-}46)$$

Again, the asterisks may be omitted if there is no need for accuracy. It should also be noted that *maximum torque* and *maximum mechanical power* represent

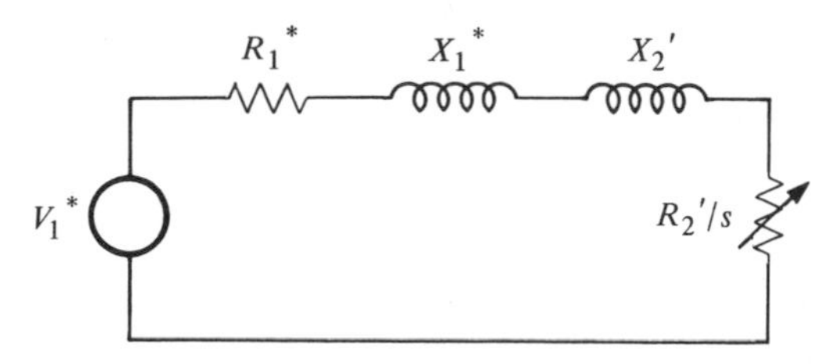

Fig. 13-13. Impedance matching for maximum torque.

two different conditions, in contrast to the synchronous machine. This is so because the speed is not constant.

Equation 13-46 permits an estimate of the speed at which the maximum torque occurs. Each of the two resistances R_1 and R_2' must be a small fraction of 1 pu (in the interest of efficiency), whereas the reactances cannot be made quite that small (because the conductors are located in slots inside the core). Thus, the square root in Eq. 13-46 should be definitely larger than R_2', which makes the slip for the condition of maximum torque less than unity. (Typical values are between 0.2 and 0.4.) Sample calculations are given in Example 13-3.

Figure 13-14 depicts the shape of a typical *torque-speed characteristic*, with a maximum torque at a slip of about one-third (i.e., at two-thirds of synchronous speed). The point of *rated operation* must be at a small value of slip, i.e., close to synchronous speed (for reasons explained in Section 13-8). At this point, the torque must be far enough below the maximum torque to provide whatever margin may be required. The *starting torque* (at $s = 1$) must be above the rated torque if the motor is expected to start at full-load, i.e., if the load exerts its full rated torque at standstill, which is not the case with all loads (see the next chapter).

The three torques exhibited in Fig. 13-14 are the main specifications on which the design of the motor for any particular purpose is based. In per-unit calculations involving these three torques, it is logical to choose rated torque as a base quantity, and consequently to choose the values of P_g and I_2' at rated conditions as base quantities. This choice further necessitates that of R_2'/s_{rated} as the impedance base, although this results in V_1^* (or V_1) not being equal to unity.

The part of the characteristic describing the range of operation between *no-load* (almost at synchronous speed but not quite so, due to rotational losses) and *full-load* (rated condition) is practically a straight line. This can be confirmed by calculating the torque

$$T = \frac{1}{\omega_{s_m}} P_g = \frac{3}{\omega_{s_m}} I_2'^2 R_2'/s \tag{13-47}$$

for small values of slip, for which

$$I_2' = \frac{V_1^*}{\sqrt{(R_1^* + R_2'/s)^2 + (X_1^* + X_2')^2}} \approx \frac{V_1^* s}{R_2'} \tag{13-48}$$

Substituting the last equation into the previous one leads to

$$T \approx \frac{3}{\omega_{s_m}} \frac{V_1^{*2} s}{R_2'} \tag{13-49}$$

which says that the torque is approximately proportional to the slip.

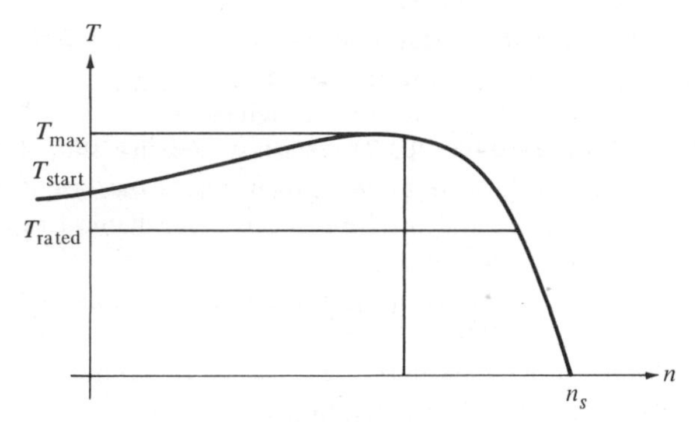

Fig. 13-14. Torque-speed characteristic.

Note the contrast of all this to the synchronous motor, which has only one steady-state speed and responds to a change of the load torque by a change of its torque *angle.* The induction motor adjusts its *speed* to meet load changes. On the other hand, these speed changes between no-load and full-load are comparatively slight, which explains why the characteristic of the induction motor is sometimes called a *constant-speed characteristic.* This is not to be taken literally but only as a contrast to the *variable-speed* property of certain other types of motors that will be encountered in later chapters.

13-10 OUTSIDE THE RANGE OF MOTOR OPERATION

The two special conditions of zero slip and unity slip are the extreme limits of motor operation, but it is possible for the induction machine to operate outside this range. In particular, *negative slip* means operation at more than synchronous speed. This can never occur as a result of the electromagnetic torque but only due to an outside torque, i.e., when the machine is *driven* beyond its synchronous speed.

As an example, let us envision a vehicle or train driven by an induction motor (not the normal choice for that purpose), and let this train run *downhill*, without shutting off the motor. The force of gravity accelerates the train and may eventually drive it beyond the motor's synchronous speed. What happens to the motor in such a case?

It must be kept in mind that the magnetic field always rotates at synchronous speed as long as the stator is connected to its source (in our case, presumably over trolley wires). Thus, the relative motion between field and rotor is reversed, which reverses the rotor voltages and currents. This, in turn, causes a reversal of the stator currents (apart from their magnetizing and core loss components) and thereby of the power at the stator terminals. So the machine operates as a *generator*, delivering electric power to its source.

It is remarkable how simply the same conclusions can be drawn from the equivalent circuit (Figs. 13-6 and 13-7). With a negative slip, the resistances R_2'/s and $(1 - s)R_2'/s$ are both negative, which makes the torque and the mechanical power negative, i.e., the machine receives mechanical power and produces a torque that opposes its motion.

It is perfectly feasible to operate an induction machine as a power generator, driven by a prime mover. But such an *induction generator* needs something extra, namely a source for its magnetic field. The synchronous machine, as the reader remembers, obtains its field excitation from a separate source; the induction generator has nothing of this sort. So there must be at least one synchronous generator operating in parallel to it, to supply it with currents that produce its magnetic field.

Another drawback of the induction generator is the nature of its power factor. The impedance of the secondary branch of the equivalent circuit (Fig. 13-6), with a negative slip, is in the second quadrant. So the current I_2' (with the voltage in the axis of reference) is in the third quadrant. Figure 13-15 shows this current and its phasor addition to the exciting current I_ϕ to obtain the stator current I_1. Reversing the current arrow for generator operation (taking the negative of I_1) shows that the output current of an induction generator is inevitably *leading*, like that of an underexcited synchronous generator. The fact that the loads of most power systems need a supply of lagging reactive power constitutes, therefore, a serious handicap for the induction generator.

One more contingency deserves some study, namely operation at positive values of slip above unity, corresponding to *negative speeds*, i.e., motion in the direction opposite to that of the magnetic field. To help visualizing this possibility, consider again our train driven by an induction motor but, this time, let it try to climb up a steep grade, one requiring a torque larger than the maximum motor torque. Whenever a load torque exceeds the motor torque, the motor and

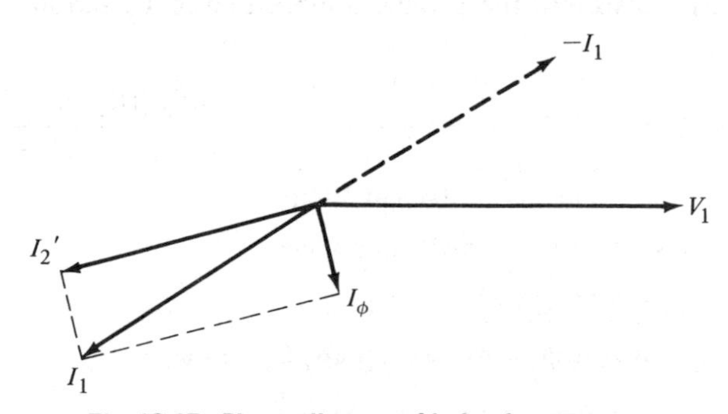

Fig. 13-15. Phasor diagram of induction generator.

its load are slowed down. When the speed of our train is reduced to zero, however, it does not stop but begins to roll backward.

For this condition, the resistance $(1 - s)R_2'/s$ is negative, indicating that the motor receives mechanical power (is driven by the force of gravity). But this time, the resistance R_2'/s remains positive, and so does, therefore, the electromagnetic torque, although its magnitude decreases as the train is being accelerated backward. So the motor acts as a (rather ineffective) electric *brake*, straining desperately and futilely to stem the downhill motion. It receives both mechanical and electrical power and converts it all into heat.

Clearly, this is not an operating condition of any practical significance. It is, however, theoretically meaningful and will be encountered again in a later chapter. This is why the curve of Fig. 13-14 has been extended into the region of negative speed. The torque vanishes asymptotically as the negative speed is increased toward infinity.

13-11 EXAMPLES

Example 13-1 (Section 13-8)

A 60-Hz, four-pole, Y-connected induction motor is rated 5 HP, 220 v (line to line). The equivalent circuit parameters are:

$$R_1 = 0.48\ \Omega \qquad R_2' = 0.42\ \Omega \qquad B_m = -1/30 \text{ mhos}$$

$$X_1 = 0.80\ \Omega \qquad X_2' = 0.80\ \Omega$$

The motor is operating with slip of 0.04. Find the input current and power factor. Find the air gap power, the mechanical power, and the electromagnetic torque.

Solution

Use the equivalent circuit in Fig. 13-6. The shunt conductance, G_c, is negligible in determining I_1. Let $\mathbf{Z}_f$ be the parallel combination of $\mathbf{Y}_\phi$ and $\mathbf{Z}_2'$. Let $X_m = -1/B_m = 30\ \Omega$

$$\mathbf{Z}_f = R_f + jX_f = \frac{1}{G_c + jB_m + \dfrac{1}{(R_2'/s) + jX_2'}} \approx \frac{jX_m\,[(R_2'/s) + jX_2']}{(R_2'/s) + j(X_2' + X_m)}$$

All of the quantities are given. Substitution yields

$$\mathbf{Z}_f = 9.71\,\underline{/23.3^\circ} = 8.92 + j\,3.84\ \Omega$$

$$\mathbf{Z}_{in} = \mathbf{Z}_1 + \mathbf{Z}_f = (0.48 + j\,0.80) + (8.92 + j\,3.84)$$

$$= 9.40 + j\,4.64 = 10.5\,\underline{/26.3^\circ}\ \Omega$$

The phase voltage is $220/\sqrt{3} = 127$ v. Use V_1 as the reference. $\mathbf{V}_1 = 127\underline{/0^\circ}$ v. The input current is

$$\mathbf{I}_1 = \mathbf{V}_1/\mathbf{Z}_{\text{in}} = (127\underline{/0^\circ})/(10.5\underline{/26.3^\circ}) = 12.1\underline{/-26.3^\circ} \text{ amp}$$

The power factor is $\cos(-26.3^\circ) = 0.90$. The air gap power is given by

$$P_g = 3(R_2'/s)\,I_2'^2 = 3R_f I_1^2 = 3(8.92)\,(12.1)^2 = 3920 \text{ w}$$

The mechanical power is given by

$$P_{\text{mech}} = (1 - s)\,P_g = (.96)\,(3920) = 3760 \text{ w}$$

Synchronous speed is $\omega_{s_m} = 4\pi f/p = 4\pi 60/4 = 188.5$ rad/sec. The operating speed is $\omega_m = (1 - s)\,\omega_{s_m} = 181$ rad/sec. The electromagnetic torque is

$$T = P_{\text{mech}}/\omega_m = 3760/181 = 20.8 \text{ nm}$$

The electromagnetic torque can also be found from

$$T = P_g/\omega_{s_m} = 3920/188.5 = 20.8 \text{ nm}$$

Example 13-2 (Section 13-8)

The induction motor of Example 13-1 has stator core loss of 51 w and rotational loss of 138 w. For operation with slip of 0.04, find the output power and the efficiency.

Solution

The output power is

$$P_0 = P_{\text{mech}} - P_{\text{rot}} = 3760 - 138 = 3620 \text{ w}$$

Tabulate all of the losses

Stator core loss $= P_c = 51$ w

Stator copper loss $= P_{R_1} = 3R_1 I_1^2 = 3(0.48)\,(12.1)^2 = 211$ w

Rotor copper loss $= P_{R_2} = 3R_2' I_2'^2 = sP_g = (0.04)\,(3920) = 157$ w

Rotational losses $= P_{\text{rot}} = 138$ w

$\sum$ losses $= P_c + P_{R_1} + P_{R_2} + P_{\text{rot}} = 557$ w

The efficiency is given by

$$\eta = \frac{P_{\text{out}}}{P_{\text{out}} + \sum \text{losses}} = 1 - \frac{\sum \text{losses}}{P_{\text{out}} + \sum \text{losses}} = 0.87$$

This concludes the solution of the problem, as stated. We can enlarge our perspective by also finding the output power in horsepower, the air gap voltage E_1, and the current I_2'

$$P_{out} = 3620 \text{ w}/(746 \text{ w/HP}) = 4.85 \text{ HP}$$

This shows that operation with slip of 0.04 does not quite deliver the rated output power for this machine. The air gap voltage is the voltage generated in a stator winding by the air gap flux. From the equivalent circuit, we can find E_1

$$\mathbf{E}_1 = \mathbf{Z}_f \mathbf{I}_1 = (9.71 \angle 23.3^\circ)(12.1 \angle -26.3^\circ) = 117.5 \angle -3^\circ \text{ v}$$

We can find the current I_2'

$$\mathbf{I}_2' = \mathbf{E}_1/\mathbf{Z}_2' = (117.5 \angle -3^\circ)/(10.53 \angle 4.36^\circ) = 11.2 \angle -7.4^\circ \text{ amp}$$

Example 13-3 (Section 13-9)

Using the induction motor of Example 13-1, (a) find the value of slip at which maximum torque is developed, (b) find the maximum torque, the corresponding mechanical power and rotor copper loss, (c) find the starting torque for operation with rated input voltage.

Solution

(a) Use the equivalent circuit of Fig. 13-13. We can neglect G_c.

$$V_1^* = \frac{(jX_m)\,V_1}{R_1 + j(X_1 + X_m)} = 123.7 \angle 0.9^\circ \text{ v}$$

$$\mathbf{Z}_1^* = R_1^* + jX_1^* = \frac{(R_1 + jX_1)(jX_m)}{R_1 + j(X_1 + X_m)} = 0.909 \angle 59.9^\circ = 0.46 + j\,0.79\ \Omega$$

The value of R_2'/s for maximum torque is given by Eq. 13-46

$$R_2'/s = \sqrt{R_1^{*2} + (X_1^* + X_2')^2} = [(.46)^2 + (0.79 + 0.80)^2]^{1/2} = 1.66\ \Omega$$

The slip is $s = R_2'/1.66 = 0.42/1.66 = 0.253$.

(b) From the equivalent circuit of Fig. 13-13, we can find the current I_2' and the air gap power

$$\mathbf{I}_2' = \mathbf{V}_1^*/(\mathbf{Z}_1^* + \mathbf{Z}_2') = (123.7 \angle 0.9^\circ)/(2.65 \angle 36.9^\circ) = 46.7 \angle -36^\circ \text{ amp}$$

$$P_g = 3(R_2'/s)(I_2')^2 = 3(1.66)(46.7)^2 = 10{,}860 \text{ w}$$

$$T_{max} = P_g/\omega_{sm} = 10{,}860/188.5 = 57.6 \text{ nm}$$

Compare T_{max} with the torque of 20.8 nm for slip of 0.04

$$P_{mech} = (1 - s) P_g = 8110 \text{ w}$$

$$P_{R_2} = 3R_2' I_2'^2 = 3(0.42)(46.7)^2 = 2750 \text{ w}$$

Compare this rotor copper loss with 157 w when the slip was 0.04 to realize that this heating can be tolerated only for very brief intervals.

(c) For starting, the slip is unity. Use the equivalent circuit of Fig. 13-13. Now we have

$$\mathbf{Z}_2' = R_2' + jX_2' = 0.42 + j\,0.80\ \Omega$$

$$\mathbf{I}_2' = \mathbf{V}_1^*/(\mathbf{Z}_1^* + \mathbf{Z}_2') = (123.7\,\underline{/0.9^\circ})/(1.82\,\underline{/61^\circ}) = 68\,\underline{/-60.1^\circ} \text{ amp}$$

The air gap power is

$$P_g = 3R_2'(I_2')^2 = 3(0.42)(68)^2 = 5830 \text{ w}$$

The starting torque is

$$T_{st} = P_g/\omega_{sm} = 5830/188.5 = 31 \text{ nm}$$

13-12 PROBLEMS

13-1. A three-phase, 50-Hz voltage is applied to a three-phase, six-pole induction motor. The motor is running at a constant speed with slip of 0.04. Find the following: (a) speed in rpm of the rotating field in the air gap relative to the stator winding. (b) speed of the rotor in rpm. (c) frequency of the rotor currents. (d) speed in rpm of the rotor mmf relative to the rotor winding.

13-2. A three-phase induction motor runs at almost 900 rpm at no-load and at 864 rpm at rated load when supplied with power from a 60-Hz, three-phase source. (a) How many poles does this motor have? (b) Find the slip at rated load. (c) Find the frequency of the rotor voltage when running at rated load.

13-3. A three-phase, 25-Hz induction motor runs at a speed of 360 rpm at rated load. (a) Find the synchronous speed, such that the slip is a reasonable value. (b) Find the frequency of the rotor currents.

13-4. The induction motor of Example 13-1 is operated with rated voltage, but with increased load that makes the slip become 0.05. Find the input current, the air gap power, and the electromagnetic torque. (Compare the ratio of torque to slip in this problem with the same ratio in Example 13-1.)

13-5. The induction motor of Example 13-2 is operated with rated voltage, but with increased load to make the slip become 0.05. Find the output power and the efficiency.

13-6. A four-pole, three-phase, Y-connected induction motor is rated 15 HP, 60 Hz, 220 v (line to line). The parameters of the equivalent circuit are:

$$R_1 = 0.23\ \Omega \qquad R_2' = 0.15\ \Omega \qquad X_m = 14\ \Omega$$

$$X_1 = 0.47\ \Omega \qquad X_2' = 0.47\ \Omega$$

Rotational loss is 300 w. Stator core loss is 200 w. For a slip of 0.05, find the stator current, the air gap power, and the electromagnetic torque.

13-7. For the machine in Problem 13-6, find the output power and the efficiency for operation with slip of 0.05.

13-8. For the machine in Problem 13-6, find (a) the maximum torque and the slip at which it occurs, (b) the starting torque and the input stator current for the starting condition.

13-9. An eight-pole, three-phase, Y-connected, 60-Hz, 440-v (line to line) induction motor is rated 100 HP. The equivalent circuit parameters are:

$$R_1 = 0.086\ \Omega \qquad R_2' = 0.063\ \Omega \qquad X_m = 6.8\ \Omega$$

$$X_1 = 0.19\ \Omega \qquad X_2' - 0.16\ \Omega$$

The rotational loss is 1.6 kw. The stator core loss is 1.0 kw. For operation with slip of 0.03, find the stator current, the air gap power, and the electromagnetic torque.

13-10. For the machine in Problem 13-9, find the output power and the efficiency for operation with slip of 0.03.

13-11. For the induction motor in Problem 13-9, find the maximum torque and the slip at which it occurs.

13-12. A three-phase, 60-Hz, four-pole induction motor is operating at a speed of 1728 rpm. The output power is 57 kw. The rotational loss is 1540 w. The stator copper loss is 2900 w. The stator core loss is 1920 w. Find the efficiency.

13-13. A three-phase, 60-Hz, four-pole induction motor takes 5100 w from the line. The rotational loss is 73 w. The stator core loss is 130 w. The stator copper loss is 230 w. The rotor copper loss is 170 w. Find the following: (a) the air gap power, (b) the slip, (c) the efficiency.

13-14. A three-phase, 60-Hz, six-pole induction motor delivers its rated power of 50 HP with a slip of 0.05. The rotor copper loss is 2 kw. The stator copper loss is 2 kw. The stator core loss is 1.3 kw. Find the efficiency.

13-15. A simplifying approximation in the equivalent circuit of Fig. 13-13 is to assume R_1^* equal to zero. An induction motor has $R_2' = 0.04$, $(X_1^* + X_2') = 0.24$, $V_1^* = 1.028$, all in per-unit. Rated torque is developed for slip of 0.04. (a) Find the maximum torque in per-unit and the slip at which it occurs. (b) Find the starting torque in per-unit.

13-16. Use the equivalent circuit of Fig. 13-13 with $R_1^* = 0$, $R_2' = 0.04$, $(X_1^* + X_2') = 0.12$, $V_1^* = 1.007$, all in per-unit. Rated torque is produced for slip of 0.04. (a) Find the maximum torque in per-unit and the slip at which it occurs. (b) Find the starting torque in per-unit.

13-17. Use the equivalent circuit of Fig. 13-13 with $R_1^* = 0$, $R_2' = 0.08$, $(X_1^* + X_2') = 0.24$, $V_1^* = 1.028$, all in per-unit. Rated torque is produced for a slip of 0.08. (a) Find the maximum torque in per-unit and the slip at which it occurs. (b) Find the starting torque in per-unit.

13-18. For the induction motor of Problem 13-6, the shunt conductance, G_c, is 0.00545 mho per phase. Solve Problem 13-6 using the equivalent circuit of Fig. 13-7 and Eq. 13-23. (Compare these answers with Problem 13-6.)

14 Application of Induction Motors

14-1 LOAD CHARACTERISTICS

An electric motor drives its load in a steady-state condition when its electromagnetic torque is just equal and opposite to the torque exerted by the load. (This statement assumes that the torque corresponding to the rotational losses is either negligible or that it has been incorporated into the load torque.) In other words, the motor and its load are settling on that speed at which this condition of *equilibrium* is satisfied. With the aid of a given torque-speed characteristic (like that of Fig. 13-14), the steady-state speed is determined as that speed at which the magnitude of the motor torque equals that of the load torque.

This is simple enough when the load torque is a given constant. But many loads exert torques that depend on the speed. In other words, the load may have its own torque-speed characteristic, which must be known. The steady-state speed is then obtained by intersecting the graphs of the motor and load characteristics, or by the corresponding analytical operation.

Figure 14-1 shows the torque-speed characteristics of a typical induction motor, plus two possible load characteristics. The horizontal line describes the property of a *constant-torque* load. For the other curve, the torque increases with the speed. This is a quite frequently encountered phenomenon, which is generally referred to as *viscous friction*. As many readers know, viscosity is a property of nonideal liquids and gases by which they resist being moved, to an extent depending on the speed of the motion. One of the most familiar manifestations of viscous friction is the *wind resistance* felt by a person who extends a hand outside of a moving vehicle. On a calm day, this force becomes zero when the vehicle stops; yet it can be quite strong at high speeds. For a simple approximation, forces of viscous friction are often assumed to be proportional to the speed, and sometimes to its square.

Typical examples of viscous-friction loads driven by electric motors are centrifuges and fans. Many types of loads exert both constant and viscous-friction torques; i.e., their torques are nonzero even at the lowest speeds and increase further at higher speeds. Load characteristics with negative slopes can also occur but are much less frequently encountered.

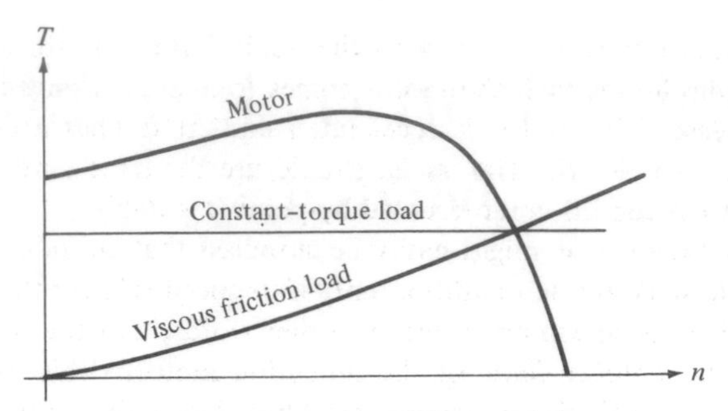

Fig. 14-1. Torque-speed characteristics.

14-2 STABILITY

The stability problem of the induction motor is different from that of the synchronous motor. The reason for this difference is the fact that the electromagnetic torque of the induction motor is a function of the slip, whereas the torque of the synchronous motor is a function of the time integral of the slip (the torque angle δ).

In Fig. 14-2, we have again the typical torque-speed characteristic of an induction motor, and a constant-torque load characteristic. There are two points of intersection, each representing a condition of equilibrium. The motor "chooses" the condition corresponding to point B; that corresponding to point A is unstable.

To confirm this statement, we shall investigate what would happen if the motor were running at the speed of point A and the load torque were undergoing a slight change. (The most minute variation in temperature or air pressure, perhaps caused by a human voice or any other sound, would be enough to produce such an effect.) If the change in load torque is an increase, then it

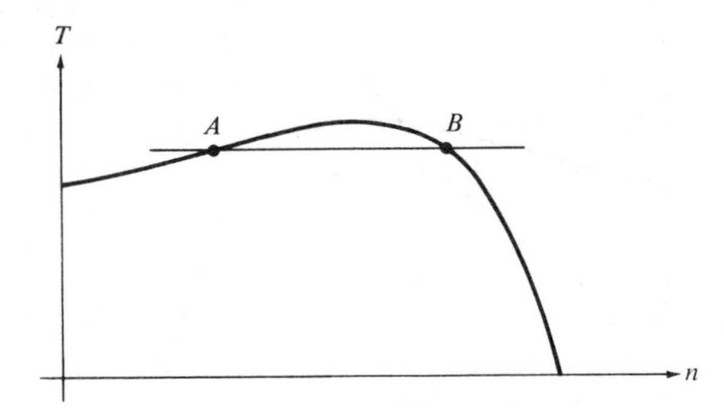

Fig. 14-2. Illustrating stability.

causes the motor to slow down, with the result that the motor torque decreases and the equilibrium is lost; the motor comes to a stop. Conversely, if the load torque decreases, the motor is accelerated until it reaches a new equilibrium at the speed of point B. The reader should use the same kind of reasoning to confirm that the condition represented by point B is *stable.*

From this example, it might easily be surmised that an induction motor can be in a stable steady-state condition only if its speed is larger than that at which the torque has its maximum value; in other words, that the maximum torque constitutes the *stability limit* of the induction motor. This would also be in good agreement with the previously established fact that, for the sake of efficiency, rated operation should occur at a small value of slip, certainly at a point to the right of the maximum torque.

Nevertheless, this formulation of the stability limit is valid only for a constant-torque load. Figure 14-3 shows once again the familiar motor characteristic, and also the torque-speed characteristic of a load with viscous friction. There is only one point of intersection, and it happens to be to the left of the maximum torque. Does it represent a stable or an unstable equilibrium?

To answer this question, let the motor operate at the condition described by this point, and conisder the effect of a slight increase of the load torque. It leads to a lower speed, and at such a speed, the motor torque exceeds the load torque, thereby accelerating the motor to restore the equilibrium. Conversely, a slight decrease of the load torque leads to a higher speed, at which the load torque exceeds the motor torque, again with the result that the motor regains its equilibrium condition, according to the point of intersection in the diagram.

A general criterion for stability of an induction motor may thus be formulated as follows: a point of intersection between motor and load characteristics represents a stable equilibrium, if, to the left of this point, the motor torque is larger than the load torque, and, to the right of it, the load torque is larger than the motor torque.

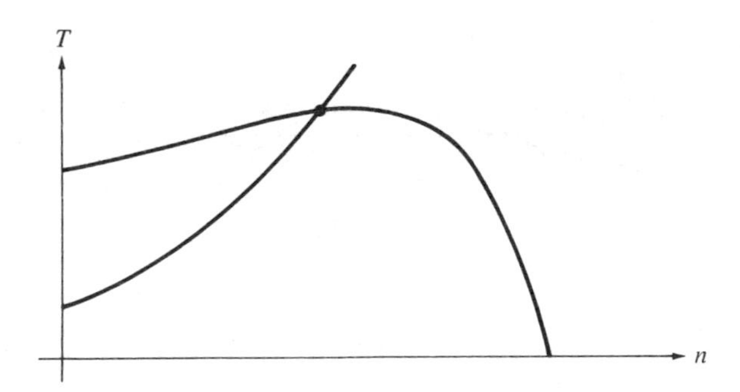

Fig. 14-3. Stability in a case involving viscous friction.

14-3 SPEED CONTROL

The possibility of controlling the speed of a motor, i.e., of being able to change the speed arbitrarily, is of crucial importance for many motors, whether they are driving vehicles, appliances, or industrial machines. In the familiar example of the gasoline-powered automobile engine, the main mechanism of speed control is the throttle, which regulates the supply of fuel to the engine.

From the outset, the distinction between speed control and *speed variation* should be clearly kept in mind. By speed variation, we mean the way in which many motors (including the induction motor but not the synchronous motor) respond to load changes by changing their steady-state speed. To return to the example of the automobile: when the road begins to go uphill, the engine slows down (speed variation), but it does the same thing when the driver "steps off the gas" (speed control). In speed variation, the operating point moves along the torque-speed characteristic; in speed control, this characteristic itself is changed.

In the case of the *induction motor*, there are several different methods by which the purpose of speed control can be accomplished. Since normal operating speeds (between no-load and full-load) are close to the *synchronous speed*, the idea of changing the synchronous speed itself may be the first to come to mind. But it is also possible to control the motor speed without changing the synchronous speed. Ways to do this include changing the magnitude of the input voltage, adjusting elements of the equivalent circuits, or even inserting additional elements into it. Each of these methods will be discussed in the following sections.

14-4 CHANGING THE SYNCHRONOUS SPEED

As the reader knows, the synchronous speed is the speed at which the magnetic field in the air gap rotates. It depends entirely on two factors: the frequency of the input voltage, and the number of poles. (See Eq. 9-3 or 9-4.)

It could be surmised that the *number of poles* is a constant property of a motor. Actually, it *can* be changed, by reconnecting parts of the stator winding. In that case, the rotor, being of the squirrel-cage type, assumes the same number of poles as the stator, by itself without any switching. The idea is illustrated by Fig. 14-4, which shows the developed stator and rotor surfaces and two stator coils belonging to one phase.

Let the two little circles in the left half of each diagram represent the conductors of one coil, the other two circles the other coil. The two coils (or two groups of such coils) may be connected in series or in parallel to each other. In either case, if their current directions are as indicated by the crosses and

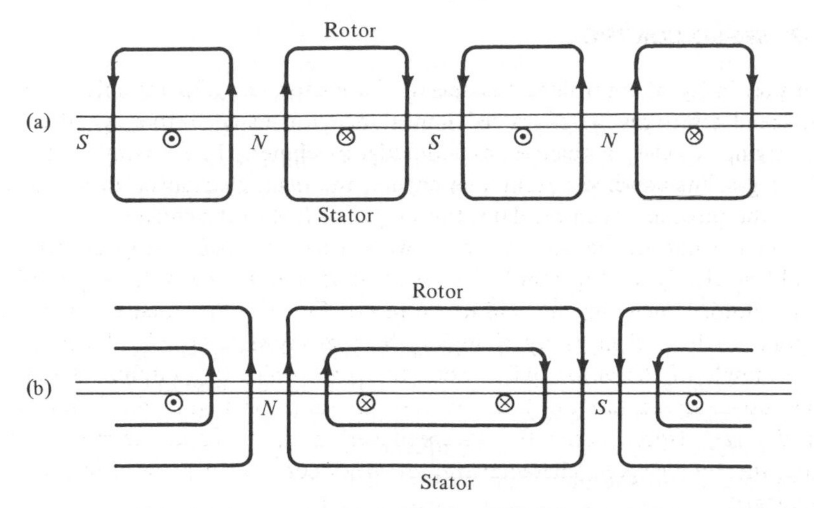

Fig. 14-4. Pole switching: (a) four poles; (b) two poles.

dots in Fig. 14-4a, then the magnetic field has four poles (indicated by the letters *N* for north and *S* for south). On the other hand, current directions according to Fig. 14-4b produce only two poles. Thus, all that is needed in order to change the number of poles is to alter the connections between the two coils (or coil groups) of each phase. Many induction motors, in situations where the need for speed control is satisfied by two distinct speeds, at a two-to-one ratio, are equipped with switches based on this principle. For a more detailed analysis, the interested reader is referred to the specialized literature.*

The other quantity determining the synchronous speed is the *frequency* of the supply voltage. Until fairly recent advances in solid-state electronics, the idea of adjusting that frequency was not really practical. It could have been done only by having the induction motor supplied by a generator of its own, and since that generator would have needed an adjustable-frequency output, it would have had to be driven by an adjustable-speed prime mover. But where such a prime mover had been available, why should it not have been used to drive the motor load itself, dispensing with the other two machines?

Nowadays, changing the frequency is done electronically, by first *rectifying* the available constant-frequency voltage, i.e., by converting a-c into d-c, and then using *inverters* to reconvert d-c into a-c, at the desired frequency. This important method will be discussed in the following section.

It must be pointed out that, in general, any substantial reduction of the

*For example, P. L. Alger, *The Nature of Induction Machines*, Gordon & Breach, 1965.

synchronous speed must be accompanied by a similar reduction in the *magnitude* of the supply voltage. This is so because the voltage induced in the stator winding (which differs from the terminal voltage by the effect of imperfections only) is, like any speed voltage, proportional to the product of flux and speed. Thus, if the speed is much reduced, without a similar reduction of the voltage, the flux must increase, far above saturation, which raises the magnetizing current to intolerably high values.

14-5 ELECTRONIC FREQUENCY CONTROL

A block diagram of one method of obtaining variable frequency is shown in Fig. 14-5. The converter output is d-c. Its operation was discussed in Section 10-10. The magnitude of this d-c voltage can be adjusted by changing the firing angle of the converter. The inverter absorbs power from the d-c bus and delivers it as three-phase a-c to the induction motor. The frequency of this a-c output is adjusted by spacing the firing angles of the switching devices used in the inverter. By this method, the torque-speed characteristics of an induction motor can be changed in an infinite variety of ways, since both the magnitude and the frequency of its input voltage are adjusted.

First, consider a single-phase inverter as shown in Fig. 14-6. The switching devices shown in the figure are power transistors, which may be turned ON or turned OFF by controlling the base current. The diodes protect the power transistors from ever being required to carry reverse current, which would ruin them. The diodes provide a freewheeling role, but they must be located in the d-c part of the circuit, since the load terminals have a-c voltage.

Consider A, B, C, and D to function as ideal switches. A and D (B and C) must never be ON simultaneously as this would short the d-c supply. When A and C are ON, the output voltage, v_{kn}, is equal to $+V_B$. When B and D are ON, the output voltage, v_{kn}, is equal to $-V_B$. When A and B (or C and D) are ON, the output voltage is zero. Controlling the time of switching enables the output voltage to be a-c whose frequency is controlled by adjusting the interval between switching times. Figure 14-7 shows that the output voltage is composed of alternating rectangular pulses whose magnitude is V_B.

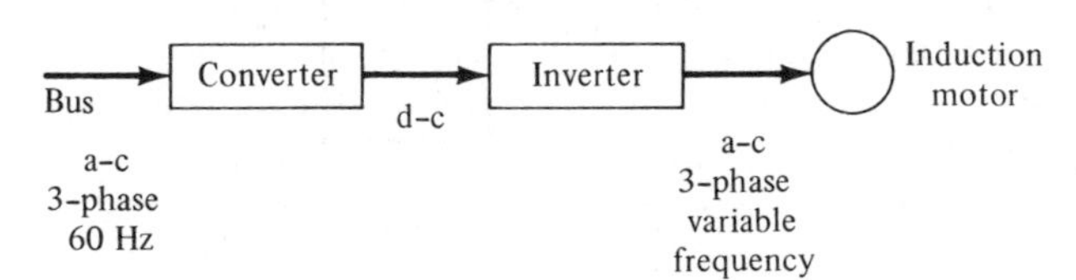

Fig. 14-5. Variable frequency control of an induction motor.

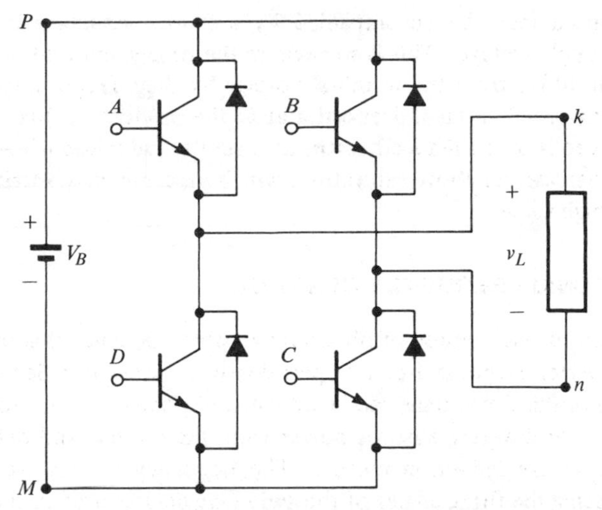

Fig. 14-6. Full-bridge, single-phase inverter.

A full-bridge, three-phase inverter circuit is shown in Fig. 14-8. Any two switches in series across the d-c bus must never be ON simultaneously. The six switches provide the means of connecting the terminals a, b, c, of the load to either of the terminals, P or M, of the d-c source. Figure 14-9 shows the operation when each switch is closed for a duration of 180°. First, investigate the voltage v_{ab}. From 0° to 60°, A is ON, a is tied to P. F is ON, so b is tied to M. Therefore $v_{ab} = V_{PM} = +V_B$. From 60° to 120°, both A and F are still ON, so $v_{ab} = +V_B$. At 120°, F is turned OFF and C is turned ON. A is ON, so a is tied to P. C is ON, so b is also tied to P. Therefore $v_{ab} = 0$. At 180°, A is turned OFF and D is turned ON. From 180° to 240°, a is tied to M. C

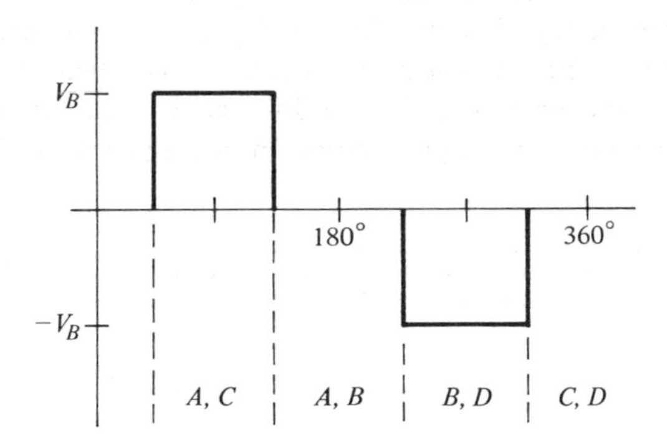

Fig. 14-7. Output voltage of a single-phase inverter.

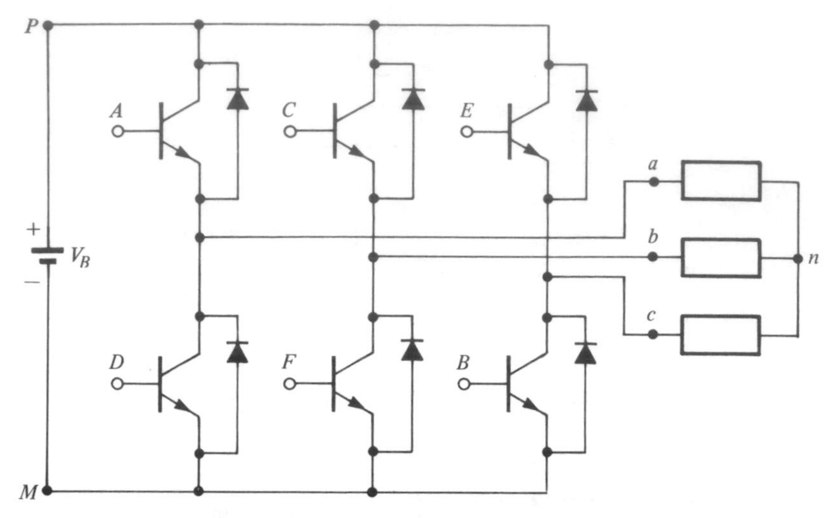

Fig. 14-8. Full-bridge, three-phase inverter.

is still ON, so b is tied to P. Therefore, $v_{ab} = V_{MP} = -V_B$. From 240° to 300°, C and D are ON, so $v_{ab} = -V_B$. At 300°, C is turned OFF and F is turned ON. From 300° to 360°, both a and b are tied to M, so $v_{ab} = 0$. The voltage v_{ab} is shown in Fig. 14-9. The voltage v_{bc} is also shown, and we see that it lags behind v_{ab} by 120°. The frequency of the output voltage is controlled by adjusting the interval between firing pulses. The magnitude of the output voltages can be controlled by changing the input d-c voltage, V_B. Another means is available by shortening the duration of the ON time for each switch.

Thyristors are also used in inverters. However, additional circuit components are required to provide the reverse voltage condition that must be achieved to allow the thyristor to turn OFF. In a converter, reverse voltage occurs in due course as the circuit functions. In the simple inverter circuit, reverse voltage is never achieved across a switch that is conducting. Therefore, means must be provided to have available a reverse voltage that can be connected to the thyristor and enable control to be established so the gate can then hold it in the OFF condition.

The a-c load current function will depend upon the load circuit. An induction motor has the air gap voltage, E_1, generated in its stator windings. An induction motor can be viewed either as a circuit containing a voltage source E_1, or as a purely passive circuit (the equivalent circuit). Another possible load on the inverter output is a synchronous motor. Here we also have an equivalent circuit that enables problems to be solved. In any case, the voltage functions and the current functions may have harmonics. Appropriate filter circuits can be used with a converter and with an inverter to help reduce troublesome harmonics.

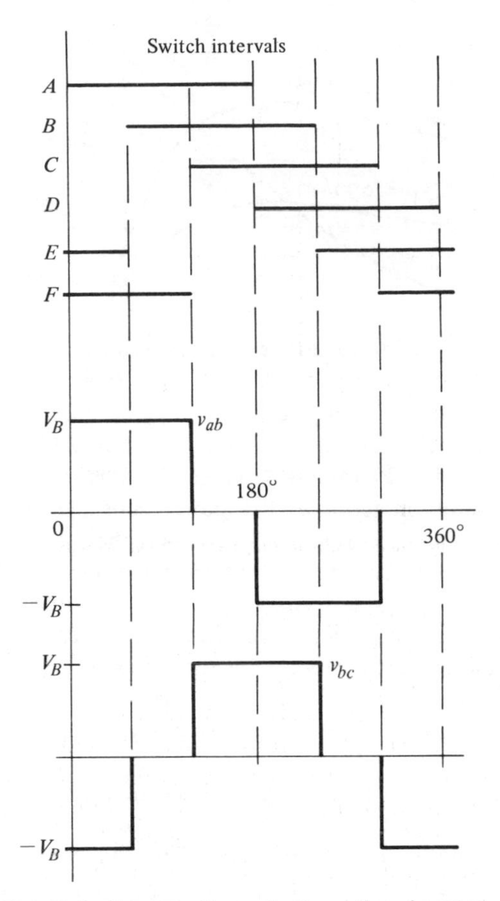

Fig. 14-9. Output voltage of a three-phase inverter.

14-6 CHANGING THE INPUT VOLTAGE

To change the magnitude of the voltage supply, leaving the frequency constant, requires merely the use of transformers. The question is how such a change affects the torque-speed characteristic of an induction motor.

The equivalent circuit of Fig. 13-6 indicates that, for a given value of slip, all currents ought to be proportional to the voltage V_1. Such a statement of proportionality does, however, ignore the fact that the elements G_c and B_m (especially the latter) are inherently nonlinear. For instance, if the voltage were raised substantially above rated value, the magnetic circuit would become more highly saturated, and the current through the susceptance B_m would increase much more than the voltage. For voltages below rated magnitude, such nonlinear

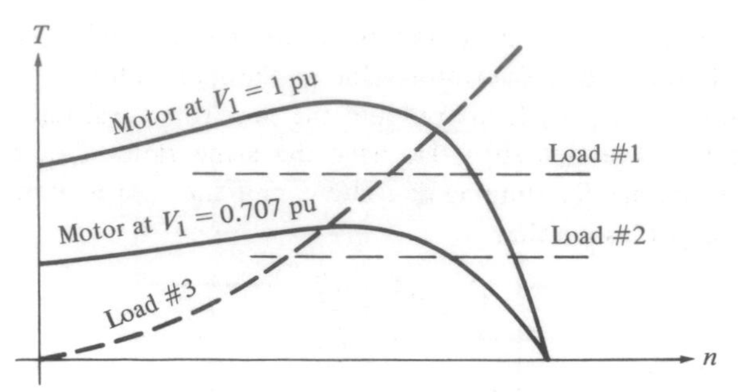

Fig. 14-10. Effect of voltage change.

effects are much less drastic, and at any rate, they affect mainly the current I_ϕ. The other elements of the equivalent circuit are linear, and so the current I_2' *is* practically proportional to V_1 for a wide range of voltage variation.

Expressing the torque in terms of the power consumed in the main branch of the equivalent circuit, according to Eq. 13-47, then leads to the conclusion that, for a given slip, the electromagnetic torque is proportional to the *square* of the voltage. If the torque-speed characteristic of a motor is known for rated voltage (say, $V_1 = 1$ pu), then the characteristic for any other voltage magnitude is found by multiplying each ordinate by the square of the voltage ratio. For instance, Fig. 14-10 shows two such characteristics, for a voltage ratio of $0.707 = 1/\sqrt{2}$. At the reduced voltage, the torque for any given speed is one-half of that for full voltage.

The diagram also contains three plausible load characteristics. For load No. 1, the voltage reduction is not practical at all; the motor just could not drive the load at the lower voltage. For the much smaller load, No. 2, the voltage reduction would be feasible, but it would produce only a slight decrease in speed, and that at an operating condition rather close to the stability limit.

Only for the pure viscous-friction load, No. 3, does voltage reduction constitute a sensible scheme for speed control. The two steady-state speeds are substantially different from each other, and there is no stability problem, the load torque being zero at zero speed. So the use of transformers for speed control of induction motors is practically limited to viscous-friction loads. The typical application is for *fans* with two or three speeds.

14-7 ROTOR CIRCUIT CONTROL METHODS

Of the several circuit elements of the induction motor, the rotor resistance R_2 has a special significance for the steady-state speed. This might be guessed from the fact that the equivalent circuit contains the element R_2'/s, and it can also be readily demonstrated.

Let the rotor resistance of an induction motor be changed from R_2 to R_2^*, which changes its referred-to-the-stator value at the same ratio, from R_2' to $R_2'^*$. Consider this motor, with both the old and the new rotor resistance, operating at such speeds that the two slips also have the same ratio: i.e., consider the motor with the original R_2 running at a slip s, and the motor with the altered R_2^* running at a slip s^*, such that

$$\frac{s^*}{s} = \frac{R_2^*}{R_2} = \frac{R_2'^*}{R_2'} \tag{14-1}$$

This makes

$$\frac{R_2'^*}{s^*} = \frac{R_2'}{s} \tag{14-2}$$

This means that the equivalent circuits for these two conditions are identical. Thus, the rotor current, the air gap power, and, most pertinently, the electromagnetic torque have exactly the same values. For instance, if the rotor resistance is doubled ($R_2^* = 2R_2$), the torque is unchanged if the slip is also doubled. This relation makes the construction of the new torque-speed characteristic easy, as Fig. 14-11 illustrates. This diagram shows the original characteristic and what happens to it if the rotor resistance is doubled and if it is tripled.

It is well worthwhile to study such a set of characteristics. They are all anchored at the same synchronous speed, like those for changes of the voltage magnitude (Section 14-6), but unlike those for changes of frequency (Section 14-5). For small loads, the speed changes due to the added rotor resistance may be slight, but they increase with increasing load torques. Most significantly, the maximum torque remains the same, which eliminates problems of stability. The more the rotor resistance is increased, the smaller is the speed at which the maximum torque is obtained.

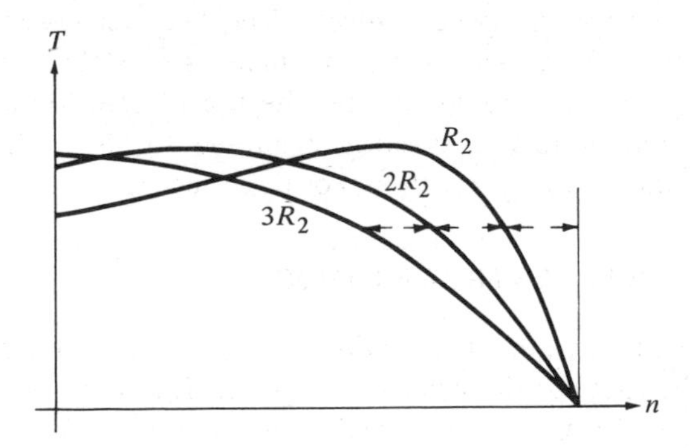

Fig. 14-11. Effect of changing rotor resistance.

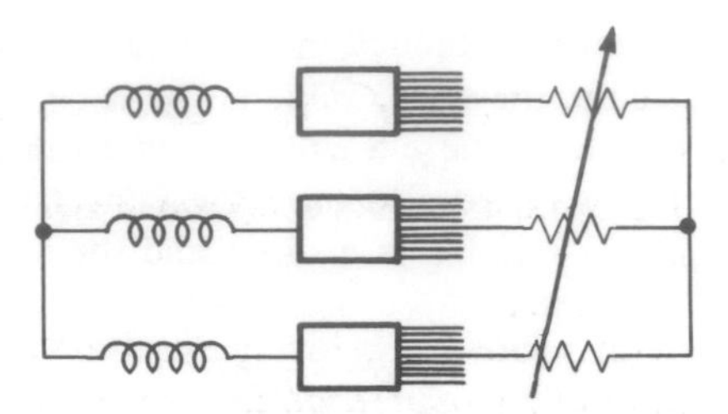

Fig. 14-12. Rotor circuit with slip rings.

The implementation of this scheme requires the motor to be of the wound-rotor type, with each of three rotor terminals connected to a *slip ring*. (These construction elements were described, in the context of the synchronous machine, in Section 9-1). The brushes in turn are connected to three adjustable resistors (rheostats), which are thereby inserted into the rotor circuits. Figure 14-12 illustrates the idea schematically for a wye-connected rotor. The symbols drawn for each phase represent, from left to right, the rotor windings, the slip rings, the brushes, and the resistors. The arrow through the resistors suggests that they are adjusted together, keeping the three-phase circuit balanced.

Speed control by means of rotor rheostats has certain drawbacks. It can be used only to decrease the speed, never to increase it beyond the *base speed* (i.e., the speed without added resistance). In particular, the synchronous speed remains an impenetrable barrier. Also, the extent to which the speed is altered by the insertion of a certain amount of resistance varies greatly with the load. Most importantly, inserting resistance into a main circuit branch always increases the copper losses in that branch and, thus, reduces the *efficiency*.

It should not be surprising that any method of increasing the slip results in greater power losses in the rotor circuit. Equation 13-31 stated clearly that the slip is associated with such losses. In the case of a viscous-friction load, the extent of these losses depends on the shape of the load characteristic, but for a constant-torque load, slip and rotor copper losses are strictly proportional to each other.

Nevertheless, it is possible to recover this "lost" power and thereby actually to avoid its loss. The way to accomplish this is to insert into the rotor circuit not a resistance but a *voltage source*. If the source voltage is in phase with the rotor current, then the source is equivalent to a resistance, except that the power "consumed" by it is not converted into heat but into some other form of power from which it can be reconverted into electric power and returned to the power system. Furthermore, the voltage of this "injected" source can be in phase *opposition* to the rotor current. In this case, it represents a negative resistance; instead of consuming power, it supplies additional power to the motor, resulting in higher speeds. Depending on the magnitude of this voltage,

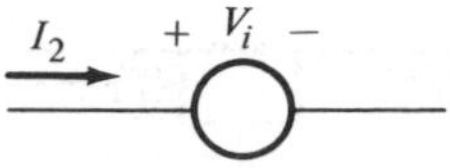

Fig. 14-13. Definition of injected voltage.

the speed may then be either above or below its synchronous value. Incidentally, the words *in phase* and *in phase opposition* used in this paragraph are meant to be in accordance with the polarity marks of Fig. 14-13.

Injected voltages can accomplish even more than that. By changing their phase angles, the *power factor* of the motor can be adjusted. Considering that the "inevitable" lagging power factor is a major disadvantage of the induction motor (the power factor of the synchronous motor is adjustable), this constitutes an attractive feature. In addition, the speed change brought about by the injected source, regardless of its phase orientation, is nearly independent of the load.

No wonder, then, that great efforts have been made for many years to design devices capable of injecting voltages into the rotor circuits of induction motors. The problem is that, in any steady state, the injected voltage must be of the same frequency as the rotor current. As the reader knows, this is *slip frequency*, which is a variable. Nevertheless, many different injected source devices have been developed and successfully built. There are mechanical devices coupled to the motor shaft; some are incorporated into the construction of the motor itself (the Schrage motor). They have one construction element in common, the *commutator*, which is also used in d-c machines and which the reader will encounter in that context.

It sometimes happens in engineering that a great deal of inventiveness and ingenuity is ultimately wasted, when its fruit is superseded by later developments. With all their attractiveness, mechanical injected source devices are expensive and, being rotating machines, they add to the maintenance needs of the motor. Nowadays, they can no longer compete economically with electronic speed control devices.

The idea of speed control by injected sources can also be realized by electronic means. Refer to Fig. 14-14. The rotor energy is transferred through the

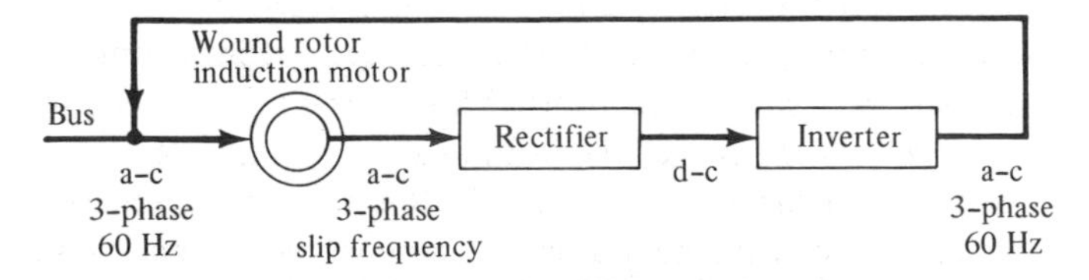

Fig. 14-14. Speed control using devices in the rotor circuit.

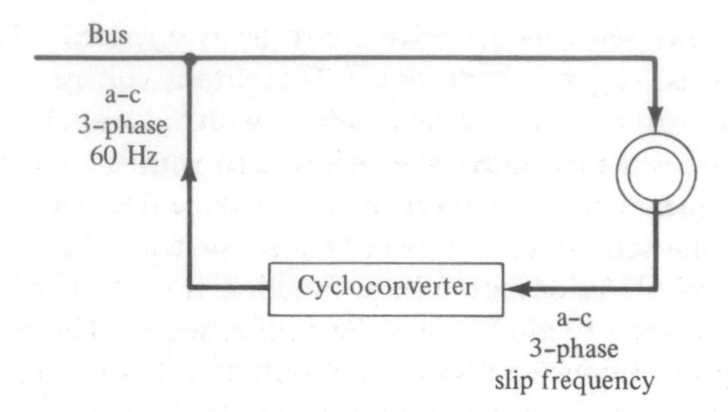

Fig. 14-15. Control by injected rotor voltage.

rectifier to a d-c bus. An inverter transfers the energy back to the a-c source. Speed control is effected by adjusting the voltage of the d-c bus. The motor can produce torque only if current flows in the rotor windings. Current can flow in the rotor circuit only if the magnitude of the rotor voltage, sE_1/a, exceeds the voltage of the d-c bus. An increase of the d-c bus voltage must result in increased slip (slower speed) in the induction motor.

Another method of rotor circuit control is shown in Fig. 14-15. An electronic frequency converter called a *cycloconverter* changes a-c at the source frequency into a-c of a lower frequency. Energy may be transferred in either direction. The output voltage may be controlled in magnitude, frequency, and phase angle. This is what is needed in the rotor circuit of an induction motor to control its speed, as well as to control its power factor on the stator side.

To illustrate the synthesis of a low frequency voltage function, consider Fig. 14-16. This dual converter consists of two three-phase converters connected with opposite polarity to a single load. Compare this circuit with Fig. 10-21. Positive load current flows when the positive converter is operated. Negative

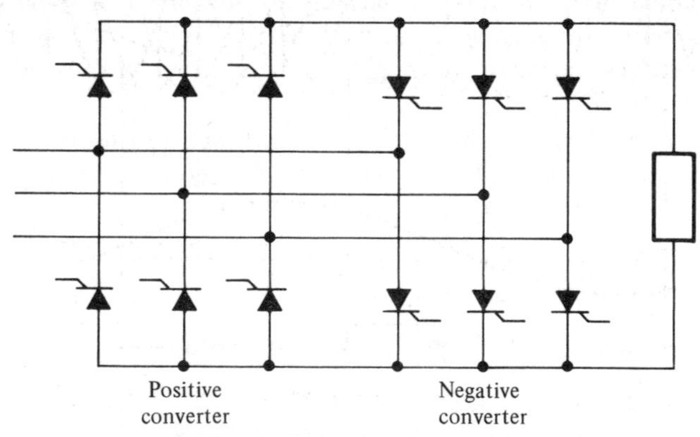

Fig. 14-16. Three-phase dual converter.

load current flows when the negative converter is operated. The average value of load voltage is $V_{ave} = (3V/\pi) \cos \alpha$. The output voltage is zero for α equal to 90°. If the firing angle is varied slowly, with values below and above 90°, then the output will vary above and below zero with the same low frequency. The output frequency must be lower than the source frequency.

We consider the solid state devices to be ideal switches that may be turned ON or OFF arbitrarily. An output voltage function is synthesized from pulses obtained by the choice of switches and their firing angles. Figure 14-17 illustrates one possibility for producing a sinusoidal output voltage. The load current depends upon the circuit connected to the output. The load current is low frequency a-c. It may still be considered to have nearly a constant value during the short interval that one pair of thyristors is conducting. The waveshape, magnitude, frequency, and phase of the output voltage are controlled by the choice of firing angles and the separation of the firing pulses. The choice of firing angles is not a simple matter. If the output current is positive, the positive converter must be used. If the output current is negative, the negative converter must be used. Both converters may operate in either rectifier or inverter mode.

It is also possible to use a frequency converter to inject voltages of constant frequency into the rotor circuit of an induction motor. The resulting operation is like that of a synchronous motor, and its speed corresponds to the difference between stator and rotor frequencies. Changes in load torque will be accom-

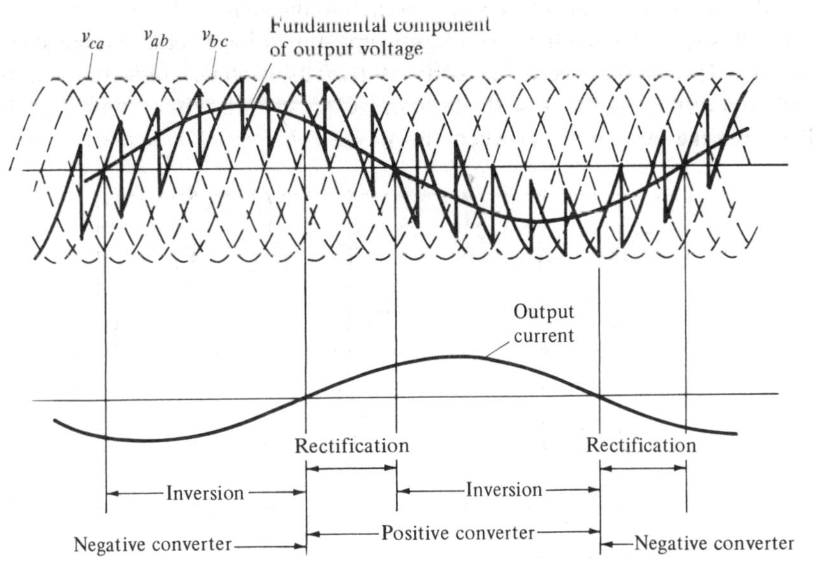

Fig. 14-17. Single-phase output of a three-phase dual converter.

modated by changes of the torque angle associated with the relative positions of the stator field and the rotor field. If the power factor angle of the stator current is to be controlled, the phase of the injected voltage must be set in relation to the voltage generated in the rotor windings. It is necessary to have some means of sensing the rotor position and to use this information in the control of the firing pulses in the cycloconverter.

14-8 THE PROBLEMS OF STARTING

The torque-speed characteristic of an induction motor, first sketched in Fig. 13-14, exhibits not only the portion between full-load (rated) and ideal no-load (synchronous) speed, but also, as features of particular interest, the maximum torque, and the *starting torque*, i.e., the torque at zero speed. Whether or not the value of the starting torque is above that of rated torque, depends on the design of the motor, which in turn can be made to suit the nature of the load to be driven by the motor. For instance, a motor intended to drive a pure viscous-friction load needs only a small starting torque (compared to rated value), etc.

The torque is not the only quantity whose value for the starting condition must be studied. The *currents*, both in the stator and in the rotor windings, have inevitably much larger values at starting than at rated operation. This can be readily seen from the equivalent circuit of Fig. 13-6 where the resistance R_2'/s at the slip $s = 1$ is only a fraction of its value at rated conditions, and even more drastically from Fig. 13-7, in which the resistance $(1 - s)R_2'/s$ becomes zero at standstill, reducing the motor to a *short-circuited transformer*. So it is hardly surprising that *starting currents* may exceed rated values by factors like 5 or 10 or even more.

The question whether such excessive currents may be tolerated (and if not, what is to be done about them) has several aspects. One absolute limitation lies in the electromagnetic *forces* between conductors located on the same part (stator or rotor) of the machine. Such forces do not contribute anything to the motor torque, and they are normally contained and rendered harmless by the mechanical strength of the materials used in the design. But such forces are proportional to the square of the current magnitude (except for some attenuating and, therefore, beneficial effect of saturation), and at certain excessive current values, some motors can be severely damaged or even mechanically destroyed by these forces unless protective devices (fuses or automatic circuit breakers) disconnect the motor in time from its source.

Another aspect of excessive currents is the problem of *heat* dissipation. In the long run, any current exceeding rated value by more than a limited safety margin leads to excessively high temperatures, i.e., temperatures too high to be tolerated by the insulating materials used in the design of the motor. But since the temperature always rises gradually, never suddenly (due to the energy-storing property of the heat capacity of the body being heated), excessive current values are permissible for limited time intervals. This principle is the basis of *intermittent-duty* motor ratings (previously mentioned at the end of Section 1-6), and it is of vital importance in the starting operation of all electric motors.

The temperature rise of a body subjected to heating depends directly on the heat *energy* transmitted to this body. Therefore, to check whether a motor overheats during the starting interval (without actually finding out by experience), the power losses must be known as a function of time from the instant of starting to the reaching of the steady state. Like all other operating quantities of the induction motor, these losses are functions of the slip, so that the first step must be to find the speed as a function of time for the starting interval.

The procedure, illustrated in Fig. 14-18, begins with points of the torque-speed characteristics of both the motor and the load (the more points the better for accuracy) being tabulated or fed into the computer memory. (The torque corresponding to the rotational losses may be taken into consideration by being either included in the load torque or deducted from the electromagnetic torque of the motor.) The difference between these two functions is the *accelerating torque*, expressed as a function of the speed

$$T_{acc}(n) = T_{motor} - T_{load} = J\frac{d\omega_m}{dt} = J\frac{2\pi}{60}\frac{dn}{dt} \tag{14-3}$$

where J is the familiar symbol for the moment of inertia of the motor and the load coupled to it.

The next step is the separation of variables

$$dt = J\frac{2\pi}{60}\frac{1}{T_{acc}(n)}\,dn \tag{14-4}$$

Figure 14-18b illustrates this step, being a plot of both the accelerating torque and its reciprocal against the speed. Integrating the latter function (or adding small finite area elements in the diagram) from zero speed to each intermediate point up to the steady state, and multiplying the results by $J(2\pi)/60$, yields the function t versus n, the inverse of which is the desired result, sketched in Fig. 14-18c. Note that the curve of $1/T_{acc}$ versus n goes to infinity for $n \rightarrow n_{ss}$. Therefore, the last point of the t-versus-n-curve cannot be obtained. This merely indicates that the steady-state speed is reached asymptotically.

After this, it remains to determine the power losses as a function of time, to subtract their rated steady-state value, and to integrate the remainder over the starting time interval. The result is the extra amount of heat energy generated during this interval.

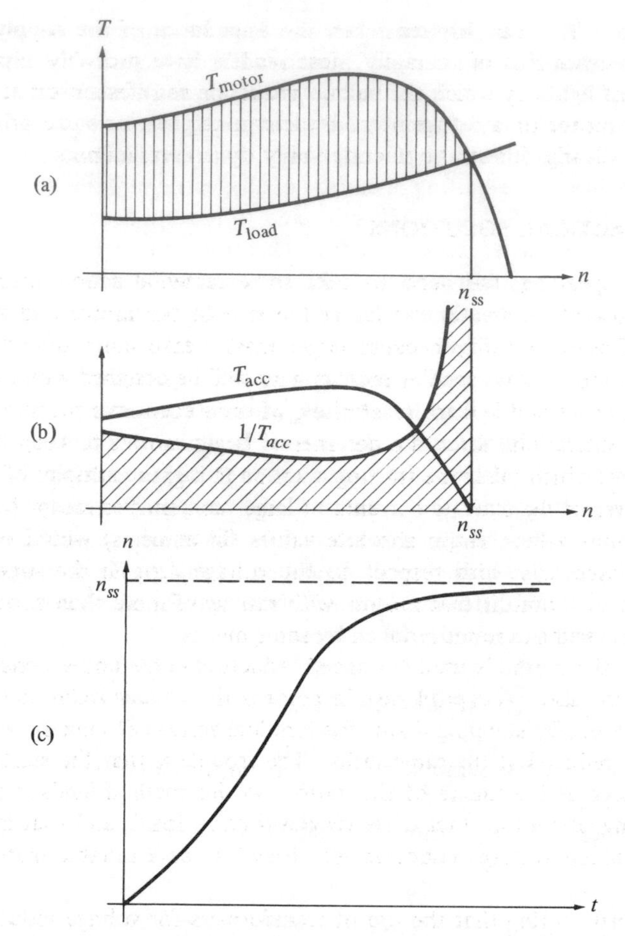

Fig. 14-18. Dynamics of starting: (a) motor and load characteristics; (b) accelerating torque and its reciprocal; (c) speed as a function of time.

The reader recognizes that this analysis was another example of *mechanical transients*, often also called dynamics. It was pointed out in the context of synchronous machines (Section 11-9) that the *electrical transient* phenomena that inevitably accompany all nonperiodic changes of current magnitudes are often assumed to occur in much shorter time intervals and, therefore, to be irrelevant to such studies.

Yet another problem to be considered in any discussion of motor starting is the effect of the current magnitude on the supply voltage. There are cases in which the current drawn by a motor in its starting interval would not pose any threat, either mechanically or thermally, to the motor, and yet would not be acceptable. This can happen when the impedance of the supply line would cause too much loss of voltage. Most readers have probably experienced the dimming of lights by which the voltage reduction manifests itself at the moment when the motor of a refrigerator, a vacuum cleaner, or some other appliance starts up. Clearly this can be tolerated only up to certain limits.

14-9 PRACTICAL SOLUTIONS

Generally speaking, the need to take some remedial action against excessive starting currents is greater the larger the size of the motor and the higher its rating. This is so partly because larger masses take more time to accelerate. Another reason is that smaller motors may well be designed with comparatively large resistances and leakage reactances, whereas economic pressures and operational considerations force the designer of larger motors to keep the imperfections down, which raises the starting currents to higher multiples of rated values. Finally, even if the starting currents of larger and smaller motor had about the same per-unit values, larger absolute values (in amperes) would be more of a problem, especially with respect to the voltage drop in the supply lines. A crude rule of thumb is that motors with ratings of more than about 5 HP must have their starting currents reduced by some means.

Most of the methods used for speed reduction come under consideration for this purpose also. A typical case in point is the *voltage reduction* discussed in Section 14-6. By stepping down the terminal voltage at some ratio, the starting current is reduced at the same ratio. The trouble is that the starting torque is then reduced at the square of this ratio. So this method lends itself mostly to the starting of motors that drive *viscous-friction* loads, and that is the type of load for which voltage reduction was found to be a suitable method of speed control.

It is worth noting that the use of transformers for voltage reduction is more effective in reducing the supply current than the motor current. To see this, refer to the single-phase diagram of Fig. 14-19. If the autotransformer has a step-down ratio a:1, the starting motor current is reduced at the same ratio,

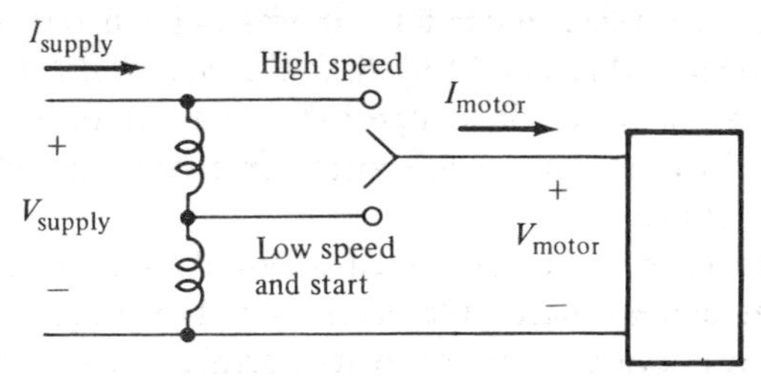

Fig. 14-19. Step-down autotransformer (single-phase diagram).

but the starting supply current (the primary current of the autotransformer) is reduced by the square of this ratio, as is the motor torque.

Another scheme is the *wye-delta switch*, a three-pole double-throw switch by which the three stator windings are connected either in delta for running, or in wye for starting. This changes the phase voltages supplied to the three stator windings at the ratio $\sqrt{3}:1$, and thereby the starting currents in the motor windings at the same ratio, whereas the starting currents in the supply lines, and also the starting torque, are all reduced at the ratio $3:1$. The absence of any transformer or other device except for the switch makes this scheme appear economically attractive, but it has its limitations and drawbacks. It does not make possible any other ratios than those named, it is applicable only to motors whose stator windings are "normally" connected in delta, and the switching operation causes both arcing at the switch contacts and a momentary interruption of power (comparable to that caused by gear-shifting in manual automobile transmissions).

If a reduction of the starting currents is to be obtained without a reduction of the starting torque, other methods, also familiar to the reader from the discussion of speed control, are available. They are the use of electronic devices for the reduction of both the magnitude and the *frequency* of the supply voltage (see Section 14-5), and the insertion of additional *resistance* or the injection of a *voltage* into the *rotor* circuit of wound-rotor motors (see Section 14-7).

By any one of these methods, the starting currents can be reduced to any extent desired. In addition, as can be deduced from Fig. 14-11, a wide range of such reductions is accompanied not by any loss, but rather by a gain of starting torque.

The insertion of additional circuit elements is not feasible for squirrel-cage rotors. This brings about a conflict between the design requirements for starting and for running. Higher values of rotor resistance would be desirable for starting, since they would reduce the starting currents and, at the same time, increase the starting torque. On the other hand, it is fundamental that winding resistances are responsible for power losses and should be kept down for the sake of good steady-state efficiencies.

Quite remarkably, it is possible to some extent to design squirrel-cage rotors in such a way that their resistances are larger for starting than for running. This have-your-cake-and-eat-it-too feature is based on the dependence of the *effective* resistance of a conductor on the frequency, commonly known as *skin effect*. Its principles are briefly explained in the next section.

14-10 MAKING USE OF SKIN EFFECT

When a constant current flows in a cylindrical conductor, it is uniformly distributed; in other words, the *current density* has the same value at every point of the conductor's cross-section. On the other hand, a *time-varying* current

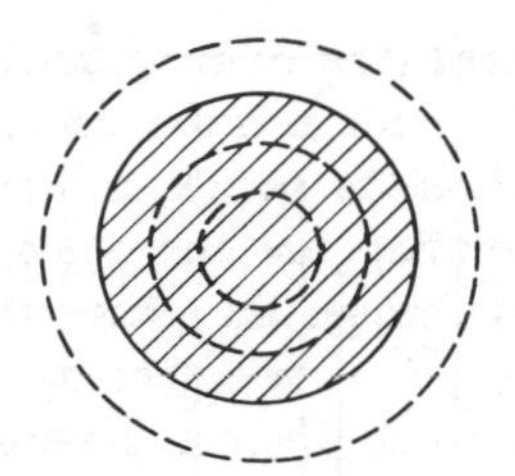

Fig. 14-20. Skin effect.

tends to concentrate in those parts of the cross-section that have fewer flux linkages than the others.

For instance, in the case of a single metallic wire of circular cross-section, all lines of magnetic flux are concentric circles, some of them inside the conductor surface (Fig. 14-20). Points nearer the circumference are surrounded by fewer lines, thus they have fewer flux linkages than points close to the center. It is as if the conductor consisted of parallel fibers, among which those closer to the center have higher inductances than those farther outside. As a result, the current density is highest at the surface. The higher the rate of change of the current, the more the current is crowded toward the conductor surface (its "skin").

In the sinusoidal steady state, skin effect becomes more pronounced with larger current magnitudes, higher frequencies, and increased conductor sizes. The outward manifestation of skin effect lies in increased power losses caused by the nonuniform current distribution. This corresponds to an *effective* resistance

$$R_{\text{eff}} = \frac{P_R}{I^2} = \mathfrak{Re}\,(\mathbf{Z}) \tag{14-5}$$

which is larger than the d-c (zero frequency) resistance, and increases with the frequency. A quantitative treatment of this subject is found in the literature on electromagnetic field theory.*

For motor windings made of round conductors, the skin effect at power frequencies turns out to amount to very little, the difference between effective and d-c resistance rarely being more than a few percent. But it is possible to take advantage of the location of the rotor conductors in slots inside the ferromagnetic core, to increase the nonuniformity of the current distribution. Figure 14-21a shows one slot of a *deep-bar* rotor, with a few typical lines of flux. The wedge above the conductor, made of nonconducting, nonmagnetic material, merely serves to secure the conductor in its place.

The upper part of the conductor (nearer to the rotor surface) has fewer flux linkages than the lower part. A time-varying current is thus crowded toward

*See, for example, W. Hayt, *Engineering Electromagnetics*, McGraw-Hill, 1967.

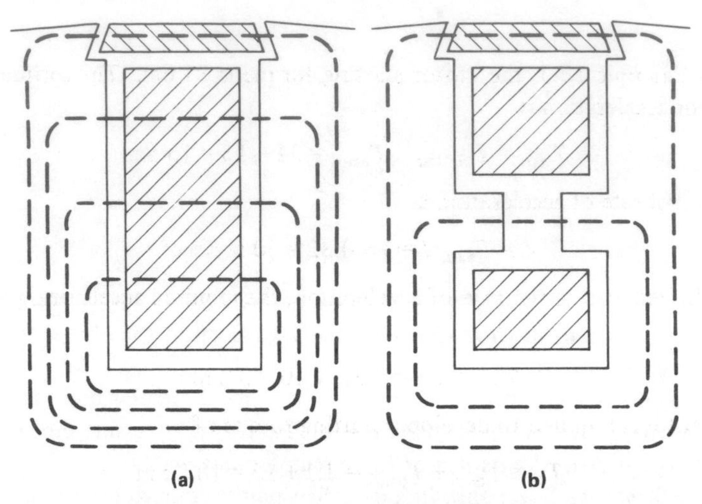

Fig. 14-21. (a) Deep bar; (b) double squirrel cage.

the rotor surface. The resulting increase of resistance can be sufficient at 60 Hz to produce a noticeable improvement in the starting properties of the motor, whereas in normal running conditions, when the rotor frequency is only a small fraction of 60 Hz, the effective resistance of the rotor circuit practically equals its d-c resistance.

Another widely used idea based on the same principle is the *double squirrel cage* illustrated by the sketch of one rotor slot in Fig. 14-21b. The upper conductor has fewer flux linkages and thus takes on most of the starting current, even though it is designed to have a larger resistance than the lower conductor. When the motor is running, it is the lower conductor that carries most of the current, due to its lower resistance, because at the small slip frequency the flux linkages don't matter. So the rotor circuit has a larger effective resistance for starting than for running.

14-11 EXAMPLES

Example 14-1 (Section 14-9)

The 5-HP induction motor of Example 13-1 is started with its load connected. The combined moment of inertia of the motor and the load is 0.32 kgm^2. The torque-speed characteristic of the load is a constant value of 15 nm. (a) If the motor is started with rated voltage, find the initial rate of acceleration. (b) If the motor is started with reduced voltage that makes the initial rate of acceleration be 10 rad/sec^2, find the value of the voltage at the stator terminals.

Solution

(a) From Example 13-3, the motor starting torque is 31 nm. The torque available for acceleration is

$$T_{acc} = T_{motor} - T_{load} = 31 - 15 = 16 \text{ nm}$$

The initial rate of acceleration is

$$\alpha = T_{acc}/J = 16/0.32 = 50 \text{ rad/sec}^2$$

(b) For the reduced initial rate of acceleration, the required accelerating torque is

$$T_{acc} = J\alpha = 0.32 \times 10 = 3.2 \text{ nm}$$

The motor is required to develop a starting torque of

$$T_{st} = T_{acc} + T_{load} = 3.2 + 15 = 18.2 \text{ nm}$$

With the rated voltage of 127 v, the starting torque is 31 nm. The reduced starting torque is proportional to the square of the impressed voltage

$$T'_{st} = kV_1'^2$$

$$V_1'^2 = V_1^2(T'_{st}/T_{st}) = (127)^2 \times (18.2/31) = 9470$$

$$V_1' = 97.3 \text{ v per phase}$$

Observe that reducing the voltage to 75 percent has resulted in a reduction of the initial rate of acceleration to 20 percent of what it could be with rated voltage.

Example 14-2 (Section 14-9)

Let the motor in Example 13-1 have a wound rotor. (a) Find the amount of resistance that must be added to the rotor circuit to enable this machine to develop maximum torque at starting, with all quantities referred to the stator side. (b) With rated voltage and the resistance of part (a), find the initial rate of acceleration with the same load as in Example 14-1.

Solution

(a) When operated with the rotor shorted, maximum torque is developed for slip of 0.253, from Example 13-3. To develop maximum torque for slip $s^* = 1$ requires

$$R_2'^* = R_2'(s^*/s) = 0.42(1/0.253) = 1.66 \ \Omega$$

The amount to be added is

$$R_2'^* - R_2' = 1.66 - 0.253 = 1.41\ \Omega \text{ per phase}$$

(b) The maximum torque is 57.6 nm. The torque available for acceleration is

$$T_{acc} = T_{motor} - T_{load} = 57.6 - 15 = 42.6 \text{ nm}$$

The initial rate of acceleration is

$$\alpha = T_{acc}/J = 42.6/0.32 = 133 \text{ rad/sec}^2$$

Example 14-3 (Section 14-7)

The motor in Example 13-1 is to be controlled by connecting an external voltage source to the rotor circuit. The electromagnetic torque is to be 20.8 nm, and the current $\mathbf{I}_2'$ is to be $11.2\ \underline{/-7.4^\circ}$ amp, as in Examples 13-1 and 13-2. However, the slip is to be 0.3. Find the injected voltage, with all quantities referred to the stator side.

Solution

The equivalent circuit of Fig. 13-13 has to be modified to include the injected voltage source, as shown in Fig. E-14-3. The voltage is changed in magnitude and frequency in reflecting it to the stator side. From Examples 13-2 and 13-3 we have

$$\mathbf{V}_1^* = 123.7\underline{/0.9^\circ} \text{ v} \qquad \mathbf{E}_1 = 117.5\underline{/-3^\circ} \text{ v}$$

$$\mathbf{Z}_1^* = 0.46 + j\,0.79\ \Omega \qquad \mathbf{I}_2' = 11.2\underline{/-7.4^\circ} \text{ amp}$$

Let

$$\mathbf{Z} = (R_1^* + R_2'/s) + j(X_1^* + X_2')$$
$$= (0.46 + 0.42/0.3) + j(0.79 + 0.80) = 2.45\underline{/40.5^\circ}\ \Omega$$

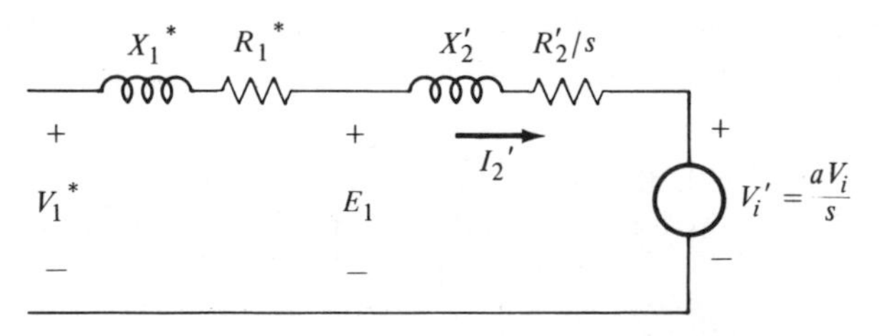

Fig. E-14-3. Equivalent circuit with injected voltage.

Write Kirchhoff's voltage law for the equivalent circuit

$$\mathbf{V}_1^* = \mathbf{Z}\mathbf{I}_2' + \mathbf{V}_i'$$

Solving for the injected voltage

$$\mathbf{V}_i' = 100.7 - j\,13.1 = 101.5\,\underline{/-7.4^\circ}$$

Note well! The air gap power is not $P_{R_2}/s = 3R_2' I_2'^2/s$. The air gap power is $P_g = 3E_1 I_2' \cos\theta_2$, where θ_2 is the angle between E_1 and I_2'. The mechanical power is $P_{mech} = (1 - s)\,P_g$. The rotor circuit power is $P_{rot.cct} = sP_g = P_{R_2} + P_X$, where P_X is the power absorbed from the rotor circuit by the injected voltage source.

14-12 PROBLEMS

14-1. The motor in Problem 13-6 is to be started with a constant-torque load of 20 nm. The motor and load have combined moment of inertia of 0.4 kgm^2. Find the initial rate of acceleration when started with: (a) rated voltage, (b) 80 percent of rated voltage.

14-2. Let the motor in Problem 13-6 have a wound rotor. The load torque is constant at 63 nm. If resistance is added to the rotor circuit to make $R_2'^* = 3R_2' = 1.26\ \Omega$ per phase, find the speed at which the motor runs.

14-3. How much resistance must be added to the rotor circuit of the motor in Problem 13-6 to develop maximum torque at starting (all quantities referred to the stator side)?

14-4. A four-pole, 60-Hz, 220-v (line to line), three-phase squirrel cage induction motor develops maximum electromagnetic torque of 2.5 per-unit at a slip of 0.16 when operating at rated voltage and frequency. This motor is to be operated at 50 Hz and 200 v (line to line). Under these new conditions, find the maximum electromagnetic torque and the speed in revolutions per minute at which it is developed.

14-5. The motor of Problem 13-6 is connected to a 50-Hz, three-phase source. The voltage is 106 v per phase. The load is adjusted to make the slip be 0.06. Find the stator current, the air gap power, and the electromagnetic torque.

14-6. The full-bridge, three-phase inverter is operated with the switching sequence as shown in Fig. 14-9. (a) Find and sketch the voltage function v_{ca}. (b) Find and sketch the phase voltages v_{an}, v_{bn}, v_{cn}. Assume the balanced three-phase load consists of three identical resistances.

14-7. A four-pole, 60-Hz wound-rotor induction motor is controlled by voltage injected into the rotor circuit. Both rotor and stator have three phases. All values are per-phase. $R_1 = 0$. $X_1 = 0$. $R_2' = 0.1\ \Omega$. $X_2' = 0.2\ \Omega$. $X_m = 6\ \Omega$. The electromagnetic torque is 26 nm. The injected voltage V_i' is adjusted to make the slip to be 0.3 and the stator power factor to be unity. (a) Find the injected voltage V_i', with values referred to the stator side. (b) If the standstill voltage ratio is $a = 2$, find the magnitude of the actual injected voltage.

14-8. The motor in Problem 14-7 is operated with torque of 20 nm. The injected voltage is adjusted to make the slip be 0.5 and the stator power factor to be 0.9 lagging. (a) Find the injected voltage V_i', with values referred to the stator side. (b) If the standstill voltage ratio is $a = 2$, find the injected voltage magnitude referred to the rotor side.

14-9. An eight-pole wound-rotor induction motor has its stator connected to a 220-v (line to line) three-phase, 60-Hz source. When the rotor is standing still, the open-circuit rotor voltage is 180 v (line to line). This machine is to be used as a frequency changer by having its rotor driven at a speed of 1200 rpm in the opposite direction of rotation from the stator field. Find the frequency and magnitude of the open-circuit rotor voltage.

14-10. For the machine and conditions in Problem 14-5, (a) find the maximum torque and the slip at which it occurs, (b) find the starting torque. (Compare these answers with those to Problem 13-8.)

15
Symmetrical Components

15-1 INTRODUCTION

This chapter constitutes a necessary prelude to the study of the following two. Up to this point, all rotating machines discussed in this book were connected to *balanced three-phase* systems, and accordingly, their analysis was based on the rotating magnetic field produced by balanced polyphase currents. But the two-phase servomotor to be studied in Chapter 16 operates with unbalanced supply voltages, and as far as the single-phase motor of Chapter 17 is concerned, it is certainly not obvious that it has anything at all to do with rotating fields.

The theory of *symmetrical components* permits the study of these devices on the basis of rotating fields. It was established in 1918 by Fortescue,* and it may be viewed as one of those methods, occasionally encountered in engineering studies, by which a more complicated problem can be reduced to a set of simpler ones. For instance, by means of Fourier analysis, the study of circuits with periodic but nonsinusoidal currents and voltages can be reduced to sinusoidal steady-state problems. Another example, closer to our subject, is the analysis of all problems of balanced three-phase circuits on a per-phase basis, whereby they are reduced to single-phase problems. Similarly, symmetrical components serve to reduce the problems of unbalanced polyphase circuits to a set of balanced ones.

The word *polyphase* is used here, just as it was in Fortescue's historic paper, to indicate that this theory can be applied to systems with any number of phases. Specifically, *three-phase* symmetrical components have long since become a well-established tool for the analysis of unbalanced conditions in power systems, particularly the drastic imbalances arising from accidental single-phase short circuits or transmission line breaks. They need not be studied as prerequisites to anything in this book. Nevertheless, a section of this chapter will be devoted to them because they have become a part of the general education of an electric power engineer.

*C. L. Fortescue, "Method of Symmetrical Coordinates Applied to the Solution of Polyphase Networks," *AIEE Transactions*, vol. 37, pp. 1027–1140, 1918.

15-2 THREE-PHASE SYMMETRICAL COMPONENTS

The term *symmetrical*, as applied to systems of three or more phases, refers to a set of voltage (or current) phasors that are equal in magnitude and differ from each other by one phase angle

$$\varphi = \frac{2\pi}{m} \tag{15-1}$$

where m is the number of phases. This is the familiar 120° angle for three phases, 90° for four phases, etc.

There are two possible ways for three phasors, say $\mathbf{V}_a$, $\mathbf{V}_b$, and $\mathbf{V}_c$, to form a symmetrical, three-phase system. With $\mathbf{V}_a$ as the axis of reference, they can be

$$\begin{aligned} \mathbf{V}_a^+ &= V^+ \underline{/0^\circ} \\ \mathbf{V}_b^+ &= V^+ \underline{/-120^\circ} \\ \mathbf{V}_c^+ &= V^+ \underline{/120^\circ} \end{aligned} \tag{15-2}$$

or they can be

$$\begin{aligned} \mathbf{V}_a^- &= V^- \underline{/0^\circ} \\ \mathbf{V}_b^- &= V^- \underline{/120^\circ} \\ \mathbf{V}_c^- &= V^- \underline{/-120^\circ} \end{aligned} \tag{15-3}$$

The superscripts (+ and −) have been chosen to indicate the ***phase sequence***, which makes the difference between the two systems. In Eqs. 15-2, the voltage of phase b follows that of phase a by one-third of a period, that of phase c follows that of phase b, and so on ad infinitum. In other words, the phase sequence is a-b-c-a-b . . . , or simply a-b-c for short. This will be called the ***positive*** phase sequence, which explains the choice of the superscript + for this sequence. In Eqs. 15-3, the phase sequence is a-c-b (or c-b-a, for anyone who prefers that), to be called the ***negative*** phase sequence, as indicated by the superscript −. No other phase sequence than these two is possible, and either one of them is changed into the other one by interchanging any two of the three phasors.

The following mathematical operations are greatly facilitated (and, at the same time, made more elegant and "symmetrical") by the introduction of the complex number

$$\boldsymbol{\alpha} = 1 \underline{/120^\circ} \tag{15-4}$$

With the aid of this symbol, any positive-sequence set of voltages (regardless of axis of reference) can be written

$$\begin{aligned} \mathbf{V}_a^+ &= \mathbf{V}^+ \\ \mathbf{V}_b^+ &= \boldsymbol{\alpha}^2 \mathbf{V}^+ \\ \mathbf{V}_c^+ &= \boldsymbol{\alpha} \mathbf{V}^+ \end{aligned} \tag{15-5}$$

and any negative-sequence set can be written

$$\mathbf{V}_a^- = \mathbf{V}^- \qquad \mathbf{V}_b^- = \boldsymbol{\alpha}\mathbf{V}^- \qquad \mathbf{V}_c^- = \boldsymbol{\alpha}^2\mathbf{V}^- \tag{15-6}$$

Suppose now these two sets are added to each other, meaning that a new set is formed by adding the phasors belonging to the same phase. Thus

$$\begin{aligned} \mathbf{V}_a &= \mathbf{V}_a^+ + \mathbf{V}_a^- = \mathbf{V}^+ + \mathbf{V}^- \\ \mathbf{V}_b &= \mathbf{V}_b^+ + \mathbf{V}_b^- = \boldsymbol{\alpha}^2\mathbf{V}^+ + \boldsymbol{\alpha}\mathbf{V}^- \\ \mathbf{V}_c &= \mathbf{V}_c^+ + \mathbf{V}_c^- = \boldsymbol{\alpha}\mathbf{V}^+ + \boldsymbol{\alpha}^2\mathbf{V}^- \end{aligned} \tag{15-7}$$

The three-phase system obtained by this summation is not symmetrical. Neither the magnitudes nor the phase differences of these three phasors are equal. So an unsymmetrical set of three-phase quantities can be obtained by the addition of two symmetrical ones of different phase sequences.

Can this statement be reversed? That is, can any unsymmetrical set of three-phase quantities be decomposed into two symmetrical *components*? In general, the answer is *no*, because three independent equations cannot be satisfied by a set of only two quantities (V^+ and V^-). So the theory of three-phase symmetrical components introduces a *third* quantity, and it does so quite logically by raising the powers of $\boldsymbol{\alpha}$ by one more notch from those of Eqs. 15-5, just as those powers are one notch above those of Eqs. 15-6 (as can be verified by substituting $\boldsymbol{\alpha}^4$ for $\boldsymbol{\alpha}$ in the third of Eqs. 15-5). Thus

$$\begin{aligned} \mathbf{V}_a^0 &= \mathbf{V}^0 \\ \mathbf{V}_b^0 &= \boldsymbol{\alpha}^3\mathbf{V}^0 = \mathbf{V}^0 \\ \mathbf{V}_c^0 &= \boldsymbol{\alpha}^6\mathbf{V}^0 = \mathbf{V}^0 \end{aligned} \tag{15-8}$$

These three phasors are identical, having the same magnitude and phase angle. They form what is called a *zero-sequence* system.

Now let these three symmetrical systems be added, just as two of them were added in Eqs. 15-7.

$$\begin{aligned} \mathbf{V}_a &= \mathbf{V}_a^+ + \mathbf{V}_a^- + \mathbf{V}_a^0 = \mathbf{V}^+ + \mathbf{V}^- + \mathbf{V}^0 \\ \mathbf{V}_b &= \mathbf{V}_b^+ + \mathbf{V}_b^- + \mathbf{V}_b^0 = \boldsymbol{\alpha}^2\mathbf{V}^+ + \boldsymbol{\alpha}\mathbf{V}^- + \mathbf{V}^0 \\ \mathbf{V}_c &= \mathbf{V}_c^+ + \mathbf{V}_c^- + \mathbf{V}_c^0 = \boldsymbol{\alpha}\mathbf{V} + \boldsymbol{\alpha}^2\mathbf{V}^- + \mathbf{V}^0 \end{aligned} \tag{15-9}$$

Since these three equations can be solved for the three unknowns $\mathbf{V}^+$, $\mathbf{V}^-$, and $\mathbf{V}^0$, it follows that any set of three phasors $\mathbf{V}_a$, $\mathbf{V}_b$, and $\mathbf{V}_c$ can be resolved into a positive-sequence, a negative-sequence, and a zero-sequence system. This can be done quite systematically by means of Cramer's rule, or by writing Eqs. 15-9 as a single matrix equation and finding the inverse matrix (which amounts to the same thing).

Some readers may prefer the following method, based on the relation

$$1 + \boldsymbol{\alpha} + \boldsymbol{\alpha}^2 = 0 \tag{15-10}$$

which can be confirmed by performing this addition either analytically (using rectangular form) or graphically. The method consists of eliminating $\mathbf{V}^-$ and $\mathbf{V}^0$ by multiplying the second of Eqs. 15-9 by $\boldsymbol{\alpha}$ and the third one by $\boldsymbol{\alpha}^2$, and then adding these two to the first one. With the aid of Eq. 15-10, this procedure leads directly to the solution

$$\mathbf{V}^+ = \tfrac{1}{3}(\mathbf{V}_a + \boldsymbol{\alpha}\mathbf{V}_b + \boldsymbol{\alpha}^2\mathbf{V}_c) \tag{15-11}$$

A similar operation yields the result

$$\mathbf{V}^- = \tfrac{1}{3}(\mathbf{V}_a + \boldsymbol{\alpha}^2\mathbf{V}_b + \boldsymbol{\alpha}\mathbf{V}_c) \tag{15-12}$$

and finally

$$\mathbf{V}^0 = \tfrac{1}{3}(\mathbf{V}_a + \mathbf{V}_b + \mathbf{V}_c) \tag{15-13}$$

Needless to say, these results can be applied to any set of three voltages or currents, whether they are phase or line quantities.

Equation 15-13 deserves closer attention. It shows that the zero-sequence component has a special property: it exists only if the sum of the three voltages (or currents) is *not* zero. There are quantities that cannot have a zero-sequence component. For instance, the sum of three line voltages $\mathbf{V}_{ab}$, $\mathbf{V}_{bc}$, and $\mathbf{V}_{ca}$ is *always* zero (this follows from Kirchhoff's voltage law), no matter how unbalanced they may be. The same goes for the line currents in a circuit without a neutral (return) conductor.

Solving an unbalanced three-phase problem by means of symmetrical components amounts to a transformation of variables. Actually, the original three quantities (e.g., $\mathbf{V}_a$, $\mathbf{V}_b$, and $\mathbf{V}_c$) are replaced by no less than nine new ones ($\mathbf{V}_a^+$, $\mathbf{V}_b^+$, $\mathbf{V}_c^+$, $\mathbf{V}_a^-$, etc.) but only three of them are independent. The basic idea is that the response of a three-phase circuit to *symmetrical* sets of voltages (or currents) can be found on a per-phase basis, and the three responses to the symmetrical components can then be added (superposed).

Responses to positive- and negative-sequence excitation are not the same, due to the presence of rotating machines (generators and motors) in the circuit. For instance, let a positive-sequence set of voltages be supplied to an induction motor. The direction of the rotating magnetic field may then be called the *forward* direction, and this is also the direction in which the motor runs. Now, if there is, at the same time, also a negative-sequence component in the supply voltages, then there must also be a magnetic field rotating in the *backward* direction, opposite to the motion of the rotor. This field induces additional voltages in the rotor conductors, and their different frequency makes the rotor reactance different. Without going any further at this point, the reader can already see that a rotating machine responds differently to supply voltages of different sequences. There will be more on this subject in the next chapter.

The response of a circuit, even without rotating machines, to zero-sequence excitation is something else again. It depends on such factors as delta or wye connections and the presence or absence of neutral conductors. It lies outside the scope of this book.

15-3 TWO-PHASE SYSTEMS

For two phases, Eq. 15-1 is not valid. It is easy enough to see why: if two coils are placed on a stator at a distance of 180° from each other, and if they carry currents whose phase difference is 180° (currents in phase opposition), then they constitute a *single-phase*, two-coil winding, and they can never produce a rotating field.

A two-phase winding must consist of coils located at a space difference of 90 electrical degrees, and they must carry currents with a 90° phase difference. Under these conditions, a rotating magnetic field is produced, just as it is with three or more phases. (See Problem 8-9).

Figure 15-1 shows two possible phasor diagrams of two-phase voltages. These diagrams are not symmetrical stars, and the phasors do not really deserve being called symmetrical. Nevertheless, the theory of symmetrical components can be adapted to such voltages or currents. For one thing, they could be viewed as

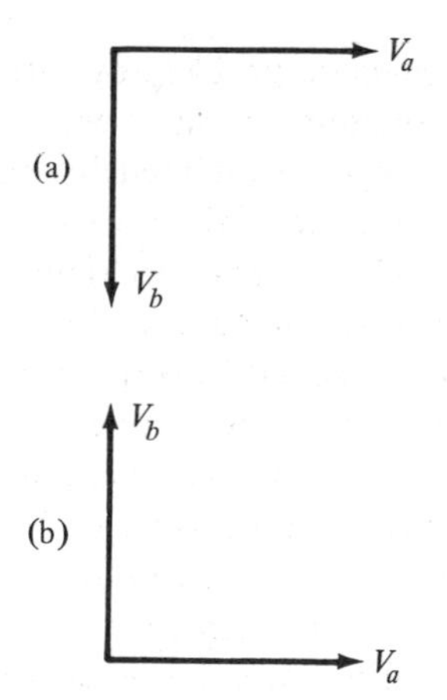

Fig. 15-1. Two-phase voltages: (a) positive sequence; (b) negative sequence.

truncated four-phase quantities, with two of the four phases having zero magnitudes. The following method is simpler.

Let the phase sequence of Fig. 15-1a be called the *positive* one (V_b following V_a by a quarter-period), and that of Fig. 15-1b the *negative* one. Thus the following equations, similar to the three-phase relations of Eqs. 15-5 and 15-6, may be written (regardless of the choice of the axis of reference).

$$\mathbf{V}_a^+ = \mathbf{V}^+ \qquad \mathbf{V}_b^+ = -j\mathbf{V}^+ \tag{15-14}$$

and

$$\mathbf{V}_a^- = \mathbf{V}^- \qquad \mathbf{V}_b^- = j\mathbf{V}^- \tag{15-15}$$

Note that the imaginary number j takes the place of the complex number $\boldsymbol{\alpha}$. (In fact, j would be the value of $\boldsymbol{\alpha}$ for *four* phases.)

Adding the positive- and negative-sequence, two-phase systems results in a new two-phase system whose two phasors are not equal to each other in magnitude and whose phase difference is not 90°.

$$\mathbf{V}_a = \mathbf{V}_a^+ + \mathbf{V}_a^- = \mathbf{V}^+ + \mathbf{V}^- \qquad \mathbf{V}_b = \mathbf{V}_b^+ + \mathbf{V}_b^- = -j\mathbf{V}^+ + j\mathbf{V}^- \tag{15-16}$$

Since these two equations can be solved for the two unknowns $\mathbf{V}^+$ and $\mathbf{V}^-$, it follows that any arbitrary set of two phasors (as here $\mathbf{V}_a$ and $\mathbf{V}_b$), can be resolved into a positive-sequence and a negative-sequence system, each capable of producing a rotating magnetic field when supplied to a two-phase winding.

The simplest way to find these two *components* is to multiply the second of Eqs. 15-16 by j, and either to add it to the first one, or to subtract it from the first one. The results are (using the relation $j^2 = -1$)

$$\mathbf{V}^+ = \tfrac{1}{2}(\mathbf{V}_a + j\mathbf{V}_b) \tag{15-17}$$

and

$$\mathbf{V}^- = \tfrac{1}{2}(\mathbf{V}_a - j\mathbf{V}_b) \tag{15-18}$$

Again, this applies to any set of two phasors, whether they represent voltages or currents. See Example 15-1 for an illustration worked out both analytically and graphically.

In the next two chapters, two types of rotating machines will be studied on the basis of this theory.

15-4 EXAMPLES

Example 15-1 (Section 15-3)

A two-phase, four-wire system has unbalanced voltages: $\mathbf{V}_a = 120\,\underline{/0^\circ}$ v, $\mathbf{V}_b = 80\,\underline{/130^\circ}$ v. Find the positive sequence components and the negative sequence components.

Solution

Express the given voltages in rectangular coordinates

$$\mathbf{V}_a = 120\,\underline{/0^\circ} = 120 + j\,0 \text{ v}$$

$$\mathbf{V}_b = 80\,\underline{/130^\circ} = -51.4 + j\,61.3 \text{ v}$$

Multiply $\mathbf{V}_b$ by j

$$j\mathbf{V}_b = 80\,\underline{/220^\circ} = -61.3 - j\,51.4 \text{ v}$$

The positive sequence component is found by using Eq. 15-17

$$\mathbf{V}_a^+ = \mathbf{V}^+ = \tfrac{1}{2}(\mathbf{V}_a + j\mathbf{V}_b) = \tfrac{1}{2}(58.7 - j\,51.4) = 39\,\underline{/-41.2^\circ} \text{ v}$$
$$\mathbf{V}_b^+ = -j\mathbf{V}^+ = 39\,\underline{/-131.2^\circ} \text{ v}$$

The negative sequence component is found by using Eq. 15-18

$$\mathbf{V}_a^- = \mathbf{V}^- = \tfrac{1}{2}(\mathbf{V}_a - j\mathbf{V}_b) = \tfrac{1}{2}(181.3 + j\,51.4) = 94.2\,\underline{/15.8^\circ} \text{ v}$$
$$\mathbf{V}_b^- = j\mathbf{V}^- = 94.2\,\underline{/105.8^\circ} \text{ v}$$

A phasor diagram illustrating these quantities is shown in Fig. E-15-1.

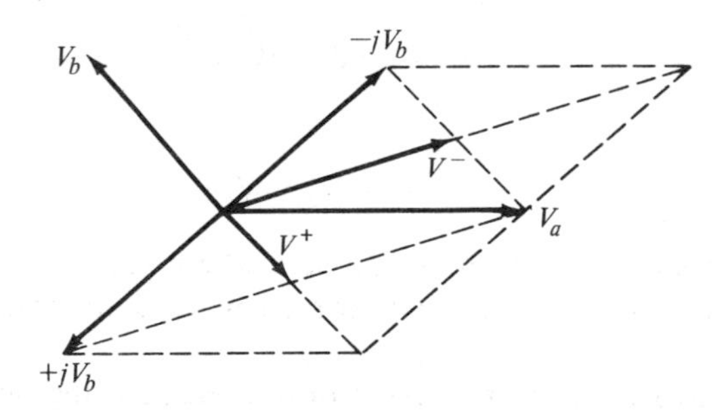

Fig. E-15-1. Graphical solution for symmetrical components.

15-5 PROBLEMS

15-1. A two-phase, four-wire system has unbalanced voltages: $\mathbf{V}_a = 210 \angle 50^\circ$ v, $\mathbf{V}_b = 90 \angle 150^\circ$ v. Find the positive sequence component and the negative sequence component.

15-2. A two-phase, four-wire system has unbalanced voltages: $\mathbf{V}_a = 130 \angle -10^\circ$ v, $\mathbf{V}_b = 90 \angle 40^\circ$ v. Find the positive sequence component and the negative sequence component.

15-3. A two-phase, four-wire system has unbalanced voltages: $\mathbf{V}_a = 100 \angle 80^\circ$ v, $\mathbf{V}_b = 120 \angle 0^\circ$ v. Find the positive sequence component and the negative sequence component.

15-4. A three-phase, three-wire system has unsymmetrical voltages to neutral: $\mathbf{V}_{an} = 100 \angle -135^\circ$ v, $\mathbf{V}_{bn} = 100 \angle 135^\circ$ v, $\mathbf{V}_{cn} = 100 \angle 45^\circ$ v. Find the zero sequence component, the positive sequence component, and the negative sequence component.

15-5. A three-phase, three-wire system has unsymmetrical voltages to neutral: $\mathbf{V}_{an} = 100 \angle 0^\circ$ v, $\mathbf{V}_{bn} = 100 \angle -120^\circ$ v, $\mathbf{V}_{cn} = 100 \angle -60^\circ$ v. Find the zero sequence component, the positive sequence component, and the negative sequence component.

15-6. A three-phase, three-wire system has unsymmetrical voltages to neutral: $\mathbf{V}_{an} = 100 \angle 0^\circ$ v, $\mathbf{V}_{bn} = 85 \angle -120^\circ$ v, $\mathbf{V}_{cn} = 50 \angle 120^\circ$ v. Find the zero sequence component, the positive sequence component, and the negative sequence component.

16

Two-Phase Servomotors

16-1 PURPOSE AND DESIGN PRINCIPLES

The machines to be studied in this chapter are basically induction motors, but their purpose is different from that of the motors discussed so far in this book. The name *servomotors* is meant to imply that they should be considered as auxiliary parts, namely, as components of automatic control systems. Their essential purpose is not power conversion per se, but rather the translation of a time-varying *signal* into a corresponding time-varying *motion*. Some examples may help to illustrate the idea to those readers who have not yet studied feedback control systems.

(a) The direction of a radar beam is to be adjusted in response to the ever-changing position of an airplane or space vehicle.

(b) A valve is to be turned in response to the changing level of a liquid in a container or to the changing pressure of a gas in a tank.

(c) The position of a ship's rudder, or of a tool, is to be adjusted in response to the command of an operator or to a planned schedule.

(d) An example from the realm of the electric power engineer: the setting of a field rheostat is to be corrected in response to the fluctuating voltage of a power system, or to the changing speed of a motor shaft.

In all these cases and in many others, what is desired of the motor is a burst of instantaneous action. Sometimes, a servomotor may be in continuous motion; at other times, it may have to wait at standstill, always ready for a quick short dash in one or the other direction. In any case, the concept of steady-state operation is irrelevant for a servomotor, and so is that of efficiency. What matters is its ability to follow a signal faithfully and with only minimal time delay.

The two-phase servomotor is one of several devices that have been developed specifically for this purpose. (Another such device, the d-c servomotor, will be discussed in a later chapter.) It is basically a two-phase induction motor with a squirrel-cage rotor. In contrast to the ordinary (not servo-) induction motor, however, it is only rarely, if ever, operating at speeds close to synchronous speed. Its torque at such speeds need not be nearly as large as that which it develops at standstill and, particularly, at speeds in the opposite direction. In other words, the torque-speed characteristic of a servomotor should look like that sketched in Fig. 16-1, a diagram in which a typical characteristic of an

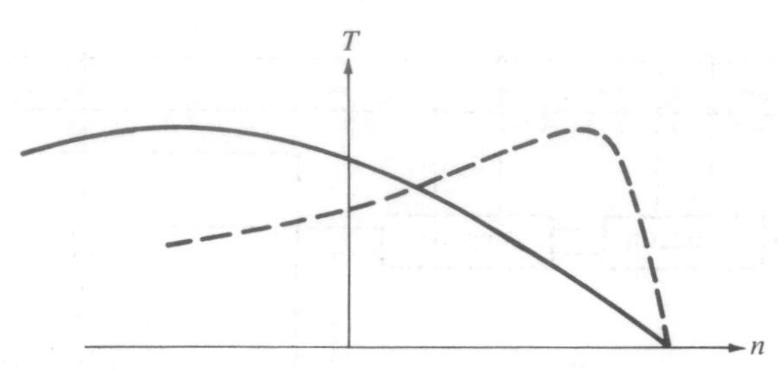

Fig. 16-1. Comparison of torque-speed characteristics.

ordinary induction motor, having the same maximum torque, is also drawn, as a broken line, for comparison.

This diagram may look familiar to the reader. It is the insertion of resistance into the rotor circuit that produces this kind of change in the motor characteristic (see Fig. 14-11). Accordingly, servomotors are designed to have much more rotor resistance than ordinary motors of comparable size. The increased rotor copper loss caused by this design is not a serious drawback for a servomotor.

16-2 THE TWO STATOR WINDINGS

The characteristic of Fig. 16-1 is valid for balanced two-phase operation of the servomotor, i.e., with the two stator windings carrying currents of equal magnitude and a 90° phase difference. This corresponds to just one magnitude and direction of the signal quantity; to different signals, the motor must respond differently. Specifically, it must develop torques whose magnitudes and directions change with those of the signal.

The way to accomplish this is to supply an alternating voltage of constant frequency and magnitude to one of the two stator windings (called the *reference* phase), whereas the voltage supplied to the other stator winding (the *control* phase) has a variable magnitude reflecting the time variation of the signal. The two voltages originate from the same a-c power source, since they must maintain their same frequency; there must also be a constant phase difference of 90° between these two voltages, which may be obtained by means of a phase-shifting network in the control phase circuit, or, more crudely, by a capacitor inserted into the circuit of the reference winding.

Figure 16-2 is a schematic description of the circuits and devices involved. The two coil symbols represent the stator windings, and they are drawn in a way suggestive of their locations, around the squirrel-cage rotor and in *space quadrature* (i.e., at a 90° distance from each other). V_a is the constant-amplitude voltage supplied to the reference winding, and V_b is the control voltage, which is also alternating, but has amplitudes determined by the signal. Using the

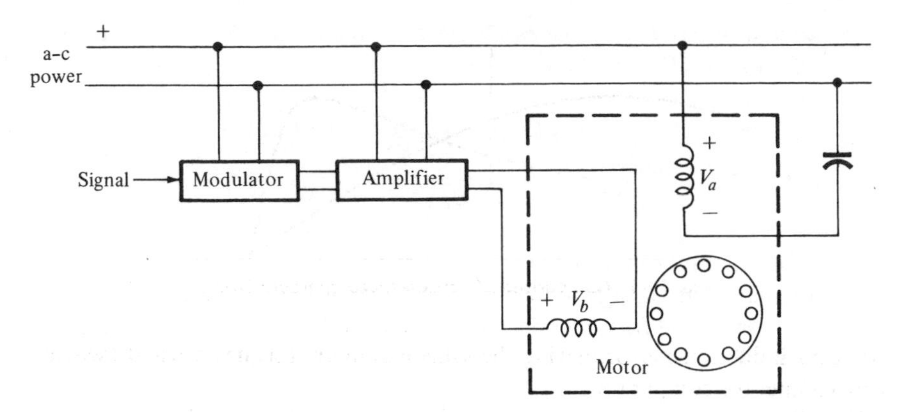

Fig. 16-2. Two-phase servomotor.

terminology of electric communications theory, V_b is the result of having the amplitude of the supply voltage *modulated* by the signal. Devices acting as modulators must be chosen depending on the nature of the signal quantity (whether it is electrical, mechanical, or whatever). The output of the modulator is *amplified* before being supplied to the control winding of the servomotor. This permits a proper adjustment of its range of magnitudes, and it also avoids *loading down* (i.e., drawing too much current from) the modulator.

It now remains to be shown just how the servomotor responds to the changing signal. For this purpose, the method of symmetrical components (applied to two-phase systems as in Section 15-3) will be used.

16-3 ANALYTICAL STUDY

We begin by choosing the reference voltage as the axis of reference

$$\mathbf{V}_a = V\underline{/0^\circ} \tag{16-1}$$

and exprėssing the control voltage in terms of the reference voltage

$$\mathbf{V}_b = kV\underline{/-90^\circ} \tag{16-2}$$

The factor k takes both positive and negative values. In writing Eq. 16-2, the choice has been made arbitrarily that k is positive when V_b *lags* V_a by 90°. The reason for this choice is consistency with the terminology of the previous chapter. That is, the two voltages V_a and V_b are to form a positive-sequence system when $k = 1$, and a negative-sequence-system when $k = -1$.

To investigate the behavior of the servomotor when k has neither of these two extreme values, the two voltages V_a and V_b will now be split into their positive-sequence and negative-sequence *components*. This is done by substituting Eqs. 16-1 and 16-2 into Eqs. 15-17 and 15-18

$$\mathbf{V}^+ = \tfrac{1}{2}(\mathbf{V}_a + j\,\mathbf{V}_b) = \frac{V}{2}(1 + k) \tag{16-3}$$

$$\mathbf{V}^- = \tfrac{1}{2}(\mathbf{V}_a - j\,\mathbf{V}_b) = \frac{V}{2}(1 - k) \tag{16-4}$$

Both of these components are in the axis of reference.

We now have a total of four new voltages replacing the original two, because V^+ and V^- each stand for a two-phase system, according to Eqs. 15-14 and 15-15. Figure 16-3 illustrates a servomotor being supplied from these four fictitious voltage sources.

If only the two positive-sequence sources V_a^+ and V_b^+ were acting, (i.e., with the other two voltages set at zero), then there would be a magnetic field rotating in what may be called the *positive direction*, and the response of the motor to these sources could be found by the methods of analysis introduced in Chapter 13. The same methods can also be used to find the response of the motor to the other two sources V_a^- and V_b^-, with the magnetic field rotating in the *negative direction*. Finally, the two responses thus obtained are to be superposed to give the actual response of the servomotor.

For the *positive sequence*, the equivalent circuit of Fig. 13-6 will be used. It is redrawn in Fig. 16-4, with the familiar complex elements $\mathbf{Z}_1 = R_1 + jX_1$ and $\mathbf{Y}_\phi = G_c + jB_m$. There is also a new symbol

$$\mathbf{Z}^+ = R_2'/s + jX_2' \tag{16-5}$$

which is used to emphasize that this impedance appears in the equivalent circuit for the positive sequence. Let the reader be reminded that all quantities repre-

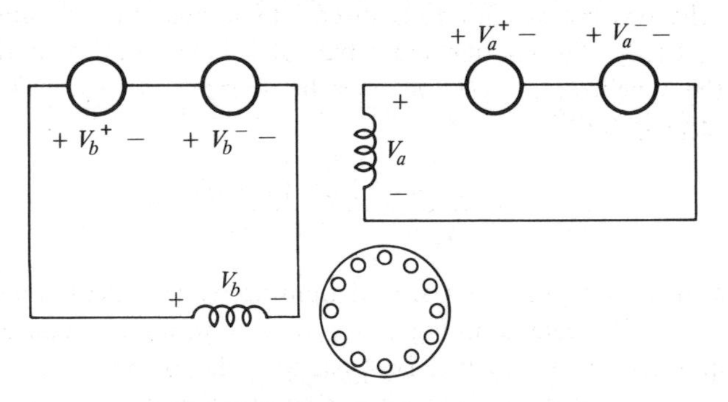

Fig. 16-3. Four fictitious sources.

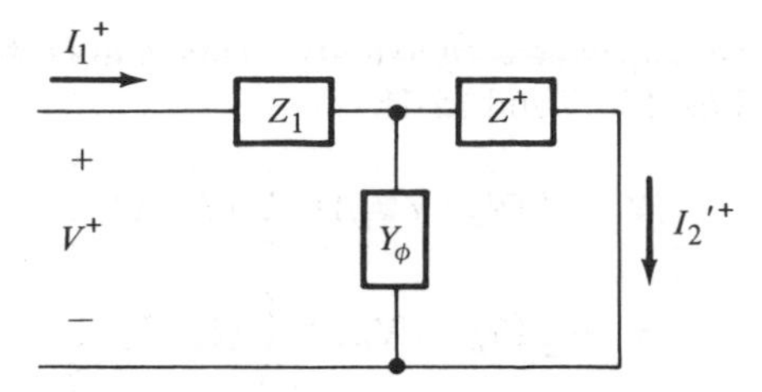

Fig. 16-4. Equivalent circuit for positive sequence.

sent one phase of a balanced two-phase system. The currents can now be calculated as

$$\mathbf{I}_1^+ = \frac{\mathbf{V}^+}{\mathbf{Z}_1 + \dfrac{1}{\mathbf{Y}_\phi + \dfrac{1}{\mathbf{Z}^+}}} \tag{16-6}$$

and (by current division)

$$\mathbf{I}_2'^+ = \mathbf{I}_1^+ \frac{\dfrac{1}{\mathbf{Z}^+}}{\mathbf{Y}_\phi + \dfrac{1}{\mathbf{Z}^+}} \tag{16-7}$$

The reader may be wondering why the negative-sequence response should be any different. The key to the answer lies in the fact that the slip s, on which every response depends, is defined in reference to the positive synchronous speed, i.e., to the speed of the rotating field produced by a set of positive-sequence currents. To be specific, slip is the normalized speed of the rotor relative to the positive synchronous speed. (The analytical definition of the slip is Eq. 13-13.) To describe the effect of a field rotating in the negative direction, the synchronous speed n_s must be replaced by $-n_s$, making the normalized relative speed

$$\frac{-n_s - n}{-n_s} = \frac{n_s + n}{n_s} = \frac{2n_s - (n_s - n)}{n_s} = 2 - s \tag{16-8}$$

Figure 16-5 illustrates this important relationship by showing a comparison of a speed scale and a slip scale (as in Fig. 13-5) for both positive and negative speeds.

So the equivalent circuit of Fig. 16-4 has to undergo just one change in order to be applicable to negative-sequence excitation: the resistance R_2'/s must be

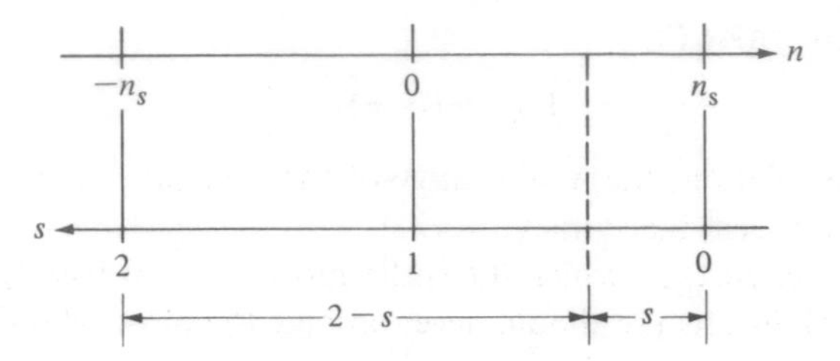

Fig. 16-5. Speed and slip.

replaced by $R_2'/(2 - s)$, or the impedance $\mathbf{Z}^+$ of Eq. 16-5 by

$$\mathbf{Z}^- = \frac{R_2'}{2 - s} + jX_2' \tag{16-9}$$

This change has been made in Fig. 16-6.

When Eqs. 16-6 and 16-7 are rewritten for the negative-sequence response, all that has to be done is replace Z^+ by Z^-. Thus

$$\mathbf{I}_1^- = \frac{\mathbf{V}^-}{\mathbf{Z}_1 + \cfrac{1}{\mathbf{Y}_\phi + \cfrac{1}{\mathbf{Z}^-}}} \tag{16-10}$$

and

$$\mathbf{I}_2'^- = \frac{\cfrac{1}{\mathbf{Z}^-}}{\mathbf{Y}_\phi + \cfrac{1}{\mathbf{Z}^-}} \mathbf{I}_1^- \tag{16-11}$$

The *actual* currents can now be found by *superposition*. For instance, in the reference phase (a)

$$\mathbf{I}_{1a} = \mathbf{I}_1^+ + \mathbf{I}_1^- \tag{16-12}$$

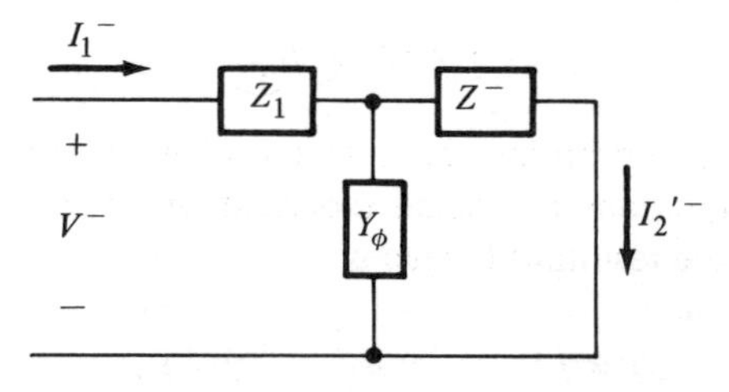

Fig. 16-6. Equivalent circuit for negative sequence.

and in the control phase (b)

$$\mathbf{I}_{1b} = -j\mathbf{I}_1^+ + j\mathbf{I}_1^- \tag{16-13}$$

In this whole calculation, the nonlinearity of the element Y_ϕ causes some (fortunately, not too much) inaccuracy.

The equivalent circuit, on which the entire procedure was based, is a most useful analytical tool, but its use should never obscure the physical reality behind it. The basic fact is that balanced two-phase currents produce a rotating magnetic field whose direction depends on the phase sequence of these currents. Since unbalanced stator currents can be replaced by their positive- and negative-sequence components, the actual field configuration at any instant can be viewed as the equivalent of *two rotating fields* with opposite directions of rotation.

When the motor happens to be running in the positive direction, then that rotating field that rotates in the same direction induces voltages of slip frequency sf in the rotor. This accounts for the flow of slip-frequency rotor currents whose interaction with the rotating field produces a torque in the positive direction. At the same time, due to the presence of another field rotating in the negative direction, there are also voltage and currents whose frequency is $(2 - s)f$ and to which the rotor impedance has, therefore, a different value than to the slip-frequency currents. These currents interact with their field to produce a torque in the negative direction. So the actual motor torque is the resultant of two torques in opposite directions.

16-4 THE RESULTANT TORQUE

Both component torques, the one in the positive direction that is due to the positive-sequence component currents, and the one in the negative direction that is due to the negative-sequence component currents, will be determined by the method given in Chapter 13 (Eq. 13-37), i.e., in terms of the air gap power P_g. Thus

$$P_g^+ = 2(I_2'^+)^2\, R_2'/s \tag{16-14}$$

and

$$P_g^- = 2(I_2'^-)^2\, R_2'/(2 - s) \tag{16-15}$$

where the factor 2 is appropriate because the diagrams (Figs. 16-4 and 16-6) from which these equations are taken represent one phase of a balanced two-phase circuit. Then the resultant torque is

$$T = T^+ - T^- = \frac{1}{\omega_{s_m}}\,(P_g^+ - P_g^-) \tag{16-16}$$

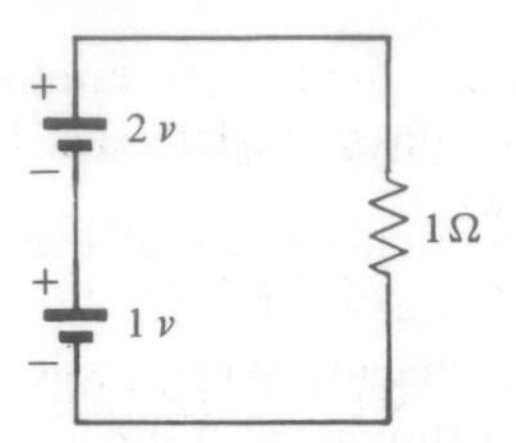

Fig. 16-7. Illustrating superposition.

At this point, a serious objection could be raised: how can the method of superposition be applied to P_g and T which are not linear functions of the currents? To appreciate the relevance of this question, just try to calculate power by superposition in the elementary circuit of Fig. 16-7. The current is 3 amp, and the power consumed by the resistance is 9 w. Now set each one of the source voltages equal to zero, one at a time; the currents are 1 and 2 amp, respectively, and the resistance consumes 1 and 4 w, respectively, suggesting a wrong total power consumption of 5 w. Clearly, the current, which is a linear function of the source voltage, can be found correctly by superposition, but the power, a quadratic function of the source voltage, can *not*.

To see the justification of the use of superposition in the special case of the torque calculation from positive- and negative-sequence components, let the reader first be reminded that power in the sinusoidal steady state is the average, taken over one period, of the product of instantaneous voltage and current

$$P = \text{ave}\,(vi) = \frac{1}{\tau}\int_0^{\tau} vi\,dt = VI\cos\theta \tag{16-17}$$

where τ is the period, V and I are the rms values of voltage and current, and θ is their phase difference.

Now let the voltages and currents in each of the two phases be expressed in terms of their components. For the reference phase,

$$v_a = v_a^+ + v_a^- \tag{16-18}$$

$$i_a = i_a^+ + i_a^- \tag{16-19}$$

and the instantaneous power

$$p_a = v_a i_a = (v_a^+ + v_a^-)\,(i_a^+ + i_a^-) \tag{16-20}$$

Similarly, in the control phase, the instantaneous power is

$$p_b = v_b i_b = (v_b^+ + v_b^-)\,(i_b^+ + i_b^-) \tag{16-21}$$

So the total average power is the sum of *eight* product averages. To get meaningful expressions for them, all phase angles must be defined. Let the voltage phasors be expressed as follows

$$
\begin{aligned}
\mathbf{V}_a^+ &= V^+ \underline{/\varphi} \\
\mathbf{V}_b^+ &= V^+ \underline{/\varphi - 90^\circ} \\
\mathbf{V}_a^- &= V^- \underline{/\psi} \\
\mathbf{V}_b^- &= V^- \underline{/\psi + 90^\circ}
\end{aligned}
\tag{16-22}
$$

Each current lags its voltage by an angle θ^+ or θ^-. Thus

$$
\begin{aligned}
\mathbf{I}_a^+ &= I^+ \underline{/\varphi - \theta^+} \\
\mathbf{I}_b^+ &= I^+ \underline{/\varphi - 90^\circ - \theta^+} \\
\mathbf{I}_a^- &= I^- \underline{/\psi - \theta^-} \\
\mathbf{I}_b^- &= I^- \underline{/\psi + 90^\circ - \theta^-}
\end{aligned}
\tag{16-23}
$$

Now, the eight averages can be expressed. They are

$$
\begin{aligned}
P_1 &= \text{ave}\,(v_a^+ i_a^+) = V^+ I^+ \cos\theta^+ \\
P_2 &= \text{ave}\,(v_a^+ i_a^-) = V^+ I^- \cos(\varphi - \psi + \theta^-) \\
P_3 &= \text{ave}\,(v_a^- i_a^+) = V^- I^+ \cos(\psi - \varphi + \theta^+) \\
P_4 &= \text{ave}\,(v_a^- i_a^-) = V^- I^- \cos\theta^- \\
P_5 &= \text{ave}\,(v_b^+ i_b^+) = V^+ I^+ \cos\theta^+ \\
P_6 &= \text{ave}\,(v_b^+ i_b^-) = V^+ I^- \cos(\varphi - 90^\circ - \psi - 90^\circ + \theta^-) \\
P_7 &= \text{ave}\,(v_b^- i_b^+) = V^- I^+ \cos(\psi + 90^\circ - \varphi + 90^\circ + \theta^+) \\
P_8 &= \text{ave}\,(v_b^- i_b^-) = V^- I^- \cos\theta^-
\end{aligned}
\tag{16-24}
$$

Note that P_6 is the negative of P_2, and P_7 is the negative of P_3. This means that, when all the products of positive-sequence voltages and negative-sequence currents, and vice versa, are added, their average values cancel each other out. What remains is

$$P = 2V^+ I^+ \cos\theta^+ + 2V^- I^- \cos\theta^- \tag{16-25}$$

which is *positive-sequence power plus negative-sequence power*. Thus, the use of superposition is justified for obtaining the *average* values, but not the *instantaneous* values of power and torque.

A further comment is appropriate concerning those four terms whose averages vanished when they were added. They represent a *pulsating* component of the total power. The existence of such a component was to be expected since there is a uniform power flow in *balanced* polyphase systems only. The corresponding torque fluctuations are absorbed by the inertia of the motor and of the load. Thus, they do not produce any substantial speed fluctuations during a period. (If they did, it would interfere with the function of the servomotor.) On the other hand, the motor cannot respond to "fast" signal variations occurring during a period. This can be a serious limitation to the use of servomotors operating at 60 Hz. There are indeed cases in which shorter response times are required, and for such cases, higher supply frequencies are used.

16-5 CHARACTERISTICS

The electromagnetic torque of a two-phase servomotor, as calculated in the two previous sections, depends on two independent parameters: the signal (expressed by the factor k introduced in Eq. 16-2), and the motor speed. Such relations can be expressed graphically by *families of curves*; for instance, the torque can be plotted as a function of the speed, each curve belonging to one value of the signal. Figure 16-1 is such a curve, that for a signal corresponding to $k = 1$. It describes the motor response to a balanced positive-sequence system of voltages. It is redrawn as the top curve in Fig. 16-8, and it will now be shown how all the other curves, for arbitrary values of k, can be obtained from that top curve.

The first step is to get the characteristic for $k = -1$, i.e., the motor response to a balanced negative-sequence supply. On that curve, all torques are negative,

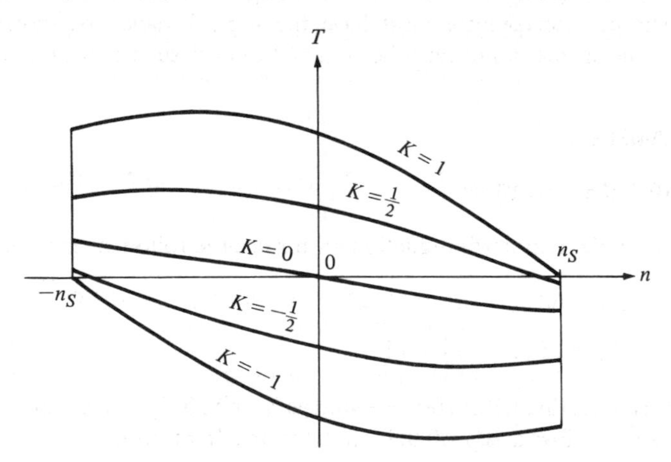

Fig. 16-8. Torque-speed characteristics.

and their magnitudes for a speed n are the same as those on the other curve for a speed $-n$. In other words, the slip s is replaced by $2 - s$, which is exactly what was done in the analytical procedure of Section 16-3.

For other values of k, there are both positive-sequence and negative-sequence torques to be determined and added (i.e., their magnitudes are to be subtracted), in accordance with Eq. 16-16. The way to find the component torques from the two curves for $k = \pm 1$ is based on the fact that, for the linear circuits of Figs. 16-4 and 16-6, the air gap power (and, thus, the torque) is proportional to the square of the supply voltage. So, referring to Eqs. 16-3 and 16-4,

$$T^+ = T_{(k=1)} \left(\frac{V^+}{V}\right)^2 = T_{(k=1)} \left(\frac{1+k}{2}\right)^2 \tag{16-26}$$

and

$$T^- = T_{(k=-1)} \left(\frac{V^-}{V}\right)^2 = T_{(k=-1)} \left(\frac{1-k}{2}\right)^2 \tag{16-27}$$

For instance, for $k = \frac{1}{2}$, the torque at any speed is found as the algebraic sum of $\frac{9}{16}$ times the value of the top curve and $\frac{1}{16}$ times the value of the bottom curve, both values taken at that same speed. When this has been done for several speeds, it is seen how the smaller signal produces the smaller torque.

Figure 16-8 shows the characteristics for $k = \frac{1}{2}$, 0, and $-\frac{1}{2}$, in addition to the top and bottom curves, and any additional curves could be added for in-between values. It should be noted that, for signals near $k = 0$, the torque is opposite to the instantaneous motion. This is just what is required, since the motor should be brought to standstill in response to a zero signal. To produce this result, the characteristic of a servomotor must have the typical shape first shown as Fig. 16-1, which means the motor must be designed to have such a characteristic.

16-6 EXAMPLES

Example 16-1 (Section 16-4)

A 60-Hz, two-pole, two-phase induction motor has the following parameters:

$$R_1 = 60\ \Omega \qquad R_2' = 160\ \Omega \qquad X_m = -1/B_m = 200\ \Omega$$

$$X_1 = 15\ \Omega \qquad X_2' = 15\ \Omega$$

This motor is operating with reference voltage $\mathbf{V}_a = 110\,\underline{/0^\circ}$ v and control voltage $\mathbf{V}_b = 55\,\underline{/-90^\circ}$ v. For a slip of 0.8, find the resultant torque in synchronous watts. Neglect G_c.

Solution

The given voltages are for $k = 0.5$. Using Eqs. 16-3 and 16-4, we can find $\mathbf{V}^+ = 82.5\,\underline{/0^\circ}$ v and $\mathbf{V}^- = 27.5\,\underline{/0^\circ}$ v. For positive sequence, use the equivalent circuit in Fig. 16-4.

$$\mathbf{Z}_f = \frac{(jX_m)(R_2'/s + jX_2')}{R_2'/s + j(X_m + X_2')} = \frac{(j\,200)(200 + j\,15)}{200 + j\,215} = 137\,\underline{/47.2^\circ}$$

$$= 93.1 + j\,101 = R_f + jX_f$$

$$\mathbf{Z}_1 = 60 + j\,15$$

The total positive sequence impedance is

$$\mathbf{Z}_t^+ = \mathbf{Z}_f + \mathbf{Z}_1 = 153.1 + j\,116 = 192\,\underline{/37.2^\circ}\ \Omega$$

The positive sequence component of stator current is

$$\mathbf{I}_1^+ = \mathbf{V}^+/\mathbf{Z}_t^+ = (82.5\,\underline{/0^\circ})/(192\,\underline{/37.2^\circ}) = 0.43\,\underline{/-37.2^\circ}\text{ amp}$$

The positive sequence air gap power is

$$P_g^+ = 2(I_1^+)^2\,R_f = 2(0.43)^2\,(93.1) = 34.4\text{ w}$$

For negative sequence, use the equivalent circuit in Fig. 16-6

$$\mathbf{Z}_b = \frac{(jX_m)(R_2'/(2-s) + j\,X_2')}{R_2'/(2-s) + j(X_m + X_2')} = \frac{(j\,200)(133 + j\,15)}{133 + j\,215} = 106\,\underline{/38.1^\circ}$$

$$= 83.4 + j\,65.4 = R_b + jX_b$$

The total negative sequence impedance is

$$\mathbf{Z}_t^- = \mathbf{Z}_b + \mathbf{Z}_1 = 143.4 + j\,80.4 = 164\,\underline{/29.3^\circ}$$

The negative sequence component of stator current is

$$\mathbf{I}_1^- = \mathbf{V}^-/\mathbf{Z}_t^- = (27.5\,\underline{/0^\circ})/(164\,\underline{/29.3^\circ}) = 0.17\,\underline{/-29.3}\text{ amp}$$

The negative sequence air gap power is

$$P_g^- = 2(I_1^-)^2\,R_b = 2(0.17)^2\,(83.4) = 4.8\text{ w}$$

The resultant torque is

$$T\omega_{sm} = P_g^+ - P_g^- = 34.4 - 4.8 = 29.6\text{ synchronous watts}$$

For a given machine and rated frequency, the conversion of resultant air gap-power $P_g^+ - P_g^-$ to torque involves a constant ratio. Engineers have found it convenient to work with power and to call it torque measured in synchronous watts.

16-7 PROBLEMS

16-1. The machine of Example 16-1 is operated with the same voltages, but with slip of 1. Find the resultant torque in synchronous watts.

16-2. The machine of Example 16-1 is operated with the same voltages, but with slip of 0.9. Find the resultant torque in synchronous watts.

16-3. For the machine and the conditions in Problem 16-2, find the currents in the stator windings.

16-4. A 60-Hz, two-pole, two-phase servomotor has the following parameters:

$$R_1 = 260\ \Omega \qquad R_2' = 830\ \Omega \qquad X_m = 960\ \Omega$$

$$X_1 = 60\ \Omega \qquad X_2' = 60\ \Omega$$

The reference winding voltage is $\mathbf{V}_a = 120\underline{/0^\circ}$ v. The control winding voltage is $\mathbf{V}_b = 48\underline{/-90^\circ}$ v. The motor is running with slip of 0.6. Find the resultant torque in synchronous watts.

16-5. For the motor and the conditions in Problem 16-4, find the currents in the stator windings.

16-6. A 60-Hz, two-pole, two-phase servomotor has reference voltage $\mathbf{V}_a = 1\underline{/0^\circ}$ per unit. The control winding voltage is $\mathbf{V}_b = k\underline{/-90^\circ}$ per unit. When this motor is operated with $k = +1$, the direction of the resultant torque is counterclockwise, and points on the torque-speed curve are:

n(pu)	-1.0	-0.7	-0.4	0	0.4	0.7	1.0
T(pu)	1.1	1.25	1.15	1.0	0.8	0.48	0

(a) Find the torque for $k = 0.5$ and slip of 0.6.

(b) Find the value of k for which the "stalled torque" (i.e., the torque at zero speed) is 0.7 per unit.

(c) Find the value of k for which the no-load speed will be 0.7 per unit counterclockwise.

16-7. The servomotor in Problem 16-4 is operating with a slip of 0.7. The reference winding current is $\mathbf{I}_{1a} = 0.125\underline{/31^\circ}$ amp. The control winding current is $\mathbf{I}_{1b} = 0.036\underline{/-38^\circ}$ amp. Find the voltages impressed on the windings for these conditions.

17
Single-Phase Motors

17-1 PURPOSE

At this point, the reader is reminded of the basic reason why electric energy is mostly generated and transmitted in *polyphase* form: only in this form can the energy be generated and transmitted at a *uniform* rate. This fact manifests itself in rotating generators and motors by the presence of the *rotating magnetic field*. In the steady-state operation of polyphase generators and motors, the stator and rotor currents produce rotating mmf waves of constant magnitude. Since these waves rotate at constant (synchronous) speed, they maintain a constant torque angle, and thereby a constant torque and constant power.

None of this can be said of *single-phase* circuits and machines. The power is *pulsating*, and the mmf wave of a single-phase stator winding is stationary, with a continuously changing magnitude. So it should come as no surprise that single-phase machines are distinctly inferior to their polyphase counterparts. Incidentally, the performance of single-phase machines also turns out to be more difficult to understand and to analyze.

Why, then, are single-phase machines used at all? In most cases, the reason is the easier availability of single-phase power in homes, apartments, schools, offices and stores; in other words, in almost every room except those in industrial plants. The point is that the economic facts of life don't justify the wiring of three-phase power in such rooms. Instead, the output of the three-phase transmission and distribution system is divided into three single-phase outputs, as evenly balanced as possible. The familiar outlets for lamps, appliances, business machines, small tools, etc. have two "hot" contacts, not three. (In some cases, there is a third contact for grounding; its purpose is not to carry current in normal operation, only to provide safety in cases of damage or malfunction.)

So the great majority of single-phase machines are *motors* driving such loads as fans, compressors for refrigerators or air conditioners, washing machines, or phonograph turntables. Compared to industrial machinery, these loads are small, generally in the *fractional horsepower* range. Even smaller loads include countless moving toys, electric razors, and many others. For all such purposes, the drawbacks of single-phase power are not too high a price to pay for the simplicity and economy of single-phase wiring.

There is one major application for *large* single-phase motors: *electric locomotives* have their power supplied either through overhead trolley wires, or through *third rails*, with the two regular rails serving as grounded return conductors. In either case, the power is either single-phase a-c, or d-c. Three-phase supply systems have been tried but lead to difficulties with crossings and switches. Moreover, induction motors have less desirable characteristics (for this purpose) than certain types of commutator motors that will be discussed in a later chapter and require a single-phase a-c or a d-c supply.

The following sections of this chapter deal only with single-phase induction and synchronous motors.

17-2 THE TWO ROTATING FIELDS

The mmf wave produced by the current in a single-phase stator winding has been studied in Section 8-8. Specifically, there was the equation of the standing wave (Eq. 8-54) and its subsequent discussion. In the present context, the subscript a (indicating phase a, one of the three phases under discussion in that section) can be omitted.

$$\mathfrak{F}(\theta, t) = \mathfrak{F}_{\text{peak}} \sin \theta \cos \omega t \tag{17-1}$$

where ω is the radian frequency of the current. The origin of the space scale (the point where $\theta = 0$) has been chosen arbitrarily 90 electrical degrees away from the magnetic axis of the winding, and the origin of the time scale (the instant when $t = 0$) has been chosen at the time when the current has its positive maximum value.

A straightforward method of analysis would now proceed to find the voltages induced at any instant in each of the conductors of a squirrel-cage rotor. Instead, only the essential relations will be derived here. For the sake of simplicity, let the flux distribution be considered proportional to the mmf wave (i.e., let the space harmonics due to saturation be disregarded). Thus

$$B(\theta, t) = B_{\text{peak}} \sin \theta \cos \omega t \tag{17-2}$$

Now the voltage induced in that rotor conductor whose location at the time $t = 0$ happens to be $\theta = 0$, will be found. This conductor rotates at the angular velocity ω_m. Thus, its location at the time t is

$$\theta = (p/2)\, \omega_m t \quad \text{electrical radians} \tag{17-3}$$

When this is substituted into Eq. 17-2, the induced voltage can be found as Blu, and it turns out to be a *nonsinusoidal* function of time

$$e = E_{\max} \sin\, [(p/2)\, \omega_m t] \cos \omega t \tag{17-4}$$

It should be kept in mind that the two ωs in this equation are different quantities, except in the case of a two-pole machine running at synchronous speed.

Many readers know (for instance, from communications theory) that the *product* of two sinusoids with different frequencies can be converted into the *sum* of two other sinusoids whose frequencies are the sum and the difference of the original ones. For the case of Eq. 17-4, the applicable formula is

$$\sin x \cos y = \tfrac{1}{2}\left[\sin(x+y) + \sin(x-y)\right] \tag{17-5}$$

Thus, the voltage induced in this rotor conductor can be viewed as consisting of *two components* whose radian frequencies are the sum and the difference of the radian frequencies ω and $(p/2)\,\omega_m$. If the *slip* is now defined as in Eq. 13-13

$$s = \frac{\omega_{s_m} - \omega_m}{\omega_{s_m}} = \frac{(2/p)\,\omega - \omega_m}{(2/p)\,\omega} = \frac{\omega - (p/2)\,\omega_m}{\omega} \tag{17-6}$$

then it can be seen that the radian frequencies of the two component voltages are

$$\omega_1 = \omega - (p/2)\,\omega_m = s\omega \tag{17-7}$$

and

$$\omega_2 = \omega + (p/2)\,\omega_m = 2\omega - [\omega - (p/2)\,\omega_m] = (2-s)\,\omega \tag{17-8}$$

Or the radian frequencies may be divided by 2π to obtain the two rotor frequencies in Hertz. They are sf and $(2-s)f$, exactly what they were found to be in the rotor of the two-phase servomotor (Section 16-3). This leads to a physical explanation for the presence of the two voltages in the rotor of a single-phase motor, which is the same as it was in the case of the servomotor: the magnetic field can be viewed as the sum of *two rotating fields*, both moving at synchronous speed but in opposite directions.

Now that the similarity between single-phase and unbalanced two-phase excitation has been established, a more suitable method of analysis for the single-phase induction motor comes to mind easily: consider the single-phase stator current as an extreme case of an unbalanced two-phase system of currents, namely the case of one out of two phase currents being zero. In other words, a fictitious second winding is to be thought added to the stator, in space quadrature to the actual stator winding. As long as this second winding is *open circuited,* it carries zero current, and its actual presence or absence makes no difference to the rest of the machine. (This situation should not be confused with that of a two-phase servomotor whose signal is zero. In that case, the control winding has zero voltage, not zero current; the winding is short circuited, not open circuited.)

17-3 ANALYSIS BASED ON SYMMETRICAL COMPONENTS

The method of symmetrical components is to be applied to the two-phase system of currents

$$\begin{aligned} \mathbf{I}_a &= \mathbf{I}_1 \\ \mathbf{I}_b &= 0 \end{aligned} \tag{17-9}$$

where $\mathbf{I}_1$ is the actual stator current. According to Eqs. 15-17 and 15-18, the components are

$$\begin{aligned} \mathbf{I}^+ &= \tfrac{1}{2}(\mathbf{I}_a + j\mathbf{I}_b) = \tfrac{1}{2}\mathbf{I}_1 \\ \mathbf{I}^- &= \tfrac{1}{2}(\mathbf{I}_a - j\mathbf{I}_b) = \tfrac{1}{2}\mathbf{I}_1 \end{aligned} \tag{17-10}$$

The reader may want to verify these results, in the manner of Eq. 15-16, by forming the sums: $\mathbf{I}_a = \mathbf{I}_a^+ + \mathbf{I}_a^- = \mathbf{I}^+ + \mathbf{I}^-$ and $\mathbf{I}_b = \mathbf{I}_b^+ + \mathbf{I}_b^- = -j\mathbf{I}^+ + j\mathbf{I}^-$. The results are in agreement with Eq. 17-9.

Each of these two sets of component currents accounts for the presence of one of the two rotating fields. In effect, by the introduction of these components, the single-phase motor is replaced by an equivalent set of two fictitious two-phase motors, each carrying balanced stator currents. Each of these two motors can now be analyzed by the methods derived earlier for the balanced operation of polyphase motors.

The applicable equivalent circuits, for operation at an arbitrary slip s, are that of Fig. 16-4 for the positive-sequence components and that of Fig. 16-6 for the negative-sequence components. From these circuits, the terminal voltages (per phase) of the two fictitious motors are seen to be

$$\begin{aligned} \mathbf{V}^+ &= \left(\mathbf{Z}_1 + \frac{1}{\mathbf{Y}_\phi + 1/\mathbf{Z}^+}\right)\mathbf{I}^+ \\ \mathbf{V}^- &= \left(\mathbf{Z}_1 + \frac{1}{\mathbf{Y}_\phi + 1/\mathbf{Z}^-}\right)\mathbf{I}^- \end{aligned} \tag{17-11}$$

Each of these two voltages contains the drop across the stator impedance and a voltage induced by the motion of its own rotating field. These two induced voltages will be called E^+ and E^-, respectively. Thus, using Eqs. 17-10,

$$\begin{aligned} \mathbf{E}^+ &= \frac{1}{\mathbf{Y}_\phi + 1/\mathbf{Z}^+}\,\frac{\mathbf{I}_1}{2} \\ \mathbf{E}^- &= \frac{1}{\mathbf{Y}_\phi + 1/\mathbf{Z}^-}\,\frac{\mathbf{I}_1}{2} \end{aligned} \tag{17-12}$$

which permits rewriting Eqs. 17-11

$$\mathbf{V}^+ = \frac{\mathbf{Z}_1\mathbf{I}_1}{2} + \mathbf{E}^+$$
$$\mathbf{V}^- = \frac{\mathbf{Z}_1\mathbf{I}_1}{2} + \mathbf{E}^- \tag{17-13}$$

The terminal voltage across the *actual* stator winding of the single-phase motor consists of the impedance drop $\mathbf{Z}_1\mathbf{I}_1$ and the total voltage induced by the two moving magnetic fields. Thus

$$\mathbf{V}_1 = \mathbf{Z}_1\mathbf{I}_1 + \mathbf{E}^+ + \mathbf{E}^- \tag{17-14}$$

It is worth noting that this also equals

$$\mathbf{V}_1 = \mathbf{V}^+ + \mathbf{V}^- \tag{17-15}$$

(from Eqs. 17-13). Thus, V^+ and V^- can be viewed as the components of the actual terminal voltage V_1.

17-4 THE EQUIVALENT CIRCUIT

Equations 17-14 and 17-12 can be interpreted as describing a two-terminal network with the terminal quantities V_1 and I_1. For this purpose, the factor $\frac{1}{2}$ in Eqs. 17-12 is associated to the impedances instead of the current. Thereby the actual stator current I_1 appears as flowing through a series connection of three impedances, namely the stator impedance Z_1, one-half the impedance of the parallel combination of $1/Y_\phi$ and Z^+, and one-half that of the parallel combination of $1/Y_\phi$ and Z^-.

In halving the impedances of these parallel combinations, their branch impedances are halved, or their branch admittances doubled, leading to the equivalent circuit of the single-phase induction motor shown in Fig. 17-1. The arrows drawn through the impedances $\frac{1}{2}Z^+$ and $\frac{1}{2}Z^-$ are to indicate that these elements are dependent on the operating condition of the motor. Specifically, they are

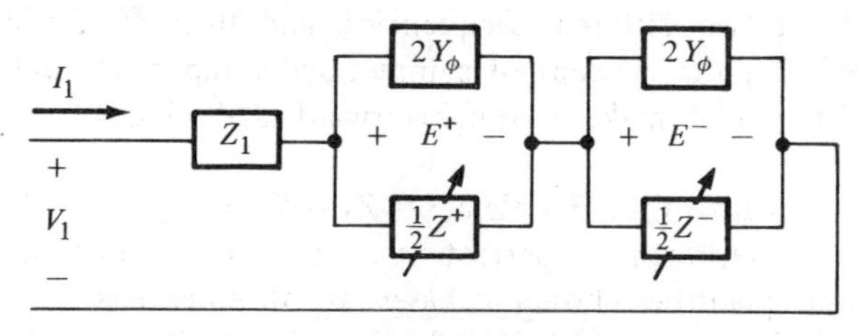

Fig. 17-1. Equivalent circuit of single-phase induction motor.

functions of the slip. According to Eqs. 16-5 and 16-9, they are

$$\begin{aligned} \frac{1}{2}\mathbf{Z}^+ &= \frac{1}{2}\frac{R_2'}{s} + j\frac{1}{2}X_2' \\ \frac{1}{2}\mathbf{Z}^- &= \frac{1}{2}\frac{R_2'}{2-s} + j\frac{1}{2}X_2' \end{aligned} \tag{17-16}$$

The differences between these two impedances constitutes the key to the analysis. To begin with, they are equal *at standstill*, when $s = 1$. In this condition, the two induced voltages E^+ and E^- are equal, which leads to the physical explanation that the two rotating fields have the same magnitude. Furthermore, the two rotor frequencies sf and $(2 - s)f$ are equal to each other as well as to the stator frequency f.

For another way to look at the condition of standstill, the stator and rotor windings are the primary and secondary, respectively, of a single-phase transformer. All mmf waves and the flux distribution are standing waves that may be written in the form of Eqs. 17-1 and 17-2. (These equations were originally meant to describe the mmf wave and flux distribution of the stator current only. But at standstill, all waves have the same form, only different peak values.) Application of the trigonometric formula (Eq. 17-5) shows that such a standing wave is equal to the sum of two traveling waves, each having a peak value equal to one-half the maximum peak value of the standing wave, and each rotating at synchronous speed, one in each direction.

It may well be surmised at this point (and will be confirmed in the next section) that there is *no torque* at standstill, since each of the two rotating fields accounts for a torque in its direction, resulting in two equal and opposite torques when the two fields are equally strong and moving in opposite directions. It is a fact that the "pure" single-phase induction motor cannot start by itself. One might say it would not know in which direction to turn. This constitutes a serious handicap of such a motor; remedies to overcome it will be shown in later sections of this chapter.

The situation changes radically once the motor has been started, no matter how this has been accomplished. At any speed other than zero, the rotor carries currents of two different frequencies, and their mmf waves are not standing waves. But the concept of symmetrical components and the equivalent circuit of Fig. 17-1 make a straightforward analysis of such a condition possible.

It is helpful to realize that the elements $\mathbf{Z}_1 = R_1 + jX_1$, $\mathbf{Z}_2' = R_2' + jX_2'$ and $\mathbf{Y}_\phi = G_c + jB_m$ all represent imperfections. In other words, their per-unit values, with rated quantities chosen as bases, are all much less than unity. The voltage across Z_1 is always much less than rated magnitude, and the currents in the exciting branches are only small fractions of rated magnitude. On the other hand, the currents in the other two branches may well be of rated magnitude or even more, depending on the slip.

When the motor runs *forward* (in the positive direction), the slip is less than unity, and so the resistance R_2'/s becomes larger than the resistance $R_2'/(2 - s)$, which in turn makes $|\mathbf{Z}^+| > |\mathbf{Z}^-|$ and thereby $|\mathbf{E}^+| > |\mathbf{E}^-|$. The conclusion is that, when the motor runs forward, the magnetic field rotating in the positive direction is larger than the other. Conversely, when the motor runs backward, $s > 1$, which makes $|\mathbf{Z}^-| > |\mathbf{Z}^+|$ and $|\mathbf{E}^-| > |\mathbf{E}^+|$. So the flux rotating in the same direction as the rotor is always larger than the flux rotating in the opposite direction. Since the direction of the *torque* of a polyphase induction motor is the direction of its rotating flux, it may be expected (and will be confirmed in the next section) that the single-phase induction motor has a net torque in whichever direction it happens to be moving.

17-5 TORQUE CHARACTERISTICS

The electromagnetic torque of a single-phase induction motor will be determined by the method that was previously used for the two-phase servomotor: the torques due to the positive-sequence and negative-sequence components are found separately, and the resultant torque is their difference. The justification for thus using *superposition* was given in Section 16-4.

The reader is reminded that the resultant torque thus determined is the *average* torque, and that there is an additional pulsating torque that makes the total instantaneous torque alternate between positive and negative values. This is inevitable in any single-phase motor because it is a basic property of single-phase power (see Section 6-2). The inertia of the rotor keeps the steady-state speed nearly constant at the value corresponding to the average torque, but the pulsating torque component tends to cause vibration and noise which the designer of a single-phase motor must take great care to minimize.

The single-phase induction motor is analyzed, then, as if there were, on the same shaft, two two-phase motors with balanced supply voltages V^+ and V^-, respectively. The motor with the positive-sequence supply carries a stator current I^+. Its rotor current referred to the stator is found by current division (see Fig. 16-4), as it was in Eq. 16-7, which is rewritten here with reference to Eq. 17-10:

$$\mathbf{I}_2'^+ = \frac{\mathbf{I}_1}{2} \frac{1/\mathbf{Z}^+}{\mathbf{Y}_\phi + 1/\mathbf{Z}^+} \tag{17-17}$$

Note that this is one-half of the current through the impedance $Z^+/2$ in the equivalent circuit of Fig. 17-1. Similarly, the response of the motor with the negative-sequence supply components may be found in terms of the current

$$\mathbf{I}_2'^- = \frac{\mathbf{I}_1}{2} \frac{1/\mathbf{Z}^-}{\mathbf{Y}_\phi + 1/\mathbf{Z}^-} \tag{17-18}$$

which is one-half of the current through the impedance $Z^-/2$ in this equivalent circuit.

Next, the *air gap power* of each of the two fictitious two-phase motors is found, as it was in Eqs. 16-14 and 16-15:

$$P_g^+ = 2(I_2'^+)^2 \,\Re e\,(Z^+) = (2I_2'^+)^2 \,\Re e\,(Z^+/2) \tag{17-19}$$

which equals the power consumed by the real part of $Z^+/2$ in the equivalent circuit. Likewise,

$$P_g^- = 2(I_2'^-)^2 \,\Re e\,(Z^-) = (2I_2'^-)^2 \,\Re e\,(Z^-/2) \tag{17-20}$$

and this equals the power consumed by the real part of $Z^-/2$ in this same circuit.

All of this proves the remarkable fact that the equivalent circuit of Fig. 17-1 describes correctly not only the terminal quantities V_1 and I_1 of the single-phase induction motor and, thus, its input power

$$P_{\text{in}} = V_1 I_1 \cos\theta \tag{17-21}$$

but also the division of this power into the stator copper loss (the power consumed in the real part of the impedance Z_1), the core losses (consumed in the real parts of the two exciting admittances), and the total air gap power

$$P_g = P_g^+ + P_g^- = P_{\text{in}} - P_{R1} - P_c \tag{17-22}$$

Finally, the *torques* of the two fictitious two-phase motors are

$$T^+ = \frac{1}{\omega_{sm}} P_g^+ \qquad T^- = \frac{1}{\omega_{sm}} P_g^- \tag{17-23}$$

and the *resultant torque* of the actual single-phase motor (counted as positive in the forward direction) is

$$T = T^+ - T^- = \frac{1}{\omega_{sm}} (P_g^+ - P_g^-) \tag{17-24}$$

Example 17-1 goes through this entire calculation for a motor with realistically given constants. Repeating the procedure for several speeds yields a *torque-speed characteristic* whose typical shape is shown in Fig. 17-2. The main features of the curve are what the reader has already been led to expect: there is

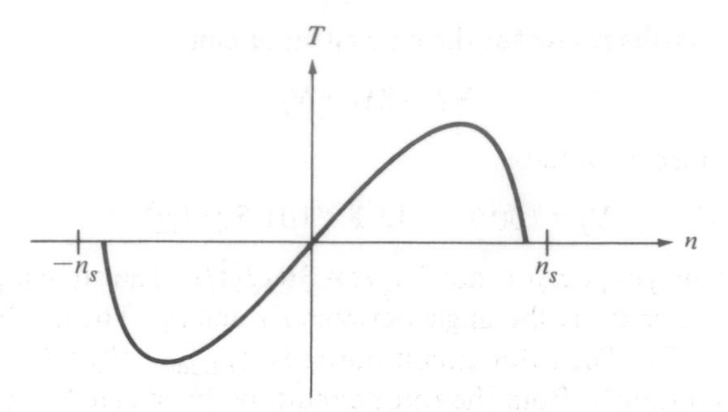

Fig. 17-2. Torque-speed characteristic.

no torque at zero speed, and otherwise, the torque is always in the direction of the motion. Also, in the vicinity of positive synchronous speed, the forward torque T^+ approaches zero, just as it does for the polyphase motor, but there is a substantial backward torque T^-. As a consequence of this, the single-phase induction motor at no-load does not come as close to synchronous speed as the polyphase induction motor. Finally, the curve is symmetrical inasmuch as the torque at any speed n has the same magnitude as at the opposite speed $-n$.

17-6 POWER LOSSES AND EFFICIENCY

All the following power calculations are based on the division of the air gap power P_g into the two parts P_g^+ and P_g^-. Consequently, all results are valid as expressions of *average power*, according to the findings of Section 16-4.

After the stator copper loss and the core losses are subtracted from the input power, the remaining power is thought of as "crossing the air gap" and going to the rotor. What happens to this power, the air gap power? Just as in the polyphase induction motor, a part of it is lost (converted into heat) in the rotor resistance, and the rest is converted into *mechanical power* which, in turn, can be subdivided into rotational losses (mostly friction and windage) and the *output power.*

The mechanical power can be obtained from the torque as

$$P_{\text{mech}} = \omega_m T = \frac{\omega_m}{\omega_{s_m}} (P_g^+ - P_g^-) = (1 - s)(P_g^+ - P_g^-) \tag{17-25}$$

Note that it is proportional to the difference between P_g^+ and P_g^- whereas the air gap power P_g is the sum of these two quantities.

The rotor copper losses are the difference between the air gap power and the mechanical power

$$P_{R_2} = P_g - P_{\text{mech}} = P_g^+ + P_g^- - (1 - s)(P_g^+ - P_g^-) = sP_g^+ + (2 - s)P_g^- \tag{17-26}$$

The last expression gives the rotor copper losses as the sum of two terms whose values are in exact agreement with the concept of the two rotating fields. The first term is the resistance loss of the slip-frequency currents induced by the forward rotating field; the other term is likewise associated with the backward rotating field. An equally valid explanation of the two terms is by reference to the two fictitious two-phase motors of the previous section. In the motor with the positive-sequence supply, the rotor copper loss at the slip s equals sP_g^+, and in the motor with the negative-sequence supply, at the same slip s, it is $(2 - s)P_g^-$.

Example 17-2 shows power loss calculations for the motor of Example 17-1. The resulting *efficiency* is considerably lower than what should be expected from other types of electric motors. This should not be surprising in view of the fact that each of the two rotating fields is responsible for its own core losses and its own rotor copper loss. It is one of several disadvantages of the single-phase motor in comparison to the polyphase motor. The most serious of them is the need for some additional construction element in order to obtain a *starting torque.*

17-7 THE SPLIT-PHASE PRINCIPLE

Most single-phase induction motors are equipped with a second stator winding, called the *auxiliary* or *starting* winding, which is placed in space quadrature to the *main* stator winding. The idea is somehow to obtain currents in the two stator windings that are not in phase with each other, even though they are drawn from the same source. In such a case, their positive- and negative-sequence components are not equal, and so they do produce a starting torque.

There is no problem in finding room for the auxiliary winding on the stator. It was explained in Section 8-6 that single-phase windings should not occupy the entire stator surface because this would result in a low distribution factor. A reasonable compromise is using two-thirds of the stator surface for the main winding and the remaining third for the starting winding. The presence of the starting winding may also be considered as giving some physical reality to the method of analysis in which the single-phase motor is viewed as an unbalanced two-phase motor.

The following calculation serves as an indication of how the *starting torque* depends on the magnitude and phase angle of the current in the starting winding. It is based on the concept of the two rotating fields produced by the positive- and negative-sequence current components. It ignores the fact that, in general, the two stator windings have different numbers of turns and different winding factors, so that balanced currents actually produce different mmf wave magnitudes. Thus, the result obtained is not strictly correct, only qualitatively significant.

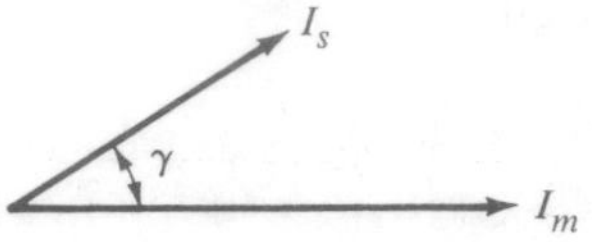

Fig. 17-3. Phasor diagram of stator currents.

Let the current in the main winding be called I_m, and that in the starting winding I_s. Use the symbol c for the ratio of their magnitudes, and γ for their phase difference. Since the voltage is irrelevant, the current I_m will be chosen as the axis of reference. Thus

$$\begin{aligned} I_s &= cI\,\underline{/\gamma} = cI(\cos\gamma + j\sin\gamma) \\ I_m &= I\,\underline{/0} \end{aligned} \tag{17-27}$$

Note that the angle γ is called positive when I_s leads I_m. This happens to be the case in all practical applications, as will be made clear in the next section. The two current phasors are sketched in the diagram of Fig. 17-3.

Now the positive- and negative-sequence *components* are found in accordance with Eqs. 15-17 and 15-18, with the phase subscripts a and b now replaced by s and m, respectively.

$$\begin{aligned} I^+ &= \tfrac{1}{2}(I_s + jI_m) = \tfrac{1}{2}I(c\cos\gamma + j\,c\sin\gamma + j) \\ I^- &= \tfrac{1}{2}(I_s - jI_m) = \tfrac{1}{2}I(c\cos\gamma + j\,c\sin\gamma - j) \end{aligned} \tag{17-28}$$

Since this derivation is aiming for the *starting* torque, the applicable equivalent circuits of Figs. 16-4 and 16-6 can be simplified by neglecting the core loss conductances (the real parts of the exciting admittances). The reader remembers from the theory of the transformer that in the short-circuit condition (which corresponds to the starting condition of the induction motor), the core losses are negligible compared to the copper losses. Thus the admittances assume pure imaginary values $\mathbf{Y}_\phi \approx jB_m$, and the *air gap power* for the positive- and negative-sequence components can be written as

$$\begin{aligned} P_g^+ &= 2I^{+2}\,\mathcal{R}e\left(\frac{1}{\mathbf{Y}_\phi + 1/\mathbf{Z}_{\text{start}}^+}\right) \\ P_g^- &= 2I^{-2}\,\mathcal{R}e\left(\frac{1}{\mathbf{Y}_\phi + 1/\mathbf{Z}_{\text{start}}^-}\right) \end{aligned} \tag{17-29}$$

At this point, it is helpful to realize that

$$\mathbf{Z}_{\text{start}}^+ = \mathbf{Z}_{\text{start}}^- = R_2' + jX_2' \tag{17-30}$$

and to introduce the abbreviation

$$\mathcal{R}e\left(\frac{1}{\mathbf{Y}_\phi + 1/\mathbf{Z}_{\text{start}}^-}\right) = \mathcal{R}e\left(\frac{1}{\mathbf{Y}_\phi + 1/\mathbf{Z}_{\text{start}}^+}\right) = R_{\text{start}} \tag{17-31}$$

Returning to Eqs. 17-29, they also contain the *magnitudes* of the component currents. Form Eqs. 17-28, these magnitudes are

$$\begin{aligned} I^+ &= \tfrac{1}{2} I \sqrt{(c\cos\gamma)^2 + (1 + c\sin\gamma)^2} \\ I^- &= \tfrac{1}{2} I \sqrt{(c\cos\gamma)^2 + (-1 + c\sin\gamma)^2} \end{aligned} \tag{17-32}$$

Now Eqs. 17-31 and 17-32 are substituted into Eqs. 17-29:

$$\begin{aligned} P_g^+ &= \tfrac{1}{2} I^2 \left[(c\cos\gamma)^2 + (1 + c\sin\gamma)^2\right] R_{\text{start}} \\ P_g^- &= \tfrac{1}{2} I^2 \left[(c\cos\gamma)^2 + (-1 + c\sin\gamma)^2\right] R_{\text{start}} \end{aligned} \tag{17-33}$$

and so the starting torque is, according to Eq. 17-24

$$\begin{aligned} T_{\text{start}} &= \frac{1}{2\omega_{s_m}} I^2 R_{\text{start}} \left[(1 + c\sin\gamma)^2 - (-1 + c\sin\gamma)^2\right] \\ &= \frac{1}{2\omega_{s_m}} I^2 R_{\text{start}} (4\, c \sin\gamma) \end{aligned} \tag{17-34}$$

Finally, by using the magnitude relations from Eqs. 17-27, the starting torque can be written as

$$T_{\text{start}} = \frac{2R_{\text{start}}}{\omega_{s_m}} I_m I_s \sin\gamma \tag{17-35}$$

This result indicates, for one thing, that the starting torque is positive for a positive angle γ. This was to be expected because the system consisting of the currents I_s and I_m is closer to a positive-sequence than to a negative-sequence system. This also agrees with the facts that $I^+ > I^-$ (Eqs. 17-32) and that $P_g^+ > P_g^-$ (Eqs. 17-33).

Otherwise, the essential message of Eq. 17-35 is that the torque increases with the magnitude of the current in the starting winding and with the phase angle between the two stator currents. Stated in such general, nonspecific terms, this result remains valid even if the assumption of two identical stator windings (on which the derivation was based) is dropped.

Incidentally, for a machine with two identical stator windings, the two stator currents may be made to form a balanced two-phase system by setting $c = 1$ and $\gamma = 90°$. In that case, the torque Eq. 17-35 gives exactly the same result as the method used in the study of polyphase induction motors.

17-8 THE PRINCIPAL TYPES OF SINGLE-PHASE INDUCTION MOTORS

With the stator of a single-phase induction motor equipped with a starting winding, the question is how to obtain a phase difference between the two stator currents drawn from the same supply voltage. Since each of the two stator windings is the equivalent of an RL circuit, any method of changing the ratio of resistance over inductance for one of them will serve this purpose. It is reasonable to give the starting winding the larger value of this ratio, for a number of reasons.

(a) The resistance of a coil of a given wire size is proportional to its number of turns, the inductance to the square of this number. So a starting winding with fewer turns than the main winding already has a larger R/L ratio.

(b) To increase the resistance of the starting winding further does not require the series connection of an additional resistor. Just use a thinner wire, or a wire made of a metal with a higher resistivity.

(c) The higher power loss due to the increased resistance of the starting winding can be tolerated for the short starting interval (a few seconds at the very most). They do not affect the running efficiency of the motor, provided the auxiliary winding is *disconnected* after the motor has come close to its steady-state speed.

A large number of single-phase induction motors are built in accordance with these principles. They are equipped with starting windings whose R/L ratios are higher than those of their main windings, and with switches to disconnect their starting windings. These switches operate *automatically*, based on either the speed (i.e., in response to the centrifugal force) or on the time elapsed after the closing of the main switch. Such motors are generally known as *split-phase motors.*

Larger phase differences between the two stator currents and thereby larger starting torques can be obtained by connecting a *capacitor* in series with the starting winding. As an example, Fig. 17-4 shows typical values of current phasors. I_m is the current in the main stator winding, and I_{s_R} is the current in the starting winding (assumed here to be identical with the main winding) with resistance added. The phase difference between these two currents is only about 14°. It is true that this angle could be increased by adding more resistance. But this attempted remedy would at the same time decrease the magnitude of the current in the starting winding and so it would hardly improve the torque.

By contrast, connecting a suitable amount of capacitance in series with the starting winding not only raises the phase angle γ (in the diagram of Fig. 17-4 to $\gamma_c = 90°$), but actually increases the magnitude of I_s. Consequently, much larger starting torques are obtained by this method. Even though it is another application of the split-phase principle, the established term for a motor with a capacitor in its starting circuit is not split-phase motor but *capacitor motor.*

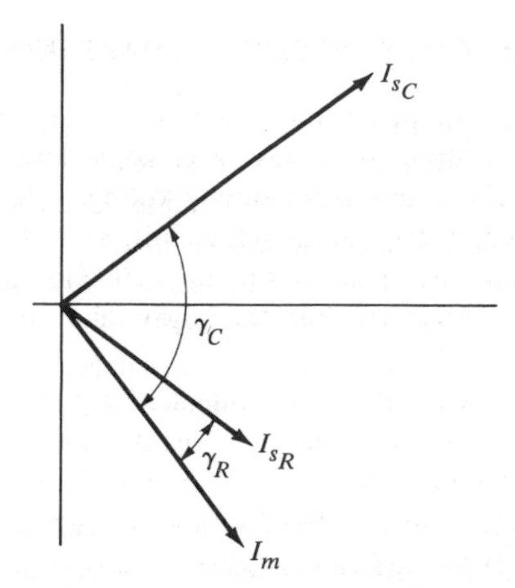

Fig. 17-4. Phasor diagram of split-phase and capacitor motors.

Since capacitor motors have so much larger starting torques than split-phase motors, the reader might wonder why split-phase motors are so popular. The answer lies in the economics of the two motor types. Capacitors in the order of magnitude required for this purpose are very expensive, whereas an increase in the resistance of the starting winding actually reduces its cost, as was explained earlier. So capacitor motors are basically used in those cases in which the starting torques of split-phase motors would not be sufficient, e.g., motors that have to start under full rated load.

Many capacitor motors have automatic switches to disconnect the starting circuit after starting (*capacitor-start* motors), just like split-phase motors but for a different reason. The relatively inexpensive type of capacitor that makes its use for the starting of a single-phase motor economically justifiable, the so-called *electrolytic* capacitor, would have too short a life expectancy if it were left permanently in the circuit. So its use must be limited to starting.

It is quite desirable, however, to operate a single-phase induction motor with its starting winding and a capacitor connected permanently (*capacitor-run* motor). Such a motor operates more nearly like a two-phase motor. It has a more nearly uniform power flow and torque. Due to the reduction of the backward rotating field, it is capable of delivering a higher output, and both the efficiency and the power factor are improved, compared to a motor running with only one stator winding. In an ideal case, the two stator mmf waves are equal in magnitude and out of phase by 90°; the motor operates as a pure polyphase motor; there is no backward rotating field; mechanical power and torque are constant; the pulsating component of the power drawn from the single-phase supply line is continuously stored and restored by the capacitor.

The amount of capacitance needed to approach this ideal case as much as possible depends on the operating condition but is much less, in any case, than the capacitance that produces the highest possible starting torque. This can be appreciated without any calculation by taking a look at the equivalent circuit of a polyphase motor. At running, the impedance of each phase is much larger than at starting; thus, a much larger capacitive reactance is needed to turn the phase angle of the current in the starting winding than for starting, and a larger capacitive reactance means a smaller capacitance.

Thus, a capacitor-run motor may be equipped with a permanent-duty capacitor that produces optimal performance for a certain (e.g., rated) operating condition, but only a weak starting torque compared to that of a capacitor-start motor. It is also possible to use a compromise value of capacitance to obtain a somewhat higher (still far from the best possible) starting torque at some sacrifice in running performance.

Finally, the conflicting requirements of starting and running operation may be reconciled by using two different capacitors in parallel to each other. The larger one, of the electrolytic type, is disconnected after starting, while the other, of a type suitable for permanent duty, remains in the circuit.

All the types of single-phase induction motors discussed above obtain their starting torque by the split-phase method. Several other possibilities of obtaining starting torques are occasionally used. *Shaded-pole* motors constitute the least expensive type but they produce only a weak starting torque. At the opposite extreme, there are *repulsion-start* motors whose starting torque is far above that of other single-phase induction motors, but whose construction is far costlier. Details about these motor types are considered as being of interest mainly to the specialist, and will not be discussed here.

Regardless of the method by which the starting torque is obtained, mention must be made of the problem of the starting *current*, so important for many polyphase induction motors. For most single-phase motors, these problems are of little concern. This is so mainly because the starting time interval is too short for the current to cause any damage, due to the relatively small size and, consequently, small moment of inertia of these motors. But the fact that the starting current is highly in excess of rated value is often quite noticeable, inasmuch as the voltage drop in the supply line may cause a momentary dimming of all lights operating in parallel with the motor during its starting interval.

Once started, single-phase induction motors have operating characteristics similar to those of polyphase induction motors. For speed control, the methods described in Chapter 14 are available if needed.

17-9 SINGLE-PHASE SYNCHRONOUS MOTORS

There is a limited but significant field of application for single-phase synchronous motors. They are used to drive small-scale devices required to move at a strictly constant predetermined speed. Typical examples are electric clocks and precision phonograph turntables. (Most "ordinary" turntables are driven by induction motors, which are less expensive. They take advantage of the fact that most people's ears are not sensitive enough to notice the slight pitch variations caused by the motor's slight speed variations.)

By itself, the single-phase synchronous motor has no *starting torque*, for two separate reasons. First, as with every *synchronous* motor, its average torque is zero at any but synchronous speed, and it must start and be brought up to near synchronous speed by induction motor action; this action is obtained by the use of damper windings (see Section 10-7) for most large-scale synchronous motors; for the small motors under discussion, however, the eddy currents induced in the rotor are quite sufficient. Second, as with every *single-phase* motor, it has no torque at zero speed, unless an additional design element is provided, e.g., an auxiliary winding based on the split-phase principle, or shaded poles.

Another distinction of these motors is that the use of field windings carrying currents from separate (d-c) sources, common for larger polyphase synchronous motors, is neither practical nor economical; fortunately, it is not necessary either. The motors under discussion can have *permanent magnets* as their rotors, or they can have salient poles and operate as *reluctance motors* (see Section 12-6). In addition, there is the interesting possibility of obtaining a synchronous torque by means of the phenomenon of *hysteresis*.

Since eddy currents in the rotor produce a torque, it might be expected that hysteresis is capable of the same thing. It is, but with an important difference: eddy currents are caused by the relative speed between the rotor and the magnetic field, and, thus, they cease to exist at synchronous speed, whereas hysteresis remains present as long as the field distribution lags behind the mmf wave. Figure 17-5 illustrates the principle. It is a "snapshot" showing the axes of the forward rotating components of the stator mmf wave and of the flux distribu-

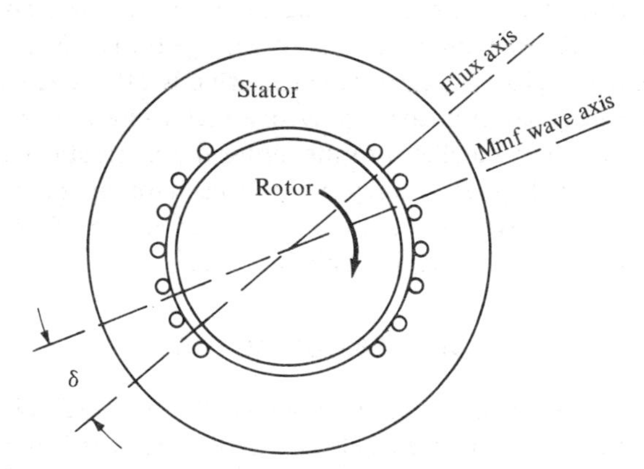

Fig. 17-5. Hysteresis motor.

tion, taken at some arbitrary instant. At synchronous speed, the torque angle formed by these two axes remains constant as long as the load is constant, and this angle changes in response to any load change in the same way as it does for any synchronous motor, thereby maintaining the steady-state equilibrium.

Hysteresis motors have no rotor windings. Their rotors are solid cylinders made of steel alloys distinguished by wide hysteresis loops. Since the hysteresis torque is determined by the angle δ, it is independent of the speed and remains constant from start to synchronous speed, which is a unique property of this type of motor.

17-10 EXAMPLES

Example 17-1 (Section 17-5)

A single-phase induction motor is rated $\frac{1}{4}$ HP, 100 v, 60 Hz, four poles. The parameters of the equivalent circuit are

$$R_1 = 2\ \Omega \qquad R_2' = 4\ \Omega \qquad X_m = -\frac{1}{B_m} = 66\ \Omega$$

$$X_1 = 2.4\ \Omega \qquad X_2' = 2.4\ \Omega$$

The rotational losses, core losses combined with windage and friction, are 37 w. The motor is operating with rated voltage and rated frequency and with a slip of 0.05. Find the input current and the resultant electromagnetic torque.

Solution

Use the equivalent circuit shown in Fig. 17-1. When numerical values are substituted, we find that the impedance across E^- is much smaller than that across

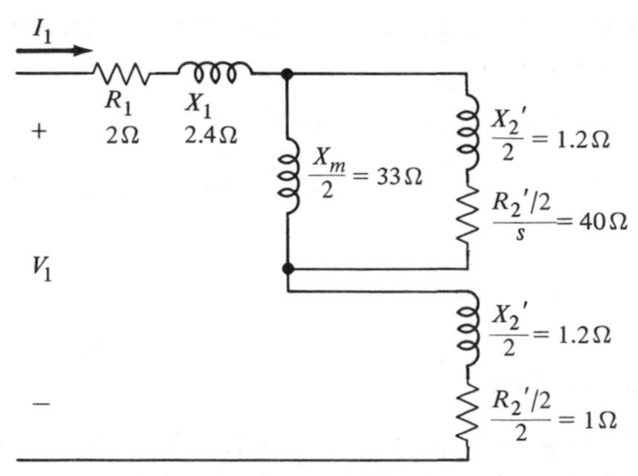

Fig. E-17-1. Approximate equivalent circuit of single-phase induction motor.

E^+ to which it is to be added. Therefore, great accuracy is not required in its calculation. Thus

$$0.5\mathbf{Z}_b = 0.5 \frac{(jX_m)\,[R_2'/(2-s) + jX_2']}{R_2'/(2-s) + j(X_m + X_2')} \approx 0.5(R_2'/2 + jX_2')$$

This approximation is reasonable for values of slip less than 0.1. The corresponding equivalent circuit is shown in Fig. E-17-1. For this example, we have

$$0.5\mathbf{Z}_b = 1 + j\,1.2\ \Omega = 0.5R_b + j\,0.5X_b$$

Next, compute $0.5\mathbf{Z}_f$

$$0.5\mathbf{Z}_f = \frac{(j\,33)\,(40 + j\,1.2)}{40 + j\,34.2} = 25.1\ \underline{/51.2^\circ}$$

$$= 15.7 + j\,19.6\ \Omega = 0.5R_f + j\,0.5X_f$$

$$\mathbf{Z}_1 = 2 + j\,2.4\ \Omega$$

The total impedance is

$$\mathbf{Z}_t = \mathbf{Z}_1 + 0.5\mathbf{Z}_f + 0.5\mathbf{Z}_b = 18.7 + j\,23.2 = 29.8\ \underline{/51.1^\circ}\ \Omega$$

The stator current is

$$\mathbf{I}_1 = \mathbf{V}_1/\mathbf{Z}_t = (110\ \underline{/0^\circ})/(29.8\ \underline{/51.1^\circ}) = 3.69\ \underline{/-51.1^\circ}\text{ amp}$$

The positive sequence air gap power is the power in the positive sequence portion of the equivalent circuit

$$P_g^+ = I_1^2(0.5R_f) = (3.69)^2(15.7) = 213.8\text{ w}$$

The negative-sequence air gap power is the power in the negative-sequence portion of the equivalent circuit

$$P_g^- = I_1^2(0.5R_b) = (3.69)^2(1) = 13.6\text{ w}$$

Synchronous speed is $\omega_{s_m} = 4\pi f/p = 188.5$ rad/sec. The resultant torque is found from Eq. 17-24

$$T = (1/\omega_{s_m})\,(P_g^+ - P_g^-) = 1.06\text{ nm}$$

Example 17-2 (Section 17-6)

The machine in Example 17-1 is operated with rated voltage and frequency with a slip of 0.05. Find the output torque, the output power, and the efficiency.

Solution

These are the same operating conditions as in Example 17-1. Use Eq. 17-25 to find the mechanical power

$$P_{mech} = (1 - s)(P_g^+ - P_g^-) = 0.95(213.8 - 13.6) = 190 \text{ w}$$

Subtract the rotational losses to find the output power

$$P_{out} = P_{mech} - P_{rot} = 190 - 37 = 153 \text{ w}$$

The shaft speed is $\omega = (1 - s)\,\omega_{s_m} = 179$ rad/sec. The output torque is

$$T_{out} = P_{out}/\omega = 153/179 = 0.85 \text{ nm}$$

Tabulate the losses

$$P_{R_1} = I_1^2 R_1 = (3.69)^2(2) = 27.2 \text{ w}$$

$$P_{R_2}^+ = sP_g^+ = (0.05)(213.8) = 10.7 \text{ w}$$

$$P_{R_2}^- = (2 - s)P_g^- = (1.95)(13.6) = 26.5 \text{ w}$$

$$P_{rot} = 37 \text{ w}$$

$$\sum \text{losses} = P_{R_1} + (P_{R_2}^+ + P_{R_2}^-) + P_{rot} = 101 \text{ w}$$

The efficiency is

$$\eta = 1 - \frac{\sum \text{losses}}{P_{out} + \sum \text{losses}} = 1 - \frac{101}{153 + 101} = 0.60$$

17-11 PROBLEMS

17-1. The single-phase induction motor of Example 17-1 is operated with rated voltage and rated frequency, but with increased load so that the slip is increased to 0.06. Find the input current and the resultant electromagnetic torque.

17-2. For the machine in Problem 17-1, with slip of 0.06, find the output power, the output torque, and the efficiency.

17-3. A single-phase induction motor is rated $\frac{1}{2}$ HP, 155 v, 60 Hz, four poles. The parameters of the equivalent circuit are:

$$R_1 = 0.8\ \Omega \qquad R_2' = 1.6\ \Omega \qquad X_m = 42\ \Omega$$

$$X_1 = 1.8\ \Omega \qquad X_2' = 2.2\ \Omega$$

The rotational losses, friction and windage combined with core losses, are 43 w. For operation with rated voltage, rated frequency, and with slip of

0.04, find (a) the input current and the resultant electromagnetic torque, (b) the output power and the output torque, and (c) the efficiency.

17-4. A six-pole, two-phase, 60-Hz induction motor has the following parameters:

$$R_1 = 0.43\ \Omega \qquad R_2' = 0.32\ \Omega \qquad X_m = 30\ \Omega$$

$$X_1 = 0.84\ \Omega \qquad X_2' = 0.84\ \Omega$$

This motor is operated as a single-phase motor with $\mathbf{V}_a = 220\,\underline{/0^\circ}$ v and with winding "b" open circuited. The load is adjusted to make the slip be 0.05. (a) Find the current in the stator winding. (b) Find the voltage in winding "b."

17-5. A small two-phase, two-pole, 60-Hz induction motor has the following parameters:

$$R_1 = 38\ \Omega \qquad R_2' = 20\ \Omega \qquad X_m = 540\ \Omega$$

$$X_1 = 30\ \Omega \qquad X_2' = 30\ \Omega$$

Find the starting torque when this motor is operated with $\mathbf{V}_a = 110\,\underline{/0^\circ}$ v and $\mathbf{V}_b = 110\,\underline{/-90^\circ}$ v.

17-6. The machine in Problem 17-5 is to be operated as a single-phase motor. For starting, a resistance of 80 Ω is connected in series with winding "b," and both this circuit and winding "a" are energized from one line with $\mathbf{V}_a = 110\,\underline{/0^\circ}$ v. At standstill ($s = 1$), the current in winding "b" can be calculated by using the impedance of the total circuit, while the current in winding "a" is not affected by adjustments in winding "b." Find the starting torque.

17-7. For single-phase starting, the resistor in Problem 17-6 is replaced by a capacitor with a reactance of 110 Ω. Find the starting torque.

17-8. For a two-phase induction motor, starting torque may be calculated from $T_{\text{start}} = k_s I_a I_b \sin\alpha$, where I_a is the current in winding "a," I_b is the current in winding "b," and α is the phase angle between the currents I_a and I_b. Use the results of Problems 17-5, 17-6, and 17-7 to check the validity of this formula and to evaluate k_s for this motor.

17-9. The two-phase induction motor of Problem 16-4 is to be used as an a-c tachometer by having its shaft rigidly coupled to some larger machine. Winding "a" is connected to a 60-Hz voltage, $\mathbf{V}_a = 120\,\underline{/0^\circ}$ v. Winding "b" is open circuited ($I_b = 0$). The motor can be analyzed as if it were a single-phase motor. Find the voltage in winding "b" for the following: (a) slip of 0.96, (b) slip of 0.98, (c) slip of 1.00, (d) slip of 1.04.

18
Commutator Machines

18-1 D-C POWER COMPONENTS

The machines discussed up to this point, whether they are single-phase or polyphase, synchronous or asynchronous, generators or motors, are all operating with alternating currents and voltages. Their most important mode of operation is the sinusoidal steady state.

Yet there is also a distinct place for direct-current (d-c) components in power systems. For instance, many chemical processes essential to our industrial society are based on *electrolytic* action, which only unidirectional currents can produce. Such currents are not necessarily supplied by d-c generators; they may be drawn from a-c power systems through rectifiers.

An entirely different example of the use of d-c in power systems is the *transmission* of large blocks of power over long distances. Although a-c is the most commonly used form of power for this purpose, there are cases when the use of d-c is competitive or even preferable, due largely to the absence of stability problems. Such cases do not involve d-c generators or motors (which are not suitable for the high voltages of transmission lines); where power is to be transmitted in d-c form, it is generated as a-c, rectified for transmission, and inverted (i.e., converted back to a-c) at the receiving end of the line. This form of power transmission has become practical and economical only with the introduction and steady improvement of electronic rectifiers and inverters, first in the form of mercury-arc tubes but nowadays as solid-state semiconductor devices.

Thus, the reader can appreciate that d-c generators and motors, which were, historically, the first rotating electric machines built, have now only a limited field of application, even in the presence of d-c transmission lines or d-c loads. Actually, there remains one major need for d-c machines: *motors* whose speed is either to change sharply with changing loads, or to be controlled quickly and over a wide range.

18-2 HOMOPOLAR MACHINES

The very nature of electromagnetic induction favors a-c machines against d-c machines. If Faraday's law is expressed in the form $e = d\lambda/dt$, it appears unlikely, to say the least, that voltages can be induced continuously in any but alternating form. Surely, flux linkages cannot keep increasing (or decreasing) all the time. Even when the equation is written as $E = Blu$, the question remains whether it is at all possible to induce unidirectional voltages. Can a conductor "cut" lines of flux continuously in the same direction? In other words, can a conductor move relative to a magnetic north pole without subsequently also moving relative to a south pole?

The question can be answered affirmatively by considering the *homopolar* type of machine, which is rather different from all other types of electric machines. The principle is illustrated by the sketch of Fig. 18-1. The right side of the figure represents a cut through the plane marked *a-a* in the left view.

The stator carries a *field winding* that is supplied with d-c but is not located outside (surrounding) the rotor like the stator winding of any other machine. The idea is that all lines of flux cross the plane air gap in the same direction (in the figure from left to right, i.e., from rotor to stator), whereas they cross the cylindrical air gap (the one surrounding the rotor) in radial direction. Thus, there are no alternating poles along the circumference, and all *armature* (rotor) conductors have voltages induced in the same direction as they move past a single pole, not past a pair of poles. There are two *slip rings*, not shown in the figure, to which the two ends of each armature conductor are connected and to which a pair of fixed *brushes* makes continuous sliding contact.

The trouble with this design is that there is no way to connect the armature conductors in series to form coils. Any attempt to use a *return conductor* leads to the formation of a turn in which the total induced voltage is zero. Thus, all the armature conductors must be connected in parallel. It is impossible, with one pair of slip rings and brushes, to obtain voltages higher than those induced

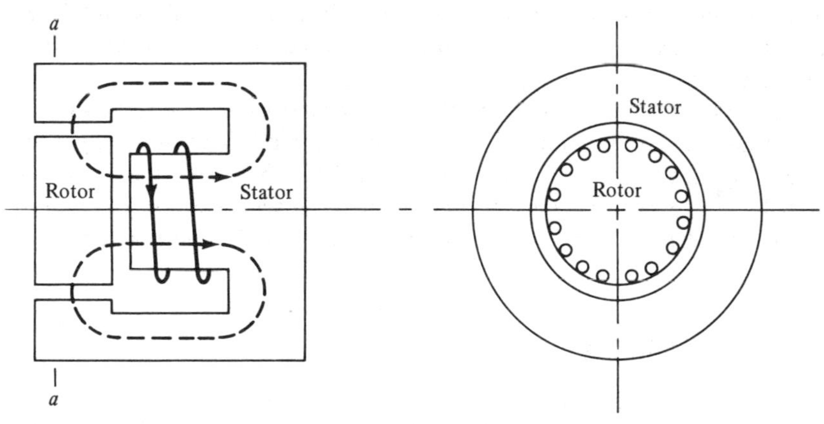

Fig. 18-1. Homopolar machine.

in a single armature conductor. The homopolar machine is used only as a generator supplying *low-voltage* d-c loads.

18-3 THE IDEA OF THE COMMUTATOR

To understand how it is possible to build rotating electric machines with unidirectional voltages, without having to accept the limitations of the homopolar design, let the first step be a look at an "inside-out" synchronous machine, i.e., one whose rotor carries the armature conductors, whereas the field winding is placed on the stator. (It was explained in Section 9-2 why the preferred design for actual synchronous machines is just the opposite.)

Figure 18-2 is a sketch of such a machine, one with two salient poles. It shows the (d-c excited) field winding on the stator, and two mean lines of flux. The two indicated positive (arrow) directions are consistent with each other; i.e., when the field current is positive, the same thing is true of the flux; in that case, the flux crosses the air gaps from the top to the bottom of the picture; in other words, the stator pole at the top is a magnetic north pole, the one at the bottom a south pole.

On the rotor, the figure shows (in the interest of clarity) only two armature conductors, or one turn of an armature coil, in an arbitrary position. When the rotor moves, an *alternating* voltage is induced in the armature conductors, with one cycle for every revolution of the rotor. To transfer this alternating voltage to a pair of stationary terminals, each of the two ends of the turn is connected to a *slip ring* (the same would be true of the two ends of a coil consisting of many turns, or of the two ends of a full distributed winding). A stationary

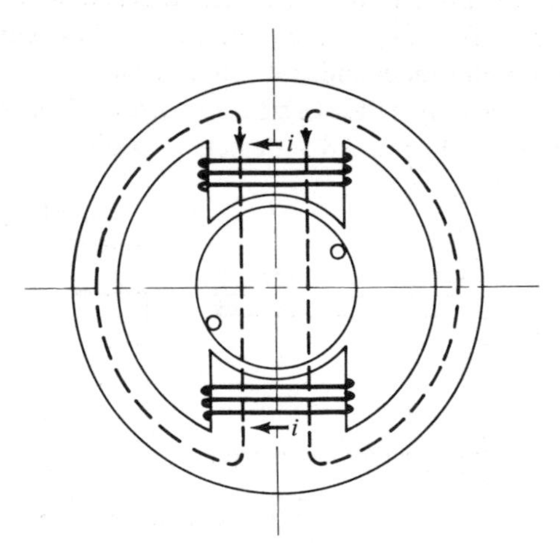

Fig. 18-2. "Inside-out" synchronous machine.

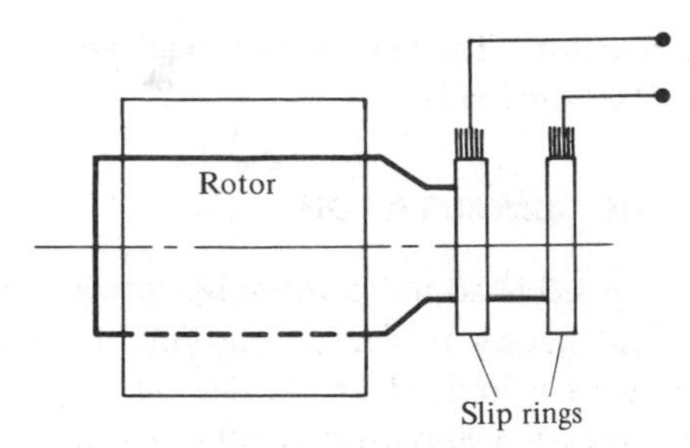

Fig. 18-3. Rotor, slip rings, and brushes.

brush is pressed against each of the two slip rings. To be able to show the slip rings and brushes clearly, another view of the rotor of the machine of Fig. 18-2 is needed; this is given in Fig. 18-3, in which the stator is omitted.

The left slip ring is connected to that conductor that happens to be, at the instant of the snapshot (Figs. 18-2 and 18-3), in the upper half of the picture, facing the north pole. Half a period later, this same conductor is facing the south pole, which is why the polarity of the slip ring and the terminal connected to it is alternating.

To obtain a unidirectional, not alternating, voltage, one terminal has to be in direct contact with whichever conductor happens to be facing the north pole at any time, and the other terminal with whichever conductor faces the south pole. This can be done by omitting the slip rings, locating the two brushes at the top and bottom, and pressing them against the bared conductor ends, as shown in Fig. 18-4. With such a scheme, the voltage between the two brushes always has the same direction, even though the voltage induced in each conductor remains alternating. To improve the "bumpy ride," the conductor ends can be widened to present a smooth cylindrical blank surface.

The next step is to extend this idea to a full *armature winding* distributed over the entire rotor surface. Figure 18-5 depicts a rotor carrying a number of

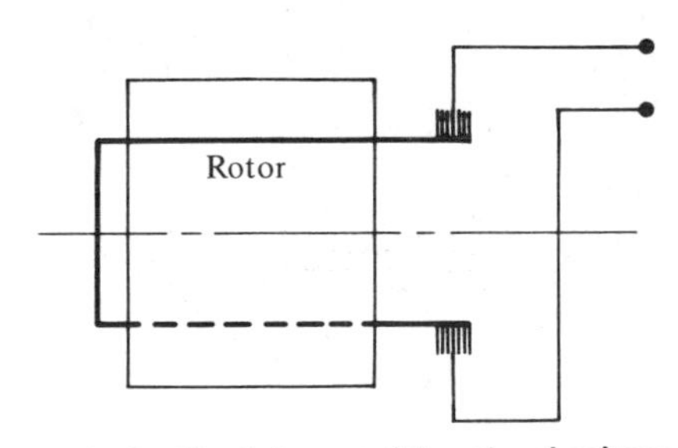

Fig. 18-4. Obtaining a unidirectional voltage.

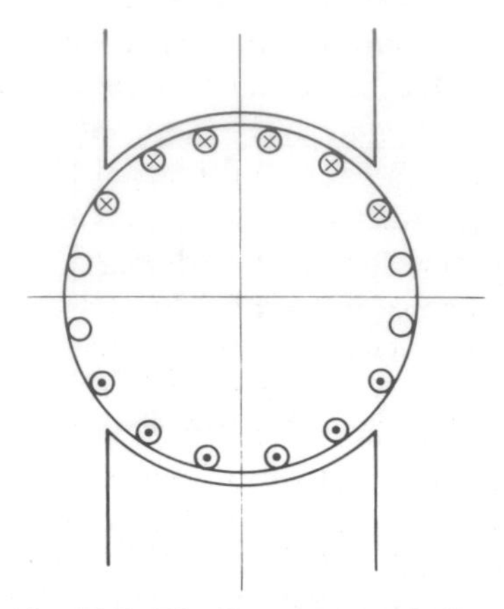

Fig. 18-5. Distributed rotor winding.

armature conductors. Each of the small circles may represent one or several such conductors, all located in slots as usual. When the rotor moves, all conductors facing one of the stator poles have induced voltages directed "into the paper," all those facing the other pole have induced voltages directed "out of the paper." The way to determine the voltage direction, from the polarity of the stator poles and the direction of rotation, was given in Section 7-1.

Let all armature conductors be connected to form a *continuous* winding, i.e., one without beginning and end, just closed in itself. (A discussion of how this is done is left for the next section.) Figure 18-6 describes such a winding as a chain of sources, using arrows to indicate the directions of the induced voltages. This diagram also indicates the location of the two brushes. They divide the

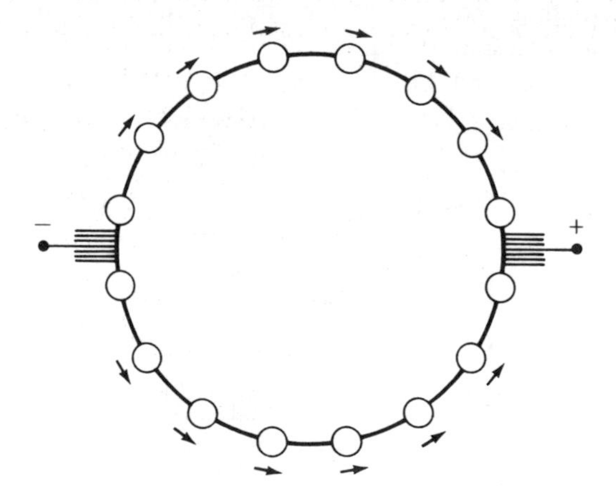

Fig. 18-6. Armature winding as a chain of sources.

winding into two halves that consitute two parallel paths through the armature. As the armature rotates, each source moves from one half into the other and back again. So, again, each individual source voltage is alternating, but the voltage between the brushes is unidirectional, practically constant, in accordance with the polarity markings at the terminals.

At any moment, there must be some armature conductors that do not face either one of the stator poles. They are said to be in the *neutral zones*, i.e., between the poles, and no voltage is induced in these conductors by their motion, as is suggested in Fig. 18-6 by the absence of arrows. These are the conductors with which the brushes are in contact.

Once again, the brushes need not jump from one bare conductor end to the next. This is avoided by connecting the end of each armature coil to a metallic piece with a smooth cylindrical surface, called a *commutator bar*. These pieces, together with thin insulating strips (usually made of mica) between them, constitute the *commutator*. In Fig. 18-7, the cylinder on the left side represents the armature, with only one conductor sketched. From the end of each armature coil, there is a lead to a commutator bar. The smaller cylinder to the right is the commutator. The rotor consists of both the armature and the commutator. Figure 18-7 also shows the stationary brushes. (They are drawn at the top and bottom, for the sake of clarity. Their actual location must be such that the bars on which they press are the ones that are connected to the conductors in the neutral zones.)

The commutator is the distinctive construction element that makes a d-c machine (except for a homopolar one) possible. Its name indicates that it changes (commutates) the alternating voltages induced in the armature conductors into direct voltages between the terminals. In effect, it serves as a mechanical rectifier. It does so by enabling the brushes to be in contact with points of the armature that are *fixed in space*, not rotating with the armature.

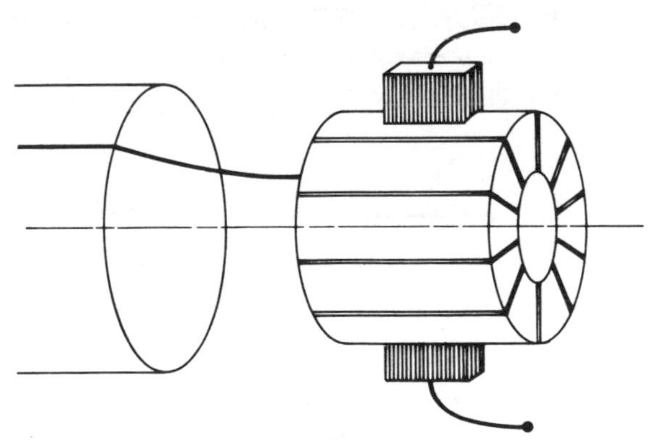

Fig. 18-7. Commutator.

The commutator may be considered the weak point of the d-c machine. It adds to its cost, and its maintenance requires special attention. There are problems peculiar to the process of commutation; a later section will be devoted to them and their solution. Such considerations restrict the area of application of d-c machines, but there are still cases in which their advantages outweigh all those drawbacks.

18-4 ARMATURE WINDINGS

Just as in a-c machines, each armature conductor must be connected to one that is approximately 180 electrical degrees distant, as shown in Fig. 18-8 for a two-pole machine. This ensures that, whatever the direction of the induced voltage in conductor a, the induced voltage in conductor b has the opposite direction, and the two conductors form one *turn* whose voltage is twice that of the individual conductor.

The two *slots* in which conductors a and b are located may (and usually do) contain several conductors. So several turns whose conductors are located in the same pair of slots can be connected in series to form a *coil.* For that case, the two lines marked a and b in Fig. 18-8 represent one *coil side* each.

Now the coils must be connected to each other to form a *d-c armature winding.* The problem is that such a winding, in contrast to those of a-c machines, must be *continuous*, without beginning or end. When the armature rotates by one slot pitch (i.e., the total width of one slot and one tooth), the armature must look exactly as it did before that motion, even though each individual conductor has changed its position.

Refer again to Fig. 18-8, and call, arbitrarily and for the purpose of this explanation only, coil side a the "forward" and b the "return" coil side. It follows that one-half of all coil sides must be "forward" coil sides, the other half "return" coil sides, and since both of these two "kinds" of coil sides appear all the way around the circumference, the two must alternate. For instance, if adjacent coil sides are numbered 1, 2, 3, etc., and if all coil sides

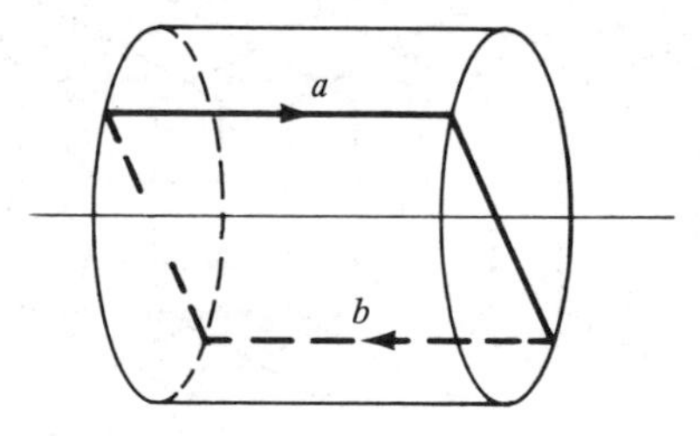

Fig. 18-8. Turn of an armature coil for a two-pole machine.

with odd numbers are considered "forward" coil sides, then the even-numbered ones are the "return" coil sides.

As an example, let there be 16 coil sides and let coil sides No. 1 and No. 8 form the first coil. In this case, the second coil consists of coil sides No. 3 (not No. 2, which is a "return" coil side) and No. 10. The whole sequence of coil sides is 1-8-3-10-5-12-7-14-9-16-11-2-13-4-15-6, and back to No. 1. Thus, the need for a continuous winding is fully satisfied.

These numbers could be entered into Fig. 18-6 (which happens to have 16 source symbols, one for every coil side), starting at any point. Another description of this winding, one that shows the true location of the coil sides with respect to each other, is given in Fig. 18-9. In this scheme, the armature is developed (as it was so frequently in Chapter 8), and each coil side is represented by a vertical line. The connections between the individual turns within a coil are not shown. Tracing the complete circuit in the diagram, beginning with No. 1 upward and No. 8 downward, one has to go back to No. 1 and repeat this cycle as many times as there are turns per coil. After the last turn, one follows the shown connection to the first conductor of coil side No. 3, etc.

The reader may have noticed that, in this example at least, the distance between the coil sides forming a coil is not exactly 180°. This is true if each coil side occupies its own slot. In the armature design that is generally preferred, however, each slot contains two coil sides, arranged in two *layers*. Figure 18-10 depicts such a slot, with four conductors per coil side (i.e., four turns per coil). The upper (outside) layer may contain the odd-numbered coil sides, the lower (inside) layer the even-numbered ones, or vice versa.

This two-layer arrangement also has the more important advantage that the connection *crossings*, which are seen in Fig. 18-9, can be taken care of in an orderly way. Figure 18-11, which is a redrawing of a part of Fig. 18-9, illustrates the idea. The full lines represent coil sides and connnections located in

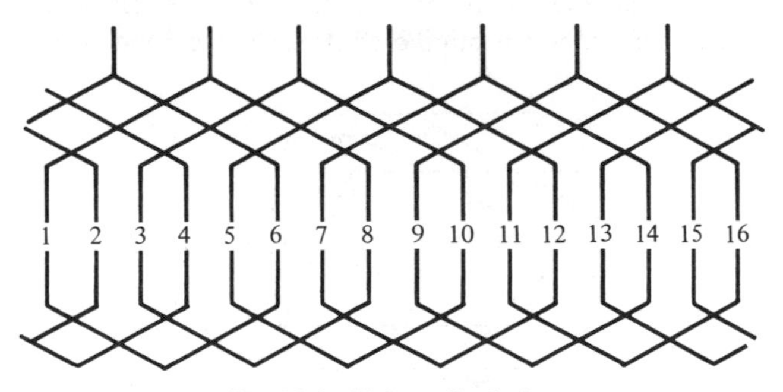

Fig. 18-9. Eight-coil winding.

Fig. 18-10. Slot with two coil sides.

the upper layer, i.e., their distance from the rotor axis is the radius of the outside ("forward," odd-numbered) coil sides. The broken lines represent coil sides and connections in the lower layer (that of the inside, "return," even-numbered coil sides). In every turn, there are two short radial pieces leading from one layer to the other. In other words, at the places appearing as dots in Fig. 18-11, the wires of each turn are twice bent at right angles.

To complete the armature winding, the connections to the commutator bars must be added. In the example of Fig. 18-9, they are shown as the short vertical lines on the top of the diagram. Since this winding consists of eight coils, there are eight commutator bars, each connected to the end of one coil.

For the armature winding of a d-c machine with *more than two poles*, each coil consists of coil sides located at a distance of 180 *electrical* degrees. To connect the coils to each other for such an armature, there are two basic possibilities, illustrated in Fig. 18-12. In a *lap* winding (the left diagram), all the coils "belonging" to one pair of poles follow each other before, in tracing the sequence of conductors through the armature, the other coils are reached. In a *wave* winding, on the other hand, the tracing proceeds in the same direction around

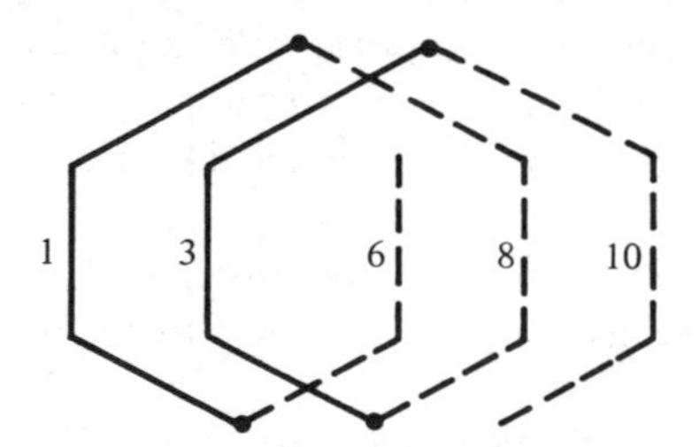

Fig. 18-11. Two layers.

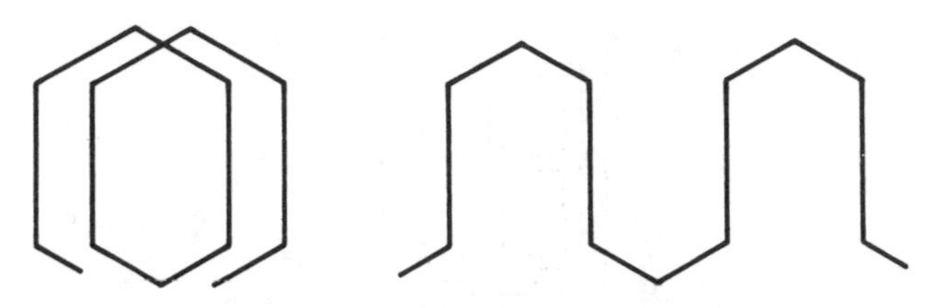

Fig. 18-12. Lap and wave winding.

the circumference, going through one coil for each of the pairs of poles, before returning to a coil near the first one.

The distinction between lap and wave windings has its significance. A closer investigation would reveal that, for a machine with p poles, a lap winding has p parallel paths through the armature from brush to brush, whereas a wave winding has two such parallel paths, regardless of the number of poles. In other words, if the total number of conductors in the armature winding is called C, a lap winding has C/p conductors in series in each path, a wave winding $C/2$ conductors. Thus, the lap winding is the choice for higher current magnitudes, the wave winding for higher voltages. For a machine with two poles, this distinction becomes irrelevant, as there are two parallel paths in either case, as a glance at Fig. 18-6 confirms. For a more detailed study of d-c armature windings, the interested reader is referred to the literature.*

18-5 THE RESULTANT MMF WAVE

The commutator not only performs the "miracle" of combining alternating coil voltages to produce a direct voltage between the terminals. It also produces, no less miraculously, a *stationary rotor mmf wave*, even though each rotor conductor rotates.

Return to the schematic diagram of Fig. 18-6, in which there are arrows to indicate the directions of the voltages induced in each coil side. When the machine operates as a *generator* (the rotor driven by a prime mover, the armature terminals connected to an electric load), then *currents* flow in the armature through the two circuit branches from brush to brush, and the direction of these currents is the same as that of the voltages, i.e., from the negative to the positive terminal and back through the outside load. When the machine operates as a *motor* (driving a mechanical load, the armature terminals connected to an electric source), then the current direction is opposite to that of the voltage arrows, from the positive to the negative terminal. In either case, the currents in all con-

*For instance: Kloeffler, Kerchner, and Brenneman, *Direct-current Machinery*, The Macmillan Co., New York, 1950.

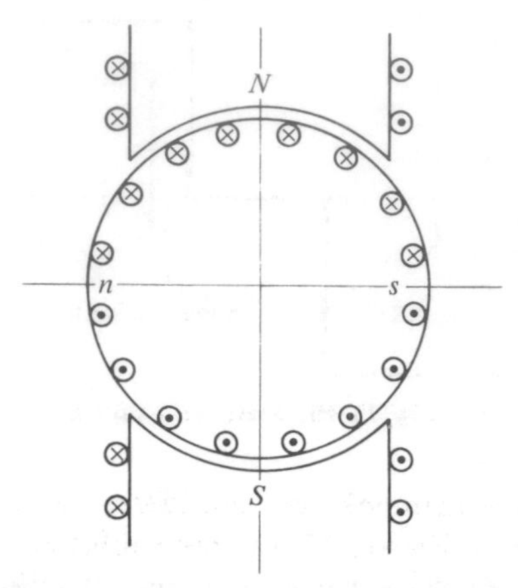

Fig. 18-13. Armature and field current directions.

ductors facing a north pole have one direction (e.g., "into the paper,") and in all conductors facing a south pole, they have the other direction ("out of the paper"). This is shown, for a two-pole machine, in Fig. 18-13, which also indicates (arbitrarily chosen) directions for the stator currents. The letters N and S mark the stator north and south poles, respectively, and the letters n and s the rotor north and south poles.

The remarkable result of this whole chain of thought is that the rotor poles remain stationary no matter how fast the rotor moves. It should also be noted that the rotor poles are always in the neutral zone. Thus, the axis of the rotor mmf wave is always 90 electrical degrees away from that of the stator (field) mmf wave. Using the terminology of synchronous machines, the rotor mmf wave is always in the quadrature axis.

In Section 8-2, it was shown how the mmf wave of a distributed winding is obtained. Accordingly, Fig. 18-14 contains the rotor winding of Fig. 18-13 redrawn in developed form, and its mmf wave, in the form of a sequence of steps and also by its characteristic triangular approximation. (There is no need to introduce sinusoidal waveshapes in the study of d-c machines, thus, no space

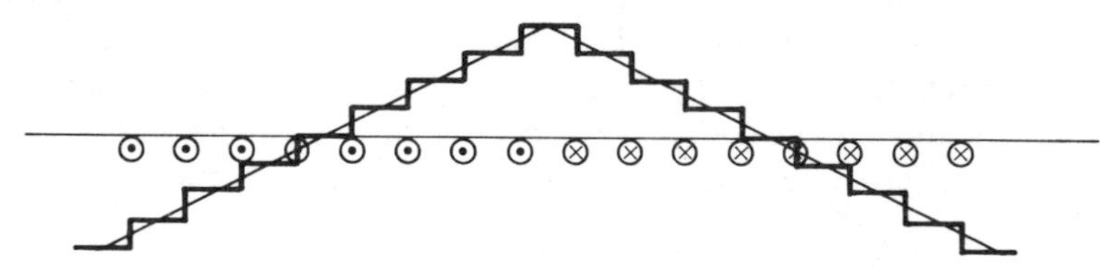

Fig. 18-14. Rotor mmf wave.

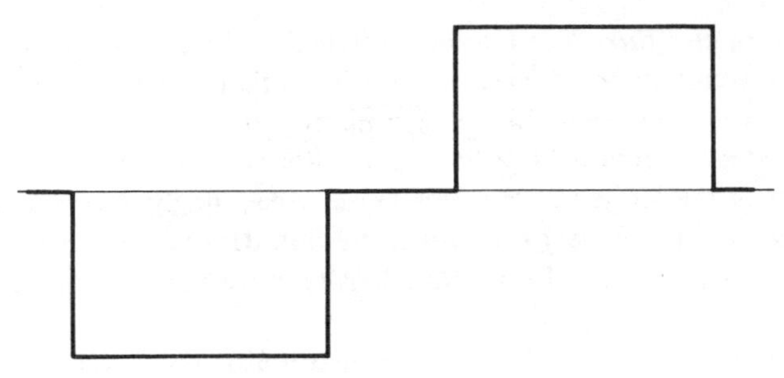

Fig. 18-15. Stator mmf wave.

fundamentals and space harmonics are considered here.) For this diagram, the left-side neutral zone of Fig. 18-13 has been arbitrarily chosen as the origin, and counterclockwise as the positive direction. Positive values are assigned to an mmf wave where it would produce flux lines pointing toward the rotor (as in Section 8-2).

By contrast, the *stator* carries a concentrated winding, since it has salient poles. Thus, the stator mmf wave is rectangular, with poles in the direct axis, i.e., at the center of the pole faces. The sketch of this wave (Fig. 18-15) is based on the same directional choices as that of the rotor mmf wave (Fig. 18-14) and on the current directions of Fig. 18-13. The diagram neglects fringing at the pole edges.

The next picture, Fig. 18-16, is the *resultant mmf wave.* It is obtained by adding the ordinates of the two previous figures, using the triangular approximation for the rotor mmf wave. It is this wave that is responsible for the *flux*, just as in synchronous and induction machines.

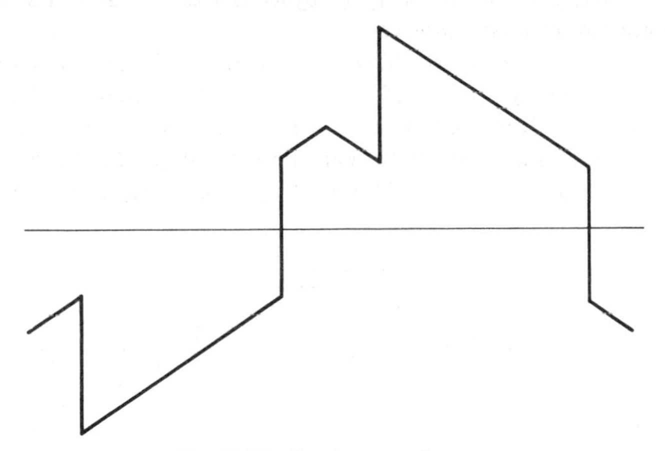

Fig. 18-16. Resultant mmf wave.

The shape of the *flux distribution*, sketched in Fig. 18-17, differs from the shape of the resultant mmf wave in two important respects. One of the differences is due to the fact that, in the neutral zones, i.e., between the stator poles, the distance between the stator and rotor cores is a high multiple of the radial length of the air gap at the pole faces. Accordingly, the reluctance in the neutral zone is much larger, and, thus, the flux density much smaller, than at the pole faces. In Fig. 18-17, the flux density in the neutral zone is assumed to be negligible.

At the pole faces, the shape of the flux distribution tends to be similar to that of the mmf wave because a major portion of the reluctance of the magnetic circuit is that of the air gap whose length is constant. Only when the iron portion of the magnetic circuit is *saturated*, does the shape of the flux distribution at the pole faces differ noticeably from that of the resultant mmf wave. In Fig. 18-17, the full line shows the curvature due to saturation; the broken line is what the flux distribution would be without saturation.

The reader is reminded of the concept of the *flux per pole* introduced in Section 8-4. The flux per pole is proportional to the area under the flux distribution curve, for the part of the circumference corresponding to one pole. It is characteristic for the d-c machine that this flux is the same as it would be if it were produced by the stator mmf wave alone, except for the effect of saturation. This is seen by comparing the broken line in the flux-distribution diagram of Fig. 18-17 with what it would be in the absence of rotor currents. This comparison is shown in Fig. 18-18. Clearly, the integral for each half of the broken line is equal to that represented by the shaded area.

The term *armature reaction* was previously (in Section 9-6, in the context of synchronous machines) introduced for the effect of the armature current upon the flux distribution. In a d-c machine, armature reaction causes a distortion of the flux distribution, leading to a higher state of saturation near one edge of

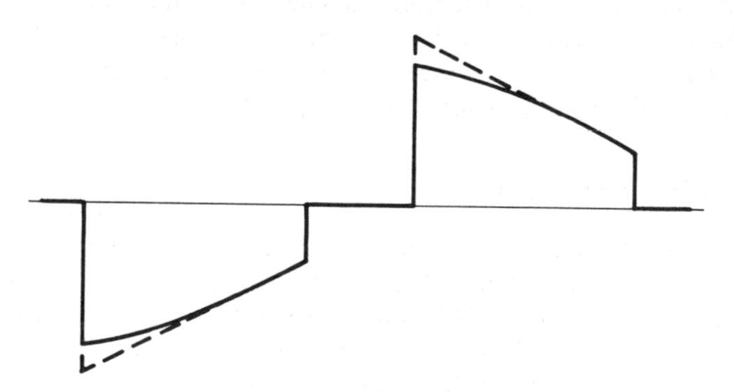

Fig. 18-17. Flux distribution (with and without saturation).

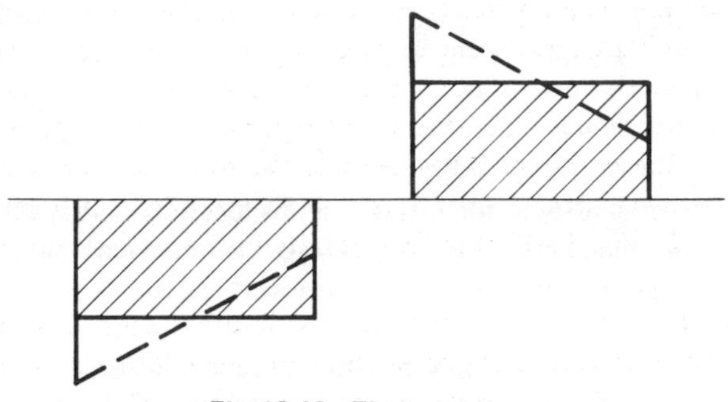

Fig. 18-18. Flux per pole.

each pole and, thereby, to a reduction of the flux per pole. One speaks of a *demagnetizing effect* of armature reaction.

For an elementary study of the operation of d-c machines, it is best to disregard armature reaction completely. When it has to be taken into consideration, in the interest of greater accuracy, then the extent of the flux reduction has to be determined, usually on an empirical basis. This reduction depends, for a given field excitation, on the value of the armature current. A good approximation consists of subtracting from the stator mmf (i.e., from the height of the rectangular wave in Fig. 18-15) an amount proportional to the armature current.

It is possible to counteract the demagnetizing effect of armature reaction by adding another winding to the stator, consisting of conductors embedded in the pole faces. This winding, called a *compensating winding*, has an mmf wave with its axis in the neutral zone, just like the armature winding. It is connected in series with the armature winding, so that these two windings produce equal and opposite mmf waves over the pole faces. This winding adds slightly to the resistance of the armature circuit, but the self-inductance of the whole armature circuit is reduced. Since compensating windings add to the cost, small d-c machines seldom have them. Large d-c machines usually are equipped with compensating windings. They are especially needed for machines subjected to overloads or rapidly changing loads. See Example 18-6.

18-6 VOLTAGE AND TORQUE

The importance of determining the flux per pole lies in the fact that both the *voltage* induced in the armature and the electromagnetic *torque* exerted on the rotor are proportional to it. This was shown in Chapter 8, but the specific equations derived in that chapter were based on the space fundamental of the flux distribution and, thus, on sinusoidal wave shapes, which are not relevant to the study of d-c machines.

Let the calculation of the induced voltage be started by considering an individual armature conductor. The voltage induced in that conductor is a function of time

$$e_{\text{cond}} = Blu = Blr\omega \tag{18-1}$$

where B is the flux density at the instantaneous location of that conductor. The symbol for the mechanical angular velocity, ω, is not given a subscript because, in the study of the operation of d-c machines, there is no need to consider any radian frequency.

The entire voltage induced between the terminals is the sum of all induced voltages in any path through the armature from brush to brush. Thus, it is obtained by adding the voltages induced in all conductors connected in series.

$$e = \sum e_{\text{cond}} = \sum (Blr\omega) = lr\omega \sum B \tag{18-2}$$

Calling the number of conductors in any path C_s (the subscript standing for the series connection), the *average* flux density B_{ave} may be substituted into this equation by

$$\sum B = C_s B_{\text{ave}} \tag{18-3}$$

and this in turn may be expressed in terms of the *flux* per pole ϕ, namely as flux divided by area

$$B_{\text{ave}} = \frac{\phi p}{2\pi r l} \tag{18-4}$$

leading to the voltage equation

$$e = \frac{C_s p}{2\pi} \phi\omega \tag{18-5}$$

As the reader knows well from previous sections, this voltage is not alternating (in contrast to that of Eq. 18-1). Finally, all the constant factors in Eq. 18-5 may be lumped together by defining the *voltage constant*

$$k = \frac{C_s p}{2\pi} \tag{18-6}$$

which leaves the voltage equation in its basic form

$$e = k\phi\omega \tag{18-7}$$

Like any other motional voltage, it is proportional to flux and speed.

The electromagnetic *torque* may be calculated in a similar way, beginning with the torque exerted on a single armature conductor, expressed as force times radius

$$T_{\text{cond}} = Blir \tag{18-8}$$

and adding all such torques to obtain the total torque exerted on the rotor

$$T = \sum (Blir) = lri \sum B \tag{18-9}$$

In this case, the summation involves *all* armature conductors. For instance, if all conductors facing stator north poles carry currents directed "out of the paper," then all those facing south poles have currents "into the paper" (or vice versa), so that the torques for all armature conductors have the same direction. Thus, using the symbol C for the total number of armature conductors,

$$C = aC_s \tag{18-10}$$

where a, the number of parallel paths through the armature, equals either 2 or the number of poles p, depending on whether the armature has a wave or a lap winding (see the last paragraph of Section 18-4).

Thus, for the torque calculation,

$$\sum B = aC_sB_{\text{ave}} \tag{18-11}$$

Substituting this and Eq. 18-4 into Eq. 18-9 leads to

$$T = \frac{aC_sp}{2\pi}\,\phi i \tag{18-12}$$

One further substitution is sensible: the current i in Eqs. 18-8, 18-9, and 18-12 is the current in any of the parallel paths through the armature; for the analysis of how the entire machine operates, however, the quantity of interest is the total current at the armature terminals, always referred to as the *armature current*

$$i_a = ai \tag{18-13}$$

This simplifies the torque equation to

$$T = \frac{C_sp}{2\pi}\,\phi i_a = k\phi i_a \tag{18-14}$$

expressing the torque as the result of the interaction of armature current and flux.

It is worth noting that the armature current in the torque equation, Eq. 18-14, as well as the induced voltage in Eq. 18-7, is "spelled" with a lower-case symbol. This is meant to indicate that these equations are not limited to the steady state (in which case capital letters would be appropriate). They are perfectly valid under many *transient* conditions, e.g., during changes of load and speed, except only that changes of *flux* would induce additional voltages (transformer voltages) not included in the speed voltage equation, Eq. 18-7.

Another comment on Eq. 18-14 is that the *torque constant* k is identical with the voltage constant introduced in Eq. 18-6, provided consistent units are used (newton-meters for the torque, radians per second for the angular velocity, webers for the flux, etc.). This fact makes it simple to relate the two basic equations derived in this section (Eqs. 18-7 and 18-14) to each other, for instance by forming the mechanical *power*

$$T\omega = k\phi i_a \frac{e}{k\phi} = ei_a \tag{18-15}$$

This expresses the *conversion of power* accomplished in the d-c machine. For *generator* operation, the mechanical input power must be somewhat more than $T\omega$, to provide for rotational losses, and the electric output power must be somewhat less than ei_a, because of copper losses. In *motor* operation, the input power is more than ei_a, and the output power less than $T\omega$.

Finally, let it be repeated, in the context of Eq. 18-15, that the symbols e and i_a are used to indicate validity under transient conditions. Again, the only exception to this statement would be the case (not too important a case, from a practical point of view) of a flux changing at a sufficiently high rate to induce a noticeable transformer voltage in the armature. It is clear enough that, in such a case, the change of magnetic energy storage would have to be entered into any equation expressing a balance sheet of power conversion.

18-7 FIELD WINDING CONNECTIONS

In a d-c machine, the currents at the terminals of both the armature winding and the field winding are of the same kind, namely, d-c. This is in contrast to the synchronous machine in which the field current is unidirectional but the armature current is alternating. Thus, there is no difficulty, in a d-c machine, in interconnecting the two windings. The various ways in which this can be done give rise to a remarkable variety of operating characteristics.

For the following discussion, it must be kept in mind that the field winding constitutes only an auxiliary circuit, one in which no electromechanical energy conversion occurs. In the steady state, all the power needed by the field winding is lost in the form of heat, due to the resistance of that winding. Consequently, it is necessary for the designer to see to it that this power loss is held down to a small fraction of the power that the machine is capable of converting between electrical and mechanical forms. This accounts for the significant difference between field windings to be connected in series and in parallel to the armature.

A *parallel* connection of armature and field windings is generally referred to as a *shunt* connection, and a machine with this connection is called a *shunt motor* or a *shunt generator*, as the case may be. Since the voltage across the two circuit branches is the same, the field current must have a much smaller steady-state value than the rated armature current. Using the conventional capital letters for steady-state values, $I_f << I_{a\ \mathrm{rated}}$. Or, with rated values as bases for per-unit notation, $I_f << 1$ pu. Consequently, the resistance of the field circuit is $R_f >> 1$ pu. Typically, the field winding of a shunt machine consists of many thousands of turns of a wire whose cross-sectional area is much smaller than that of the armature winding.

By contrast, in a *series motor* (there is no need to consider series generators, as will be seen later), the same current i_a flows through the armature and field windings. The steady-state voltage across the field winding must be held to a value much below 1 pu, which leads to the statement $R_f << 1$ pu, the direct opposite of the one for shunt machines. The typical field winding of a series machine consists of just a few turns per pole, made of a wire with a cross-section sufficient for the largest current expected to be carried.

Finally, it is possible and sometimes useful to combine series and parallel connections. A machine with both types of field windings is referred to as a *compound motor* or a *compound generator.*

18-8 PROBLEMS OF COMMUTATION

The operation of the commutator was explained in principle, earlier in this chapter. At a closer look, there are some peculiar problems associated with it.

To begin with, the *sliding contact* between brush and commutator surfaces can never be as good as a permanent motionless connection between two conducting parts. No surface can be ideally flat or ideally cylindrical, no matter how meticulously it may have been smoothed, buffed, or polished. To be at all satisfactory, the contact between brush and commutator surfaces requires that the brush be pressed against the commutator by springs. But too much spring pressure is not good either, since it results in too much friction, which, in turn, means both increased loss of power and faster wearing out of the surfaces, making it necessary to shut down the machine frequently for maintenance.

At any rate, due to the imperfection of the surfaces, most of the current has to cross a microscopic air gap between commutator and brush. The phenomenon is similar to an electric *arc* in which an ionized gas conducts a current from one solid conducting surface to another. The voltage-current characteristic of such an arc is nonlinear, in contrast to that of a metallic conductor. For the case of the commutator-brush contact, a suitable approximation is made by assuming that there is a constant voltage of about 1 v across the arc, independent of the current. Thus, there is a total voltage of 2 v, as the contribution of the *contact resistance* to the voltage across the armature circuit resistance.

The situation described thus far in this section is essentially no different from that of the slip rings used in most synchronous machines and some induction machines. But the case of the commutator is much more critical because the current tends to be distributed over its contact surface in a nonuniform way. To understand this problem, it is necessary to investigate what is going on in an armature coil "undergoing commutation."

Figure 18-19 is a schematic representation of several armature coils (numbered 1, 2, and 3) and the commutator bars (named *a* and *b*) connected to them, at three different instants. Armature and commutator are assumed to be moving from right to left, and the brush shown in the diagrams is the one through which the current flows *out of* the machine. Remember the brushes are stationary.

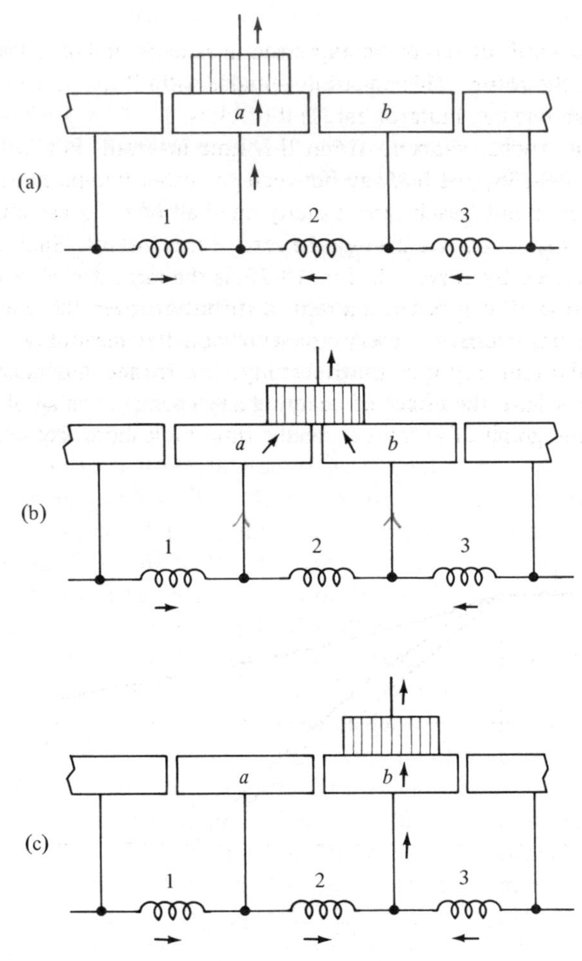

Fig. 18-19. Commutation.

Observe specifically what happens to the current in coil No. 2. At the instant of Fig. 18-19a, this current must flow from right to left; at the instant of Fig. 18-19c, from left to right. In between these two, the current direction is being reversed. That is the process of *commutation*, of nearly sudden reversal of current direction.

For coil No. 2, this process begins at the instant when the brush first touches the left edge of commutator bar b, and it ends when the brush loses contact with bar a. Disregarding the width of the thin insulating strip between bars, the time interval available for commutation is

$$\Delta t = \frac{w}{\omega} \tag{18-16}$$

where w is the width of the brush expressed in radians, and ω is again the angular velocity of the rotor. This equation remains valid if the brush is widened to cover more than one commutator bar, as it often is.

The question is what happens within this time interval. For instance, at the instant of Fig. 18-19b, just halfway between the other two pictures, do the two commutator bars a and b each carry exactly one-half of the brush current? If so, then the current i_2 in coil No. 2 would be zero at this instant. Such a *linear commutation*, described by curve 1 in Fig. 18-20, is the most desirable kind of commutation because it makes the current distribution over the contact surface uniform. But unfortunately, every armature coil has inevitably some leakage inductance. Without any quantitative study, the reader understands that this inductance must have the effect of delaying any change of value of the coil current, making the graph of current i_2 versus time look more like curve 2 in Fig. 18-20.

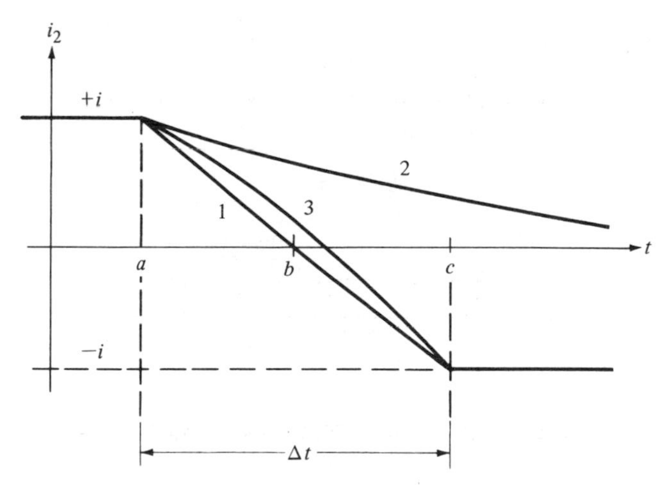

Fig. 18-20. Current in coil No. 2 of Fig. 18-19.

During the interval, Δt, the current in coil 2 is decaying toward zero. At the end of the interval, Fig. 18-19c, the current would still be flowing from right to left. The only place currents flowing into commutator bar *a* can go is to the brush through an arc. This is called *sparking*.

It has been found necessary to take radical steps against sparking. Since the fault lies with voltage induced in the coils by their leakage inductances, the only fully effective remedy is to induce a voltage that will control the rate of change of current. Since these coils are located in the neutral zone, there is room for additional salient poles called *interpoles* to serve this purpose. Figure 18-21 is a sketch of a machine with two regular poles (similar to Fig. 18-2) with the interpoles added.

The windings on the interpoles must be connected in series with the armature winding, so that their mmf is always proportional to the armature current, just like the leakage flux whose effect it is to counteract. If the interpole windings are given the right number of turns, the voltage induced in the coils undergoing commutation is just sufficient to make $di_2/dt = \pm 2i/\Delta t$, and the desired linear commutation is obtained. In actual machines, this ideal reversal of current may not be achieved. Something similar to curve 3 in Fig. 18-20 is satisfactory in eliminating sparking at the brushes. It is taken for granted that the interpole windings are correctly connected, so that the induced voltage has the right polarity; but once this is done, the connection between armature and interpoles does not have to be changed, regardless of any reversals of rotation or armature current.

Interpole windings inevitably add a little to the armature circuit resistance and, thereby, to the copper losses. Nevertheless, they are used in nearly all commutator machines as the standard remedy against the troubles of commutation.

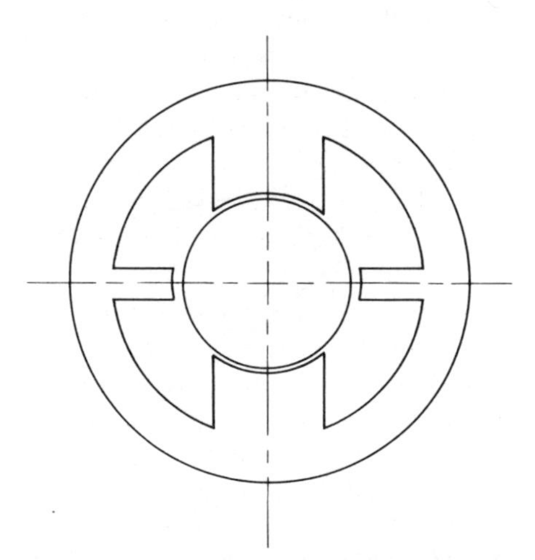

Fig. 18-21. Two-pole machine with interpoles.

18-9 EXAMPLES

Example 18-1 (Section 18-6)

A four-pole, d-c machine has a double-layer, lap winding in 29 slots. There are three coils per slot, two turns per coil, and 87 commutator segments. The flux per pole is 0.075 weber. The rotor speed is 1200 rpm. Find the d-c voltage that is generated in this machine.

Solution

The total number of conductors in the armature winding is

$$C = (29 \text{ slot})\,(3 \text{ coils/slot})\,(2 \text{ turns/coil})\,(2 \text{ conductors/turn})$$
$$= 348 \text{ conductors}$$

For a lap winding, the number of parallel paths in the armature winding is the same as the number of poles

$$a = p = 4$$

The number of conductors in series in one path is

$$C_s = C/a = 348/4 = 87 \text{ conductors}$$

Use Eq. 18-6 to find the voltage constant

$$k = C_s\, p/2\pi = (87)\,(4)/(2\pi) = 55.4$$

Thc angular velocity is

$$\omega = (n/60)\, 2\pi = (1200/60)\, 2\pi = 125.7 \text{ rad/sec}$$

The flux per pole is given

$$\phi = 0.075 \text{ weber}$$

Use Eq. 18-7 to find the d-c voltage

$$e = k\phi\omega = (55.4)\,(0.075)\,(125.7) = 522 \text{ v}$$

Example 18-2 (Section 18-6)

For the machine in Example 18-1, find the developed torque for an armature current of 24 amp.

Solution

The torque constant is the same as the voltage constant

$$k = C_s\, p/2\pi = 55.4 \text{ nm/weber amp}$$

Use Eq. 18-14 to find the torque

$$T = k\phi i_a = (55.4)\,(0.075)\,(24) = 99.7 \text{ nm}$$

Example 18-3 (Section 18-6)

Use the results of Examples 18-1 and 18-2 to demonstrate the validity of Eq. 18-15.

Solution

The mechanical power is

$$T\omega = (99.7)\,(125.7) = 12{,}530 \text{ w}$$

The electrical power is

$$ei_a = (522)\,(24) = 12{,}530 \text{ w}$$

Example 18-4 (Section 18-8)

For the machine in Example 18-1, the diameter of the commutator is 0.17 m. The width of a brush is 1.4 cm. (a) Find the duration of the commutation interval, if the speed is 1200 rpm. (b) Find the rate of change of current in a coil, during commutation, if the armature current is 24 amp.

Solution

(a) The width of the brush expressed in radians is

$$w = 0.014/(0.17/2) = 0.165 \text{ rad}$$

The angular velocity is

$$\omega = 125.7 \text{ rad/sec}$$

Use Eq. 18-16 to find the duration of the commutation interval

$$\Delta t = w/\omega = 0.165/125.7 = 0.0013 \text{ sec}$$

(b) The current through one path (in each conductor) is

$$i = i_a/a = 24/4 = 6 \text{ amp}$$

The change in current during the commutation interval is

$$\Delta i = 2i = 2 \times 6 = 12 \text{ amp}$$

The rate of change of current is

$$di/dt \approx \Delta i/\Delta t = 12/0.0013 = 9230 \text{ amp/sec}$$

Example 18-5 (Section 18-8)

For the machine in Example 18-1, use four interpoles. Find the number of turns for each interpole winding.

Solution

Refer to Fig. 18-14. Let $\mathfrak{F}_a$ denote the peak of the triangular mmf wave. The value of $\mathfrak{F}_a$ is equal to the sum of the currents in $\frac{1}{2}$ of the conductors in one pole span

$$\mathfrak{F}_a = \tfrac{1}{2}(C/p)\, i = \tfrac{1}{2}(348/4)\, 6 = 261 \text{ At}$$

It has been determined empirically that the net mmf at the interpole position must be 0.25 $\mathfrak{F}_a$ and have opposite polarity to the armature mmf. Let N_{comm} denote the number of turns in the coil around the interpole. The current through this coil is the armature current

$$\text{Net mmf} = N_{\text{comm}}\, i_a - \mathfrak{F}_a = 0.25\, \mathfrak{F}_a$$

$$N_{\text{comm}}\, i_a = 1.25\, \mathfrak{F}_a$$

$$N_{\text{comm}} = 1.25\, \mathfrak{F}_a / i_a = 1.25(261)/24 = 13.6 \text{ turns}$$

The number of turns in this coil must be an integer, so we select

$$N_{\text{comm}} = 14 \text{ turns}$$

Example 18-6 (Section 18-5)

For the machine in Example 18-1, a pole face covers 0.71 of a pole span. Find the number of conductors to be in one pole face for the compensating winding.

Solution

The purpose of the pole face winding is to compensate for the effect of the armature mmf in the region covered by a pole. See Fig. E-18-6, which shows the conductors in the pole face as well as in the armature. The mmfs of the two windings are also shown. Let y denote the ratio of a pole face to a pole span. It is given that $y = 0.71$. The number of armature conductors in one pole span is

$$C/p = 348/4 = 87 \text{ conductors}$$

The current in one conductor is

$$i = i_a/a$$

The total amperes in the armature in one pole span is

$$(C/p)\,(i_a/a) = 2\mathfrak{F}_a$$

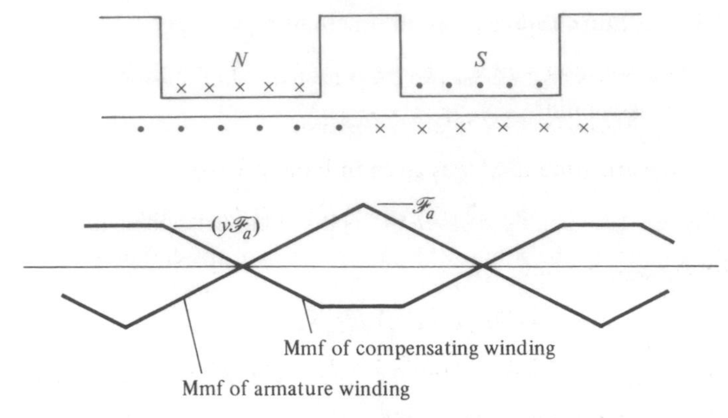

Fig. E-18-6.

Let N_{comp} denote the number of conductors in one pole face for the compensating winding. The compensating winding is connected in series with the armature circuit, so the current through each pole face conductor is the armature current. The total amperes in one pole face is

$$N_{comp} i_a = y(2\mathcal{F}_a) = y(C/p)\,(i_a/a)$$

Solving for N_{comp}, we obtain

$$N_{comp} = yC/pa = (0.71 \times 348)/(4 \times 4) = 15.4$$

Next, we select the number of slots to be in the pole face. It is desirable for the slots in the pole face to be spaced differently from the slots in the rotor. This is to avoid reluctance torques that occur if the teeth of the rotor and stator are precisely aligned. The rotor has 29 slots. The number of slots opposite one pole face is (29/4) (0.71) = 5.1 slots. We can select four slots in each pole face. This makes N_{comp} = 16 conductors.

Example 18-7 (Section 18-8)

For the compensated machine in Example 18-6, use four interpoles. Find the number of turns for each interpole.

Solution

The net mmf is required to be 0.25 $\mathcal{F}_a$. (Refer to Example 18-5.) For this machine, the compensating winding supplies a portion of the required mmf. (See

Fig. E-18-6.) At the interpole position, the mmf equation is

$$\text{Net mmf} = N_{\text{comm}}\, i_a + \tfrac{1}{2} N_{\text{comp}}\, i_a - \mathfrak{F}_a = 0.25\, \mathfrak{F}_a$$

$$N_{\text{comm}}\, i_a = 1.25\, \mathfrak{F}_a - \tfrac{1}{2} N_{\text{comp}}\, i_a$$

The peak of the armature mmf was given in Example 18-5.

$$\mathfrak{F}_a = \tfrac{1}{2}(C/p)\, i = \tfrac{1}{2}(C/p)\, (i_a/a)$$

Solving for N_{comm}, we find

$$N_{\text{comm}} = 1.25(\tfrac{1}{2})\, (C/pa) - \tfrac{1}{2} N_{\text{comp}}$$

$$= 1.25(\tfrac{1}{2})\, (348/4 \times 4) - \tfrac{1}{2}(16) = 5.6 \text{ turns}$$

We select $N_{\text{comm}} = 6$ turns.

Example 18-8 (Section 18-6)

A d-c machine has six slots in the rotor in one pole span (180°). The pole face covers 130°. See Fig. E-18-8a. One path through the armature consists of six coils of one turn each (12 conductors). The flux density under a pole face is B_0. This flux distribution is shown as a dotted line in the figure. The rotor position is indicated by an angle α, which denotes the location of slot 1 in relation to a pole tip. Show that Eq. 18-3 is only an approximation, by (a) finding B_{ave} for this machine, (b) finding ΣB as a function of angle α, (c) finding the average of this function and comparing it with $C_s B_{\text{ave}}$.

Solution

(a) We find B_{ave} over the region of one pole span

$$B_{\text{ave}} = (130°/180°)\, B_0 = (13/18) B_0$$

(b) The solid vertical lines indicate the value of flux density at the location of each conductor. This is duplicated with reverse sign under the next pole where conductors 8 through 12 are located. In Fig. E-18-8a we find $B_1 = 0$ and $B_2 = B_3 = B_4 = B_5 = B_6 = B_0$. Sum up the corresponding B for all 12 conductors to find

$$\sum B = B_2 - B_8 + B_3 - B_9 + B_4 - B_{10} + B_5 - B_{11} + B_6 - B_{12} = 10\, B_0$$

This answer holds for $20° < \alpha < 30°$. With α slightly less than $20°$, $B_2 = 0$. With α slightly greater than $30°$, $B_6 = 0$.

Figure E-18-8b shows the condition when only four conductors lie under the pole face. Summing the corresponding B for all 12 conductors now

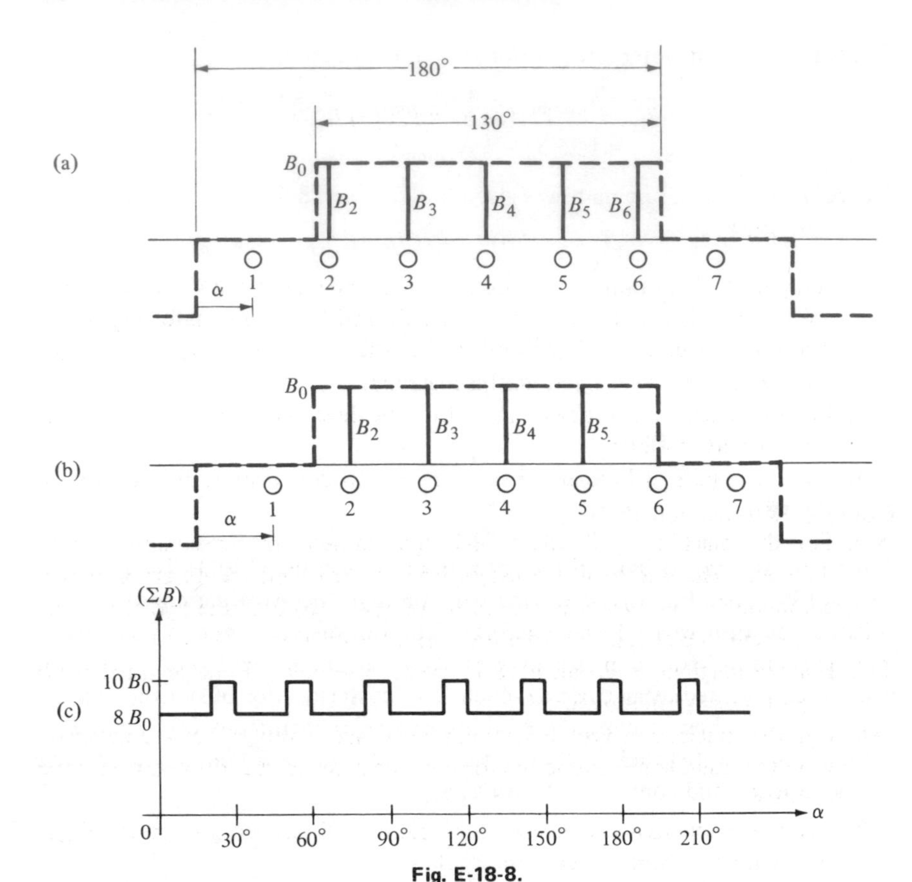

Fig. E-18-8.

shows

$$\sum B = B_2 - B_8 + B_3 - B_9 + B_4 - B_{10} + B_5 - B_{11} = 8\,B_0$$

This answer holds for $30° < \alpha < 50°$. With α slightly greater than $50°$, $B_1 = B_0$, and we go back to five conductors under the pole face. For increasing values of α this function is repetitive. An idealized sketch of this function is shown in Fig. E-18-8c.

If the rotor moves with constant velocity ($\alpha = \omega t$), then Fig. E-18-8c could also represent the voltage function. See Eq. 18-2. The voltage function has a d-c component and a ripple.

(c) The function for ΣB is periodic with $30°$ for its period. Accordingly, we can find the average of this function.

$$\left(\sum B\right)_{\text{ave}} = [(10\,B_0)(10°) + (8\,B_0)(20°)]/(30°) = (26/3)\,B_0$$

We compare this with $C_s B_{ave}$

$$C_s B_{ave} = 12(13/18)\,B_0 = (26/3)\,B_0 = \left(\sum B\right)_{ave}$$

18-10 PROBLEMS

18-1. A six-pole, d-c machine has a double layer, lap winding in 43 slots. There are three coils per slot, one turn per coil, and 129 commutator segments. The flux per pole is 0.082 weber. The rotor speed is 700 rpm. Find the d-c voltage that is generated in this machine.

18-2. For the machine in Problem 18-1, find the developed torque for an armature current of 48 amp.

18-3. Use the results of Problems 18-1 and 18-2 to demonstrate the validity of Eq. 18-15.

18-4. For the machine in Problem 18-1, the diameter of the commutator is 0.26 m. The width of a brush is 1.5 cm. (a) Find the duration of the commutation interval, if the speed is 700 rpm. (b) Find the rate of change of current in a coil, if the armature current is 48 amp.

18-5. For the machine in Problem 18-1, use six interpoles. Find the number of turns for each interpole winding.

18-6. For the machine in Problem 18-1, a pole face covers 0.7 of a pole span. Find the number of conductors to be in one pole face for the compensating winding.

18-7. For the compensated machine in Problem 18-6, use six interpoles. Find the number of turns for each interpole.

18-8. A d-c machine has four slots in the rotor in one pole span (180°). One path through the armature consists of four coils of one turn each (eight conductors). The flux density distribution is approximated by a sinusoid, $B(\theta) = B_{max} \sin \theta$. The rotor position is indicated by an angle α, which denotes the location of slot 1 in relation to the point where $\theta = 0$. Show that Eq. 18-3 is only an approximation by (a) finding B_{ave}, (b) finding ΣB as a function of angle α, (c) finding the average of this function and comparing it with $C_s B_{ave}$.

19
D-c Motors

19-1 SCHEMATIC DIAGRAMS

The operating characteristics of a d-c motor depend greatly on the type of interconnection between the armature and field windings. To study these characteristics, it is, therefore, very helpful to use so-called schematic diagrams that illustrate these connections in symbolic form. The conventional symbols for such diagrams will now be introduced.

The *armature* is described by the symbol of Fig. 19-1a. The circle is used because it is the function of the armature to act as a voltage *source*; the "ears" at both sides of the circle are to be indicative of the *brushes* at the armature terminals. By contrast, Fig. 19-1b describes the armature in terms of *circuit elements*: source, resistance, and inductance. Both descriptions omit the interpole windings, which are to be understood to be part of the armature circuit and whose resistances and inductances are included in the elements R_a and L_a.

As to the *field* winding, Fig. 19-2a gives its symbol, and Fig. 19-2b shows its description for circuit analysis. Basically, schematic diagrams show the circuit branches and their currents. They contain all the information needed for the writing of current (KCL) equations but, in general, not for voltage (KVL) equations.

As an example, Fig. 19-3 shows schematic diagrams of *compound motors*. There are two possibilities for connecting the two field windings, the shunt field winding and the series field winding. The terms *long shunt* and *short shunt* are meant to indicate whether the shunt field winding is connected across the series connection of armature and series field winding, or across the armature alone.

The reader will see later on that compound motors are essentially shunt motors with series field windings added. In other words, the main field winding is the shunt field winding; the series field winding plays only a minor, though some-

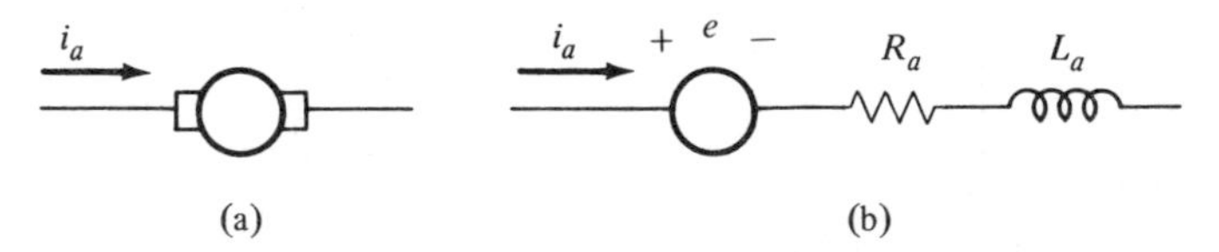

Fig. 19-1. Armature symbols: (a) schematic; (b) for circuit analysis.

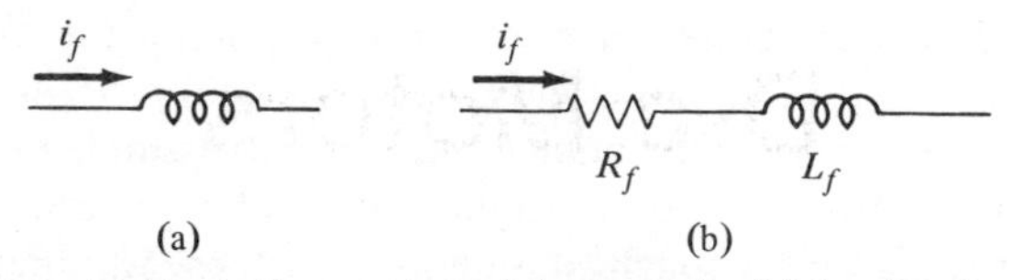

Fig. 19-2. Field winding symbols: (a) schematic; (b) for circuit analysis.

times important, role as an addition or correction with the function of adjusting the motor characteristics. For this reason, the difference between long- and short-shunt connections is practically immaterial, and for the remainder of this chapter, only the long shunt will be considered.

Observe the symbols used for the various currents. The armature current is always called i_a, the *shunt* field current i_f, and the current at the line terminals simply i (without subscript). No special symbol is needed for the current in a series field (it is either i_a or i). In a series motor, there is only one current to be considered, so there is no need for any subscripts.

The current *arrows* have been chosen so that, for motor operation, all currents have positive values, provided the terminal voltage v is positive. In other words, these arrows indicate actual current directions for motor operation, not just assumed positive directions. Consequently, the statement of Kirchhoff's current law

$$i = i_a + i_f \tag{19-1}$$

may be taken to indicate that the line current i drawn by a shunt or compound motor is always larger (though sometimes not by much) than the armature current i_a.

Returning to the circuit elements shown in Figs. 19-1b and 19-2b, the question arises whether there are any *mutual inductances* between the various windings to be considered. Here the reader is referred to Section 18-5 in which the conclu-

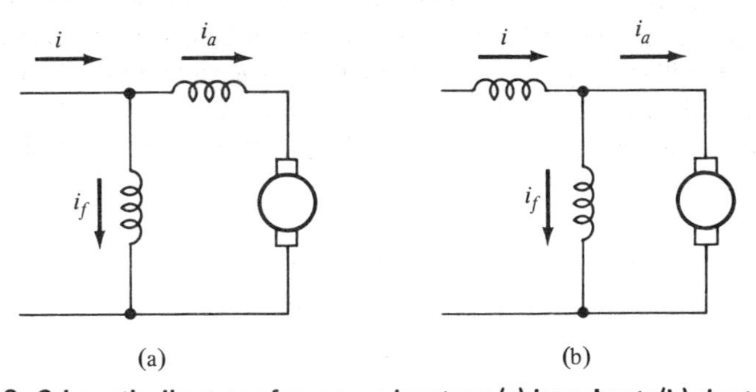

Fig. 19-3. Schematic diagrams of compound motors: (a) long shunt; (b) short shunt.

sion was reached that armature reaction has only a minor (always weakening) effect on the flux. Neglecting this effect (which is a valid procedure at low saturation, or else as a crude approximation) is the same as assuming that there is no magnetic coupling, thus, no mutual inductance, between armature and field windings, which is plausible enough considering that these windings are centered on the quadrature axis and the direct axis, respectively. Any consideration of the effect of armature reaction is equivalent to introducing some (nonlinear) mutual inductance between these two circuits.

By contrast, the two field windings of a compound machine are fully coupled to each other, and this coupling may be indicated, complete with dot notation for relative polarities, in either of the two types of diagrams.

19-2 THE SATURATION CURVE

The two basic equations derived in Section 18-6, namely the voltage equation, Eq. 18-7, and the torque equation, Eq. 18-14, both contain the magnetic flux ϕ as one of the important quantities that determine the operation of a d-c machine. This flux in turn depends on the resultant mmf of the currents in the various windings. It, thus, becomes necessary for the analysis of d-c motor operation to use the saturation curve, the relation between resultant mmf and flux.

The familiar shape of such a curve is drawn as Fig. 19-4. The abscissa is the total mmf per pole of all windings. For instance, for a (long-shunt) compound machine, it may be written as

$$\mathfrak{F} = N_f i_f \pm N_s i_a - \mathfrak{F}_d \tag{19-2}$$

In this equation, N_f stands for the number of turns per pole of the shunt field winding, N_s for that of the series field winding. The field windings may be connected so that their mmfs either aid or oppose each other, which explains the

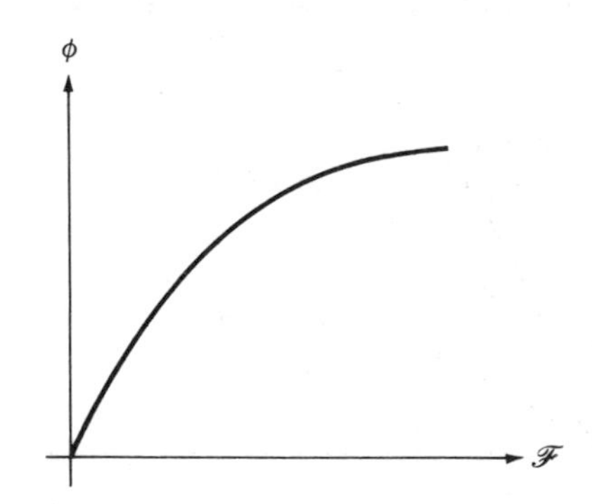

Fig. 19-4. Saturation curve.

± sign. Finally, $\mathcal{F}_d$, the demagnetizing effect of armature reaction, may be neglected in some cases; otherwise, it is most frequently assumed to be proportional to the armature current (see Section 18-5).

The reader should understand the fact, first encountered in Section 9-7 in the context of synchronous machines, that the saturation curve of a rotating machine can be interpreted, by means of scale changes, as an *open-circuit characteristic*, i.e., a relation between field current and induced voltage. Both these quantities are electrical, and the induced voltage becomes accessible and can be measured when the armature is open circuited, i.e., when it carries no current. (In an open-circuit test, the rotor is driven by a prime mover, and the field winding is energized from a separate source.)

Applied to a d-c machine, the scale change for the ordinates is essentially the same as for a synchronous machine: flux is replaced by induced voltage, since these two quantities are proportional to each other for any constant speed (Eq. 18-7). Thus, the open-circuit characteristic of a d-c machine is valid for only one speed. When variable speeds are to be considered (which is essential for the analysis of motor operation, at least), there are several ways to use the curve. One may, for instance, have a family of open-circuit characteristics for several speeds (Fig. 19-5) and interpolate for in-between values of speed. Or, one may have one curve with several ordinate scales for several speeds. Or, one may be satisfied with one curve and one scale because values of e for other speeds (at the same mmf) may be easily enough calculated by

$$e_{at\omega_2} = \frac{\omega_2}{\omega_1} e_{at\omega_1} \qquad (19\text{-}3)$$

The scale change for the abscissas is somewhat different because there may be more than one mmf-producing current involved, as Eq. 19-2 shows. One can get

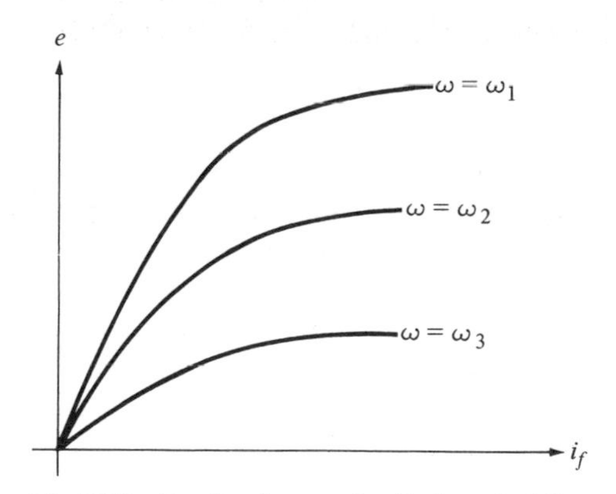

Fig. 19-5. Family of open-circuit characteristics.

around this difficulty by introducing the *effective field current*

$$i_{f\text{eff}} = \frac{\mathcal{F}}{N_f} \tag{19-4}$$

as a fictitious current flowing through the shunt field winding and producing a given mmf without aid or hindrance from the armature current. From Eq. 19-2, this current is

$$i_{f\text{eff}} = i_f \pm \frac{N_s}{N_f} i_a - \frac{\mathcal{F}d}{N_f} \tag{19-5}$$

showing that, when $i_a = 0$, there is no difference between actual and effective field current. Thus, the saturation curve e versus $i_{f\text{eff}}$ can be obtained in an *open-circuit test* by direct measurements of induced voltage and field current, for instance with the open-circuited armature driven by a prime mover.

The open-circuit characteristic also serves to describe the nature of the source voltage e as that of a *dependent source*. This concept, originally formed for the analysis of electronic devices, is also perfectly applicable to the armatures of electric motors and generators.

19-3 STEADY-STATE CHARACTERISTICS

The steady-state operation of a motor is determined by the intersection of its *torque-speed characteristic* with that of the load. At first, d-c motors will be analyzed (as were their rivals the induction motors) under the assumption of a constant supply voltage V.

The first motor type to be taken up is the *shunt motor*, because of its essentially constant flux. What this means is that, according to the equivalent circuit of Fig. 19-6, the field current in the steady state is independent of the load:

$$I_f = \frac{V}{R_f} \tag{19-6}$$

where, incidentally, capital letters are used for steady-state voltages and currents. On the other hand, the steady-state *armature* current depends very much on the load, because it is proportional to the electromagnetic torque (Eq. 18-14).

If the demagnetizing effect of armature reaction is disregarded, and the flux is, thus, considered as a constant, the torque-speed characteristic of the shunt motor can be shown to be a straight line, by substituting the voltage equation, Eq. 18-7, and the torque equation, Eq. 18-14, into the steady-state expression of Kirchhoff's voltage law for the armature circuit

$$V = R_a I_a + E = R_a \frac{T}{k\phi} + k\phi\omega \tag{19-7}$$

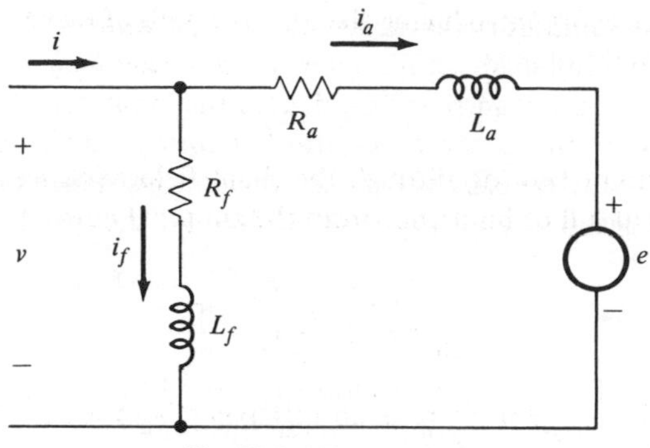

Fig. 19-6. Shunt motor.

and solving for either the torque or the speed, e.g.

$$T = \frac{k\phi}{R_a} V - \frac{k^2\phi^2}{R_a} \omega \qquad (19\text{-}8)$$

It is relevant that, in the first expression of Eq. 19-7, the term $R_a I_a$ is much smaller than either V or E, unless excessively high values of armature current are considered. Multiplying each term by the same factor does not change this relationship. Thus, the torque comes out as the relatively small difference between two much larger terms, and accordingly, the graph of the characteristic looks as it is drawn in Fig. 19-7. Its ordinate intercept (the torque at which the speed is zero, also called the *stalled torque*) is far above the range of the diagram because it would require a high multiple of rated armature current, namely V/R_a.

The abscissa intercept, on the other hand, appears clearly in the diagram. It is the *ideal no-load speed*, i.e., the speed at which the motor would run (idle, coast) in the no-load condition if it had no rotational losses.

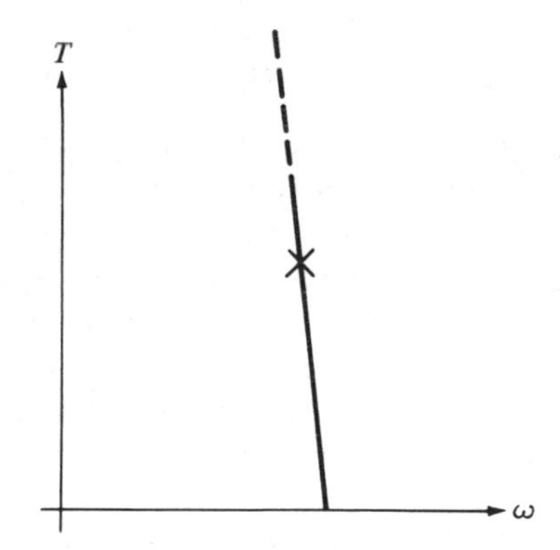

Fig. 19-7. Torque-speed characteristic of shunt motor.

The diagram also indicates the point of *rated operation*. The difference between no-load and full-load speeds (the *speed regulation*) has typical values around 5 percent of full-load speed, similar to the induction motor with which the d-c shunt motor shares the designation of *constant-speed* drive. Actually, the speed would really be constant if the armature resistance (an imperfection) were zero, as can best be seen if Eq. 19-7 is solved for the speed

$$\omega = \frac{V}{k\phi} - \frac{R_a T}{k^2 \phi^2} \tag{19-9}$$

The similarity with the induction motor ceases when the characteristic is pursued further upward past the point of rated operation: the curve does not bend over to the left, there is no maximum torque. This means that there are no stability problems, but it also means that the shunt motor is much more in need of protection against excessive loads than the induction motor. Such protection may be provided by fuses, circuit breakers, or automatic devices adjusting the armature voltage.

The steady-state analysis of the *series motor* is quite different because the flux varies widely with the load. The key lies in the fact that the same current that flows through the armature also produces the flux (Fig. 19-8). Thus, whenever an increase in load torque requires an increase in current, this also increases the flux, which in turn calls for a decrease in speed, since the product of flux and speed changes only slightly (due to the resistance drop). Accordingly, the series motor is considered a *variable-speed* motor, and its field of application is determined by this property.

There are many cases for which a variable speed is desirable. Essentially, in such cases, the load torque varies widely, and it is desired that the motor current and, thus, also its power should vary by much less. The lower speed may be viewed as cushioning the shock of the higher torque, keeping the power more nearly constant. Typical cases are electric locomotives (running uphill at a lower speed than on a level track) and cranes (lifting the loaded hook more slowly than the empty one).

To obtain the torque-speed characteristic of a series motor, one must know its

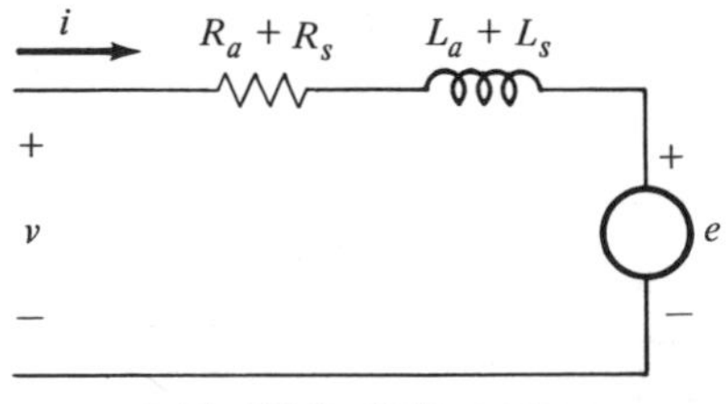

Fig. 19-8. Series motor.

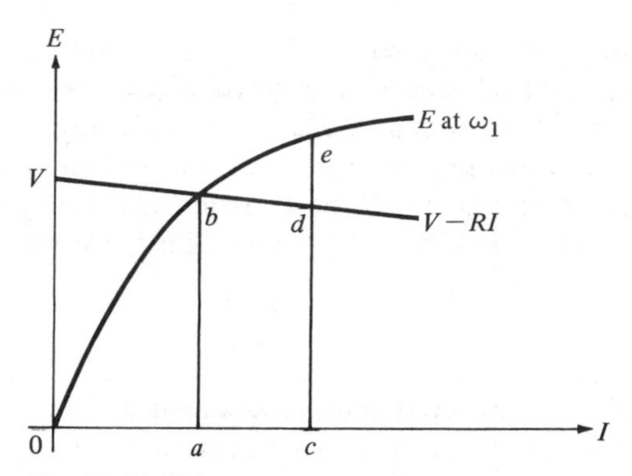

Fig. 19-9. Series motor open-circuit characteristic.

open-circuit characteristic. Refer to Fig. 19-9, in which the given curve, at the given scale, is valid at a certain speed to be called ω_1. Also drawn is the straight line representing the induced voltage $E = V - RI$, where the resistance R is the sum of the armature and series field resistances.

The intersection (point b) represents the condition at which the motor speed is ω_1. Read the corresponding current value (the distance $0a$), and calculate the corresponding torque from the power balance equation

$$T = \frac{EI}{\omega} \tag{19-10}$$

to get the first point of the torque-speed characteristic. Further points are obtained by taking arbitrary current values like the distance $0c$. The corresponding induced voltage is the distance cd, and the speed is $\omega_1\, cd/ce$. Repeating this procedure a few more times yields a curve like that of Fig. 19-10, clearly displaying the variable speed.

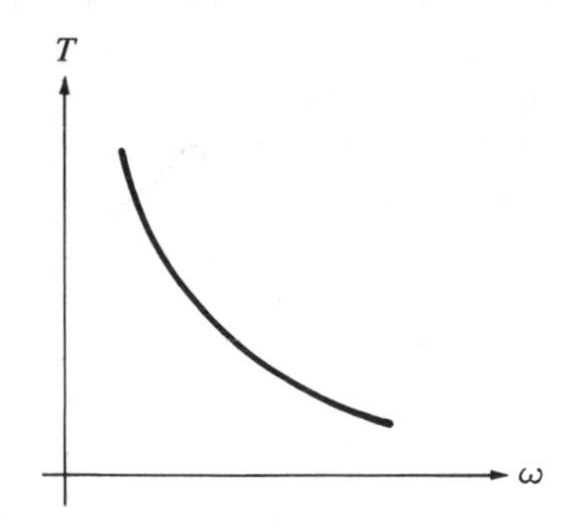

Fig. 19-10. Torque-speed characteristic of series motor.

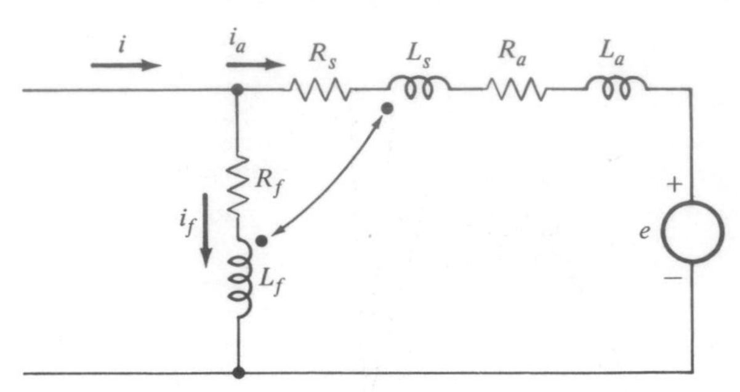

Fig. 19-11. Compound motor.

The curve has no intercepts within a reasonable range. This means that the series motor needs protection from both excessive torques and excessive speeds. The latter condition ("running away") would occur at very small load torques.

Compound motors can have their two field windings connected so that their mmfs are either aiding or opposing each other. Figure 19-11 illustrates the first-named case, the one in which the motor is said to be *cumulatively compounded*. For its study, one needs to know the open-circuit characteristic, which is viewed as a plot (valid at a speed ω_1) of the induced voltage versus the effective shunt field current

$$I_{f_{\text{eff}}} = I_f + \frac{N_s}{N_f} I_a - \frac{\mathcal{F}_d}{N_f} \tag{19-11}$$

To obtain the torque-speed characteristic of this motor, one chooses several values of armature current, and goes through the following procedure for each of them; calculate the effective shunt field current, look up the corresponding voltage E_1 on the open-circuit characteristic, calculate the actual induced voltage $E = V - (R_a + R_s)I_a$, find the speed $\omega = \omega_1 E/E_1$, and the torque $T = EI_a/\omega$.

The result of all this looks like the left curve in Fig. 19-12. For the sake of comparison, the same diagram also shows what the characteristic would be if the series field winding were reversed (*differential compounding*) or disconnected (changing the machine into a shunt motor).

As its characteristic indicates, cumulative compounding makes the motor run more slowly, to an extent depending, for a given number of series field turns, on the load. It may be said that the cumulatively compounded motor is in between the shunt and the series motor; its speed is more variable than that of the shunt motor but less so than that of the series motor. In contrast to the latter, it can idle safely without running away.

The classical case for the use of cumulative compounding is that of a motor driving a cyclical load (e.g., a metal press shaping a large number of identical automobile body parts in a row) and making use of a *flywheel*. During each

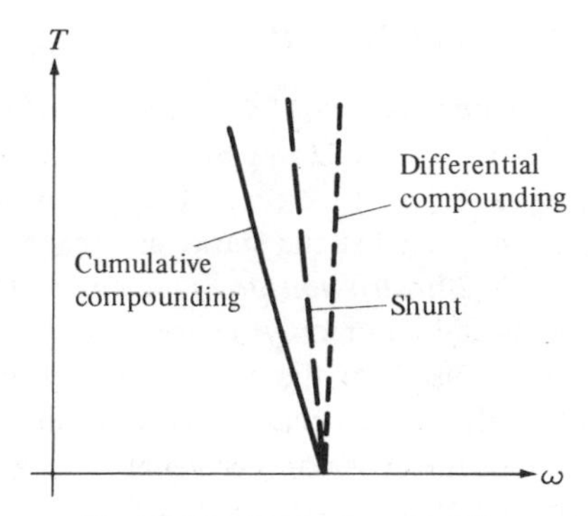

Fig. 19-12. Motor characteristics.

cycle, there is a period of peak load during which the flywheel helps the motor by contributing some of its stored kinetic energy. Then, during the period of low load torque, this energy is re-stored. The power output of the motor fluctuates much less than the power required by the load; in other words, the power of the motor stays much closer to its cyclical average than that of the load. So the power rating of the motor is less than it would be without the flywheel, which can be a considerable saving. But the cyclical change in kinetic energy stored in the flywheel can occur only if the speed is variable; hence the use of cumulative compounding.

Differential compounding is hardly ever used. The reader may refer to Section 14-2 to see that any torque-speed characteristic with a positive slope is a likely cause for instability, certainly so in case of a constant-torque load.

Some readers may have been surprised by the fact that an increase in flux (e.g., by cumulative compounding) leads to a decrease in steady-state speed, and vice versa. The explanation for this fact is that, within a reasonable range of armature currents, the product of flux and speed (the induced voltage) must remain substantially constant. But does not, one might ask, an increased flux cause an increase in torque, which would lead to an increase in speed? The answer to that is that the torque is proportional to the product of flux and armature current, and the increase in flux is accompanied by a relatively larger decrease in armature current.

This seemingly paradoxical relationship between flux and steady-state speed will be encountered again in the context of speed control. It also accounts for the slight correction of the torque-speed characteristic (as against Fig. 19-7) that arises when the demagnetizing effect of armature reaction is taken into consideration. The actual curve is a little steeper than it would be without armature reaction; it is even possible for a shunt motor with too much armature reaction to become unstable, like a differentially compounded motor.

19-4 POWER LOSSES AND EFFICIENCY

The various power conversions occurring in a d-c motor do not differ in essence from those in a synchronous or induction motor. *Core losses* (hysteresis and eddy-current losses) are caused in the rotor of a d-c machine (but not in its stator) by its relative motion against the stationary magnetic field. They may be considered together with the *friction and windage losses*, since they all are basically no-load losses and are commonly determined together in no-load tests. The term *rotational losses* is used for the combined friction and windage losses and core losses, indicating that they occur in the machine that rotates even if it delivers no output power. This does not mean that they are independent of the load; they are functions of flux and speed, and both flux and speed change with a changing load, particularly in the case of a series motor.

Copper losses are caused by the resistances of all current-carrying windings. The copper losses in shunt field windings are constant, i.e., independent of the load, whereas the copper losses in armature and series field windings are proportional to the square of the armature current. Figure 19-13 is a power flow diagram for a *compound motor*. It shows agreement with Kirchhoff's laws by displaying the relations

$$P_{in} = VI = V(I_f + I_a) = VI_f + VI_a \tag{19-12}$$

and

$$VI_a = (E + RI_a)I_a = EI_a + RI_a^2 \tag{19-13}$$

where the symbol R once more stands for the combined armature and series field resistance.

It was mentioned above that the standard experiment by which the rotational losses are determined is a *no-load test*. The idea is that any arbitrary values of flux and speed can be duplicated in such a test without the armature having to carry any current. But it is true that the rotational losses determined by such

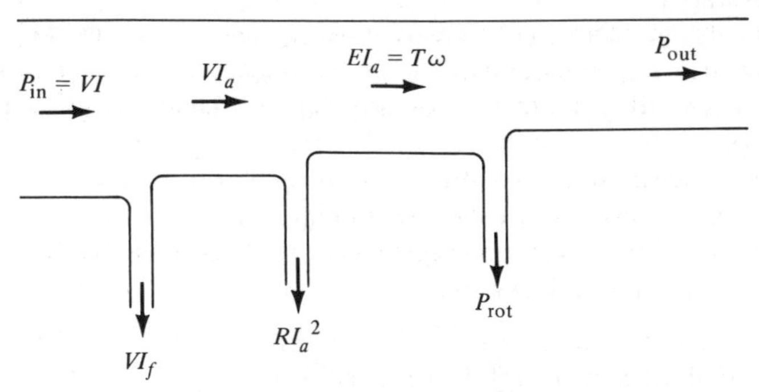

Fig. 19-13. Power flow diagram of compound motor.

a test do not quite equal those under corresponding load conditions. Even though the magnitude of the flux in the no-load test is equal to that under load, the flux *distortion* due to armature reaction and to leakage fluxes around the armature slots causes an increase in core losses under load. This difference is often accounted for by adding a small empirically determined value (usually one percent of the output power) to the no-load rotational losses. This correction is then referred to as the stray load losses.

Finally, even though the division of power losses into constant and variable (*load*) losses is not as clear-cut as for a-c machines, it still remains substantially true that the efficiency $\eta = P_{\text{out}}/P_{\text{in}}$ has its highest value at that load at which the pure load losses (RI_a^2) amount to about one-half of the total losses.

19-5 SPEED CONTROL

Due to its commutator, a d-c motor is more expensive than an induction motor of comparative ratings, both in the price of purchase and the cost of maintenance. In addition, it needs a supply of d-c power, which is not always easily available; it may either require a source of its own or, more likely, rectifiers to convert the available a-c into d-c. In spite of these handicaps, the d-c motor has managed to hold on to its field of application. There are two reasons: the possibility of variable-speed characteristics (for series and compound motors), and the availability of several desirable methods of speed control, which are the subject of this section.

Speed control means the shifting of the torque-speed characteristic (see Section 14-3). A straightforward analytic expression of this relation is available for the shunt motor only. (For other d-c motors, the characteristic depends on the saturation curve.) Equations 19-8 or 19-9 show that three quantities may be considered as parameters: the armature resistance R_a, the flux ϕ, and the terminal voltage V. Incidentally, this statement is also true for the series or compound motor.

Changing the *armature resistance* is probably the oldest method of speed control, but it is still in use, especially for series motors. When a resistor is connected in series with the armature, the armature resistance is in effect increased. Equation 19-9 indicates that this results in a reduction of the steady-state speed, except for the ideal no-load condition. A family of characteristics, for different values of armature resistance, is sketched in Fig. 19-14.

This method of speed control is simple and fairly inexpensive, but it has the following limitations and drawbacks:

(a) The speed can only be reduced, never raised, against the *base speed* (i.e., the speed without any resistance added).
(b) The method is relatively ineffective at no-load.

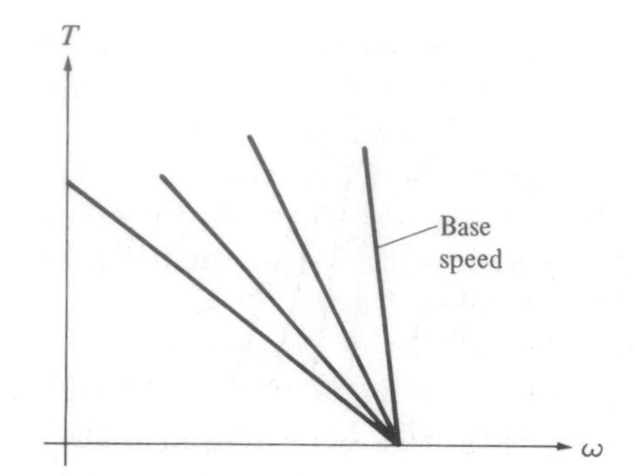

Fig. 19-14. Armature-resistance control of shunt motor.

(c) The motor loses its "constant-speed" property.
(d) The method is grossly wasteful of energy. Roughly speaking, the percentage of speed reduction equals the percentage of input power consumed by the added resistance.

Speed control by changing the *flux* is an important method for shunt motors, where it can be accomplished without touching the main (armature) circuit. One way is to reduce the field current by connecting additional resistance in series with the field winding. An adjustable resistor used in this manner is called a *field rheostat*. Figure 19-15 is a family of torque-speed characteristics of a shunt motor with field-rheostat control.

This method of speed control is economically sound (the power loss in the field circuit is actually reduced when a field rheostat is added), but it has its limitations too:

(a) The speed can only be raised, not reduced.
(b) More speed generally means more power. If a motor is operating at rated conditions, an increase in speed without a corresponding decrease in load torque means an overload. This can also be seen by checking what happens to the armature current $I_a = T/k\phi$, when the flux is reduced. The broken-line hyperbola in Fig. 19-15 (representing $T\omega$ = constant) shows how higher speeds are permissible for lower torques only.

Instead of field rheostats, *thyristors* can be used to control the field current and thereby the speed. In this case, the shunt field winding is disconnected from the armature and energized from the single-phase or three-phase a-c power system, just like the field winding of a synchronous machine (see Section 10-10).

Whether field rheostats or thyristors are used, the base speed is the lowest obtainable for a given load. It is the speed that results from turning the field rheostat resistance down all the way to zero, or from setting the firing angle to

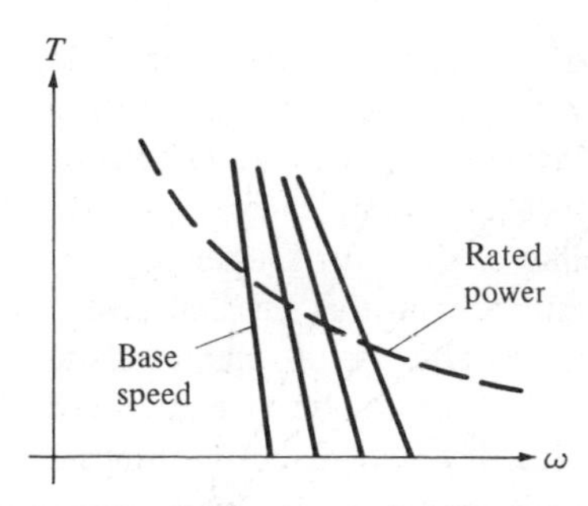

Fig. 19-15. Flux control of shunt motor.

zero. Adding resistance to the field circuit, or delaying the firing of the thyristors, reduces the field current and thereby the flux. As the reader known, this increases the steady-state speed, as long as the load torque remains within reasonable limits. For extreme overloads, the response to a reduction of flux would be the opposite, namely a decrease in speed, as can be seen from Eq. 19-9 or by extending the curves of Fig. 19-15 far upward.

The third basic method of speed control is by changing the *armature voltage V*. This terminology refers to the most frequent application of this method, namely for shunt motors, when the field winding is disconnected from the armature and energized from a constant d-c source, while an adjustable voltage is supplied to the armature. The result, in accordance with Eq. 19-8, is seen in Fig. 19-16.

It is plausible to consider the speed obtained with the full available voltage the base speed. Accordingly, the method may be viewed as a way to obtain lower speeds, just like the armature resistance method, but without its drawbacks. Armature-voltage control reduces the no-load speed just as much as the full-load speed (all the way down to zero, if desired), it does not change the slope of the characteristics, and it does not waste any substantial amount of power.

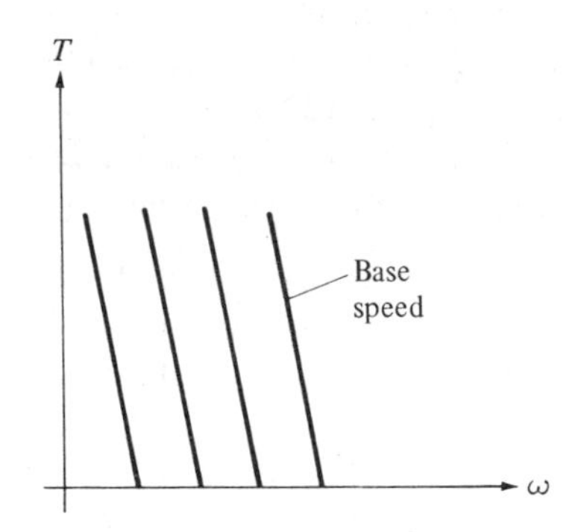

Fig. 19-16. Armature-voltage control of shunt motor.

There are several ways to obtain the adjustable armature voltage. Older installations use a separately excited d-c generator with adjustable field excitation (see Chapter 20). This generator has to be driven by a prime mover running at substantially constant speed. Examples for what this prime mover can be are induction motors, gasoline engines, or Diesel engines. (The typical example is the so-called Diesel-electric locomotive; in that case, the d-c motor is normally a series motor, since its variable-speed characteristic is suitable for traction.) Regardless of the type of prime mover, this method of speed control is comparatively expensive, having three full-sized machines doing the job of one.

Modern industrial installations use electronic control of the armature voltage by means of *thyristors*. The analysis of an armature circuit with electronic control is more involved than that of a field circuit, not so much because of the difference in power ratings, but rather because of the induced speed voltage e in the armature circuit. This study will be undertaken in Sections 19-10 to 19-12.

Finally, it should be mentioned that two different methods of speed control (e.g., field control and armature-voltage control) may be combined to give speeds both above and below the base speed.

19-6 THE FOUR QUADRANTS

When the speed of a d-c shunt motor is controlled by an adjustable armature voltage, the family of torque-speed characteristics (Fig. 19-16) may be meaningfully extended to include negative values of both torque and speed. Refer to Fig. 19-17 in which the symbol V_1 represents the highest available voltage magnitude.

The *third quadrant*, in which both speed and torque are reversed, describes motor operation in the backward direction, something that is required not only

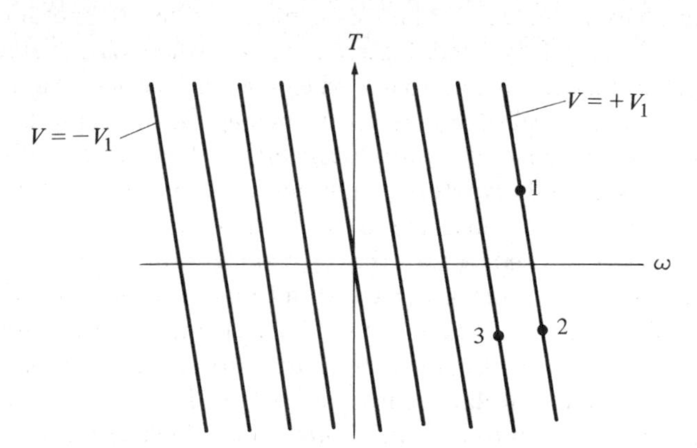

Fig. 19-17. Positive and negative values of torque and speed.

for motors driving vehicles (trains, trucks, etc.), but also for many industrial processes. As might have been expected, this kind of operation is obtained by reversing the armature voltage. (Incidentally, the same result could be achieved by reversing the flux, but not both, since the electromagnetic torque $T = k\phi I_a$ would remain unchanged if both the armature and field currents were reversed.) The availability of an armature voltage that is adjustable in both directions makes a smooth transition between forward and backward running possible, in contrast to the abrupt "throwing" of a reversing switch.

In the *second and fourth quadrants* of the diagram, the electromagnetic torque is opposed to the motion. The reader knows very well that this represents generator action. The reason to study it in the context of motor operation is the fact that it also describes a motor temporarily acting as a *brake*.

Visualize an electrically driven train reaching a downhill portion of its track. With the force of gravity added to that of its motor, the train is accelerated downward. When its speed exceeds the ideal no-load speed of the motor, the torque becomes negative, i.e., opposing the motion; the motor acts as a brake. If the previous steady-state condition (on the level part of the track) was described by point 1 in Fig. 19-17, the new downhill steady state may be at point 2. Quite possibly, the speed at this point might be considered too high for safety; with an adjustable armature voltage, all that is needed is to reduce this voltage to reach a lower steady-state speed at, say, point 3.

Elevators are often operated in accordance with this principle. The same d-c machine acts as a motor when the elevator is going up, and as a brake when it is going down. Other applications include cranes.

A brake can be defined as a device that converts unwanted mechanical energy into something else. The familiar friction brake converts this energy into heat, which means in general that it is lost; in addition, friction surfaces are wearing off and, thus, have a limited lifetime. By contrast, an electric motor can, as was shown above, do the job by operating as a generator whose output is returned to the same system that serves as a source when the machine acts as a motor. This mode of operation is called *regenerative braking*. In the case of the motor armature supplied through thyristors, regenerative braking is possible only if another complete set of thyristors, connected in an inverter circuit, is used to return the output of the d-c machine to the a-c system.

Often it is desired to use some form of braking, not to oppose a force like gravity, but to slow down the motor and its load more quickly than if they were left to themselves. In such a case, the mechanical energy to be converted is the kinetic energy of the moving masses. In contrast to the cases considered before, there is no steady state and so the curves of Fig. 19-17 are not applicable. But it remains true that, in order to have the motor act as a brake, its armature current and its torque must be reversed, and that this can be accomplished by reducing the armature voltage, making it smaller than the induced voltage. If the motor and its load are to be brought to a complete standstill by this method, the armature voltage must be turned down to zero, gradually or in several steps.

Electric braking need not be regenerative, especially when the amount of energy to be saved would not be economically significant. In such a case, the motor armature may be disconnected from its source, and, while the machine is acting temporarily as a generator driven by inertia, its electric output may be consumed in a resistor. This energy is then lost but, at least, the problems of wear and tear associated with friction are avoided.

19-7 TRANSIENT CONDITIONS

In discussing the d-c motor up to this point, none of its *energy-storing elements* were taken into consideration. Thus, all equations and diagrams obtained are limited to the steady state. But, whenever any of the quantities determining a steady state is changed, the motor must undergo a transient period before it can reach a new steady state.

The energy-storing elements are partly mechanical and partly electrical: the *inertia* of the rotor and the load makes sudden changes of speed impossible, and the *inductances* of the various windings keep their currents from changing suddenly. (The effect of capacitances is generally ignored in the study of electromagnetic devices, because the energy storage in electric fields is much smaller than that in magnetic fields.) The quantities whose changes constitute "disturbances" are the voltages across the windings, and the load torque.

The following discussion is simplified by the assumption that the field current is held constant, an assumption that is valid for the most important, though certainly not for all, cases; in particular, it is valid for the use of d-c motors as *servomotors*. In addition, the effect of armature reaction will be disregarded. This eliminates the magnetic coupling between the armature and field circuits, and, together with the assumption of a constant field current, it makes the magnetic flux constant. This leaves two energy-storing elements to be considered: the moment of inertia of the rotating mass, and the armature inductance.

Accordingly, there is one mechanical and one electrical equilibrium equation to be written:

$$T = T_{\text{load}} + J\,\frac{d\omega}{dt} \tag{19-14}$$

where the torque for the rotational losses is either neglected or included in T_{load}, and

$$v = e + R_a i_a + L_a\,\frac{di_a}{dt} \tag{19-15}$$

(There is an inconsistency in this notation: the torques T and T_{load} may be time-varying, just like the speed, the currents, and the voltages. Nevertheless, capital letters must be used for them.)

These two equations contain mechanical and electrical quantities that are related to each other by the basic equations, Eqs. 18-7 and 18-14. For a constant flux, these two relations can be simplified by defining a new constant

$$k' = k\phi \tag{19-16}$$

so that they read

$$e = k'\omega \tag{19-17}$$

and

$$T = k' i_a \tag{19-18}$$

When these two relations are substituted into the equilibrium equations, Eqs. 19-14 and 19-15, two differential equations for the two time functions ω and i_a are obtained:

$$k' i_a = T_{\text{load}} + J \frac{d\omega}{dt} \tag{19-19}$$

and

$$v = k'\omega + R_a i_a + L_a \frac{di_a}{dt} \tag{19-20}$$

One of these two variables, ω or i_a, can be eliminated. Thus, one either differentiates Eq. 19-20, solves for $d\omega/dt$, and substitutes into Eq. 19-19:

$$k' i_a = T_{\text{load}} + \frac{J}{k'} \left(\frac{dv}{dt} - R_a \frac{di_a}{dt} - L_a \frac{d^2 i_a}{dt^2} \right) \tag{19-21}$$

or one solves Eq. 19-19 for i_a, and substitutes it and its derivative into Eq. 19-20:

$$v = k'\omega + \frac{R_a}{k'} \left(T_{\text{load}} + J \frac{d\omega}{dt} \right) + \frac{L_a}{k'} \left(\frac{dT_{\text{load}}}{dt} + J \frac{d^2 \omega}{dt^2} \right) \tag{19-22}$$

Finally, the terms of these two equations are collected and placed in their traditional sequence

$$\frac{L_a J}{k'} \frac{d^2 i_a}{dt^2} + \frac{R_a J}{k'} \frac{di_a}{dt} + k' i_a = \frac{J}{k'} \frac{dv}{dt} + T_{\text{load}} \tag{19-23}$$

and

$$\frac{L_a J}{k'} \frac{d^2 \omega}{dt^2} + \frac{R_a J}{k'} \frac{d\omega}{dt} + k'\omega = v - \frac{R_a}{k'} T_{\text{load}} - \frac{L_a}{k'} \frac{dT_{\text{load}}}{dt} \tag{19-24}$$

The reader should notice that the coefficients on the left sides of these two equations are the same, whereas the "forcing functions" on the right sides are different linear functions of the two independent variables v and T_{load}.

It is worthwhile to ask, at this point, how the transient responses would change if it were somehow possible to ignore the armature inductance. Setting L_a equal to zero in Eq. 19-23 or 19-24 leads to a first-order differential equation. This means that the free component of the response (i_a or ω) would then be an exponential function of time, with the (mechanical) time constant

$$\tau_m = \frac{R_a J}{k'^2} \tag{19-25}$$

If, on the other hand, the armature circuit existed by itself, without any relation to moving masses, then it would be an RL circuit and its free response would be an exponential function of time, with the (electrical) time constant

$$\tau_a = \frac{L_a}{R_a} \tag{19-26}$$

The differential equations, Eqs. 19-23 and 19-24 can be rewritten in terms of these two time constants:

$$k'\tau_a\tau_m \frac{d^2 i_a}{dt^2} + k'\tau_m \frac{di_a}{dt} + k' i_a = \frac{J}{k'}\frac{dv}{dt} + T_{load} \tag{19-27}$$

and

$$k'\tau_a\tau_m \frac{d^2\omega}{dt^2} + k'\tau_m \frac{d\omega}{dt} + k'\omega = v - \frac{R_a}{k'} T_{load} - \frac{L_a}{k'}\frac{dT_{load}}{dt} \tag{19-28}$$

The reader familiar with second-order linear, constant-coefficient differential equations (e.g., from circuit or system theory) knows that the waveshape of the free response is found from the *characteristic equation* which, in this case, is

$$\tau_a\tau_m s^2 + \tau_m s + 1 = 0 \tag{19-29}$$

Its roots are

$$s = -\frac{1}{2\tau_a} \pm \sqrt{\frac{1}{4\tau_a^2} - \frac{1}{\tau_a\tau_m}} \tag{19-30}$$

So the free response can be either overdamped or oscillatory, depending on the relation between the two time constants τ_a and τ_m. A comparison of Eq. 19-30 with the general expression

$$s = -\alpha \pm \sqrt{\alpha^2 - \omega_0^2} \tag{19-31}$$

shows that the *undamped natural frequency* is

$$\omega_0 = \sqrt{\frac{1}{\tau_a \tau_m}} \tag{19-32}$$

and the *damping ratio* is

$$\zeta = \frac{\alpha}{\omega_0} = \frac{\dfrac{1}{2\tau_a}}{\sqrt{\dfrac{1}{\tau_a \tau_m}}} = \sqrt{\frac{\tau_m}{4\tau_a}} \tag{19-33}$$

Accordingly, the free response is oscillatory if $\tau_m < 4\tau_a$. Two cases with realistic values of time constants are worked out in examples 19-4 and 19-5.

A complication arises when the load torque depends on the speed, as it does in the case of viscous friction. In such cases, Eqs. 19-21 through 19-33 are no longer strictly valid. Furthermore, a *change of load* then means that one torque-speed characteristic of the load is replaced by another. Very likely, these characteristics are not even linear, in which case a study of the transient response lies outside the haven of linear analysis and beyond what is expected of the reader.

19-8 STARTING

A different problem of transient operation arises when a motor is being started. Just to connect it to its source by closing a switch is not permissible because this would result in a highly excessive armature current. The same problem was discussed for induction motors in Section 14-8. For d-c motors, it is even more pronounced because the current drawn by the armature of a stationary d-c motor would be limited by its resistance only, whereas for a comparable a-c motor, there is also the much larger inductive reactance.

An easy way out is available when the motor armature is supplied from an *adjustable source*, e.g., thyristors, for the purpose of speed control. In this case, the motor is started by connecting its armature to a voltage that is adjusted to a value sufficiently low to keep the current within its permissible limit. As the rotor is gaining speed, the armature current decreases, and the voltage can be gradually raised to its full rated value.

Another speed control device, the armature circuit *resistor*, is also well suited for the purpose of starting. When the switch connecting the armature to its source is closed, enough resistance is connected in series with the armature to limit the current to the desired value, and this resistance is turned down to zero in several steps as the rotor accelerates.

To start a d-c motor not equipped for speed control, a *starting resistor* (*starting box*) can be used. It is also connected in series with the armature, and its difference from an armature resistor for speed control lies in its smaller heat-dissipating capacity. The starting process is usually a matter of a few seconds, whereas speed control may be used for a long time. So a starting resistor may be of a much less expensive design than one intended for speed control.

The value of the resistance needed for starting is, in most cases, a high multiple of that of the armature itself. Adding this resistance changes the time constants τ_m and τ_a substantially. Equations 19-25 and 19-26 indicate that the mechanical time constant is increased, and the electrical one decreased at the same time. If they originally had the same order of magnitude (as they often do), the addition of the starting resistor makes $\tau_m >> \tau_a$. For this case, the roots of the characteristic equation take on significant values. Since $1/4\tau_a^2 >> 1/\tau_a\tau_m$, the square root in Eq. 19-30 approaches the value $1/2\tau_a$. Then the root with the minus sign becomes

$$s_2 \approx -\frac{1}{2\tau_a} - \frac{1}{2\tau_a} = -\frac{1}{\tau_a} \tag{19-34}$$

The other root is the difference of two almost equal terms. A good approximation is obtained by using the binomial theorem:

$$\begin{aligned} s_1 &= -\frac{1}{2\tau_a} + \left(\frac{1}{4\tau_a^2} - \frac{1}{\tau_a\tau_m}\right)^{1/2} = -\frac{1}{2\tau_a} + \left(\frac{1}{4\tau_a^2}\right)^{1/2} - \frac{1}{2}\left(\frac{1}{4\tau_a^2}\right)^{-1/2}\frac{1}{\tau_a\tau_m} + \cdots \\ &= -\frac{1}{2\tau_a} + \frac{1}{2\tau_a} - \frac{1}{2}(2\tau_a)\frac{1}{\tau_a\tau_m} = -\frac{1}{\tau_m} \end{aligned} \tag{19-35}$$

What this means is that the free response consists of two exponential functions, a "fast" electrical one (due to the inductance) and a "slow" mechanical one (due to the inertia). In the first short time interval after a disturbance, the current comes close to its steady-state value before the speed can undergo any substantial change. This is why a starting resistance is chosen as the value $V/I_a - R_a$, where I_a is the highest permissible value of the armature current (typically one and one-half to two times rated value).

The steps by which the starting resistance is reduced should be chosen so that after each step the current again comes close to, but does not exceed, its limit. The switching may be automatic, e.g., in response to the increasing speed or to the decreasing current. The duration of the starting process depends partly on the load.

19-9 A-C COMMUTATOR MOTORS

It was pointed out earlier that a simultaneous reversal of the armature and field currents does not change the operation of a d-c motor, since the torque $T = k\phi i_a$ remains unchanged. It follows that, in principle, a d-c motor can be supplied from an a-c source. In practice, this possibility is restricted to *series motors*, because the field winding of a shunt motor would have too much inductive reactance.

The development of solid-state power rectifiers has greatly reduced the field of application of a-c commutator motors; but such motors are still in use for two specific purposes. One of them is for small tools and appliances that work with either a-c or d-c (*universal motors*). The other is electric railroad locomotives for which a variable-speed (series) characteristic is desired and whose trolley wires carry high-voltage, single-phase a-c.

The a-c operation of commutator motors has its problems. In most cases, it turns out to be necessary to equip the motor with a compensating winding (see Section 18-5) to counteract the effect of armature reaction. In addition, there is a greater likelihood of commutator sparking than with a d-c supply.

19-10 SINGLE-PHASE, HALF-WAVE THYRISTOR DRIVE

Small d-c motors may be operated from a single-phase, a-c source using the system shown in Fig. 19-18. The thyristor permits control of the armature voltage by adjustment of the firing angle α. See Fig. 19-19. The thyristor may be turned ON at angle α provided that the source voltage v_1 exceeds the voltage e_a generated in the motor. (The words *generated voltage* and *induced voltage* are interchangeable.) The equation of the armature current $i_a(t)$ will be determined below. It increases until v_1 approximately equals e_a. Then, the armature current decreases until zero current occurs at angle β. We call β the extinction angle, and we call the conducting interval λ. After the thyristor is OFF, the motor coasts for an interval μ until $2\pi + \alpha$ is reached and the thyristor is turned ON again. Both the armature current and the armature terminal voltage are functions that have an a-c component with the frequency of the source and harmonics of that frequency. Both functions have a d-c component, however, for which the performance of the motor is described in Section 19-3. A pulse of torque, corresponding to the pulse of armature current, accelerates the motor during the major portion of the conducting interval. During the coasting interval,

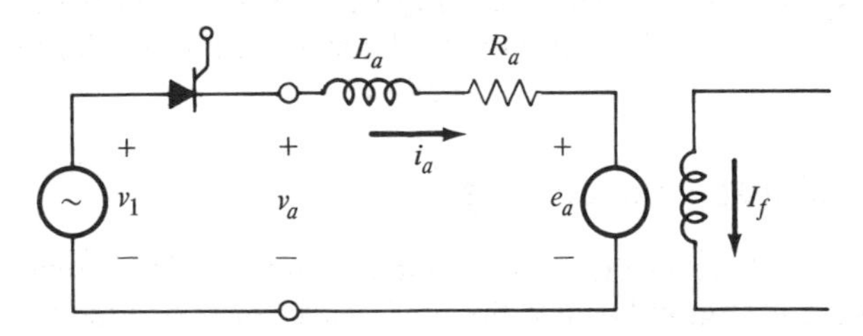

Fig. 19-18. Circuit of single-phase, half-wave thyristor drive.

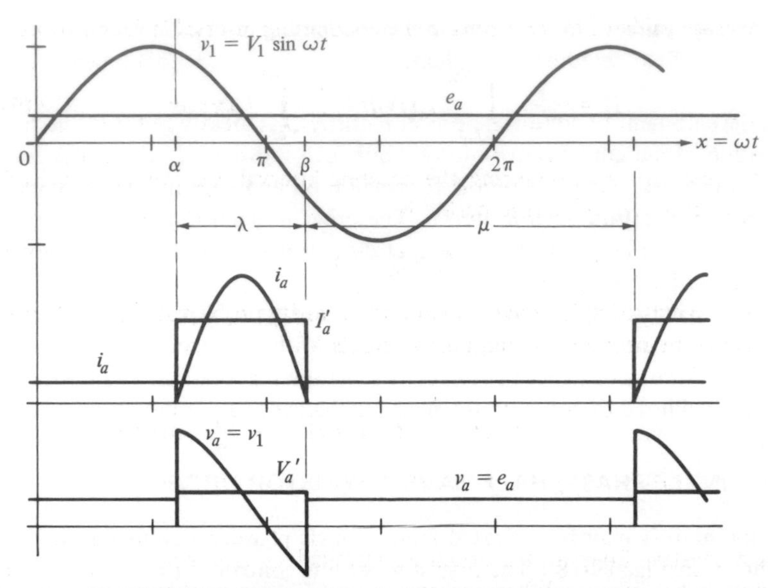

Fig. 19-19. Waveshapes of single-phase, half-wave thyristor drive.

the load torque causes the motor to slow down. This is satisfactory, provided these periodic changes in speed are small compared with the average speed.

During a conducting interval ($\omega t_1 = \alpha < x < \beta = \omega t_2$), the armature circuit is described by the following differential equation

$$R_a i_a + L_a(di_a/dt) = v_1 - e_a = V_1 \sin \omega t - e_a \tag{19-36}$$

This can be solved for the armature current i_a during the conducting interval.

$$i_a(t) = (V_1/Z_a) \sin(\omega t - \theta_a) - E_a/R_a + I_3 e^{-t/\tau_a} \tag{19-37}$$

where the armature impedance at frequency ω is

$$\mathbf{Z}_a = R_a + j\omega L_a = Z_a \angle \theta_a \tag{19-38}$$

the generated voltage with the speed assumed to be constant is

$$E_a = k\phi\omega = k'\omega \tag{19-39}$$

the armature circuit time constant is $\tau_a = L_a/R_a$, and I_3 is a constant to be found from $i_a(t_1) = 0$. Then β can be found by solving the equation for the next zero value, because we know $i_a(t_2) = 0$. Since this is a transcendental equation, graphical or computational means may be used to solve it; or a few guesses at the answer for β will lead to a reasonable evaluation.

The average value of current during the conducting interval is found from

$$I_a' = \frac{1}{\lambda/\omega} \int_{\alpha/\omega}^{\beta/\omega} i_a(t)\, dt = \frac{1}{\lambda} \int_{\alpha}^{\beta} i_a(x)\, dx \tag{19-40}$$

Since the current is zero during the coasting interval, the average value of the armature current for the entire period is

$$I_a = I_a' \lambda/2\pi \tag{19-41}$$

During a conducting interval, the armature voltage v_a equals v_1. The average of this function during a conducting interval is V_a'.

$$V_a' = \frac{1}{\lambda/\omega} \int_{\alpha/\omega}^{\beta/\omega} v_1(t)\, dt = \frac{1}{\lambda} \int_{\alpha}^{\beta} v_1(x)\, dx = \frac{V_1}{\lambda} [\cos\alpha - \cos\beta] \tag{19-42}$$

During a coasting interval, the armature terminal voltage is equal to the voltage generated in the motor. When the motor speed is assumed to be constant, $e_a = E_a = k'\omega$. The average value of the terminal voltage for the entire period is

$$V_a = V_a'(\lambda/2\pi) + E_a(\mu/2\pi) \tag{19-43}$$

Considering d-c components only, Kirchhoff's voltage law is

$$V_a = R_a I_a + E_a \tag{19-44}$$

This law must also be satisfied by the average values during the conducting interval

$$V_a' = R_a I_a' + E_a' \tag{19-45}$$

where E_a' is the average value of e_a during the conducting interval

$$E_a' = \frac{1}{\lambda/\omega} \int_{t_1}^{t_2} e_a(t)\, dt \tag{19-46}$$

As long as the speed is considered constant, $E_a' = E_a = e_a$.

Equation 19-45 could also be derived by integrating Eq. 19-36 over the conducting interval

$$R_a \int_{t_1}^{t_2} i_a(t)\, dt + L_a \int_0^0 di_a = \int_{t_1}^{t_2} v_1(t)\, dt - \int_{t_1}^{t_2} e_a(t)\, dt \tag{19-47}$$

where $t_1 = \alpha/\omega$ and $t_2 = \beta/\omega$. Since $i_a(t_1) = 0$ and $i_a(t_2) = 0$, the integral $\int_0^0 di_a = 0$. Divide the other integrals by the duration of the interval $t_2 - t_1 = \lambda/\omega$. This puts them into the form of Eqs. 19-40 and 19-42.

At any instant, the developed *torque* of the motor must equal the sum of the load torque and the accelerating torque

$$k' i_a = T(t) = T_L(t) + J\, d\omega_m/dt \tag{19-48}$$

where J is the moment of inertia of the system, k' is the torque constant with a fixed value of flux, $T(t)$ is the motor torque, and $T_L(t)$ is the load torque. We can integrate all of the terms in Eq. 19-48 over a full period and divide by the period

$$\begin{aligned} k' \frac{\omega}{2\pi} \int_0^{2\pi/\omega} i_a(t)\, dt &= \frac{\omega}{2\pi} \int_0^{2\pi/\omega} T(t)\, dt \\ &= \frac{\omega}{2\pi} \int_0^{2\pi/\omega} T_L(t)\, dt + J \frac{\omega}{2\pi} \int_{\omega_1}^{\omega_1} d\omega_m \end{aligned} \tag{19-49}$$

Since the speed is the same at the beginning and at the end of the period, $\int_{\omega_1}^{\omega_1} d\omega_m = 0$. The remaining terms are in the form of average values

$$I_a = \frac{\omega}{2\pi} \int_0^{2\pi/\omega} i_a(t)\, dt \tag{19-50}$$

$$T = \frac{\omega}{2\pi} \int_0^{2\pi/\omega} T(t)\, dt \tag{19-51}$$

$$T_L = \frac{\omega}{2\pi} \int_0^{2\pi/\omega} T_L(t)\, dt \tag{19-52}$$

Substituting in Eq. 19-49, we get

$$k' I_a = T = T_L \tag{19-53}$$

Assume the load has a constant value of T_L. During most of the conducting interval $T(t)$ exceed T_L (see i_a and I_a in Fig. 19-19), and the motor accelerates. During the coasting interval, $\mu/\omega = t_3 - t_2 = \Delta t$, the motor torque is zero. From Eq. 19-48, we can see that, for the coasting interval.

$$T_L/J = -d\omega_m/dt \approx -\Delta\omega_m/\Delta t \tag{19-54}$$

Use Eq. 19-53 and solve for $\Delta\omega_m$

$$\Delta\omega_m = -T(\Delta t)/J = -T\mu/\omega J \tag{19-55}$$

$\Delta\omega_m$ is called the speed dip. Equation 19-55 gives us a reasonable approximation to its value. If the speed dip is appreciable, then the assumption of constant speed would lose its validity.

Since operation of this half-wave circuit makes the armature current to be in pulses, the heating in the motor (which is determined by the rms value), is greater than would be caused by the average value. Consequently, the motor cannot be used continuously at its rated current. In addition, the d-c component taken from an a-c source may cause saturation in transformers, so this system is restricted to small motors. However, the system allows wide speed range by controlling the armature voltage. When operating with a given firing angle, an increase in load torque makes the motor adjust to a slower speed, and, thus, to increase the armature current. With these changes, the extinction angle β adjusts to a larger value and reduces the average value of the armature voltage V_a. This operation has a more drooping speed-load characteristic than for a shunt motor with fixed armature terminal voltage. If it is desired to maintain the same speed, the firing angle must be decreased with increasing load.

19-11 THREE-PHASE, HALF-WAVE THYRISTOR DRIVE

For d-c motors of 5 HP and larger, a three-phase system is used. One such system is shown in Fig. 19-20. This circuit makes it possible to have three conducting intervals during one cycle of the source voltage. This improves the situation in regard to harmonics because their frequencies are higher than for the single-phase system. For small currents, the conducting interval can be less than 120°, and the performance is the same as for the single-phase

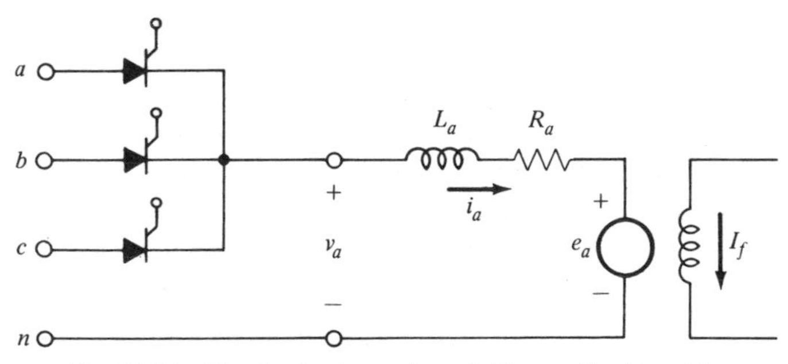

Fig. 19-20. Circuit of a three-phase, half-wave thyristor drive.

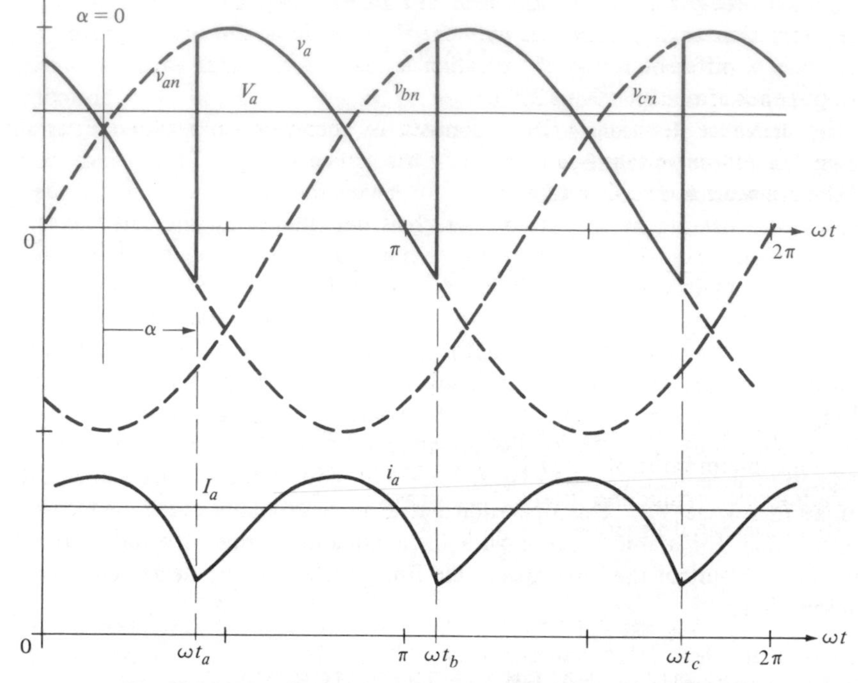

Fig. 19-21. Waveforms of a three-phase, half-wave thyristor drive.

system in section 19-10. Refer to Fig. 19-21. The reference for the firing angle α is selected to be the point in the cycle at which forward voltage is available to a thyristor, when v_{an} equals and subsequently exceeds v_{cn}, and similarly for the remaining phases. The value of the average armature voltage during a conducting interval is given by

$$V_a' = \frac{1}{\lambda} \int_{\alpha+\pi/6}^{\beta+\pi/6} v_a(x)\, dx \tag{19-56}$$

with $\lambda < 2\pi/3$.

Every period consists of three conducting intervals and three coasting intervals. The average armature voltage over the entire period reflects both kinds of intervals, like that of Eq. 19-43 for the single-phase system.

$$V_a = \frac{\lambda}{2\pi/3} V_a' + \frac{\mu}{2\pi/3} E_a \tag{19-57}$$

In this equation, speed changes are assumed to be negligible.

In this three-phase system, the load can increase beyond the point at which the extinction angle β coincides with the firing angle of the next thyristor. In this case, a different mode of operation is reached, in which each conducting interval ends after 120° ($\lambda = 120°$).

The armature terminal voltage consists of successive, repetitious intervals when the motor voltage is a segment of the source voltage. The average value of the armature terminal voltage is

$$V_a = \frac{1}{(2\pi/3)} \int_{\alpha+\pi/6}^{\alpha+5\pi/6} v_{an}(x)\, dx = \frac{1}{(2\pi/3)} \int_{\alpha+\pi/6}^{\alpha+5\pi/6} V_1 \sin x\, dx$$

$$= \frac{3\sqrt{3}}{2\pi} V_1 \cos\alpha = 0.827\, V_1 \cos\alpha \tag{19-58}$$

The d-c voltage at the armature terminals is adjustable between zero (for $\alpha = 90°$) and a maximum value of $0.827\, V_1$ (for $\alpha = 0°$).

During one conducting interval, the armature circuit is described by the following differential equation

$$R_a i_a + L_a(di_a/dt) = v_1 - e_a \tag{19-59}$$

where v_1 denotes the applicable source phase voltage. The solution of this equation has been studied in Section 19-10 with Eq. 19-37. It will be helpful in analyzing this mode of operation to consider the components of this answer. Accordingly, let

$$i_a(t) = i_1(t) + i_2(t) + i_3(t) \tag{19-60}$$

where i_1 is the component due to the source voltage. Choosing $v_1 = v_{ab} = V_1 \sin \omega t$,

$$i_1(t) = (V_1/Z_a) \sin(\omega t - \theta_a) \tag{19-61}$$

and

$$\mathbf{Z}_a = R_a + j\omega L_a = Z_a \underline{/\theta_a} \tag{19-62}$$

The component due to the motor generated voltage, E_a, is i_2. Assume the motor speed is constant. Then $E_a = k'\omega$ is constant, and so is

$$i_2(t) = -E_a/R_a \tag{19-63}$$

The remaining component, i_3, is the free response. This describes any required transient that allows the operation to adjust to various initial conditions that may be encountered.

$$i_3(t) = I_3\, e^{-t/\tau_a} \tag{19-64}$$

where τ_a is the time constant of the armature circuit

$$\tau_a = L_a/R_a \tag{19-65}$$

When this system operates in periodic steady state, the current at the end of a conducting interval must equal the current at the beginning of a conducting interval. $i_a(t_b) = i_a(t_a + 2\pi/3\omega) = i_a(t_a)$. See Fig. 19-22. Since i_2 is constant, any change in i_1 during the interval must be equal and opposite to the change in i_3

$$\Delta i_1 = i_1(t_b) - i_1(t_a) \tag{19-66}$$

$$\Delta i_3 = i_3(t_b) - i_3(t_a) \tag{19-67}$$

$$\Delta i_3 = -\Delta i_1 \tag{19-68}$$

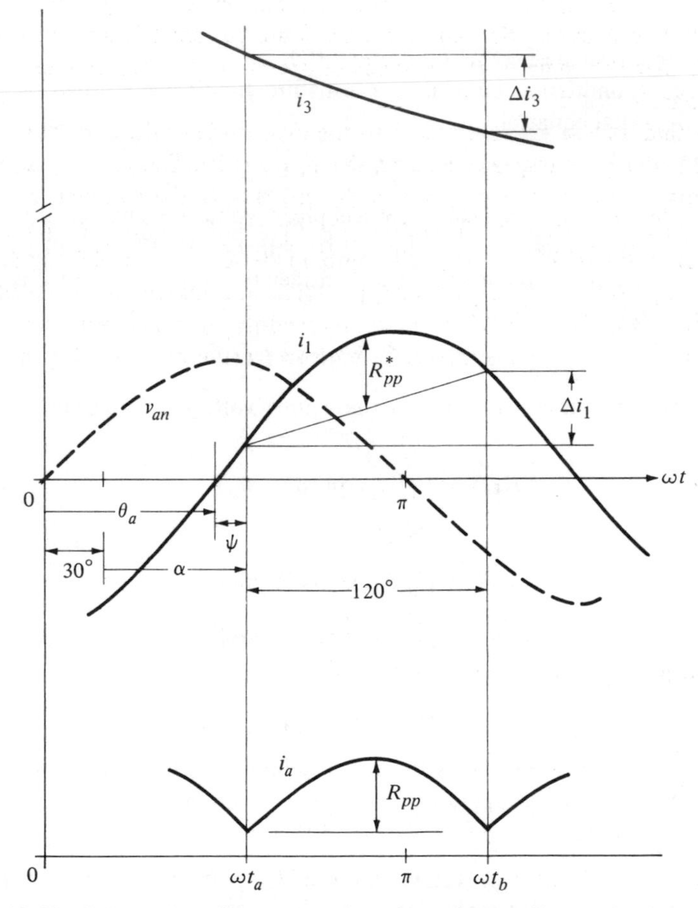

Fig. 19-22. Components of the current for a three-phase, half-wave thyristor drive.

Equation 19-68 along with Eqs. 19-61, 19-64, 19-66, and 19-67 can be solved to evaluate I_3. The average value of the armature current is given by

$$I_a = \frac{1}{(2\pi/3)} \int_{\alpha+\pi/6}^{\alpha+5\pi/6} i_a(x)\, dx \tag{19-69}$$

Kirchhoff's voltage law must hold for the d-c quantities in the armature circuit.

$$V_a = R_a I_a + E_a \tag{19-70}$$

This describes the operation of a d-c motor in the manner in which it was analyzed in Section 19-3. However, for this circuit both armature voltage and armature current have sizeable ripples of frequency 3ω. If we can estimate the magnitude of the ripple in the current, we can make a judgment about the validity of this analysis. So long as the armature current has positive value, we can know that the analysis of this mode of operation (i.e., continuous armature current) remains valid.

We can find a close approximation to the magnitude of the current ripple. In Fig. 19-22, the peak-to-peak value of the ripple is labelled R_{pp}. Assume that the current i_3 is approximately a straight line during the conducting interval. The current i_2 is constant. Therefore, the ripple effect comes primarily from i_1. We can evaluate R_{pp}^*, which will approximate R_{pp}, and which we define as $R_{pp}^* = i_1(\omega t_a + 60°) - [i_1(\omega t_a) + \frac{1}{2}\Delta i_1]$. While the maximum of i_a does not occur at $(\omega t_a + 60°)$, it is in close proximity, and the values are very close.

For this derivation, change the time reference to the zero crossing of the current. Then

$$\alpha + 30° = \theta_a + \psi$$

$$i_1 = I_1 \sin \omega t$$

$$i_1(\omega t_a) = I_1 \sin \psi$$

$$i_1(\omega t_a + 60°) = I_1 \sin(\psi + 60°)$$

$$\Delta i_1 = I_1 \sin(\psi + 120°) - I_1 \sin \psi$$

Solving for R_{pp}^*, we find

$$R_{pp}^* = 0.5\, I_1 \cos(\psi - 30°) = 0.5\, I_1 \cos(\alpha - \theta_a) \tag{19-71}$$

We conclude

$$R_{pp} \approx 0.5\, I_1 \cos(\alpha - \theta_a) \tag{19-72}$$

Observe in Fig. 19-21 that the variation of i_a above and below I_a is not the same. A handy rule of thumb may be to assume $\frac{1}{3}$ of R_{pp} above I_a and $\frac{2}{3}$ of

R_{pp} below I_a. This enables us to check for $i_{a\,\min}$

$$i_{a\,\min} \approx I_a - (\tfrac{2}{3})\, R_{pp} \qquad (19\text{-}73)$$

and, thus, estimate whether or not $i_{a\,\min}$ is positive. If the given method results in a zero or negative estimated value of $i_{a\,\min}$, then a more detailed analysis should be done to know which mode of operation is applicable.

We also need to investigate the speed dip, as this has a bearing on our assumption of constant speed. The torque function has the same waveshape as the armature current. The moment of inertia acts as a low-pass mechanical filter. The speed will not be able to follow the higher harmonic components, so we can restrict our analysis to the fundamental component of the ripple, i.e., 3ω. We assume the armature current can be approximated by

$$i_a(t) = I_a + (R_{pp}/2) \cos (3\omega t) \qquad (19\text{-}74)$$

where we assume a new zero reference for this function. Then

$$T(t) = k' i_a(t) = k' I_a + (k' R_{pp}/2) \cos (3\omega t) \qquad (19\text{-}75)$$

The load torque is assumed to be constant and to equal the average torque

$$T_L = T_{\text{ave}} = k' I_a \qquad (19\text{-}76)$$

The torque equation for the mechanical system is

$$T(t) = T_L + J(d\omega_m/dt) \qquad (19\text{-}77)$$

Equate Eqs. 19-77 and 19-75 to find

$$J(d\omega_m/dt) = (k' R_{pp}/2) \cos (3\omega t) \qquad (19\text{-}78)$$

Solve for the speed

$$\omega_m = \frac{1}{J} \int_0^t (k' R_{pp}/2) \cos (3\omega t)\, dt$$

$$= \frac{k' R_{pp}}{2J(3\omega)} \sin (3\omega t) + \omega_{m(\text{ave})} \qquad (19\text{-}79)$$

$\omega_{m(\text{ave})}$ has constant value. We can express the motor speed in the form

$$\omega_m = \omega_{m(\text{ave})} + (\Delta\omega_m/2) \sin (3\omega t) \qquad (19\text{-}80)$$

The peak-to-peak variation in speed is the speed dip, and it is found by equating Eqs. 19-79 and 19-80.

$$\Delta\omega_m = -k' R_{pp}/J(3\omega) \qquad (19\text{-}81)$$

See Example 19-7 for some indication of how small this will be. Since the actual fundamental component of the ripple current is smaller than what was assumed in Eq. 19-74, we can conclude that the actual speed dip must be smaller than a value computed from Eq. 19-81.

For large machines, three-phase, full-wave bridge circuits are used. The principles of analysis are the same as those presented here. The solutions, however, will have ripple frequencies of 6ω, as the conducting intervals will have 60° duration. (Refer to Fig. 10-22). While these circuits are needed for their large power capability, they have the added advantages of less difficulty with harmonics and smaller value for the speed dip.

Widespread use of integrated circuits to supply the triggering pulses for the thyristors or power transistors has enabled the control systems to incorporate many desired characteristics. A feature that is needed with d-c motors is a current limit mechanism. This can function automatically to limit the armature current to a safe value no matter what operating conditions may be encountered. It is also possible to combine analog and digital techniques into the control system. Small computers are used as part of the control system where logic or calculation is required as part of the process control.

19-12 CHOPPERS

There is a need for controlling motors that operate from batteries. Examples are industrial trucks and fork lifts used inside factories for material handling. The chopper circuit shown in Fig. 19-23 uses a power transistor to connect intermittently the armature of the motor to the d-c source. This can be done at the rate of a few hundred pulses per second. Adjusting the ratio of ON time to OFF time controls the average value of the armature voltage. See Fig. 19-24.

$$V_a = V\,T_1/T_2 \tag{19-82}$$

When the switch is ON, $v_a = V$, and the armature current is increasing. When the switch is OFF, armature current flows through the freewheeling diode, $v_a = 0$, and the armature current is decreasing.

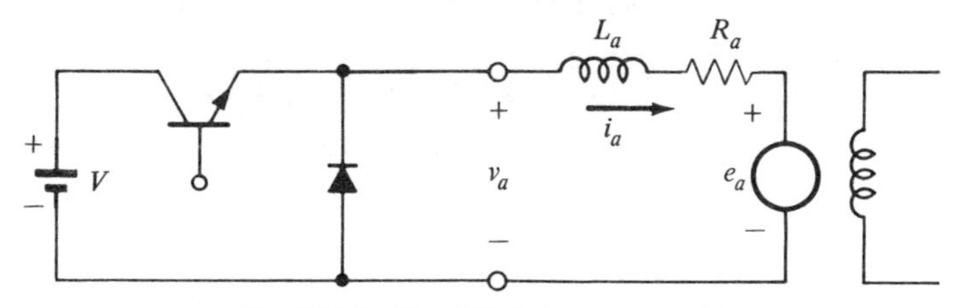

Fig. 19-23. Simplified chopper circuit.

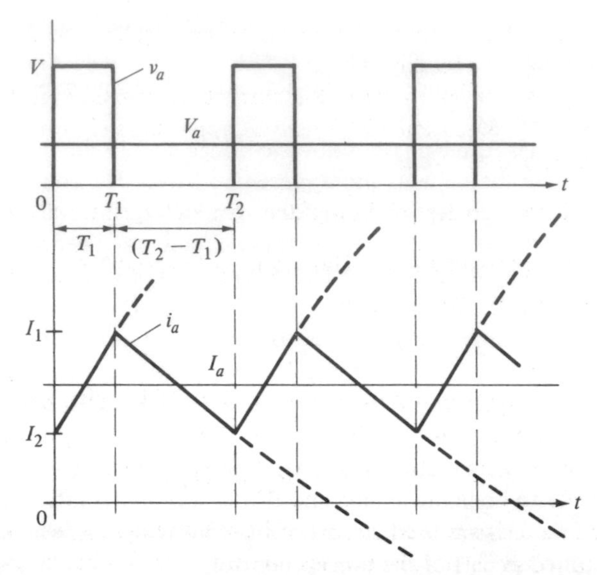

Fig. 19-24. Voltage and current waveforms for the chopper circuit.

In order to have periodic steady state it is necessary for the increase in current during T_1 to equal the decrease in current during $(T_2 - T_1)$. Kirchhoff's voltage equation must be satisfied by the average values

$$V_a = R_a I_a + E_a \tag{19-83}$$

The operating characteristics of the d-c motor with armature voltage control was studied in Section 19-5.

The frequency of the pulses is

$$f = 1/T_2 \tag{19-84}$$

The frequency must be chosen high enough to limit the peak-to-peak ripple in the current

$$R_{pp} = I_1 - I_2 \tag{19-85}$$

Assume that the speed is constant. Further, assume that $f = (1/T_2) > 5/\tau_a$. That is, $T_2 < \tau_a/5 = L_a/5R_a$. These assumptions permit the segments of current to be approximated by straight lines. The rigorous response of current is an exponential with time constant τ_a. The following are good approximations.

$$I_a = (I_1 + I_2)/2 \tag{19-86}$$

$$I_1 = I_a + R_{pp}/2 \tag{19-87}$$

$$I_2 = I_a - R_{pp}/2 \tag{19-88}$$

The simple circuit has no provision for i_a to be negative, so I_a must exceed $R_{pp}/2$ for these approximations to hold.

To find the peak-to-peak ripple in the current, remember that $R_a i_a$ is only a small fraction of V. Equation 19-83 remains approximately valid with instantaneous values of i_a and e_a. Thus

$$V - V_a \approx (e_a + R_a i_a + L_a\, di_a/dt) - (e_a + R_a i_a) = L_a\, di_a/dt$$

is a satisfactory approximation. This leads to the result $V - V_a \approx L_a\, di_a/dt \approx L_a R_{pp}/T_1$. Solving for R_{pp}, we find

$$R_{pp} \approx T_1(V - V_a)/L_a \tag{19-89}$$

With analysis similar to the derivation of Eq. 19-81, the speed dip can be found to be

$$\Delta\omega_m = -k' R_{pp}/J(2\pi f) \tag{19-90}$$

A chopper could also be used to control the current in a field winding. The chopper circuit plays the role of a series rheostat in that it has an average voltage across its terminals. It also has an average current flowing through it. In the ideal case, however, it has zero power. Using a chopper instead of a rheostat, thus, gives more efficient operation.

19-13 EXAMPLES

Example 19-1 (Section 19-3)

A d-c shunt motor has a no-load saturation curve for 1000 rpm given by

$$E = \frac{300 I_f}{1 + I_f} \qquad \text{or} \qquad I_f = \frac{E}{300 - E}$$

where E is the open-circuit voltage and I_f is the shunt field current. The demagnetizing effect of armature reaction is $(\mathfrak{F}_d/N_f) = 0.1$ equivalent shunt field amperes when the armature current is 100 amp. The armature circuit resistance is 0.12 Ω. The shunt field circuit resistance is 247 Ω. The motor is operating from a d-c source of 230 v. The load is adjusted to make the armature current to be 80 amp. Find the speed in revolutions per minute.

Solution

The actual shunt field current is

$$I_f = V/R_f = 230/247 = 0.93 \text{ amp}$$

The demagnetizing effect of armature reaction is proportional to the armature current

$$(\mathcal{F}_d/N_f) = (0.1/100) \times 80 = 0.08 \text{ equivalent shunt field amp}$$

Use Eq. 19-11 to find the effective shunt field current

$$I_{f\text{eff}} = I_f - (\mathcal{F}_d/N_f) = 0.93 - 0.08 = 0.85 \text{ amp}$$

If the speed were 1000 rpm, this field current would cause the generated voltage to be E_1. Use the no-load sat. curve to find E_1.

$$E_1 = 300(0.85)/(1 + 0.85) = 137.8 \text{ v}$$

Use Kirchhoff's voltage law for the armature circuit to find the actual generated voltage. Let E_2 denote this value

$$E_2 = V - R_a I_a = 230 - (0.12)(80) = 220.4 \text{ v}$$

For a fixed value of flux, the generated voltage is proportional to the speed

$$n_2 = E_2(n_1/E_1) = 220.4(1000/137.8) = 1600 \text{ rpm}$$

Example 19-2 (Section 19-3)

The d-c motor in Example 19-1 has a series field connected in the long shunt circuit to make the motor cumulative compound. The series field has three turns per pole and resistance of 0.02 Ω. The shunt field has 1260 turns per pole. For the same operating conditions as in Example 19-1, find the speed.

Solution

Use Eq. 19-11 to find the effective shunt field current.

$$\begin{aligned} I_{f\text{eff}} &= I_f + I_a(N_s/N_f) - (\mathcal{F}_d/N_f) \\ &= 0.93 + 80(3/1260) - 0.08 = 1.04 \text{ amp} \end{aligned}$$

If the speed were 1000 rpm, this field current would cause the generated voltage to be E_1. Use the no-load sat. curve to find E_1

$$E_1 = 300(1.04)/(1 + 1.04) = 152.9 \text{ v}$$

Use Kirchhoff's voltage equation for the armature circuit to find the actual generated voltage E_2

$$E_2 = V - (R_a + R_s)I_a = 230 - (0.12 + 0.02)(80) = 218.8 \text{ v}$$

For a fixed value of flux, the generated voltage is proportional to the speed

$$n_2 = E_2(n_1/E_1) = 218.8(1000/152.9) = 1430 \text{ rpm}$$

Example 19-3 (Section 19-3)

A d-c series motor has its no-load saturation curve for 1200 rpm given by

$$E = \frac{1400 I_s}{40 + I_s} \quad \text{or} \quad I_s = \frac{40E}{1400 - E}$$

where E is the generated voltage and I_s is the series field current. The demagnetizing effect of armature reaction is zero. The armature circuit resistance, including the series field, is 0.3 Ω. This motor is operated from a d-c source of 550 v. The load is adjusted to make the armature current to be 60 amp. Find the speed.

Solution

The effective field current is 60 amp. If the speed where 1200 rpm, this field current would cause the generated voltage to be E_1. Use the no-load sat. curve to find E_1

$$E_1 = 1400(60)/(40 + 60) = 840 \text{ v}$$

Use Kirchhoff's voltage law for the armature circuit to find the actual generated voltage E_2

$$E_2 = V - RI = 550 - (0.3)(60) = 532 \text{ v}$$

For a fixed value of flux, the generated voltage is proportional to the speed

$$n_2 = E_2(n_1/E_1) = 532(1200/840) = 760 \text{ rpm}$$

Example 19-4 (Section 19-7)

A d-c motor has the following parameters: $R_a = 0.05$ pu, $L_a = 0.0025$ pu, $k' = 1$ pu, $J = 0.8$ pu. This motor has no load connected to it, so the load torque is zero. It is at rest with the field excited, and the armature circuit is open. The motor is started by connecting the armature to a d-c source of 0.5 pu. This makes the armature voltage a step function of 0.5 pu. Find the response of the speed and the armature current.

Solution

The armature circuit time constant is

$$\tau_a = L_a/R_a = 0.0025/0.05 = 0.05 \text{ sec}$$

The mechanical time constant is

$$\tau_m = JR_a/k'^2 = (0.8)(0.05)/(1)^2 = 0.04 \text{ sec}$$

The characteristic equation is given by Eq. 19-29

$$(0.05)(0.04)\,s^2 + (0.04)\,s + 1 = 0$$

$$s^2 + 20\,s + 500 = 0$$

$$s = -10 \pm j\,20$$

The undamped natural frequency is

$$\omega_0 = \sqrt{500} = 22.36 \text{ (rad/sec)}$$

The damping ratio is

$$\zeta = \alpha/\omega_0 = 10/22.36 = 0.447$$

The damped frequency is

$$\omega_d = \sqrt{500 - (10)^2} = 20 \text{ (rad/sec)}$$

The initial conditions are: $\omega(0+) = \omega(0-) = 0, i_a(0+) = i_a(0-) = 0, (d\omega/dt)(0+) = (T - T_L)/J = 0$, because $T(0+) = k' i_a(0+) = 0$ and T_L is given to be zero. At $t = 0+$, Kirchhoff's voltage equation for the armature circuit is

$$v(0+) = R_a i(0+) + L_a(di_a/dt)(0+) + k'\omega(0+)$$

$$0.5 = 0 + (0.0025)(di_a/dt)(0+) + 0$$

$$(di_a/dt)(0+) = v(0+)/L_a = 200 \text{ (sec}^{-1}\text{)}$$

Use the roots of the characteristic equation and the initial conditions to find the solutions for speed and armature current

$$\omega(t) = 0.5 - 0.56\,e^{-10t} \sin(20t + 63.4°) \text{ pu}$$

$$i_a(t) = 10\,e^{-10t} \sin 20t \text{ pu}$$

The maximum speed of 0.604 pu occurs for $\omega t(180°/\pi) = 180°$. The speed overshoots its final value of 0.5 pu by $(0.604 - 0.5)/0.5 = 0.21$ or 21 percent. The maximum value of current is 5.14 pu, and it occurs at $\omega t(180°/\pi) = 63.4°$. Even though this motor is started with one-half of rated voltage, the maximum current is excessive. This makes clear the necessity of limiting the armature current during a starting transient.

Example 19-5 (Section 19-7)

The machine in Example 19-4 is operating with rated voltage and load torque of zero. So the speed is 1 pu, and the armature current is zero (ignore no-load losses). At $t = 0$, the load torque is changed instantaneously to be constant at rated value. So the load torque is a step function of 1 pu. Find the response of speed and armature current.

Solution

The time constants and the roots of the characteristic equation have been found in Example 19-4. For this problem, the initial conditions are as follows: $i_a(0+) = i_a(0-) = 0$, and $\omega(0+) = \omega(0-) = 1$ pu.

The torque equation is

$$J(d\omega/dt)(0+) = k' i_a(0+) - T_L$$

$$(d\omega/dt)(0+) = 0 - T_L/J = -1/0.8 = -1.25 \text{ (sec}^{-2})$$

Kirchhoff's voltage equation is

$$v(0+) = R_a i_a(0+) + L_a(di_a/dt)(0+) + k'\omega(0+)$$

$$1 = 0 + L_a(di_a/dt(0+) + 1$$

$$(di_a/dt)(0+) = 0$$

Use the roots of the characteristic equation and the initial conditions to find the solutions for speed and armature current

$$\omega(t) = 0.95 + 0.063\, e^{-10t} \sin(20t + 126.9°) \text{ pu}$$

$$i_a(t) = 1 - 1.12\, e^{-10t} \sin(20t + 63.4°) \text{ pu}$$

Example 19-6 (Section 19-10)

A 1-HP d-c motor is operated in the single-phase half-wave circuit of Fig. 19-18. The 60-Hz source voltage is 240 v (rms). The armature circuit resistance is 6.8 Ω. The self-inductance of the armature circuit is 0.20 henry. The shunt field current is set such that the voltage constant is $k' = 4.2$ v/(rad/sec) or 4.2 nm/amp. The moment of inertia is 0.068 kgm^2. The firing angle α is set to be 110°. The load is adjusted to make the speed be 22 rad/sec. (a) Find the function for armature current during the conducting interval. (b) Find the extinction angle β. (c) Find the average current during the conducting interval. (d) Find the average value of the armature current. (e) Find the speed dip.

Solution

(a) Assume the speed is constant. Then, e_a is constant

$$e_a = E_a = k'\omega_m = 4.2 \times 22 = 92.4 \text{ v}$$

The differential equation for the armature circuit during the conducting interval is given by Eq. 19-36. The impedance of the armature circuit at the source frequency is

$$\mathbf{Z}_a = R_a + j\omega L_a = 6.8 + j(377)(0.2) = 6.8 + j75.4 = 75.7 \underline{/84.8°}\ \Omega$$

The maximum value of the source voltage is

$$V_1 = \sqrt{2}\,(240) = 339.4 \text{ v}$$

The time constant of the armature circuit is

$$\tau_a = L_a/R_a = 0.20/6.8 = 0.0294 \text{ sec}$$

The current is given by Eq. 19-37

$$i_a(t) = (339.4/75.7) \sin (377t - 84.8°) - (92.4/6.8) + I_3\, e^{-t/0.0294} \text{ amp}$$

At the firing angle α, the current is zero

$$t_1 = \alpha/\omega = 110°(\pi/180°)/377 = 0.00509 \text{ sec}$$

Use $i_a(t_1) = 0$ to find $I_3 = 13.89$ amp. The armature current is

$$i_a(t) = -4.48 \cos (377t + 5.2°) - 13.59 + 13.89\, e^{-t/0.0294} \text{ amp}$$

(b) The conducting interval ends when the current reaches zero. We must find the time t_2 for which $i_a(t_2) = 0$. Refer to Fig. 19-19 for an estimate of β. We can make a first guess of $\beta = 220°$. This turns out to be too large. Subsequent guesses lead us to the answer $\beta = 211°$.

(c) The duration of the conducting interval is $\lambda = \beta - \alpha = 101°$. $\lambda/\omega = 0.00468$ sec. Use Eq. 19-40 to find the average value of current during this interval

$$I_a' = \frac{1}{\lambda/\omega} \int_{\alpha/\omega}^{\beta/\omega} i_a(t)\, dt = 1.03 \text{ amp}$$

(d) The average value of current over one cycle of the source is

$$I_a = I_a' \lambda/2\pi = 1.03(1.763/2\pi) = 0.289 \text{ amp}$$

The average torque is $T = k'I_a = 4.2 \times 0.289 = 1.21$ nm.

(e) The time duration of the coasting interval is

$$\Delta t = t_3 - t_2 = \mu/\omega = (360° - 101°)(\pi/180°)/377 = 0.012 \text{ sec}$$

Use Eq. 19-55 to compute the speed dip

$$\Delta\omega_m = -T(\Delta t)/J = -(1.21)(0.012)/(0.068) = 0.214 \text{ rad/sec}$$

The speed dip expressed as a ratio to the speed is

$$\Delta\omega_m/\omega_m = 0.214/22 = 0.0097$$

In this problem, the assumption of constant speed is quite reasonable.

As a means of checking the answers, we can find the average armature terminal voltage during the conducting interval by using Eq. 19-42.

$$V_a' = (V_1/\lambda)(\cos\alpha - \cos\beta) = (339.4/1.763)(\cos 110° - \cos 211°)$$
$$= 99.2 \text{ v}$$

Compare this with Eq. 19-45

$$V_a' = R_a I_a' + E_a = (6.8)(1.03) + 92.4 = 99.4 \text{ v}$$

This is good agreement considering that numbers have been rounded off, and especially considering the interpolation used in determining the value of β.

Example 19-7 (Section 19-11)

A 10-HP d-c shunt motor is operated in the three-phase, half-wave thyristor drive shown in Fig. 19-20. The armature circuit resistance is 0.21 Ω. The armature circuit self-inductance is 0.0105 henry. The moment of inertia is 0.432 kgm^2. The field current is set such that the voltage constant is 1.42 v/(rad/sec) or 1.42 nm/amp. The three-phase source has line-to-neutral voltage of 205 v (rms). $v_{an} = 290 \sin 377t$. The phase sequence is a-b-c. The firing angle is set at 45°. The constant-torque load requires the average electromagnetic torque to be 45 nm. (a) Find the speed. (b) Find the peak-to-peak ripple in the armature current. (c) Investigate the speed dip.

Solution

(a) The average armature terminal voltage is found from Eq. 19-58

$$V_a = 0.827\, V_1 \cos\alpha = 0.827(290)\cos 45° = 169.6 \text{ v}$$

The average current is found from Eq. 19-76

$$I_a = T/k' = 45/1.42 = 31.7 \text{ amp}$$

The generated voltage is

$$E_a = V_a - R_a I_a = 169.6 - (0.21)(31.7) = 162.9 \text{ v}$$

The speed is

$$\omega_m = E_a/k' = 162.9/1.42 = 114.7 \text{ rad/sec}$$

(b) The armature circuit impedance is

$$\mathbf{Z}_a = R_a + j\omega L_a = 0.21 + j(377)(0.0105) = 0.21 + j3.96 = 3.97 \underline{/87°}\ \Omega$$

The component of the current due to the source is

$$i_1 = (V_1/Z_a)\sin(\omega t - \theta_a) = (290/3.97)\sin(377t - 87°)$$
$$= -73\cos(377t + 3°) \text{ amp}$$

An approximation to the peak-to-peak ripple is given by Eq. 19-72

$$R_{pp} \approx 0.5 I_1 \cos(\alpha - \theta_a) = 0.5(73)\cos(45° - 87°) = 27.1 \text{ amp}$$

As a matter of interest, the current during a conducting interval is

$$i_a(t) = 73 \sin(377t - 87°) - 775.7 + 862.4\, e^{-t/0.05}$$

for $75° < \omega t < 195°$.

The minimum current is

$$i_{a\min} = i_a(t_a) = i_a(t_b) = 13.7 \text{ amp}$$

The maximum current is

$$i_{a\max} = i_a(0.00666) = 40.2 \text{ amp}$$

(c) The speed dip is given by Eq. 19-81

$$\Delta\omega_m = -k' R_{pp}/J(3\omega) = -1.42(27.1)/(0.432)(1131) = -0.079 \text{ rad/sec}$$

Expressed as a ratio to the speed, the speed dip is

$$\Delta\omega_m/\omega_m = -0.079/114.7 = 0.0007$$

Example 19-8 (Section 19-12)

The d-c motor of Example 19-7 is used in the chopper circuit of Fig. 19-23 with a d-c source of 200 v. The load requires the motor to develop an electromagnetic torque of 42.6 nm. The ON time is 2 msec. The OFF time is 4 msec. (a) Find the speed of the motor. (b) Find the peak-to-peak ripple in the armature current. (c) Find a new pulse frequency to limit the peak-to-peak ripple to 10 amp. Keep the same speed and torque.

Solution

(a) $T_1 = 0.002$ sec. $T_2 = 0.006$ sec.
Use Eq. 19-82 to find the armature terminal voltage

$$V_a = V(T_1/T_2) = 200(0.002/0.006) = 66.7 \text{ v}$$

Find the average armature current

$$I_a = T/k' = 42.6/1.42 = 30 \text{ amp}$$

Use Eq. 19-83 to find the generated voltage

$$E_a = V_a - R_a I_a = 66.7 - (0.21)(30) = 60.4 \text{ v}$$

Find the speed

$$\omega = E_a/k' = 60.4/1.42 = 42.5 \text{ rad/sec}$$

(b) Use Eq. 19-89 to find the peak-to-peak ripple

$$R_{pp} \approx T_1(V - V_a)/L_a = (0.002)(200 - 66.7)/(0.0105) = 25.4 \text{ amp}$$

(c) The peak-to-peak ripple is required to be 10 amp. Use Eq. 19-89 to find T_1

$$T_1 \approx R_{pp}L_a/(V - V_a) = 10(0.0105)/133.3 = 0.00079 \text{ sec}$$

To maintain the same speed and torque requires V_a to remain at 66.7. Use Eq. 19-82 to find T_2

$$T_2 = T_1 V/V_a = (0.00079)(200)/66.7 = 0.00237 \text{ sec}$$

The required pulse frequency is

$$f = 1/T_2 = 1/(0.00237) = 422 \text{ Hz}$$

19-14 PROBLEMS

19-1. The d-c motor of Example 19-1 is operated with the same source voltage and the same field current. The load is increased, however, to make the armature current to be 120 amp. Find the speed.

19-2. The cumulative compound d-c motor of Example 19-2 is operated with the same source voltage and shunt field current. The load, however, is increased to make the armature current to be 120 amp. Find the speed.

19-3. The d-c series motor of Example 19-3 is operated with the same source voltage, but the load is changed to make the armature current to be 20 amp. Find the speed.

19-4. A 75-HP, 500-v, 500-rpm, d-c series motor has an armature circuit resistance (including the series field) of 0.23 Ω. The rated load current at rated voltage and rated speed is 124 amp. The no-load saturation curve for 300 rpm is given by

$$E = \frac{530I}{100 + I} \quad \text{or} \quad I = \frac{100E}{530 - E}$$

where E is the generated voltage and I is the effective field current. Find the demagnetizing effect of armature reaction in equivalent series field amperes, for rated armature current (i.e., $\mathcal{F}_d/N_s = ?$).

19-5. A d-c series motor operates from a 230-v source. The armature and brush contact resistance is 0.44 Ω. The series field resistance is 0.21 Ω.

The demagnetizing effect of armature reaction is zero. The no-load saturation curve for 1000 rpm is given by

$$E = \frac{570I}{35+I} \quad \text{or} \quad I = \frac{35E}{570-E}$$

where E is the generated voltage and I is the effective field current. The rotational losses are 145 w. Stray load loss may be taken to be 1 percent of $(E \times I)$, for this problem. The load is adjusted to make the armature current to be 24 amp. (a) Find the speed. (b) Find the efficiency.

19-6. Repeat Problem 19-5, except for the armature current to be 36 amp.

19-7. A d-c shunt motor has its no-load saturation curve for 1200 rpm given by

$$E = \frac{1{,}020 I_f}{0.9 + I_f} \quad \text{or} \quad I_f = \frac{0.9E}{1020 - E}$$

where E is the generated voltage and I_f is the effective field current. The armature circuit resistance is 0.48 Ω. The shunt field current is 1.2 amp. The armature terminal voltage is 460 v. The armature current is 50 amp. (a) Find the speed. (b) Find the electromagnetic torque.

19-8. The motor in Problem 19-7 has its field current reduced to 0.6 amp. The armature terminal voltage is 460 v, and the armature current is 50 amp. (a) Find the speed. (b) Find the electromagnetic torque.

19-9. The motor in Problem 19-7 is operated with armature terminal voltage of 230 v. The field current is 1.2 amp. The armature current is 50 amp. Find the speed.

19-10. The motor in Problem 19-7 has a series field with 18 turns per pole. The resistance of the series field is 0.1 Ω. The shunt field has 1900 turns per pole. This machine is operated as a cumulative compound motor with the shunt field current set at 0.8 amp. The armature terminal voltage is 460 v. (a) The armature current is 50 amp. Find the speed. (b) Find the no-load speed.

19-11. The motor in Problem 19-7 is to have a series field added to it. The resistance of the series field is assumed to be 0.1 Ω. The armature terminal voltage is 460 v. (a) Find the value of the shunt field current to make the no-load speed be 1000 rpm. (b) With the shunt field current set at the value in (a), find the number of turns needed in the series field to make the speed be 900 rpm when the armature current is 50 amp.

19-12. A d-c shunt motor is rated 5 HP, 230 v, 18 amp, 1500 rpm. The armature circuit resistance is 0.5 Ω. The armature self-inductance is 0.03 henry. The moment of inertia is 0.196 kgm^2. The shunt field current is set to make the voltage constant be 1.4 v/(rad/sec) or 1.4 nm/amp. The motor is operating at rated voltage with no load. The speed is 164.3 rad/sec. The armature current is zero (ignore all no-load losses). A constant torque load of 21 nm is applied suddenly. (a) Find the roots of the characteristic equation. (b) Find the response of the speed and the armature current.

19-13. The motor in Problem 19-12 is at standstill with the armature circuit open. The field current is set to make $k' = 1.4$ nm/amp. The load torque is zero. The motor is started by connecting the armature to a d-c source of 80 v. Find the response of the speed and the armature current.

19-14. The motor of Problem 19-12 is initially at rest. The voltage impressed on the armature is a ramp function with a slope of 110 v/sec ($v_a(t) = 110\ tu(t)$ v). The connected load is zero. Find the response of speed and armature current.

19-15. Solve Example 19-6 for operation with $\alpha = 110°$ and the load decreased to make the speed be 26 rad/sec.

19-16. Solve Example 19-6 for operation with α set to 90° and the load such that the speed is 26 rad/sec.

19-17. Solve Example 19-6 for operation with α set to 90° and the load such that the speed is 30 rad/sec.

19-18. The d-c motor system in Example 19-7 is operated with the firing angle changed to be 70°. The torque is changed to be 50 nm. (a) Find the speed. (b) Find the peak-to-peak ripple in the armature current. (c) Investigate the speed dip.

19-19. A 100-HP d-c motor is operated in the three-phase, half-wave thyristor drive shown in Fig. 19-20. The armature circuit resistance is 0.015 Ω. The armature self-inductance is 0.0007 henry. The moment of inertia is 5.9 kgm^2. The shunt field current is set such that the voltage constant is 1.22 v/(rad/sec) or 1.22 nm/amp. The three-phase source has line to neutral voltage given by $v_{an}(t) = 315 \sin 377t$ v. The phase sequence is *a-b-c*. The load is such that the electromagnetic torque is required to be 405 nm for a speed of 170 rad/sec. (a) Find the firing angle α. (b) Find the peak-to-peak ripple in the armature current. Also make an estimate of the minimum instantaneous value of the armature current function. (c) Investigate the speed dip.

19-20. Repeat Problem 19-19, but with torque of 400 nm and speed of 136 rad/sec.

19-21. The d-c motor of Example 19-7 is used in the chopper circuit of Fig. 19-23 with a d-c source of 250 v. The load requires the motor to develop an electromagnetic torque of 23 nm. The ON time is 1.6 msec. The OFF time is 2.4 msec. (a) Find the speed of the motor. (b) Find the peak-to-peak ripple in the armature current. (c) Find the new pulse frequency that would limit the peak-to-peak ripple to 10 amp. Keep the same torque and speed.

19-22. The d-c motor of Example 19-7 is used in the chopper circuit of Fig. 19-23 with a d-c source of 200 v. The load requires the motor to develop an electromagnetic torque of 27 nm. The speed is to be 54 rad/sec. The peak-to-peak ripple in the armature current must be 12 amp. Find the pulse frequency. Find the ON time. Find the OFF time.

20
D-c Generators

20-1 SEPARATE EXCITATION

The operation of all generators and motors discussed in this book is based on Faraday's law. Voltages are induced in a system of conductors moving relative to a *magnetic field*. In the case of a motor, this magnetic field may be produced by a current drawn from the same source as the main motor current. But where is a generator to draw its field excitation from, unless its magnetic field is produced by a permanent magnet?

Three distinct possibilities exist: (1) The generator may operate in parallel with *other generators* and draw its exciting current from their output. This is the way an induction generator operates; see Section 13-10. (2) The generator may produce *its own* field current. This is quite practical for d-c generators and will be discussed in a later section. (3) The field current may come from a *separate source*. This scheme is used for most synchronous generators. Its application to d-c generators is the subject of this section.

The reader knows that, in steady-state operation, the field circuit does not take part in the electromechanical energy conversion that is the purpose of the machine. The magnetic flux and its stored energy remain constant. The source of the field circuit has to provide that energy while the flux is being built up, i.e., during a transient interval. Once a steady state is reached, the only energy drawn from that source is the one to be converted into heat in the field circuit resistance. Thus, the power rating of that source need not be more than a small fraction of that of the generator.

Just as for a synchronous generator, the source of the field circuit of a d-c generator can be a battery, an auxiliary d-c generator called an exciter (which itself may be self-excited or have a permanent magnet as its stator), or the a-c power system through rectifiers. *Thyristors* may be used to combine the function of rectification with that of current adjustment. With d-c field sources, control can be accomplished by the use of *rheostats*.

20-2 THE EXTERNAL CHARACTERISTIC

It is the job of a power generator to supply a *constant voltage* to its load, whereas the current drawn by the load may change continuously. But since a generator cannot really be an ideal voltage source, its terminal voltage does change somewhat with the load current.

The relationship between these two quantities (V versus I) can be described graphically, and is called the *external characteristic* of a generator. (The name indicates that both variables are accessible outside the machine, namely at its terminals.) For this relationship it is understood that the generator is driven at constant speed, and that its field circuit remains unchanged.

In the case of the separately excited d-c generator, there are two distinct reasons for its departure from the ideal: the ***resistance*** of the armature circuit, and the effect of ***armature reaction***. Refer to the circuit diagram of Fig. 20-1 in which the polarity marks and current arrows are drawn in agreement with actual polarities and current directions for generator operation. Accordingly, the steady-state voltage equation of the armature circuit is

$$V = E - R_a I \tag{20-1}$$

(Note that the load current I is identical with the armature current.) This is the equation of the external characteristic. If armature reaction is neglected, the induced voltage E is a constant, and the equation represents a straight line whose slope indicates the imperfection of the generator in comparison to an ideal source.

Let the effect of armature reaction be considered as a number of demagnetizing ampere-turns proportional to the armature current. (See the next-to-last paragraph of Section 18-5.) This effect can be expressed in terms of a reduction of the induced voltage, by means of the familiar no-load characteristic. In Fig. 20-2, I_f is the actual field current, and $E_{NL} = V_{NL}$ the induced voltage at no-load (point 1). Let the effective field current be $I_{f_{\text{eff}}} = I_f - \mathfrak{F}_a/N_f$; as in Eq. 19-5, with $\mathfrak{F}_d$ the number of demagnetizing ampere-turns for rated load. Point 2 shows the loss of voltage due to armature reaction. Subtracting the resistance drop $R_a I$ according to Eq. 20-1 leads to the terminal voltage (point 3).

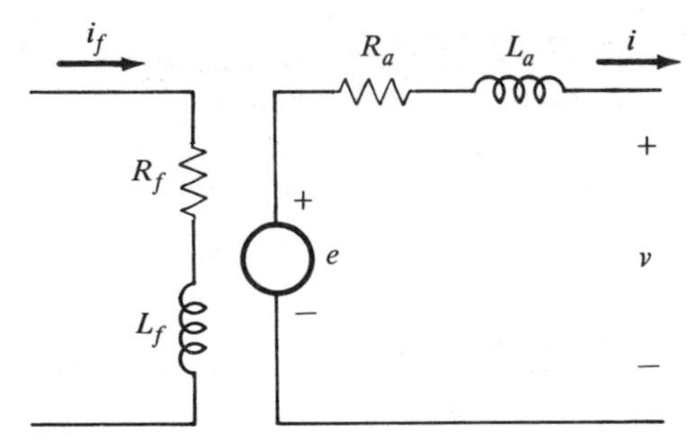

Fig. 20-1. Separately excited d-c generator.

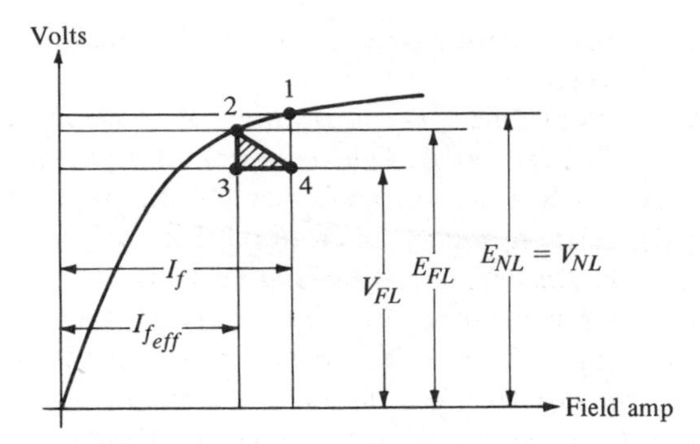

Fig. 20-2. Open-circuit characteristic.

The graphical procedure can be repeated for other values of load current. The sides of the shaded triangle are proportional to the load current. Point 2 slides along the no-load characteristic while point 4 remains on the vertical line drawn through point 1. As a result, curves of E versus I and V versus I can be plotted (see Fig. 20-3), the latter being the external characteristic.

This diagram also displays the *voltage regulation*, a quantity previously encountered in the context of transformers and synchronous generators. The definition for d-c generators is the same:

$$\epsilon = \frac{V_{\text{NL}} - V_{\text{FL}}}{V_{\text{FL}}} \tag{20-2}$$

where the full-load voltage V_{FL} is to be given rated value. In other words, the external characteristic contains rated operation as one of its points. In per-unit terms, using rated values as bases, the no-load voltage is always larger than 1 pu,

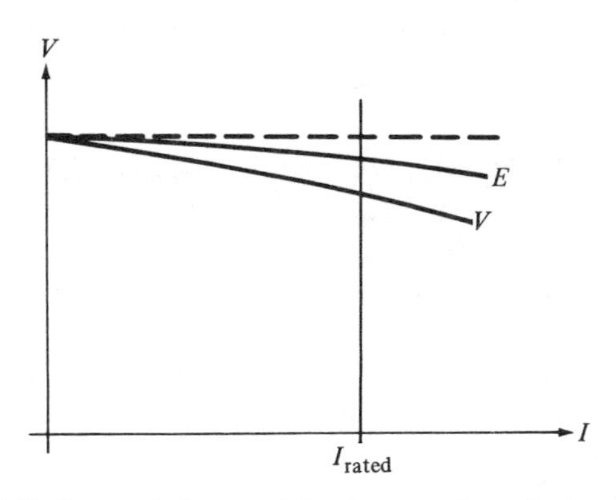

Fig. 20-3. External characteristic of separately excited generator.

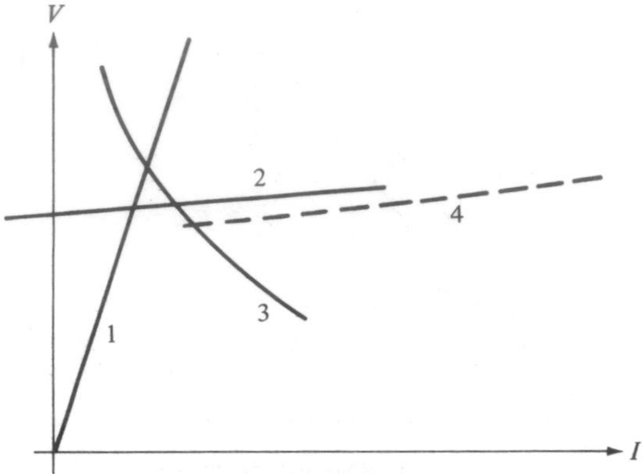

Fig. 20-4. Load characteristics.

and the regulation is

$$\epsilon = V_{NL} - 1 \tag{20-3}$$

In those cases in which this value is more than the load specifications allow, the field current of the generator must be adjusted, either manually or automatically. Specifically, the field excitation at no-load must be reduced, the external characteristic lowered for operation at no-load.

On the same diagram with the external characteristic of the generator, the voltage-current characteristic of the *load* may be drawn. Figure 20-4 shows several such characteristics: curve 1 for a pure resistance load, curve 2 for a storage battery being charged, and curve 3 for a motor driving a constant-power mechanical load. The broken curve 4 is the composite characteristic for a combined load consisting of these three loads in parallel. The intersection of this curve with the generator's external characteristic is the point of steady-state operation.

20-3 SELF-EXCITATION

It is by no means obvious that a generator can act as a source for its own field current. After all, if there is no magnetic flux, then there is no induced voltage and no field current, and if there is no field current, then there is no flux. The fact is that the possibility of self-excitation is based on the presence of *residual magnetism*.

When the field and armature circuits are properly connected to each other, and the generator is driven by a prime mover, then any amount of initial flux, no matter how small, is sufficient to let the generator *build up* its voltage. The basic idea is illustrated in Fig. 20-5. The descending branch of the hysteresis loop

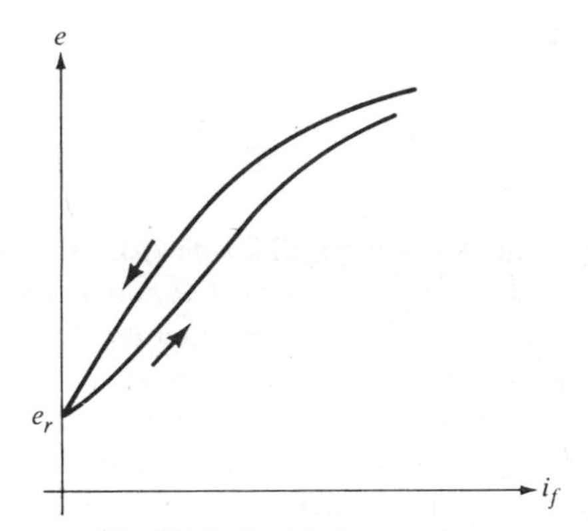

Fig. 20-5. Residual magnetism.

shows what happened when the generator was shut down after its previous run. When it is started again, the voltage e_r induced by the residual flux produces a field current. This relatively small current strengthens the flux, which in turn induces more voltage, etc.

It is only fair to ask where the residual flux is to come from if there was no "previous run." The answer is that, in its virginal state, the magnetic circuit must once be excited from a separate source.

The following more detailed explanation of the build-up process applies to the *shunt generator*. (It is also substantially valid for a compound generator. The series generator, however, which does not produce anything even distantly resembling a constant voltage under varying load, need not be considered here.)

In the circuit diagram of Fig. 20-6, the terminals of a shunt generator are left open. It is quite reasonable (though not necessary), not to connect the load until after the build-up process is completed. This reduces the circuit under consideration to an all-series loop consisting of the ideal source e, the resistance $R = R_a + R_f$, and the inductance $L = L_a + L_f$. The current in this circuit is the

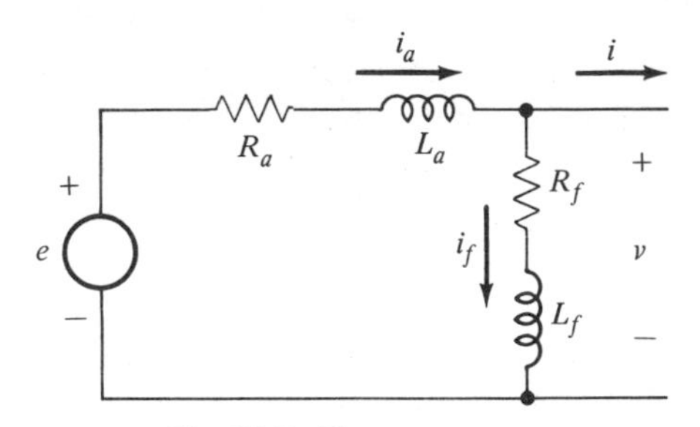

Fig. 20-6. Shunt generator.

field current i_f, and the equation of the circuit is

$$e = Ri_f + L \frac{di_f}{dt} \tag{20-4}$$

Figure 20-7 displays the open-circuit characteristic starting with residual magnetism. The voltage e_r produces a current that would increase toward the steady-state value e_r/R if this rising current did not increase the flux and, thereby, the voltage e itself. Thus, current, voltage, and flux are all rising. For instance, at the moment when the current has the value represented by the distance $0a$ in the diagram, the induced voltage is represented by the distance ac.

A straight line $0bd$, called the ***resistance line***, is drawn through the origin at a slope equal to the circuit resistance R. At the instant when e is represented by the distance ac, the distance ab represents the resistance drop Ri_f, and the distance bc, thus, is left to represent $L\, di_f/dt$, which is seen to be positive, meaning that the current is still increasing. It is now clear that the field current increases as long as $e > Ri_f$, and that the intersection (point d) represents the steady state.

The diagram also shows that the field circuit resistance (the major part of R) is chosen, or adjusted, to yield the desired no-load voltage. If R_f is increased, the slope of the resistance line is increased, which moves the point of intersection to the left and, thus, decreases the no-load voltage. But if R_f is increased beyond a certain value called the ***critical*** field circuit resistance (the one for which the resistance line is a tangent to the open-circuit characteristic), then there is no intersection (except for points very near the residual condition), and no build-up of voltage occurs.

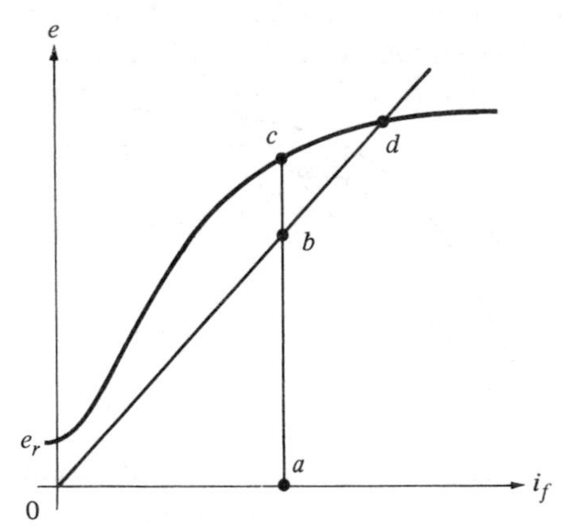

Fig. 20-7. Build-up process.

20-4 SHUNT GENERATORS UNDER LOAD

When a load is connected to the terminals of the shunt generator of Fig. 20-6, then there are three different currents to be considered, related to each other by the Kirchhoff current equation

$$i = i_a - i_f \tag{20-5}$$

Also, there is no longer a series connection of armature and field windings. Accordingly, the *field resistance line* 041 in Fig. 20-8 is drawn at the slope R_f, and its ordinates represent the steady-state values of the terminal voltage as a function of the actual field current

$$V = R_f I_f \tag{20-6}$$

On the other hand, the open-circuit characteristic represents the relation between the induced voltage and the *effective* field current, which differs from the actual field current by the effect of armature reaction. In the steady state, the induced voltage is

$$E = V + R_a I_a \tag{20-7}$$

The external characteristic of the shunt generator is now obtained in a manner similar to that used in Section 20-2 for the separately excited generator. Choose an arbitrary value of the armature current I_a, and draw a rectangular triangle whose vertical side equals $R_a I_a$ and whose horizontal side equals $\mathfrak{F}_d/N_f$, both for the chosen value of I_a. This triangle is fitted to the diagram in the way shown in Fig. 20-8: point 2 is located on the open-circuit characteristic and point 4 on the field resistance line. Thus, the coordinates of point 2 are $I_{f\,\text{eff}}$ and

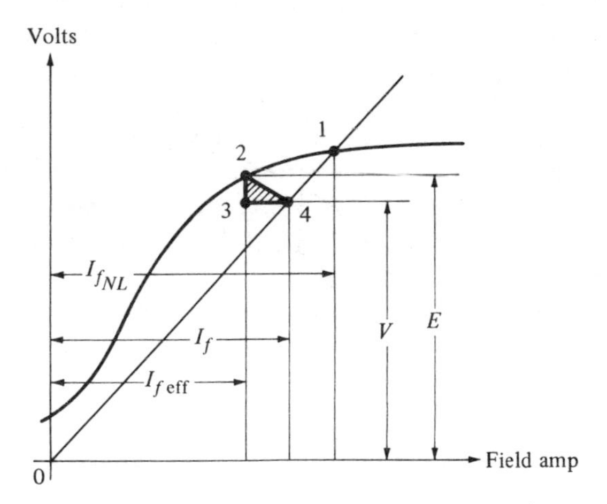

Fig. 20-8. Open-circuit characteristic and field resistance line (shunt generator).

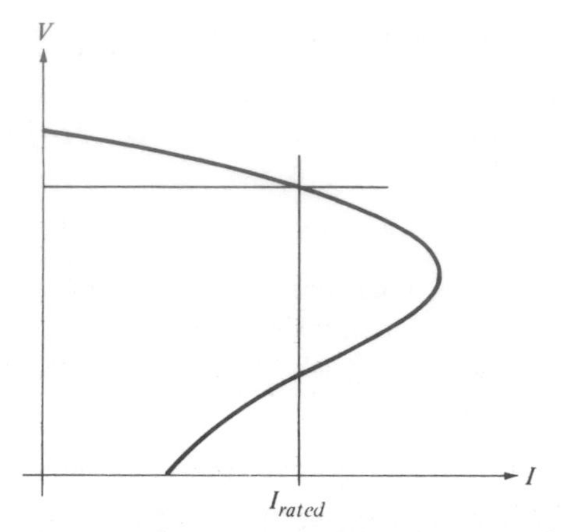

Fig. 20-9. External characteristic (shunt generator).

E, and those of point 4 are I_f and V. As to point 1, it may still be taken to represent the no-load condition, since the difference between the slopes R_f and $R = R_f + R_a$ is clearly not significant; or, to put it differently, since the triangle 234 becomes negligibly small when $I = 0$.

The external characteristic of the shunt generator, obtained by repeating this procedure (graphically or by computer) for several different values of I_a and tabulating and plotting the results, is sketched in Fig. 20-9. Its striking feature is that it reaches a maximum value of I and then turns back (to the left). The reason can be seen in Fig. 20-8: the triangle to be fitted between the open-circuit characteristic and the field resistance line must first increase in size but then decrease, when it is moved to the left. Figure 20-9 also shows that, in the case of a short-circuit at its terminals, the shunt generator delivers only a comparatively small current (in striking contrast to the separately excited generator) whose value depends on the residual magnetism.

The *voltage regulation* of the shunt generator is worse (larger) than it would be if the field winding were disconnected from the armature and supplied from a separate constant voltage source. This can be seen by comparing Figs. 20-2 and 20-8, and it can be understood from the fact that the actual field current of the shunt generator decreases with increasing load, whereas it stays constant in case of separate excitation. Thus, the shunt generator is likely to require a *field rheostat* to adjust the terminal voltage, manually or automatically, when the load changes.

20-5 COMPOUND GENERATORS

The purpose of equipping a shunt generator with an additional field winding in series with the armature, as shown in Fig. 20-10, is to improve its external char-

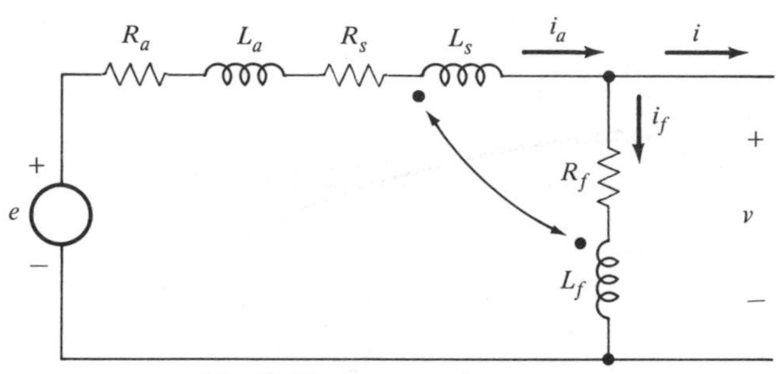

Fig. 20-10. Compound generator.

acteristic; in other words, to reduce its voltage regulation. This requires the compounding to be *cumulative*, as indicated by the coupling dots in the diagram. When the load current increases, the magnetic circuit is given additional ampere-turns, just as if the field current were increased. Such a generator may not need a field rheostat.

To be effective for this purpose, the series field winding should have enough turns to make $N_s I_a > \mathcal{F}_d$. In this case,

$$I_{f_{\text{eff}}} = I_f + \frac{N_s}{N_f} I_a - \frac{\mathcal{F}_d}{N_f} \tag{20-8}$$

is larger than I_f, and the horizontal side of the triangle to be fitted to the open-circuit characteristic and the field resistance line is reversed, as Fig. 20-11 illustrates.

According to this diagram, the terminal voltage under load is still smaller than at no-load. Correspondingly, the external characteristic of Fig. 20-12 is still

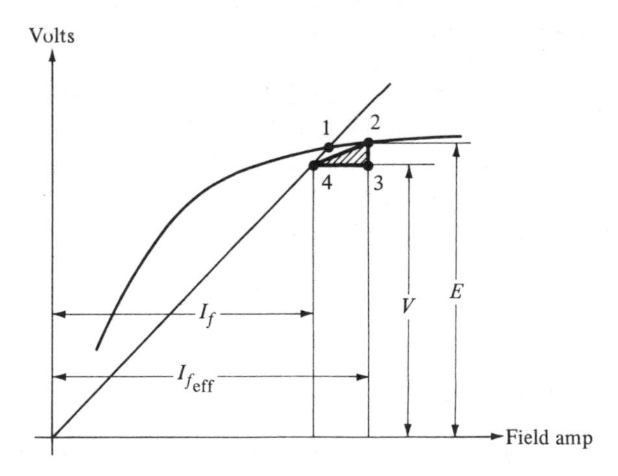

Fig. 20-11. Open-circuit characteristic and field resistance line (compound generator).

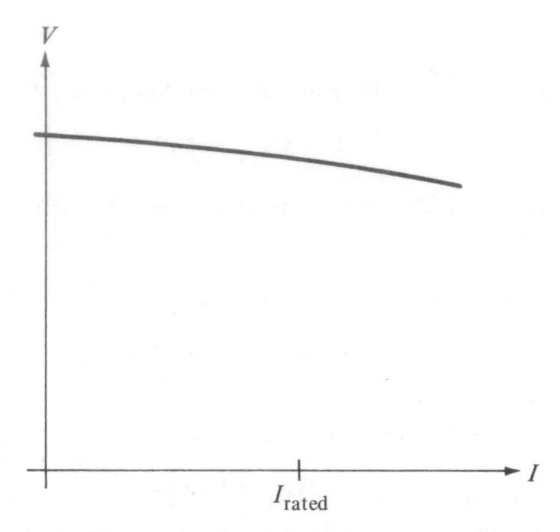

Fig. 20-12. External characteristic of compound generator.

drooping, although less than it would be without series field winding (and it does not display the short-circuit loop of Fig. 20-9). A compound generator with such an external characteristic is called *undercompounded*.

Raising the number of turns of the series field winding to such an extent that the terminal voltage at full load equals its no-load value (which makes the voltage regulation zero) is called *flat compounding* because, in this case, the external characteristic is almost a horizontal straight line. A further increase in the number of turns of the series field winding is referred to as *overcompounding*.

20-6 EXAMPLES

Example 20-1 (Section 20-4)

A d-c machine is connected in the shunt generator circuit of Fig. 20-6. When operated at rated speed of 1500 rpm., the open-circuit characteristic is approximated by

$$E = \frac{367\, I_f}{0.233 + I_f} \qquad \text{or} \qquad I_f = \frac{0.233\, E}{367 - E}$$

where E is the no-load generated voltage and I_f is the shunt field current. The armature circuit resistance is 0.14 Ω. The shunt field circuit resistance is adjusted to be 459 Ω. The demagnetizing effect of armature reaction is 0.049 equivalent shunt field amperes for an armature current of 70 amp. (a) Find the no-load terminal voltage. (b) Find the terminal voltage when the load requires the armature current to be 60 amp.

Solution

(a) At no-load, the generated voltage equals approximately the terminal voltage

$$E \approx V = R_f I_f = 459\, I_f$$

This equation describes the field resistance line. Substitute this in the open-circuit characteristic to find

$$V_{NL} = E_{NL} = 260 \text{ v}, \quad I_{f_{NL}} = 0.566 \text{ amp}$$

The intersection of the field resistance line with the open-circuit characteristic locates point 1 in Fig. 20-8.

(b) Under load conditions, the actual field current is

$$I_f = V/R_f = V/459$$

The demagnetizing effect of armature reaction is proportional to the armature current

$$\mathfrak{F}_d/N_f = (0.049)\,(60)/70 = 0.042 \text{ amp}$$

The effective field current is

$$I_{f_{\text{eff}}} = I_f - (\mathfrak{F}_d/N_f) = (V/459) - 0.042$$

Solve for V

$$V = 459 I_{f_{\text{eff}}} + 19.3$$

From Kirchhoff's voltage law for the armature circuit,

$$V = E - R_a I_a = E - (0.14)\,(60) = E - 8.4$$

Equate the two expressions for V and solve for E

$$E = 459 I_{f_{\text{eff}}} + 27.7$$

This is the equation of a line that is parallel to the field resistance line. The intersection of this line with the open-circuit characteristic is located at point 2 in Fig. 20-8. Solve the two equations to find

$$E = 246.6 \text{ v}, \quad I_{f_{\text{eff}}} = 0.477 \text{ amp}$$

The terminal voltage is

$$V = E - R_a I_a = 246.6 - 8.4 = 238.2 \text{ v}$$

Example 20-2 (Section 20-5)

The machine in Example 20-1 is to be operated as a cumulative compound d-c generator in the circuit of Fig. 20-10. The shunt field has 2180 turns per pole.

The series field has 11 turns per pole and resistance of 0.029 Ω. The resistance of the shunt field circuit is adjusted to make the no-load voltage to be 240 v. Find the terminal voltage for an armature current of 50 amp.

Solution

For $V_{\mathrm{NL}} = 240$ v, the field current must be

$$I_{f\mathrm{NL}} = 0.233(240)/(367 - 240) = 0.44 \text{ amp}$$

The resistance of the shunt field circuit must be

$$R_f = V_{\mathrm{NL}}/I_{f\mathrm{NL}} = 240/0.44 = 545\ \Omega$$

Under load conditions, the actual field current is

$$I_f = V/R_f = V/545$$

The demagnetizing effect of armature reaction is proportional to the armature current

$$\mathcal{F}_d/N_f = (0.049)\,(50)/70 = 0.035 \text{ amp}$$

The effective field current is

$$\begin{aligned} I_{f\mathrm{eff}} &= I_f - (\mathcal{F}_d/N_f) + I_a N_s/N_f = (V/545) - 0.035 + (50)\,(11)/2180 \\ &= (V/545) + 0.217 \end{aligned}$$

Solve for V

$$V = 545 I_{f\mathrm{eff}} - 118.3$$

From Kirchhoff's voltage law for the armature circuit,

$$V = E - (R_a + R_s)I_a = E - (0.14 + 0.029)\,(50) = E - 8.45$$

Equate the two expressions for V and solve for E

$$E = 545 I_{f\mathrm{eff}} - 110$$

This is the equation of a line that is parallel to the field resistance line. The intersection of this line with the open-circuit characteristic is located at point 2 in Fig. 20-11. Solve the two equations to find

$$E = 276.2 \text{ v}, \qquad I_{f\mathrm{eff}} = 0.709 \text{ amp}$$

The terminal voltage is

$$V = E - (R_a + R_s)I_a = 276.2 - 8.45 = 268 \text{ v}$$

For this machine, the series field has an effect that is strong enough to make the terminal voltage rise as the load current is increased (overcompounding).

20-7 PROBLEMS

20-1. The shunt generator in Example 20-1 has the shunt field circuit resistance set at 416 Ω. (a) Find the no-load terminal voltage. (b) Find the terminal voltage for an armature current of 90 amp.

20-2. The shunt generator in Example 20-1 is to be operated with the terminal voltage at 250 v for an armature current of 60 amp. (a) Find the field circuit resistance for this condition. (b) For the same field resistance, find the no-load voltage.

20-3. The machine of Example 20-1 is to be operated as a separately excited d-c generator in the circuit of Fig. 20-1. (a) Find the field current that makes the no-load voltage to be 250 v. (b) With the same field current, find the terminal voltage for an armature current of 80 amp.

20-4. The machine of Example 20-1 is operated as a separately excited d-c generator in the circuit of Fig. 20-1. (a) The terminal voltage is to be 250 v for an armature current of 80 amp. Find the required field current. (b) With the same field current, find the no-load voltage.

20-5. The cumulative compound d-c generator of Example 20-2 is operated with its no-load voltage at 240 v. Find the terminal voltage for an armature current of 80 amp.

20-6. The d-c generator of Example 20-1 is to have a series field added to it. It is to be operated in the circuit of Fig. 20-10. The shunt field resistance is to be adjusted to make the no-load voltage to be 250 v. Assume the resistance of the series field to be 0.014 Ω. The shunt field winding has 2180 turns per pole. Find the number of turns required in the series field winding to make the terminal voltage be 250 v for an armature current of 60 amp.

20-7. A d-c generator has its open-circuit characteristic for a speed of 1750 rpm given by

$$E = \frac{435 I_f}{1.93 + I_f} \qquad \text{or} \qquad I_f = \frac{1.93 E}{435 - E}$$

where E is the generated voltage and I_f is the shunt field current. The shunt field winding has 2100 turns per pole. The series field has two turns per pole and resistance of 0.0031 Ω. The armature circuit resistance (excluding the series field) is 0.0192 Ω. The demagnetizing effect of armature reaction is zero.

This machine is operated as a separately excited d-c generator in the circuit of Fig. 20-1. The shunt field current is set to make the no-load terminal voltage to be 250 v. Find the terminal voltage for an armature current of 400 amp.

20-8. The machine of Problem 20-7 is used with the shunt field separately excited to give a no-load terminal voltage of 250 v. The series field is connected in the armature circuit to cause cumulative ampere-turns. Find the terminal voltage for an armature current of 400 amp.

20-9. The machine of Problem 20-7 is used as a shunt generator in the circuit of Fig. 20-6. The shunt field current is adjusted to make the no-load terminal voltage to be 250 v. Find the terminal voltage for an armature current of 400 amp.

20-10. The machine of Problem 20-7 is used as a shunt generator in the circuit of Fig. 20-6. The shunt field circuit is adjusted to make the terminal voltage be 250 v for an armature current of 400 amp. Find the no-load voltage.

20-11. The machine of Problem 20-7 is used as a cumulative compound d-c generator in the circuit of Fig. 20-10. The shunt field current is adjusted to make the no-load terminal voltage to be 250 v. Find the terminal voltage for an armature current of 400 amp.

20-12. The machine of Problem 20-7 is used as a cumulative compound d-c gen- in the circuit of Fig. 20-10. The shunt field circuit is adjusted to make the terminal voltage be 250 v for an armature current of 400 amp. Find the no-load terminal voltage.

21
Synchros

21-1 SYNCHRO TRANSMITTERS

The devices to be studied in this chapter do not serve as power converters; nor can they be considered as basic components of power systems. They are comparatively small-scale instrumentlike devices used mainly in automatic control systems, often in conjunction with the servomotors studied in Chapter 16. They do have similarities with rotating electric machines: they are electromagnetic devices consisting of stators and rotors with windings that are magnetically coupled across the air gap. But they act more like transformers than like motors and generators. Even when their rotors rotate, the motional voltages are insignificant, only the transformer voltages count.

Consider first a device with a single-phase winding on the stator and a similar one on the rotor. Such a device was previously described by the sketch of Fig. 7-18 but, for the present purpose, the symbolic representation of Fig. 21-1 is preferred. The axis of the stator mmf wave is vertical, and the rotor is drawn in the position that makes the axis of its mmf wave horizontal. In this position, the mutual inductance between the two windings is zero, as the reader has been shown in Section 7-8. When the rotor is turned by 90°, then the mutual inductance has its highest value. Considering only the space fundamental of the mutual inductance M as a function of the angular rotor position θ,

$$M = M_{\max} \sin \theta \tag{21-1}$$

counting the angle θ in the "zero position" shown in the figure as zero and considering the clockwise direction as positive.

A practical device based on this principle is called an *induction regulator* (the name indicating its similarity to an induction motor). It is a transformer with an adjustable ratio of transformation. Either the stator or the rotor winding can be used as the primary, the other being the secondary. With a voltage of constant magnitude and frequency impressed on the primary terminals, the magnitude of the secondary voltage constitutes a *signal* indicating the rotor position. Such a signal can, for instance, actuate a servomotor, in response to the angular position of the rotor of the induction regulator, up to an angle of 90°. If it is desired to have a servomotor respond to the angular position of a shaft capable of continuous and unlimited motion, then more sophisticated devices are needed.

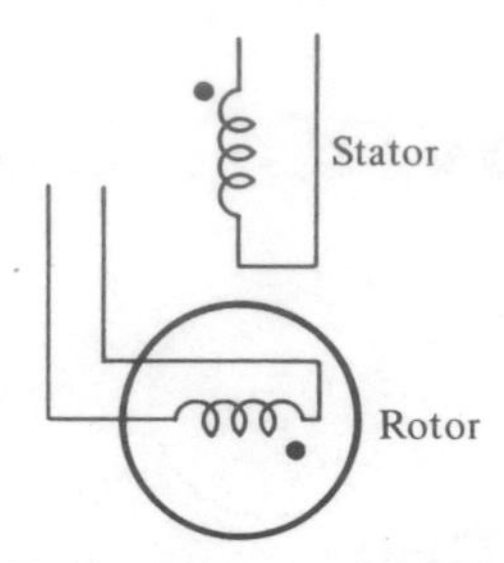

Fig. 21-1. Rotor position and magnetic coupling.

Figure 21-2 is a schematic representation of a *synchro transmitter*. Its rotor winding (which is, incidentally, a concentrated winding on a salient-pole rotor) is used as the primary. Thus, it is connected, through slip rings, to one phase of the power system, which means, in effect, to a constant a-c source. The stator carries three windings, located 120° from each other and connected to each other to form a wye. But, contrary to appearance, they are *not* three-phase windings. All induced voltages, in rotor and stator alike, are in phase or out of phase by 180° with each other, and the steady-state magnetic field is stationary, not rotating. In fact, all synchro units are strictly *single-phase* devices.

The effect of having the three stator windings located apart from each other is that secondary voltages of different magnitudes are obtained. For instance, in the *zero position* (which, for the synchro transmitter, is defined as the one shown in Fig. 21-2), the mutual inductance between the rotor winding and stator winding No. 1 has its highest possible value. According to Eq. 21-1, the magnitude of the transformer voltages induced in the three stator windings (rms values, from terminal to "neutral") are

$$\begin{aligned} E_1 &= E_0 \cos 0 = E_0 \\ E_2 &= E_0 \cos 120^\circ = -\tfrac{1}{2} E_0 \\ E_3 &= E_0 \cos(-120^\circ) = -\tfrac{1}{2} E_0 \end{aligned} \tag{21-2}$$

where E_0 is the value corresponding to the maximum mutual inductance.

Now let the rotor be turned, as an arbitrary example, by 30° in the clockwise direction. In this case, the three stator voltages are

$$\begin{aligned} E_1 &= E_0 \cos(-30^\circ) = (\sqrt{3}/2) E_0 \\ E_2 &= E_0 \cos 90^\circ = 0 \\ E_3 &= E_0 \cos(-150^\circ) = (-\sqrt{3}/2) E_0 \end{aligned} \tag{21-3}$$

The reader should have no difficulty in obtaining the values of these three voltages for any rotor position.

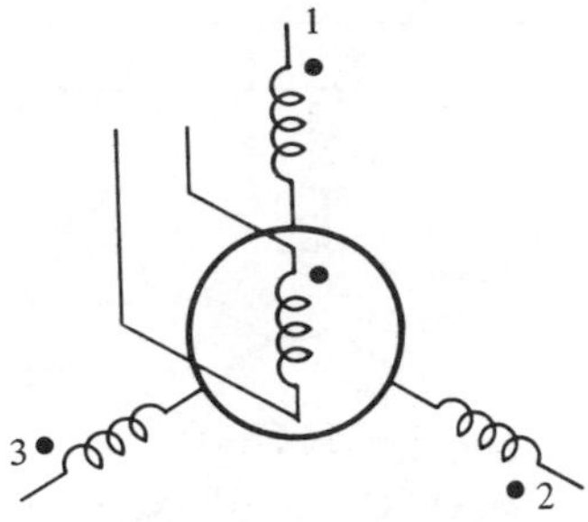

Fig. 21-2. Synchro transmitter.

These voltages are not, by themselves, suitable as signals to actuate a servomotor. They are meant to be supplied to other types of synchro units, whose description is to follow.

21-2 SYNCHRO CONTROL TRANSFORMERS

The device represented by Fig. 21-3 is known as a control transformer. (The choice of name is not very fortunate. *All* synchro units are essentially transformers, and most of them are used as components in control systems.) Its three stator windings are similar to those of a synchro transmitter, but the rotor winding is located on a rotor that has no salient poles, and its terminals are not connected to any source. Instead, the voltage across the rotor terminals constitutes the *signal* that is supplied, through an amplifier, to the servomotor. The amplifier is needed to minimize the rotor current, because the direction of the magnetic flux in this unit is to be determined entirely by the stator currents.

Figure 21-4 shows the interconnection between the two units. The stator terminals of the control transformer are connected to those of the transmitter. The power system provides the field excitation to the transmitter rotor, and also serves to supply the power needs of the amplifier and the servomotor.

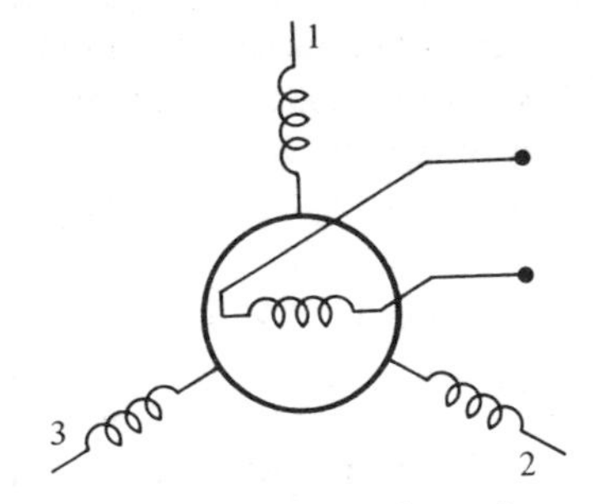

Fig. 21-3. Synchro control transformer.

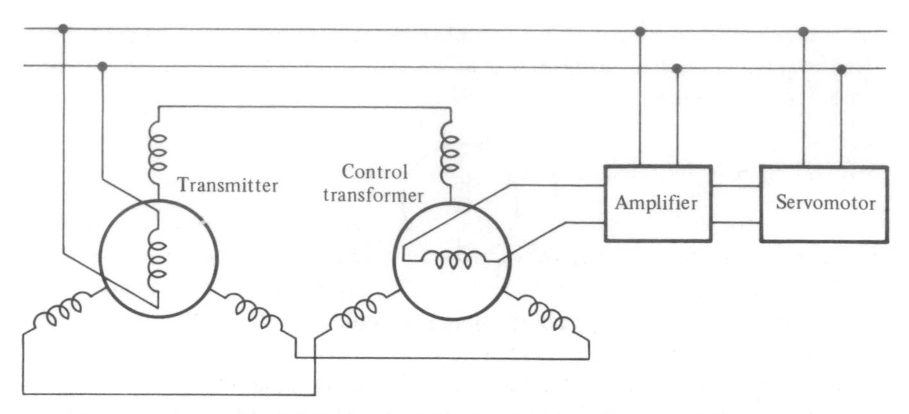

Fig. 21-4. Application of control transformer.

The key to the understanding of synchro devices lies in finding the *direction of the flux axis* in each unit. In the transmitter, this direction is determined by the rotor position, and is emphasized by its salient poles. In the control transformer, on the other hand, it depends only on its stator currents. These currents are due to the voltages induced in the transmitter stator windings, e.g., according to Eqs. 21-2 or 21-3. Their resultant flux in the control transformer must have the same direction as that in the transmitter. This is so because the sum of the voltages induced in each closed stator path must be zero (neglecting the effect of imperfections).

Finally, the magnitude of the signal voltage induced in the rotor of the control transformer depends on the angular difference between this unit's flux axis and the position of its rotor. Define the zero position for the control transformer as that shown in Fig. 21-4 (notice that this definition is different from that for the transmitter). When both units are in their zero positions, then the flux in each unit is vertical, and the signal voltage is zero. For any motion of either one or both units, in either direction, the signal voltage is proportional to the sine (again omitting the effect of space harmonics) of the *difference of the angles* by which the two units depart from their respective zero positions.

In the typical "closed loop" application of the two synchro units, the servomotor is coupled, directly or indirectly, to the shaft of the control transformer as well as to its load. Whenever the shafts of the two synchro units have equal angular positions, then the signal voltage is zero, and the servomotor does not interfere with this state of equilibrium. In all other cases, the motor develops a torque that increases with the difference between the two angular positions, up to 90°. Thus, the servomotor restores the equilibrium whenever there is a departure from it, whether this departure is caused by a disturbance acting on the load shaft, or by a command expressed as a setting of the transmitter shaft.

It is possible for the two shafts to rotate continuously in unison, (provided the speed is small enough to keep motional voltages from interfering). This performance lends credence to the name *synchro*.

21-3 OTHER SYNCHRO UNITS

In contrast to the control transformer that translates the difference of two angular positions into an electrical signal, another type of synchro unit called a *synchro receiver* produces a mechanical output signal. By its own angular position, it indicates the angular position of the transmitter, as if it were mechanically coupled to it. It serves as a *position indicator*, showing the position of a moving shaft at a distance too far to make mechanical coupling practical.

The receiver is built just like the transmitter, except that it is equipped with a mechanical viscous friction damper. This is necessary because the receiver rotor is not coupled to anything, so that, in response to a suddenly rising electromagnetic torque, it would otherwise overshoot the mark, resulting in "hunting" or even "motoring".

Figure 21-5 shows a transmitter and a receiver properly interconnected. Each unit is drawn in what is defined as its *zero position*. Their rotor windings are both excited from the power system; therefore, the flux axis in each unit is determined by the position of its rotor. The stator windings of the two units are interconnected just like those of the transmitter and the control transformer in the previous section.

The stator circuits are in a state of equilibrium when the flux axes in the two units have the same direction, so that the voltages in each closed stator path add up to zero. When the rotors are in different angular positions, then the currents resulting from this condition produce an electromagnetic torque that tends to

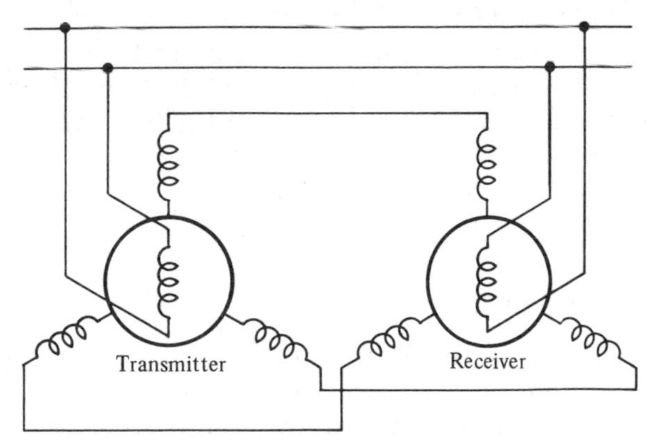

Fig. 21-5. Synchro transmitter and receiver.

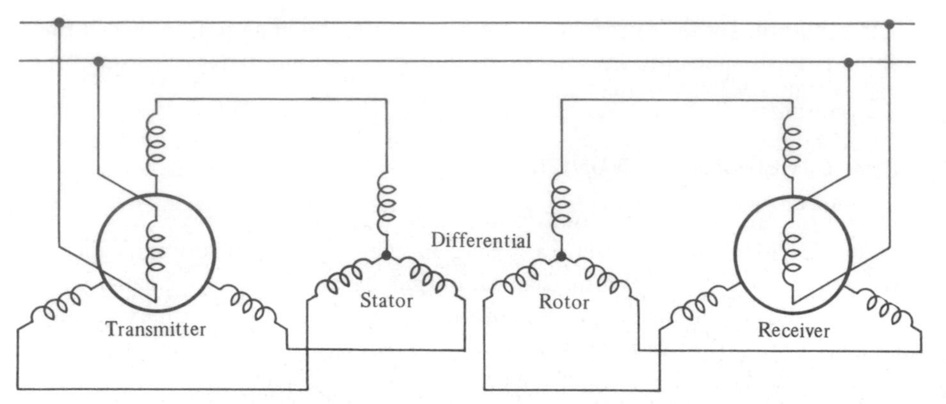

Fig. 21-6. Synchro transmitter, differential, and receiver.

reestablish the equilibrium. Thus, whenever the rotor of the transmitter is set at a new position, the rotor of the receiver follows it to the same position. The receiver may carry a pointer to indicate its position.

One more type of synchro has been developed: the *differential* unit. Its stator and rotor have each a "three-phase" winding, like the stators of all synchro units previously discussed. Thus, the differential unit resembles a small induction motor with a wound rotor. Its purpose is to generate a response indicating the sum or the difference of two angular positions. Figure 21-6 shows a differential synchro between a transmitter and a receiver. (It can also be used between a transmitter and a control transformer.) The receiver, being the unit equipped with a damper, has the function of indicating the sum or difference of the angles impressed by outside forces on the other two units. The connections shown in Fig. 21-6 are used when the difference is wanted. To obtain the sum, the roles of stator and rotor of the differential unit are interchanged.

To verify these statements, consider any arbitrary position of the transmitter and differential rotors, look for equilibrium conditions, and observe the direction of the flux in each unit. The suggested procedure is illustrated by two examples.

21-4 EXAMPLES

Example 21-1 (Section 21-3)

In the system of Fig. 21-6, let the transmitter rotor be turned 30° clockwise, and the differential rotor 60° clockwise. All angles are understood as departures from the zero positions shown in the figure. Find the position of the receiver rotor.

Solution

First the two given angles may be entered in the table below. Then the angle of the transmitter flux axis may be entered as 30° clockwise (CW), since it is determined by the transmitter rotor only. Next, the flux direction in the differential unit must be the same as that in the transmitter because the same voltages must be induced in the corresponding stator windings of these two units. This flux axis is thus located at 30° *counterclockwise* (CCW) relative to the differential rotor. To induce the same voltages in the receiver stator windings, the flux in the receiver must also be at 30° CCW. Finally, there is an electromagnetic torque acting on the receiver rotor, until the rotor position in the receiver coincides with the flux position. So the receiver turns 30° CCW, confirming the statement that it indicates the difference between transmitter and differential positions.

Transmitter		*Differential*		*Receiver*	
Rotor	Flux	Rotor	Flux	Rotor	Flux
30° CW	30° CW	60° CW	30° CW	30° CCW	30° CCW

Example 21-2 (Section 21-3)

In Fig. E-21-2, the place of the receiver is taken by a control transformer. Notice also that the stator and rotor connections of the differential unit are interchanged. Let the rotors of the transmitter and of the differential unit each be turned by 30° clockwise, while the rotor of the control transformer is held in its zero position. Find the angle indicated by the output voltage of the control transformer.

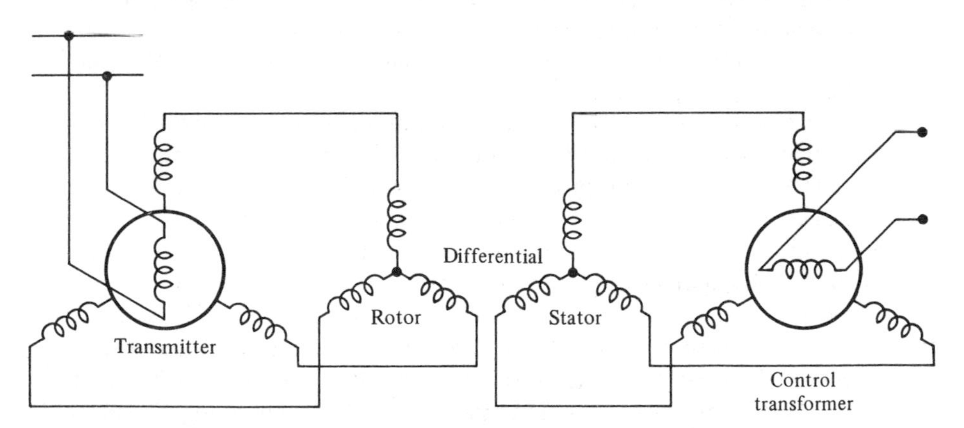

Fig. E-21-2. Synchro transmitter, differential, and control transformer.

Solution

The same tabulation as in the previous example may be used to good advantage. Thus

Transmitter		*Differential*		*Control transformer*	
Rotor	Flux	Rotor	Flux	Rotor	Flux
30° CW	30° CW	30° CW	60° CW	0	60° CW

In this case, the flux in the differential unit must be at 30° clockwise relative to the differential rotor, which puts it at 60° relative to its zero position. The flux in the control transformer then assumes the same direction. Thus, the output voltage indicates an angle of 60°, which is the sum of the two angles impressed on the other two units.

21-5 PROBLEMS

21-1. The synchro transmitter shown in Fig. 21-2 has a rotor voltage of $v_r(t) = 154 \sin 377t$. In the zero position, stator winding 1 has the voltage $v_1(t) = 98 \sin 377t$. Let α denote the angle by which the rotor is displaced clockwise from its zero position. Find the voltages in all three stator windings as functions of angle α and of time.

21-2. A transmitter and control transformer are shown in Fig. 21-4. Let α denote the angle by which the transmitter rotor is displaced clockwise from its zero position. Let τ denote the angle by which the control transformer rotor is displaced clockwise from its zero position. The transmitter rotor voltage is $v_r(t) = 154 \sin 377t$. For $\alpha = 0°$ and $\tau = 90°$, the control transformer rotor voltage is $v_c(t) = 72 \sin 377t$. Find the expression for the control transformer rotor voltage as a function of angle α, angle τ, and time.

21-3. In the system of Fig. 21-6, let the transmitter rotor be turned 73° clockwise from its zero position. Let the differential rotor be turned 51° clockwise from its zero position. The receiver rotor is allowed to turn to its position for zero torque. Find the position of the receiver rotor.

21-4. In the system of Fig. 21-6, let α denote the angle by which the transmitter rotor is displaced clockwise from its zero position. Let γ denote the angle by which the differential rotor is displaced clockwise from its zero position. Let β denote the angle by which the receiver rotor is displaced clockwise with respect to its zero position. The receiver rotor is free to turn to its position for zero torque. Find the expression for β in terms of α and γ.

21-5. For the system in Fig. 21-6, let the angles be denoted as in Problem 21-4. With all three units in the zero position, the differential rotor winding "1" has the voltage $v_1(t) = 98 \sin 377t$. Find the expression for the voltage of the differential rotor winding "1" as a function of angle α, angle γ, angle β, and time.

21-6. For the system in Fig. E-21-2, let the transmitter rotor be displaced 41° counterclockwise from its zero position. Let the differential rotor be displaced 157° clockwise from its zero position. Let the rotor of the control transformer be displaced 31° clockwise from its zero position. Find the angle indicated by the output voltage of the control transformer.

21-7. For the system in Fig. E-21-2, let α denote the angle by which the transmitter rotor is displaced clockwise from its zero position. Let γ denote the angle by which the differential rotor is displaced clockwise from its zero position. Let τ denote the angle by which the rotor of the control transformer is displaced clockwise from its zero position. With $\alpha = 0°$, $\gamma = 0°$, and $\tau = 90°$, the output voltage of the control transformer rotor winding is $v_c(t) = 72 \sin 377t$. Find the expression for the control transformer rotor voltage as a function of angle α, angle γ, angle τ, and time.

Answers to Problems

3-1. $\mu_1 = 0.0066$, $\mu_2 = 0.0073$, $\mu_3 = 0.0049$, $\mu_4 = 0.0007$ henry/m.
3-2. 5800, 1040, 112.
3-3. (a) 1500 At, (b) 2270 At/m, (c) 1.5 weber/m^2,
(d) 2.5×10^6 At/weber, (e) 0.00066 henry/m, (f) 526.
3-4. 4.9 amp.
3-5. 2.42 amp.
3-6. 3.08 amp.
3-7. 5.96 amp.
3-8. 0.00686 weber.
3-9. 0.007 m.
3-10. $A_2 = 0.0013$ m^2.
3-11. (a) 2330 At, (b) -8800 At, (c) 2810 At.
3-12. (a) 2/3, (b) 5/2, (c) 3.
3-13. $\phi_1 = 0.0006$ weber, $B_1 = 1.2$ weber/m^2, $H_1 = 0$
$\phi_2 = 0.0006$ weber, $B_2 = 2$ weber/m^2, $H_2 = 1250$ At/m.
3-14. (a) 0.00281 weber, 0 At/m, (b) 0.00562 weber, 0 At/m,
(c) 0.0076 weber, 877 At/m, (d) 0.0076 weber, 3560 At/m.

4-2. (a) 0.308 henry, 0.00221 joules, (b) 0.000352 henry, 2.53×10^{-6} joules.
4-3. (a) 0.353 amp, (b) 0.361 joule, (c) 1.12 joule, (d) 23.8 henrys.
4-4. 0.505 joules/cycle.
4-5. 25 w.
4-6. $P_{h1} = 400$ w, $P_{e_1} = 800$ w, $P_{h2} = 240$ w, $P_{e_2} = 288$ w.
4-7. 1.6.
4-8. 1.8.
4-9. 975 w.
4-10. (a) 550 v, (b) 8.2 amp.
4-11. (a) +10%, +21%, +21%, ≈+10%, (b) -9%, -9%, 0%, ≈-9%,
(c) 0%, +10%, +21%, ≈0%.
4-12. 530 v, 405 w, 3.19 amp, 0.24 lag.
4-13. (a) 110 v, (b) 3.07 amp, (c) 1.76 amp, (d) 45.6 w, (e) 1.71 amp.

4-14. (a) 16.6 amp, (b) 0.025 lag.
4-15. 220 v, 2.7 amp.
4-16. (a) $P_{c1} = 14.8$ w, $P_{c2} = 59.7$ w, (b) 320 v, (c) 2.24 amp.

5-1. (a) 171 At, (b) 30.4 amp, (c) $I_2/I_1 = 0.483$.
5-2. $\mathbf{I}_1 = 37.3\underline{/-1.43^\circ}$ amp, $\mathbf{V}_1 = 561\underline{/32.8^\circ}$ v.
5-3. $\mathbf{I}_1 = 37.3\underline{/-1.43^\circ}$ amp, $\mathbf{V}_1 = 561\underline{/32.7^\circ}$ v.
5-4. $R_{e1} = 11\ \Omega, X_{e1} = 23.4\ \Omega, G_{c1} = 13\ \mu$mhos, $B_{m1} = -23.7\ \mu$mhos.
5-5. 0.043.
5-6. 0.967.
5-7. $R_{e1} = 2.97\ \Omega, X_{e1} = 4.41\ \Omega, G_{c1} = 40\ \mu$mhos, $B_{m1} = -118\ \mu$mhos.
5-8. -0.009.
5-9. 0.963.
5-10. 0.952.
5-11. 5.15 kva.
5-12. $I_2 = 56.8$ amp.
5-13. 0.979.
5-14. 4.48 v, 56.8 w.
5-15. 40 w, 0.267 amp.
5-16. $R_{e1} = 0.0155$ pu, $X_{e1} = 0.023$ pu, $G_{c1} = 0.0077$ pu, $B_{m1} = -0.0297$ pu.
5-17. 0.024.
5-18. 0.975.
5-19. (a) 0.0125 pu, (b) 0.007 pu.
5-20. 17.1188 henrys, 0.171188 henrys, 1.7113 henrys, 0.9997.
5-21. (a) 0.9985, (b) 1.037 pu.
5-22. Fig. P-5-22(a). 0.75, 377 v, Fig. P-5-22(b). 0.995, 241 v.
5-23. (a) 34,660 v, 0.71, (b) 25,310 v, 0.943.

6-1. (a) 10 amp, (b) 40 amp, (c) 5 kva.
6-2. (a) 1/10, (b) 152 amp, (c) 15.2 amp, (d) 36.4 kva.
6-3. (a) $6\underline{/-29.4^\circ}\ \Omega$, (b) $11.3\underline{/-45^\circ}\ \Omega$.
6-4. (a) 15 kva, (b) 0.0085, (c) 0.986.
6-5. (a) 30 kva, (b) 0.0042, (c) 0.993.
6-6. Wye-Delta, 30 kva, 762/220 v, 39.3/136 amp.
6-7. $\mathbf{V}_{an} = 13{,}800\underline{/+180^\circ}$, $\mathbf{V}_{bn} = 13{,}800\underline{/-60^\circ}$, $\mathbf{V}_{cn} = 13{,}800\underline{/+60^\circ}$, $\mathbf{V}_{ab} = 23{,}900\underline{/+150^\circ}$, $\mathbf{V}_{bc} = 23{,}900\underline{/-90^\circ}$, $\mathbf{V}_{ca} = 23{,}900\underline{/+30^\circ}$, $\mathbf{I}_{na} = 7\underline{/+160^\circ} = a\mathbf{I}_{AB}$, $\mathbf{I}_{nb} = 7\underline{/-80^\circ} = a\mathbf{I}_{BC}$, $\mathbf{I}_{nc} = 7\underline{/+40^\circ} = a\mathbf{I}_{CA}$, $\mathbf{I}_C = 72.7\underline{/+70^\circ}$, $\mathbf{I}_B = 72.7\underline{/-50^\circ}$, $\mathbf{I}_A = 72.7\underline{/-170^\circ}$, phase sequence is *c-b-a*.
6-8. (a) 60 kva, (b) $P_2 = 23.6$ kw, $P_3 = 59.6$ kw, (c) 180 kva.
6-9. $\mathbf{V}_{CN} = 50.8\underline{/-90^\circ}$, $\mathbf{V}_{BN} = 50.8\underline{/+150^\circ}$, $\mathbf{V}_{AN} = 50.8\underline{/+30^\circ}$, $\mathbf{I}_{CN} = 10\underline{/-105^\circ}$, $\mathbf{I}_{BN} = 10\underline{/+135^\circ}$, $\mathbf{I}_{AN} = 10\underline{/+15^\circ}$, $\mathbf{I}_c = 4\underline{/-105^\circ}$, $\mathbf{I}_b = 4\underline{/+135^\circ}$, $\mathbf{I}_a = 4\underline{/+15^\circ}$.

7-1. (a) $W_m = (2x/5K)\,\phi^{5/2}$, (b) $W'_m = (3/5)\,(K/x)^{2/3}\,\mathfrak{F}^{5/3}$, (c) $(2/5K)\,\phi^{5/2}$, $-(2/5)\,K^{2/3}\,(\mathfrak{F}/x)^{5/3}$, (d) 205 newtons.
7-2. (a) 1.568 joules, (b) 2.744 joules, (c) 1.176 joules, (d) 0.672 joules, (e) $i(t) = 4 - 1.71\,e^{-87.5t}$ for $t > 0$, (f) 2.058 joules.
7-3. (a) 1.40 joules, (b) 2.45 joules.
7-4. 3.6 amp.
7-5. (a) $\mathfrak{R}(x) = 513{,}000(1 + 1230x)$, (b) 0.114 joules, (c) area $cbb'c'c = 0.798$ joules.
7-6. (a) $L(x) = 0.405/(1 + 1230x)$ henrys, (b) 0.684 joules, (c) 1.368 joules.
7-7. (a) $T = -11 \sin 2\theta - 14 \sin\theta$, (b) 16.5 nm, (c) -2019 nm.
7-8. (a) $T = -11 \sin 2\theta + 14 \sin\theta$, (b) +16.5 nm, (counterclockwise), (c) -16.5 nm, (clockwise), (d) 0, 180°, (also at +50.5° and $-50.5°$), (e) $W = 15.5 + 5.5 \cos 2\theta - 14 \cos\theta$ joules, (f) 180°.
7-9. 634 nm (counterclockwise).
7-10. (a) $T = 0.9 + 0.9 \cos 754t$, (b) 0.9 nm, (c) clockwise.
7-11. (a) $T = -0.9 - 0.9 \cos 754t$, (b) -0.9 nm, (c) counterclockwise.
7-12. (a) $T = 0.9\,(\sin 2\theta)\,(1 + \cos 754t)$ nm, (b) 0, 90°, 180°, 270°, (c) tends to 90° for $0 < \theta < 180°$, tends to 270° for $180° < \theta < 360°$.
7-13. (a) $v_{bb'} < 0$, (b) clockwise, (c) motor action.
7-14. (a) $v_{aa'} < 0$, (b) counterclockwise, (c) generator action.

8-1. (a) 0.958, 0.991, (b) 0.205, 0.793, (c) -0.158, 0.609.
8-2. (a) 0.957, 0.978, (b) 0.200, 0.500, (c) -0.149, 0.105.
8-3. $1560 \sin 314t$.
8-4. (a) 60 Hz, (b) $e_{a'a} = 181 \cos 377t$, $e_{b'b} = 181 \cos(377t - 120°)$, $e_{c'c} = 181 \cos(377t - 240°)$, (c) $a - b - c$, (d) $e_{a'b'} = 314 \cos(377t + 30°)$, $e_{b'c'} = 314 \cos(377t - 90°)$, $e_{c'a'} = 314 \cos(377t + 150°)$.
8-5. (a) 0.0611 weber, (b) 486 At.
8-7. $\mathfrak{F}(\theta, t) = (3/2)\,\mathfrak{F}_p \cos(\theta + \omega t - 240°)$.
8-8. $i_a(t) = I_m \cos(\omega t + 30°)$, $i_b(t) = I_m \cos(\omega t - 90°)$, $i_c(t) = I_m \cos(\omega t - 210°)$.
8-9. $\mathfrak{F}(\theta, t) = \mathfrak{F}_p \cos(\theta - \omega t)$.
8-10. (a) $\lambda = 30 \cos\theta$, (b) $e = 5655 \sin\theta$, (c) u = 37.7 m/sec, (d) $e = 5655 \sin\theta$.

9-1. 3940 v, 28,000 At.
9-2. 3690 v, 13,400 At.
9-3. 420 Hz.
9-4. 12 poles, 10 poles, 600 rpm.
9-5. 0.87 Ω.

9-6. 0.924 Ω.
9-7. 0.8, 0.025, 0.072, 0.031, 0.09, 0.9, all in pu.
9-8. 0.7, 0.012, 0.047, 0.017, 0.067, 0.77, all in pu.
9-9. 0.833 Ω.

10-1. 656 v, 5.45 amp.
10-2. 107 amp, 0.93 leading.
10-3. 373 v, 20 electrical degrees.
10-4. 57 electrical degrees.
10-5. 0.73 Ω.
10-6. 0.717 pu, 1.04 pu.
10-7. 1.25 pu.
10-8. 136 kw, 1082 nm.
10-9. 86 kw, 141 amp.
10-10. (a) 10.9 rad/sec, 0.12 (b) 5010 nm.
10-11. 15,700 nm.
10-12. 3100 v.
10-13. (a) 156 w, 156 w, (b) 69 w, 138 w.
10-14. (a) 156 w, 111 w, 267 w, (b) 69 w, 69 w, 278 w, (c) 0, 0, 313 w.
10-15. 1.12 amp, 325 v.
10-16. (a) 0.69 amp, 325 v, (b) 0.97 amp, 325 v.
10-17. 0.98 amp, 325 v.
10-18. 0.62 amp.

11-1. 3700 v, 111 amp, 0.22.
11-2. 0.28.
11-3. -0.053.
11-4. 1560 v, 10 amp, 0.26.
11-5. -0.07.
11-6. 203 amp, 1015 v.
11-7. 176 amp, 1039 v.
11-8. 0.91 Ω.
11-9. 11.3 amp.
11-10. 10.4 amp, 433 amp.
11-11. 6.7 amp, 444 amp.
11-12. 350, 200, 50, 0, 100, 250, 400 amp.
11-13. 3.1 Ω, 15.7 amp.
11-14. (a) 1785 rpm, (b) P_A = 45 kw, P_B = 60 kw, (c) Q_A = 30 kvars, Q_B = 40 kvars.
11-15. (a) 1792.5 rpm, (b) P_A = 45 kw, P_B = 30 kw, (c) PF_A = 0.99 lag, PF_B = 0.66 lag.

11-16. 0.98.
11-17. 36.9°, 0.96 lagging.
11-18. 1.02 pu, 0.97 lagging.

12-1. 1920 v.
12-2. 11.2 Ω, 6.75 Ω.
12-3. 0.73 pu.
12-4. 0.51 pu.
12-5. 1.12 pu, 1.65 pu.
12-6. 0.87 pu, 0.57 pu.

13-1. (a) 1000 rpm, (b) 960 rpm, (c) 2 Hz, (d) 40 rpm.
13-2. (a) 8 poles, (b) 0.04, (c) 2.4 Hz.
13-3. (a) 375 rpm, (b) 1.0 Hz.
13-4. 14.7 amp, 4790 w, 25.4 nm.
13-5. 4410 w, 0.856.
13-6. 38.4 amp, 11.9 kw, 63 nm.
13-7. 11 kw, 0.83.
13-8. (a) 103 nm, 0.16, (b) 36 nm, 127 amp.
13-9. 120 amp, 79.1 kw, 840 nm.
13-10. 75 kw, 0.90.
13-11. 2310 nm, 0.176.
13-12. 0.87.
13-13. (a) 4740 w, (b) 0.036, (c) 0.88
13-14. 0.86.
13-15. (a) 2.2 pu, 0.167, (b) 0.71 pu.
13-16. (a) 4.22 pu, 0.333, (b) 2.53 pu.
13-17. (a) 2.2 pu, 0.333, (b) 1.32 pu.
13-18. 38.8 amp, 11.8 kw, 62.7 nm.

14-1. (a) 40 rad/sec^2, (b) 7.5 rad/sec^2.
14-2. 160 rad/sec^2.
14-3. 5.25 R'_2.
14-4. 2.98 per unit, 1212 rpm.
14-5. 38.3 amp, 9900 w, 63 nm.
14-7. (a) $111\underline{/-7.6^\circ}$ v using V_1 as reference, (b) 16.7 v.
14-8. (a) $115\underline{/-3.6^\circ}$ v using V_1 as reference, (b) 28.8 v.
14-9. 140 Hz, 420 v.
14-10. (a) 98.6 nm, 0.187, (b) 41 nm.

15-1. $150\underline{/53^\circ}$ v, $61.2\underline{/42.7^\circ}$ v.
15-2. $42\underline{/33.4^\circ}$ v, $103.6\underline{/-26.3^\circ}$ v.

15-3. $109.6\underline{/85.4^\circ}$ v, $13.8\underline{/-51^\circ}$ v.
15-4. $\mathbf{V}_a^0 = 33.3\underline{/135^\circ}$ v, $\mathbf{V}_a^+ = 91\underline{/-105^\circ}$ v, $\mathbf{V}_a^- = 24.3\underline{/-165^\circ}$ v.
15-5. $\mathbf{V}_a^0 = 66.7\underline{/-60^\circ}$ v, $\mathbf{V}_a^+ = 33.3\underline{/0^\circ}$ v, $\mathbf{V}_a^- = 66.7\underline{/60^\circ}$ v.
15-6. $\mathbf{V}_a^0 = 13.8\underline{/-43^\circ}$ v, $\mathbf{V}_a^+ = 78.3\underline{/0^\circ}$ v, $\mathbf{V}_a^- = 13.8\underline{/43^\circ}$ v.

16-1. 34.2 synchronous watts.
16-2. 32.2 synchronous watts.
16-3. $\mathbf{I}_{1a} = 0.61\underline{/-33.6^\circ}$ amp, $\mathbf{I}_{1b} = 0.29\underline{/-126.5^\circ}$ amp.
16-4. 4.4 synchronous watts.
16-5. $\mathbf{I}_{1a} = 0.13\underline{/-39.1^\circ}$ amp, $\mathbf{I}_{1b} = 0.042\underline{/-156.3^\circ}$ amp.
16-6. (a) 0.378 pu, (b) 0.7, (c) 0.234.
16-7. $\mathbf{V}_a = 108\underline{/68^\circ}$ v, $\mathbf{V}_b = 45\underline{/3.6^\circ}$ v.

17-1. 3.99 amp, 1.26 nm.
17-2. 186 w, 1.05 nm, 0.62.
17-3. (a) 6.76 amp, 2.3 nm, (b) 374 w, 2.1 nm, (c) 0.74.
17-4. (a) $53\underline{/-33^\circ}$ amp, (b) $\mathbf{V}_b = 148\underline{/-111^\circ}$ v.
17-5. 0.17 nm.
17-6. 0.038 nm.
17-7. 0.19 nm.
17-8. 0.096.
17-9. (a) $2.54\underline{/-44.8^\circ}$ v, (b) $1.27\underline{/-44.6^\circ}$ v, (c) 0, (d) $2.54\underline{/135.2^\circ}$ v.

18-1. 247 v.
18-2. 162 nm.
18-3. $P = 11{,}850$ w.
18-4. (a) 0.00157 sec, (b) 10,160 amp/sec.
18-5. 4.5 (select 5 turns).
18-6. 5 conductors.
18-7. 2 turns.
18-8. (a) $0.637\,B_{\max}$, (b) $5.23\,B_{\max} \sin(\alpha + 67.5^\circ)$ for $0 < \alpha < 45^\circ$, (c) $5.09\,B_{\max}$.

19-1. 1605 rpm.
19-2. 1360 rpm.
19-3. 1400 rpm.
19-4. 9 amp.
19-5. (a) 925 rpm, (b) 0.90.
19-6. (a) 715 rpm, (b) 0.87.
19-7. (a) 897 rpm, (b) 232 nm.

19-8. (a) 1282 rpm, (b) 163 nm.
19-9. 424 rpm.
19-10. (a) 866 rpm, (b) 1150 rpm.
19-11. (a) 1.06 amp, (b) 4 turns per pole.
19-12. (a) $-8.3 \pm j\,16.2$
(b) $\omega(t) = 159 + 6.6\,e^{-8.3t} \sin(16.2t + 126.3°)$ rad/sec
$i(t) = 15 - 16.9\,e^{-8.3t} \sin(16.2t + 62.9°)$ amp.
19-13. $\omega(t) = 57.1 - 64.2\,e^{-8.3t} \sin(16.2t + 62.9°)$ rad/sec
$i(t) = 165\,e^{-8.3t} \sin(16.2t)$ amp.
19-14. $\omega(t) = 78.6t - 3.93 + 4.85\,e^{-8.3t} \sin(16.2t + 125.9°)$ rad/sec
$i_a(t) = 11 - 12.4\,e^{-8.3t} \sin(16.2t + 62.9°)$ amp.
19-15. (a) $i_a(t) = -4.48 \cos(377t + 5.2°) - 16.06 + 16.83\,e^{-t/0.0294}$, (b) 205°,
(c) 0.9 amp, (d) 0.238 amp, (e) −0.18 rad/sec.
19-16. (a) $i_a(t) = -4.48 \cos(377t + 5.2°) - 16.06 + 18.03\,e^{-t/0.0294}$, (b) 218°,
(c) 1.94 amp, (d) 0.690 amp, (e) −0.46 rad/sec.
19-17. (a) $i_a(t) = -4.48 \cos(377t + 5.2°) - 18.53 + 20.88\,e^{-t/0.0294}$, (b) 211°,
(c) 1.36 amp, (d) 0.457 amp, (e) −0.313 rad/sec.
19-18. (a) 52.5 rad/sec, (b) 34.9 amp, (c) −0.101 rad/sec.
19-19. (a) 35.4°, (b) 373 amp, 83 amp, (c) −0.068 rad/sec.
19-20. (a) 49.3°, (b) 472 amp, 13 amp, (c) −0.086 rad/sec.
19-21. (a) 68 rad/sec, (b) 22.9 amp, (c) 571 Hz.
19-22. 382 Hz, 1.06 msec, 1.56 msec.

20-1. (a) 270 v, (b) 240 v.
20-2. (a) 419 Ω, (b) 269 v.
20-3. (a) 0.498 amp, (b) 229 v.
20-4. (a) 0.631 amp, (b) 268 v.
20-5. 275 v.
20-6. 4 turns per pole.
20-7. 242.3 v.
20-8. 255 v.
20-9. 236.4 v.
20-10. 263 v.
20-11. 258.5 v.
20-12. 238.6 v.

21-1. $v_1(\alpha, t) = [98 \cos\alpha] \sin 377t$
$v_2(\alpha, t) = [98 \cos(\alpha - 120°)] \sin 377t$
$v_3(\alpha, t) = [98 \cos(\alpha - 240°)] \sin 377t$.
21-2. $v_c(\alpha, \tau, t) = [72 \sin(\tau - \alpha)] \sin 377t$.

21-3. 22° clockwise.

21-4. $\beta = \alpha - \gamma$.

21-5. $v_1(\alpha, \gamma, t) = [98 \cos(\alpha - \gamma)] \sin 377t$

$v_1(\beta, t) = [98 \cos \beta] \sin 377t$.

21-6. 85° clockwise.

21-7. $v_c(\alpha, \gamma, \tau, t) = [72 \sin(\tau - \alpha - \gamma)] \sin 377t$.

Index